写设计

何浩艺术书籍设计

2003—2018

何浩 著

生活·讀書·新知 三联书店

本书现名为《写设计：何浩艺术书籍设计 2003—2018》
简体版限在中国内地发行

图书在版编目（CIP）数据

写设计：何浩艺术书籍设计：2003—2018 / 何浩著. —北京：生活·读书·新知三联书店，2020.8
ISBN 978-7-108-06888-0

Ⅰ. ①写… Ⅱ. ①何… Ⅲ. ①书籍装帧－设计 Ⅳ. ①TS881

中国版本图书馆 CIP 数据核字（2020）第 111280 号

责任编辑 曾 诚
装帧设计 何 浩
责任校对 常高峰
责任印制 宋 家
出版发行 生活·讀書·新知三联书店
（北京市东城区美术馆东街 22 号 100010）
网 址 www.sdxjpc.com
经 销 新华书店
印 刷 北京雅昌艺术印刷有限公司
版 次 2020 年 8 月北京第 1 版
2020 年 8 月北京第 1 次印刷
开 本 720 毫米×1012 毫米 1/16 印张 29
字 数 50 千字 图 798 幅
印 数 0,001—2,000 册
定 价 290.00 元
（印装查询：01064002715；邮购查询：01084010542）

目 录

2007

2008

2009

2010

* 各项目依设计时间排序

序

谭 平

何浩是中央美术学院的教师，之前是我的研究生，后来又成为我招收的第一个博士生。从一开始，何浩就十分清晰自己的创作方向。很多时候，我们往往只是就他所做的工作进行探讨，这并非是一种指导与被指导的状态，而更多的是人与人之间平等的交流，是平视的感觉，犹如朋友般的交往，这同我以前的学生是不太一样的。

一般的设计师，往往会努力强调其作品的个人风格，追求鲜明的语言特色和样式。何浩与人不同之处，是他设计的每本书都并不遵循这种取向，而是真正为内容而创作，最终使每本书的呈现各不相同。因此，当人们拿到他设计的书籍时，一般并不会特别去注意其背后的设计师，而是更容易被内容本身牢牢吸引，这是何浩设计一个最重要的特点。这或许也形成了一个很有意思的现象：一般而言，艺术创作往往都要追求自己的风格，现在设计也同样有这个倾向。然而突然间，我们发现有一个设计师的作品完全看不到要塑造个人风格的企图，这反而变得非常独特。这种独特的姿态，使得何浩区别于我们所见的大部分设计师。

对当代设计而言，书籍一直是一种重要媒介。这么多年来，其功能形式也有了很多重大转变。最初，书籍主要是为了传递资讯，在现代主义时期，要求的就是设计师能最大限度地满足这种需求。但现在的情况则有所不同，我觉得，今天重要的实体书籍其实有两种类型：一种是有着能持续释放能量的内容，有必要借助纸张来固化；而另一种，如何浩一直所做的——更多呈现的是收藏品的特质，如同艺术品的延伸，其书籍实体的存在感则更为重要。人们往往需要在对书籍的摩挲和翻动中才能体验到其中的质感与温度，这种面对书籍的感觉，实际上与欣赏艺术品原作是非常相似的。未来，强化这一趋势的可能性越来越大，而对书籍总体数量的需求则可能会越来越少。与此同时，能创作这类作品的设计师也必然会越来越专精。这种专精，需要的是超越为大众快速消费品而做的那种相对简单的设计工作。这样的思考与实践，在信息泛滥拥堵的今天，重要性不言而喻。何浩同样也觉察到了这样的转变。因此，他的这十年，都在做同一件事，就是不断为中国当代艺术最具代表性的艺术家设计书籍。至今，这些作为作品延伸的书籍，总数已超过百部。

很多年以前，我们曾经做过一个项目——《中国艺术家招贴展》，当时邀请了许多重要的中国当代艺术家来设计招贴，用他们的作品与招贴设计结合，这既是设计意义上的实验，同时又是当代艺术专案。当展览推出之后，实际上人们是把它作为某种特殊视角的当代艺术展来看待的。何浩的工作同样具有这种意义，同时跨越了当代艺术与当代设计这两个领域。今天，当我们将何浩设计的所有书籍放在一起时，我们能轻易发现这同时也是中国当代艺术发展脉络的呈现。尤其是当何浩将设计师的自我表达放在设计背后时，这种现象就更为明显。或许一开始，何浩并没有意识到，他的工作将无意中记录中国当代艺术的发展状态，而历经十年的累积，他的作品恰恰成为一部重要的文献，今天看来意义非凡。

何浩的实践，其实也是一个很好的例子，来看今天的设计师如何在当代艺术与平面设计之间穿梭。在我看来，艺术与设计本是一体，实际上是一棵树的两个分叉，都是从同一个根源中生长出来的。当代艺术探讨的很多问题都极为重要，只是这种探讨并非针对实用而来，而是向着问题而去，并进一步提出未来发展的可能性。这种看法非常重要，它的形成，一方面是我们对于当代设计的独立理解；另一方面，其实也是当代平面设计的一个重要特点与重大转变。如同最初在中央美术学院的设计教学体系中所涉及的，不只是针对设计本身，而是在艺术、设计、建筑的这种大设计与大艺术方向下，来不断观察当代艺术与设计的前沿问题，来训练设计师如何观察、思维与感知。

视觉传达抑或平面设计，之前基于纸媒，今天正被网络或其他方式迅速消解。那么其真正的力量来自于何处？其实并非来自于专业性，而一定来自于思想与观念，这种思想与观念，就是当代艺术。因为，思想与观念必然会影响我们的生活方式；而生活方式，就是我们的设计。因此，设计师的工作，究竟是一种分析的途径，还是综合的方法？是外科手术式的作业，还是一种关系学的厘清？这实际上已成为艺术或设计的核心问题。何浩的设计之所以没有形式上的刻意追求，就是因为他一直在寻找设计中的准确关系——既有作者，又有自己；既在艺术系统当中，又在设计学科之内。如同何浩本人——既反思与实践（对设计），也观察与研究（对艺术）；既对西方现代设计兴趣浓厚，又亲近传统文化与东方哲学，关系与分寸把握得恰到好处。在我看来，设计师理应如此，兴趣开放，见解精微，从不简单排斥什么东西，而是把任何有价值的东西都放到自己的体系之中。

何浩的设计方法，十年间并没有太大的变化，他一直坚持这样一种角色。这种设计工作的状态，如同书法日课一般，已成为一种不断积累的过程。我觉得，这与他的成长历程和身心状态有关，也与他对这项工作的热爱有关。在这样的方式下，设计师的工作更多地转变为一种个人化的兴趣爱好，如同那些重要收藏家一样，成为一个角色特殊的研究者与观察者，这一点甚至超越了一般批评家与艺术家本人。艺术家对自身的了解其实并不容易，而作为一个特殊的研究者与观察者，设计师却有可能更为深入地了解与掌握艺术家的相关背景，去分析、比较与判断同一类艺术家的作品与概念的异同。这么多年来，从他初窥门径到成熟老练，何浩在设计中的这种角色、这份热爱，似乎从未改变。我猜想，他可能一生都会是这样，像他那样只是单纯地喜欢着自己的工作，如果恰好也能以此为业的话，这一生将多么幸福！可以预想，在20年、50年之后，他的设计所记录的历史更长，呈现的意义也必然会更为重要。也许，只有当几十年的工作摆在一起之后，我们才终有所悟，设计师的“无”其实才是真正的“有”。

2013年12月

谭　平　艺术家、教育家，曾任中央美术学院副院长，现为中国艺术研究院副院长

设计也可以是一种写作

宋晓霞

五年前，何浩送给我一部他的十年作品集。起初我并没有从书籍设计的角度去看这部集子，而倾向于把它视为一系列当代艺术的专案。当时我已在中央美院人文学院讲授“全球视野下的当代艺术研究”课程，聚焦当代艺术研究中的若干问题，引导研究生有意识地从不同的视角来审视当代艺术史。何浩的这部作品集，就是一部“设计者”视角下的中国当代艺术史。在课堂上，我借着何浩2003—2013年的工作讨论了中国当代艺术史的书写。

在现有的中国当代艺术史中，2003—2013年并非是现成的纪年。1978—2008年或者2000—2010年，通常构成中国当代艺术史的叙述单元[1]。2003年也不是一个显著的历史年份，没有发生像2001年中国成功申办奥运会、2002年中国加入WTO这样的重要历史事件。然而，世纪之交的社会转折，特别是中国与世界关系的重新定位，影响、推动了中国当代艺术的构成因素及其相互关系的错动，使之发生了内在形态的“转折”。

1942年以来的中国艺术实践已经形成了一种复杂的、充满内部矛盾的结构，所谓“转折”，首先是这个结构中各种力量的关系的重组。2003年，中国首次以国家馆的身份拟参加威尼斯双年展[2]，6月，由政府主办的“中法文化年”系

1. 例如，鲁虹：《中国当代艺术三十年：1978—2008》，长沙：湖南美术出版社，2013年；朱朱：《灰色的狂欢节：2000年以来的中国当代艺术》，桂林：广西师范大学出版社，2013年；吕澎：《中国当代艺术史2000—2010》，上海：上海人民出版社，2014年；巫鸿：《中国当代艺术史1970s—2000s》（*Contemporary Chinese Art: A History 1970s-2000s*，英文版），Thames & Hudson inc.，2014。

2. 第50届威尼斯双年展中国馆的主题“造境”（*Synthi-scapes*）着眼于“都市化”和“全球化”，选择王澍、展望、杨福东、吕胜中、刘建华五位艺术家的作品。最终因为“非典”取消了赴威尼斯的首场展览，而以“第50届威尼斯双年展中国馆在广东美术馆”的方式在国内展出。

列活动，以当代艺术展“中国怎么样”（范迪安策划，蓬皮杜文化艺术中心）作为开端。国家体制开始接纳当代艺术作为文化推广的路径，中国当代艺术不再只是地下的暗流涌动。本土与全球的对话，自20世纪90年代后期开始就是中国当代艺术发展的内驱力之一。2003年11月出版的《地之缘：亚洲当代艺术的迁徙与地缘政治》[3]，探讨了“亚洲自我表达”的多种可能性，体现了试图摆脱以西方为主导的知识体系和价值体系经验的构想与努力。随着中国艺术家愈来愈多地参与国际艺术活动，他们或在海外办展，或在世界各地创作作品，或在跨国资本的推动下参与全球艺术市场，或在双年展等国际艺术展览上获得更多交流的机会，本土与全球的对话在新世纪进入一个新的阶段。中国当代艺术的生产方式，流通方式，作品的展出、观察与批评等艺术活动方式也都出现了重大变化。商业资本对艺术的介入，既是一个国际化的趋势，在体制上也具有自己的特点。中国当代艺术的市场化过程，是与其合法化的过程以及重新工具化相伴随的。

2003年，北京举办了“再造798”综合艺术活动（徐勇、黄锐发起，邱志杰、张黎策划）。从《798艺术区大事记》[4]以及对参与798艺术区实践的当事人所做的访谈来看[5]，正是在2003年前后，北京798从一个自发形成的当代艺术聚集地进入了公共视野，在艺术全球化和中国社会变革的背景下，逐渐发展成为具有世界声誉的国际艺术区。2003年因此可以视为中国当代艺术全球化的重要节点。

思想与观念

就在中国当代艺术发生转折之际，何浩开始了他自己的艺术书籍设计历程。《戴汉志的宝丽来照片》的设计完成于2003年，虽然只是一部小书，其实颇为重要。它不仅确立了何浩后来在设计上独有的问题意识、工作方法，及其对“设计者”在书籍设计过程中的特别定位，而且留存了八九十年代中国当代艺术中的纯粹与非商业化的痕迹。戴汉志（1946—2002）是长居北京的荷兰籍策展人、学

---

3. 许江主编：《地之缘：亚洲当代艺术的迁徙与地缘政治》，杭州：中国美术学院出版社，2003年。
4. 黄锐主编：《北京798：再创造的“工厂”》，成都：四川美术出版社，2008年，第190—195页。
5. 程磊、朱其主编：《北京798：中国变革中的艺术、建筑与社会》，北京：东八时区书屋，2008年。

者和艺术经纪人，他以中国艺术史的延续和国际当代艺术实践这两条线索建立坐标轴，为奠定中国当代艺术发展的基石做出了重要贡献[6]。《戴汉志的宝丽来照片》的内容，是戴汉志生前拍摄的百余张宝丽来照片，为在戴汉志逝世周年之际寄寓哀思而结集出版。因为常常有感于当代艺术圈里狂飙突进的豪气与浮躁，何浩的设计给我的强烈印象，却是颇令人意外的那种“从容与淡然”的格调，这是一个人在“乐”过“哀”过之后，对“哀”与“乐”的“度”的把握达到了“乐而不淫，哀而不伤”程度的自然结果。何浩多年后回眸这个设计时自述：“面对这本小书，我始终非常小心地保留着照片最初在情绪上给我的感染，并以此作为设计的基调。”如何对这些拍摄时并没有既定主题的原作进行“转译”，通过书籍设计语言来强化其内在的情绪与韵味，把握与诠释中国当代艺术的往昔“味道”，既是何浩设计的灵魂，也是何浩持续性的设计艺术实验的内在逻辑，而且也是他关于中国当代艺术的写作。

对于如何处理形式和内容的关系，何浩和一般的设计师不一样。我所见的设计师通常都努力在形式上追求自己的个人风格，甚至因此造成书籍形式与内容的断裂。那些在成排的书架上一目了然的视觉形式语言，或逞勇斗狠，或运思机巧，总之设计师鲜明的语言特色和样式竭力要给人留下深刻的印象。何浩却把力气放在内容上，他花费大量精力梳理、分析图片和文字，通过反复揣摩、研究来“酿造”书籍的内在逻辑。许多时候，他不仅是书籍的“设计者”，更是书籍的“编辑者”[7]。荣荣和映里的《六里屯》（2006）就是从几百张散乱、混杂、支离的影像碎片中，从形态和内容上找出这些素材的线索，然后根据内容和形态的内在逻辑建立起相适应的形式。何浩书籍设计的过程，也是编辑的过程，甚至也是策展的过程。

---

6. 戴汉志（Hans van Dijk）1986年从荷兰来到中国，作为中国当代艺术的同行者，他关注纯粹的艺术实验，编辑中国当代艺术档案。他与艺术家联合创立的中国艺术文件仓库（CAAW），是国内最早的实验性艺术空间之一。1993年由他策划的“中国前卫艺术展”，是在欧洲举办的第一个大型、综合性的中国当代艺术展。

7. 许多经何浩设计的当代艺术书籍，编辑署名也是何浩，诸如：林天苗《不零》（2004），王蓬《王蓬》（2005），王功新、林天苗《这儿？或那儿？》（2005），缪晓春《虚拟的最后审判》（2006），张人力《第二历史》（2006），刘铮《太阳下面》（2006），林天苗《聚焦——纸上》（2007），林天苗《看影》（2007），陆亮《夜行者》（2007），唐晖《唐晖1991—2008》（2008），彭斯《抱书独行》（2009），王功新《关联——王功新录像艺术》（2010），刘晓辉《刘晓辉》（2012）等。

在何浩看来，每一个设计项目都有一个自己的内在逻辑或者设计基调。比如设计《历史中国众生相 1966—1976》的关键点，是要保留当他进入徐唯辛画室的第一瞬间作品扑面而来的“巨幅感”，接下来才是如何在形式上将空间感转译为书籍形态与语言的表现。再如《刘晓辉》，其设计基调是“暗含着一点茫然忧伤的松弛感”，但是这种感觉怎么能通过设计传达给读者呢？刘晓辉作品这样一种“有顺序但也可以没有”的调性，如何在版面设计上传达有序与无序的转换呢？

何浩将设计的关键点比喻为保险箱的密码，“破解了，一触即开，又是一番天地；破解不了，那就只能在保险箱外装饰彩绘了”。这个比喻让人对于设计师如何处理形式与内容的关系一目了然。对于何浩而言，他创作的诸多新的设计概念与方法，不是为创造而创造，更不是为了形式语言的更新而创新。他正是为了探索“保险箱的密码”而反复研究素材，所有的创新就来自于他对艺术作品不断触摸之后在理解上的不断推进，在认识上的日益深入。我想在这里列举一些何浩在设计过程中曾经遇到过的关键点：相对于陈列在美术馆空间中的艺术品原件，书籍如何成为以另一种形态构成的作品？对介于平面与立体之间的绘画，如何调整图片的“物感”与“图感”的关系以表达原作的超然出尘？怎样保留版画作品原纸张与印痕之美，而不让版画在转化为书籍的过程中由“画”沦为“图”？读者可以从书中看到他在反复揣摩研究的过程中，破解了“密码”，打开了这些关键点；同时也在这个过程中形成了自己的思想与观念，创造了自己的设计方法。

何浩以书籍设计的“无我”之境，切入设计师与作者、作者与读者、内容与形式、思想与结构之间的对话与相互关系。所谓“无我”，是指设计师聚焦于书籍内容内在的逻辑关系，而非表现上的形式变化，在对话的过程中放下自我的见解和趣味，融入作者的创作之中。对于何浩来说，这个融入的过程不限于作品，也包括对作者的品鉴，二者共同酿造出书籍的内容结构与内在气韵。设计师凭着对纸张材料、印刷工艺以及书籍设计艺术语言的实践运用，创造性地传达原作的思想与意蕴。就当代艺术书籍而言，从展览现场到书、从艺术品原作到复制品，观看空间和观看物变了，运用的媒介质料也变了。设计师运用纸张、油墨、排版、字体……各种各类的书籍设计载体、语言，对原作品的内容、意蕴进行重新模拟，用不同的表意手段，创造性地表现，比如版画纸张和油墨的质感，寻找、使用适合表现原作品的书籍设计语言表现形式，尽可能地还原从原作到书籍媒介转换中流失的内容。这个过程，不是对现实亦步亦趋的描摹，而是运用设计载体、语言在视觉、触觉以及心理上对读者的示意与触摸，诚如古人所说的“言外之意”与“象外之象”。

何浩设计的思想与观念有两个来源，一是如前所述的当代艺术，另一个是传统的艺术文脉。无论是在“无我”之境中实现设计师的“有我”，还是洞彻形势之后顺势而为的“不设计”，或是把最严苛的限制转化为最关键的提示，还有深入梳理、细致分析，临战斩钉截铁，有如庖丁解牛的“缓进速战”，从他的工作方法中不难见出古人的智慧。这一切并非出自所谓的追求，而是何浩自然而然、与其人生体验融为一体的实践。这样一种设计思维，远远超越了借助市场力量和青年亚文化所设计的“满眼的喧嚣”，也与西方现代主义风格平面设计在方法与格调上有所不同。当我们理解了何浩持续性的艺术实验的内在逻辑，看到他对书籍设计价值与意义的自省，以及书籍设计者在创作过程中的自我定位，或许不一定要着急地将它定格为东方设计观。好的设计从来不是对书籍的装饰，带有思想驱动力和历史感的书籍设计本身就是内容写作的一部分。就像艺术不是对社会变化和时代精神的记录，而本身就是社会变化和时代精神的一部分。

设计亦是一种评论

何浩关于中国当代艺术的写作，虽然并非是他有意为之，却自有其学术态度与价值判断。这些原则首先体现在他对合作对象的选择上，譬如，“当红”“权威”甚至所谓的“先锋”，都不是何浩的选择。那些有持续的问题意识、富有创造逻辑和思想智慧的艺术家，才是何浩理想的委托人。其次，在变动纷乱、许多时候个人无法预测自己命运的时代里，在这生命中呈现诸多裂痕的人生中，何浩没有将自己无保留地交付给某种时代洪流，而是确认自己当前所在的地方，明确自身的力量，再通过持续不间断的、一丝不苟的工作，以朴素而日常的劳作，完成稳定、可靠与值得信赖的设计，这也是日积月累所形成的行为方式。

具体分析起来，何浩在艺术书籍设计过程中做了大量本原性的工作。如果说何浩的设计是一种“写作”的话，那么他首先回到了以艺术家和他们的作品为中心的“写作”上来，其核心是对于人的自觉。所谓人的自觉，即自觉为具有独立精神之个体，亦视他人为独特之存在。所以，何浩在与艺术家相处的过程中，与其说是有关设计的沟通，不如说是对于艺术家之为人的体认。比如他笔下的画家张进，夫妻俩住在十平米的一间屋里，却拥有“整个世界都是自己的”自由与自在。通过大半年的深交，何浩不仅随张进了解了如何辨别古物，而且由张进的生活实践体悟了“一种有着高古之气的生活格调”。这是超越了形式的价值观，所以何浩说“设计问题的解决之道，大多来自设计之外”。

在现有的艺术评论中，其实鲜有以艺术家和他们的作品为中心的写作。何浩将自己作为设计者的身份定位为“接受者”和“评论者”，这意味着他不仅耐心地理解形成艺术家思维模式的环境和文化逻辑，还进而分析和把握构成艺术家成就的最主要的方面，即特殊的视觉方法和技术因素。这使何浩对于艺术家及其作品的认识，自然而然地超越了各种通行的标签式定位以及新奇的观念阐释。在这个过程中，作为原作品创作者的艺术家，也在“理解的对话”中和设计者共同进行“二度创作”，也成为自己作品的“接受者”和“评论者”。在何浩看来，这样一个动态的过程将以书籍为中心，在未来的读者那里重新开始、不断延续。何浩曾经引用《庄子》郢人斫垩的寓言，来描述原作者与设计者之间共同创作的理想境界。

就书籍设计而言，何浩的不同寻常之处是他的“写作”是从修整图片开始的。修图这个环节一直被何浩视为设计工作的一部分，哪怕只是去除扫描时留在画面上的脏点，或者调整一下拍摄时产生的细微的透视变形，也不会让助手代替来做。他在这个过程中一丝不苟地熟悉每一件作品，这个看上去相当缓慢的“磨墨”的过程，却构成了何浩理解作品最可靠的基础，所以他对作品的认识与评论的背后，始终有一个连贯的主体，卓然独立于艺术界流行的观念、习俗之上。例如他花了大量时间深入处理了《陈文骥》中的每一张图片，通过极尽精微的调整、取舍，提示了陈文骥作品与空间的关系，遂使《陈文骥》一书不是对原作的被动复制和简单记录，而成为能够反作用于原作的、具有独立价值的另一重阐述。

后来我与何浩交往多了才发现，伴随着他内心对个体自我与精神的珍视，是他的超越感以及对于历史的自觉。超越是超过自我，则可以进入无我的境界。设计者放下自我，才有可能放弃凭借形式变化搏出位的设计路径，转向追求书籍的内在逻辑关系。杨福东曾经开玩笑说，全世界把他的《竹林七贤》五部黑白电影从头看到尾的，大概只有他（原作者）和何浩（作为“接受者”“评论者”的设计者）两个人。正是在对于材料的分析、梳理与反复打磨的过程中，何浩逐渐把握到我们这个时代的艺术生态以及历史情境的基本逻辑，他不靠长篇大论的推衍和说理，而是凭直觉和洞见发现哪些真正对我们的文化产生了影响，哪些不过是过眼烟云。

从艺术的角度来说，何浩的艺术书籍设计是一种真正的评论，谨从以下三个方面述之。

首先，何浩对于艺术家有深切的感受。他说，“王蓬是个有着心不在焉的决绝和魂不守舍的清醒的人”，“是从骨子里特立独行的人”；何浩觉得徐冰“有点像文学作品中的早期共产党员——坚定的信仰使他能够屏蔽掉一切感知对于人的侵扰”；缪晓春“平日待人接物温良谦和与世无争，但在艺术上却无比坚忍强悍，永远在舍近求远不停歇的跋涉中”。“无论是《历史中国众生相》这组作品还是徐唯辛这个人，就如同他当时在写字楼中被公司和培训机构包围的那间画室，充满了难以一言以蔽之的层叠与复合，绝不是简单一个标签就可以归类打包的。中国当代艺术之‘当代’，亦是如斯。而设计，则是冷眼看世界。”

其次，何浩对于作品的评论言约旨远，读来令人相视而笑，莫逆于心。例如：他评论陈文骥作品的特质是“不怀旧、不愤世、不媚体制、不趋市场、始终独立于潮流之外”，说陈文骥始于 2007 年的作品远观极简节制、超然出尘，近看是毫不虚妄的厚重坚实……“几十年的艺术修为，最终化作‘提起放下’后的一笑，轻盈通透。”何浩谈朱昱的绘画，“当我第一次看到朱昱这批绘画的时候，有点吃惊——极精彩——澄明，纯净，但有一种非常内敛的‘狠’劲，直指人心”。“朱昱的绘画，既同古典绘画的格调相联结，又充满了内在的锋利和挑衅。”

何浩的设计本身也是在文字之外的另一维度的“评论”。姑以他 2006 年设计的岳敏君《被复制的偶像》为例，何浩没有因循通行的“红光亮”设计，却以与之“背道而驰的材料选择——画作在这种纸张上所呈现的不是高饱的鲜亮而是略显暗淡的灰——没心没肺的笑脸背后其实是不知所措的无奈和欲说还休的忧愁。纸张与油墨的结合不动声色地跨越了简单的复制，而成为对于画作另一维度的评论与表述”。

最后，何浩对中国当代艺术的独到见地，既有深入其里的现场感，亦有令人会心的一语中的，使当代艺术书籍的设计一扫圈中的习气与学科的负累。例如，2007年他设计巫鸿、张黎主编的《〈新摄影〉十年》，中国当代艺术的境遇较十年前已有了天壤之别。从这一认识出发，何浩没有将十年前的摄影原作重新扫描、精美印刷，而是另辟蹊径，以“复刻”的方式在全新的历史条件下再现了当年炙手的粗粝。“那因复印而产生的如刀刻般凌厉的影像，业余但饱含真诚与信仰的排版，不合理却合情的装订方式……中国观念摄影发轫之初的呐喊与勃发尽在其中”，此书也在最强烈的反差中为中国当代艺术史筑起一座具有历史感和超越其时代的纪念碑。

总之，何浩对当代艺术的“评论”，有情而无我，全然没有当代艺术评论过度文化阐释和观念套用的通病。诚如宋儒说曾皙“即其所居之位，乐其日用之常……而胸次悠然”（朱熹《论语集注》）。得益于他对当代艺术的见地及其对艺术家的认识，何浩的设计是从内容中生长出来的，以精神和意味而见长，有着古朴的士气，优雅的韵味，淡定的心性，节制的态度，慎独的威猛，以及质实的物感，在当代铺天盖地、争奇斗艳的设计中特立独行。

何浩的书籍设计，固然是基于对原作内容的体认与分析，但同时也是对自身经验的忠实、掘进与清理。何浩的每一次设计实践，都是在知与行的挑战之中，做持续的、智慧的即兴创作。在这个意义上，设计也可以是一种写作。这样的设计，既是中国当代历史的见证者，也是中国当代文化的建构者。

就在中国当代艺术因时势变迁发生又一次转折之际，何浩对于历史的感知使他敏锐地将2013年“视为中国当代艺术的一个新节点”。不过，这一次何浩已经在这十年的工作中寻求到前进的保证，形成了自己的观念与方法，拥有了历史的自觉。尤其是近年来他在学术书籍设计中的实践，使何浩的“写作”如虎添翼。万事皆有自己发生的时刻……

2019 年 9 月

宋晓霞　学者、批评家，中央美术学院人文学院教授、博士生导师

平面设计的当代转向——何浩的书籍设计实践

蒋 华

“何浩展开了一个令人兴奋的新领域实践——那就是用一种从东方文化与生活方式中生长的当代性代替了移植的现代主义，并实践于当代中国。”

何浩作为一位以书籍为媒介的独立实践者，自 2003 年至今的十多年间，以一己之力，持续地与中国当代艺术家合作，设计了超过百部的高质量书籍与画册。合作者中包括艾未未、林天苗、王功新、荣荣和映里、岳敏君、缪晓春、张大力、刘铮、展望、杨福东、徐唯辛、徐冰、喻红、叶锦添等中国当代艺术界极具特点和代表性的艺术家。这样的工作履历在世界范围内的同业中绝无仅有。

与其在当代艺术界的炽热声望相比，何浩实在可算是一位隐士。他不属于任何专业组织，与行业的热闹繁华主动保持着疏离。直至 2013 年我邀请他参加我策划的“Typojanchi 首尔国际文字设计双年展”与“第七届宁波国际设计双年展”，他才首次在平面设计界的重要展览上公开展示其项目。何浩低调隐逸，温良慎独，其“猛兽不群”的姿态，来自他对设计价值、意义的自省与自警。出于对今天流行的过度设计与滥用风格的警惕，何浩竭力将自己的工作重心从形式层面转向设计的内容与精神。正因如此，对于该如何去认识何浩的工作的重要性我以往的平面设计师朋友显然都没有做好准备。

十年的维度颇为重要，只有持续足够长的时间，实践者才有可能所思渐深并触及核心。作为同事与同行，我们始于 2006 年的交往与讨论，逐渐成为一场持续的对话。在经历了相仿的十几年设计工作之后，有幸与何浩一起共同回顾其设计历程，却无意间成就一个难得的机缘，去共同反思与清理平面设计学科层面的一些基本问题。

独立设计：一个新的起点

“这是一个最好的时代，这是一个最坏的时代。”或许每个时代的人们，都认为身处于历史中最黑暗的岁月。任何一种实践姿态都将是这种历史感下的个体选择。

经历了20世纪80年代的“85’新潮”、90年代的“平面设计在中国”、21世纪初的“大声展”，中国当代平面设计实际上肇始于先锋图像艺术，并陆续加入市场力量与青年文化，从而迅速完成平面设计新设传统的自我宣示——如90年代末深圳平面设计协会的《平面》杂志所宣称的那样，其时的深圳设计先锋称自己为中国第一代现代平面设计师，似乎之前的世界并不存在。设计师们试图移植西方现代主义设计服务体系以回应市场机遇。然而，这一新体系构建的“美丽新世界”，缺乏与中国自身久远文脉的对话，也短促得完全缺乏西方现代主义的渐进生成结构，只是直接挪用西方现代设计样式的移植性逻辑。在全球化与乡土中国、后现代与前工业的博弈中，它迅速蜕变为消费主义丛林与样式主义浪潮，成为黑暗的同谋与真实的雾霾。何浩在1998年加入这个行业，正是南方新平面设计初尝成功滋味的年代。当时的平面设计师，伴随着南方印刷业迅猛发展的隆隆机器声，比其他职业更为迅速地享受到了中国市场经济转向的果实，既让平面设计这一行业成为一种利好，同时也让设计师们愈发远离真实的世界。何浩最初的职业生涯，同样受到这种平面设计的影响，他曾谈及《平面设计在中国92’展》作品集与“平面设计师之设计历程”丛书对自己的影响。90年代中期风头正劲的现代主义风格平面设计，实际上是作为某种先锋艺术出现在媒体上，与官方主题的美术创作截然不同。何浩曾被这种样式层面的先锋性所吸引，但当时这些商业设计工作，并未给他带来太多的满足。

在何浩职业生涯的最初阶段，中国实际上还缺乏真正意义上的当代艺术画册与书籍。在国家出版体制下，出版社出版的精装画册，一般都做得厚、重、大，尽显权威感。体制内的艺术家几乎只有在取得了国家认可之后，才可能有机会出版这样的画册。如同在官方的核心美术馆举办个展一样，个人画册的出版，往往被视为一种至高无上的荣誉。正是出于这样的心理惯性，这种情形在某种程度上延续至今——比如人民美术出版社的“大红袍”系列[1]在今天依然被追捧。

对于何浩来说，2002年是一个重要的时间节点——何浩与艾未未结识并为其设计了“上海视觉艺术大学建筑设计方案”的标书，接着艾未未邀请何浩设计《戴汉志的宝丽来照片》这本小书。2002年去世的戴汉志是一个至今未被遗忘的中国当代艺术的同行者。因为某种外人不了解的原因，戴汉志拍摄了一百来张宝丽来照片，这批在他生前从未示人的照片在整理遗物时被发现。为纪念他去世周年而推出的这本小书发表了其中的30张照片，设计简单明确但情绪饱满，呈现出一种页面翻动的寂寥与肃穆。

在关键时刻，艾未未扮演了引路人的角色，无意中，这本小书引导何浩转向其事业的新方向。设计难道不就是某种工具？够用就好，但重要的是“用设计来做什么”。作为最重要的当代艺术家，艾未未这位当时最不像设计师的建筑设计师成为何浩某种意义上的导师。艾未未自己在1994年编辑出版的《黑皮书》与之后的《白皮书》《灰皮书》早已成为记录90年代当代艺术发展的重要文献。他对于内容的重视，对于形式的态度，都深深影响了何浩。

机会接踵而来，接下来就是艾未未直接邀请何浩设计其1993年回到北京之后十年作品的合集。这是个非常重要的项目，当时鲜有一个独立艺术家在体制外出版这种“正式”出版物。最初，从章法到格调，这样的工作在中国其实并没有多少现成的经验可以参照与学习，何浩也并非一开始就有明确的自觉。在很长一段时间里，设计师几乎每天骑车去艺术家工作室，与其讨论，磨合修改设计。随着内容编辑的逐步深入，设计就在其中自然生长出来，一切变得顺理成章。

艾未未《作品：北京1993—2003》确认了当代中国艺术书籍设计的一个新起点——一种从当代中国的日常生活土壤生长出来的当代设计实践，反而拥有了一种古代中国士人器物般的优雅与淡定——结实的内容，温暖的手感，郑重而坚硬，质量远超人所习见的正式出版物。而那种在编辑概念的不妥协之后呈现的出乎意料的节制，令人印象尤为深刻。

经过数年调整的何浩终于展开一个令人兴奋的新的领域——那就是用一种从

1.“中国近现代名家画集”系列，由于画册外壳使用全红的装帧，画家的名字烫金于红套之上，被俗称“大红袍”。

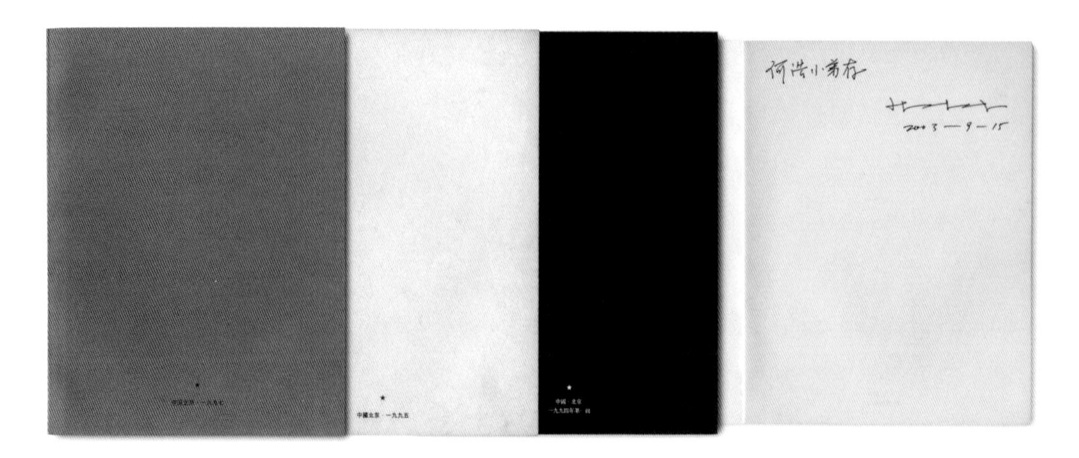

《黑皮书》《白皮书》《灰皮书》，北京，1994—1997

东方文化生活方式中生长的当代性代替移植的现代主义，并实践于当代中国。先锋与独立，实际上是一种实践形态与时代场域之间、实践主体与客观世界之间的复杂关系。独立设计是一种启蒙，既是社会批判，也是自我完善的文化建构；既是叛逆与不合作，也是新文化的书写贡献。之前是市场经济相对于院校、国企与行业协会的独立，而后是独立文化相对于市场体制的独立，都可以视为独立设计的先锋姿态与独立态度，而这正是平面设计最重要的思想资源。80 年代以来当代中国最激动人心的文化活力，集中表现在诗歌文学与当代艺术领域的先锋思想上，如同《今天》[2] 杂志，秉持了一种鲜明的当代指向。今天重估当代中国平面设计三十年，必须将先锋文学、当代艺术、独立文化与平面设计结合起来，呈现出一种新的结构。

实际上，独立设计的价值就在于一种身无媚骨、胸有方心的主体性介入。在这个波澜壮阔的时代，设计师的独立姿态成为一种设计责任，如何将一些非常精彩但缺少机会的作品传播出去，最初被商业完全忽略的独立设计正是当务之急，其最重要的价值正在于设计师的内容贡献与知识共享。这种主体性的观察、评论、介入和实践，绝非盲目的进入，而是以主体姿态“有选择地介入”和“创造性地介入”。这种当代设计的社会性与公共性，实际上是一种开放性对封闭性的

2.《今天》是 1978 年由北岛等人创刊的民间诗歌刊物，地点在北京，朦胧诗派成员会聚于此刊。后经过十年中断后，1990 年《今天》在挪威复刊，由北岛继续担任主编。

突破，这正是出版所代表的意义——揭示真相，活出真实。同样缺乏经验的设计师与艺术家当时并未意识到，他们在这个几乎空白的领域开始的工作，为平面设计，也为当代文化，出乎意料地拓展了一个新的起点。

平面设计的当代艺术史：何浩的十年

何浩2002年开始独立设计，恰好是当代艺术由地下转入地上的时期。何浩在过去的十年间完成百余部艺术书籍的设计，对于一个完全不依附于任何正规出版机构的独立设计师，可算得上是个奇迹。他的合作者在今天看来已是群星璀璨，他们在还不那么为人所了解的时候，或是在其艺术生涯最重要的阶段，与何浩合作，完成了他们第一本或者最重要的作品集。

艺术书籍是一种真实不虚的媒介，十年的积累，让这些书籍成为一条观察中国当代艺术状态的特殊线索。当展览散去，现场消失，评论过时，书籍所凝固下来的文本意义将会继续流传，成为记录时代之声的静默的黑匣子。显然，设计本身不是目标与核心，在设计语境之外的意义才是重点。面对当代中国社会、文化、艺术的实践场域，设计所承担的责任正在于此。设计师成为当代艺术一种特殊视角的观察者、评论者与批评者，并将设计视为一份礼物，代表当代投向未来。

在何浩的独立设计实践中，除了艾未未，他在中央美术学院的研究生导师谭平施加了另一种影响：一种来自纯粹艺术中复数性媒介的启示。建立于现代艺术基础上的中央美术学院的设计专业，是无可争议的当代设计先锋之地。中国最早的设计学科其实早在北洋时期就已开设，20 世纪 50 年代被取消，直到 90 年代中期才由从德国回来的当代艺术家谭平等人重设。作为当时唯一在当代艺术背景下建立的新设计学科，中央美院的设计教育从一开始就摒弃了之前那些实用主义指向的教学体系，而将先锋艺术的思想与观念注入设计。2002 年，何浩重返校园，成为谭平最早的几个研究生之一，这种学习更像是获得了一个思考与对话的开放性空间。谭老师成为何浩设计的一个重要读者，他作为一位版画家，同时也作为设计思考者，给予了何浩新的启示与鼓励。

20 世纪现代设计的发展成为艺术非常重要的催化剂，如同欧洲的早期现代主

何浩与谭平在中央美院校园中，2004

义，当代艺术也大量使用海报与工业化印刷品，其意义与同样作为复数性媒介的版画一样，都成为一种机械复制时代的艺术。这样的认识，实际上成为今天平面设计学科的建构基础。

这种当代艺术中的复数性与物质感，也集中表现在影像与媒体上。与绘画不同，这些新媒介更像是一种特殊的版画，在技术性图像生产的背后，抽离了绘画技法而依然保留着绘画般的物质感，其质感与语言的特性令人着迷。这种影像的生产方式实际上促成了包括平面设计在内的现代视觉艺术与观念艺术最重要的语言转向。与绘画相比，影像媒介所呈现的情绪更为直接，对时代的记录更为锐利。对于视觉艺术而言，影像语言的质感如同平面设计中的西文文字设计一样，展现了一种直接而清晰的现代性。

何浩早期的工作大量集中在影像，重要项目除了为戴汉志与艾未未设计的书籍，还包括荣荣和映里的《蜕》、《超越——荣荣和映里的摄影近作》与《六里屯》，张大力的《第二历史》，缪晓春的《虚拟的最后审判》、《图像 + 想象》与《“水”的艺术史研究》，刘铮的《太阳下面》与《惊梦》，张鸥的《爸爸和我》，摄影口袋书合集《外象》，重新复制的先锋摄影杂志合集《〈新摄影〉十年》，王功新的《关联》以及杨福东的《竹林七贤》等。

2007 年何浩为林天苗设计的《聚焦——纸上》，呈现的是手制纸与版画的混合结果，艺术家对于综合材料的敏锐感受成为何浩书籍设计的触发点，艺术作品媒介的物质性被小心翼翼地转换到书籍媒介中，这种以工业化印刷复制实现的喃喃细语，依然如原作般抚慰人心。

2007 年，刚刚创立“三影堂”这个非营利艺术机构的荣荣和映里，希望何浩能帮助重新设计一部限量合订本——《〈新摄影〉十年》，来纪念十年前荣荣、刘铮与朋友们一起用复印件手工制作的独立先锋摄影杂志。作为中国早期观念摄影群体旗帜的《新摄影》，曾聚集了现在看来最重要的先锋摄影力量。由于当时艺术家们的穷困以及受到诸多限制，这些摄影作品只是用复印机制作了少量的拷贝在朋友之间传播，现在市间早已无从寻觅。何浩的想法是，用最先进的印刷设备，并研制勾兑一种模拟复印碳粉的新油墨，去细腻地还原之前每一期粗糙的复印图像，保留那些原始的能量和生动的质感。荣荣与刘铮的原版设计被刻意保留，设计师精心构造了亚麻布覆盖的函套与瓦楞纸书匣，以一种持敬而克制的仪式感，将重印的四册杂志与一册文字小心翼翼地包裹起来。由此，纪念碑般的合订本构建了其与原先粗糙的手工杂志之间的微妙关系，这既是设计师向先行者的某种致敬，也是对平面设计根源的个人反思。

而在此之前，2006年荣荣和映里的《六里屯》项目中，何浩清晰地展现了设计师在内容层面上的贡献。项目命名来自艺术家曾在北京租住（后被拆毁）的村

荣荣、刘铮在手工装订复印好的《新摄影》封面和内页，北京，六里屯，1996—1998（荣荣摄）

子，荣荣和映里在这里拍摄了数量惊人的照片。最初的计划是用一本厚书来展现这些内容。在经过对于项目的深入研究之后，设计师推翻了原先的方案，提出了一个新的编辑概念，并重新组织了内容——基于作品类型，何浩将一本书变为四册，分别放置荣荣单身汉时期的个人照片、映里搬来共同生活之后拍摄的大量未挑选的整卷生活彩照，以及两人开始以合作方式拍摄的作品，外加一册薄薄的评论文集。这个新概念成为阅读的起点，通过读者随机的图像阅读，重新生成一种清晰又凌乱的图像叙述，如同记忆一般斑驳，重组后的照片观看方式，第一次变为一种非线性的叙事，一部意识流小说。函套上展现的阳光树荫照片，再一次轻轻触碰了这种情绪。这个诗意的项目，充分展现了书籍设计作为诗性媒介的惊人魅力与无穷潜能。

2008 年的北京奥运会之后，当代艺术在中国似乎有些微妙的转向。在经历一个短暂的低潮之后，何浩与从事架上绘画的艺术家合作渐多。他先后与唐晖、陈文骥、张方白、张进、靳尚谊、朱昱、喻红等不同类型的艺术家合作，这种转向实际上也是一种挑战：如何在一种相对更具古典特征的艺术中用书籍设计去捕捉那些细微的动人之处？2009 年为陈文骥设计的同名画册，可算是何浩的转型之作——格调精妙简洁，通过具空间感的图像，将陈文骥极少主义绘画的物质性成功地还原。2010 年，何浩帮助制作了《徐冰版画》，这本书全面展现了徐冰此前几乎所有的版画实践，版画在媒介上的间接性与复数性实际上是徐冰观念艺术的真正基础。何浩在书籍的设计中，保留了作品纸张和印痕的所有细节，从而将纸本作品的物质感最大限度地呈现，用“呼之欲出”形容毫不为过。2011 年，在为当代水墨艺术家中的隐者张进设计《张进画集》的时候，何浩索性与张进用了大半年的时间一起汲古访幽，谈天吃茶，书籍的编辑设计过程成为一种亦师亦友的君子之交，慢慢滋养打磨出书中的古朴格调与沉着气息，成为一种传统文脉与当代创作之间的日常生活实践。

2012 年，何浩设计了荒木经惟的《感伤之旅·堕乐园1971—2012》。这是何浩为外国艺术家在中国的展览而做的第一本书。荒木经惟最为重视书籍媒介，作为摄影师，书籍实际上是其发表作品的最重要方式，因此之前他的作品集一直保持着由其本人认可的高水准。对何浩而言，这种与不同文化传统的艺术家展开的对话成为新的挑战。这部完全创作于中国的书籍，无论是设计、制作质量，还是书籍的开阔简练，都已超越此前荒木的大部分书籍。何浩创造性地使用专色黑油墨和利用纸张肌理特质，用那些极为简单但是又有些许冒险的技术去调理设计概念，走钢丝般地实现了书籍高质量、低预算和最短完成时限的制作要求。这样的设计，大刀阔斧又轻松自在，欧美无此锋利，日本无此开阔，实际上这正是当代中国设计场域与荒木经惟作品的对话方式。这个项目也将何浩的工作推向一个更为开阔的阅读与观看空间。

与大多数设计师委托工作的被动与逼仄不同，何浩对于设计的委托关系非常谨慎，这种郑重多少亦是对于弥漫性的功利主义与机会主义设计生产关系的某种对抗。何浩很少接受来自机构的委托，也从不贸然扩大领域，大部分项目的出发点都源于其个人对这些艺术家的兴趣。他坚持对项目进行甄别与选择，仔细斟酌与委托人之间的关系，并且坚持每年只接受十个左右的委托。比起行业内许多流水线上快速工作着的设计师，这个数字实在是克制，以今天何浩纯熟技艺下的惊人效率，这无疑是设计师主动做出的一种退守姿态，是对职业设计的快速消费主义模式的摒弃。

过去十年何浩在独立出版领域的设计工作，使他成为当代出版体系之外最重要的书籍设计师。何浩以一己之力，推动平面设计以书籍这样一种特定媒介，深刻介入到中国当代文化进程中，并呈现出设计在当代文化进程中的另一种可能。这里，设计不但成为历史见证者，更重要的是，设计本身就是更为隐性却更为坚实的当代文化建构。

设计作为对话

设计是一种协同与对话的游戏，与艺术创作中的个体表现相区别，设计师的创作往往需要委托人、制作方、受众的多方协同。对话，正是何浩设计的一个关键因素，即设计姿态的主体间性（intersubjectivity）。这暗示了“之间”这一设计发生的真正场域，也意味着在设计的委托关系中，平等与主动性的姿态。东方的“物—我”，并非只是“他—我”那种西方意义上的客体与主体，而更像是“你—

左：与艺术家林天苗 / 林天苗工作室，2004　中：与艺术家徐冰 / 雅昌印刷公司，2009　右：与电影美术指导叶锦添和策展人马克·霍尔本 / 何浩工作室，2013

我”的关系，即主体间的对话，所谓的“心心相印”。何浩作品中真正的设计其实就发生于那些“之间”的关系：艺术家与设计师、作者与读者、内容与形式、思想与结构，就是这些互为主体的关系两端所发生的对立与对话、共享与协同，相互作用，相互投射。设计的目标不仅指向外在的功能，更指向创作者内在的认识。这种关系超越了主观与客观的局限，设计既是对自我主体性的发掘，也是借由设计对象发现另一主体的过程。

十年来，何浩专注于书籍设计，几乎没有扩张至其他媒介，他关注的不是如何去设计一本书，而是试图去探索书籍媒介与内容意义之间的某种对话关系。他的设计或可描述为“基于内容生长出的设计”，即将设计视为一个过程——设计即编辑——过程最终发展为设计本身。每个设计最初都源于某个编辑概念，编辑概念则基于深入的同情与体验、独到的理解与认识，而后才是恰如其分的设计描述。在此，设计师的真正价值在于如何超越装饰层面的诱惑，而为真正重要的内容贡献出合理而真实的叙事性结构，并监督把握这一结构的生产进程。

现代主义以来的设计观，过度受困于“客观性”。平面设计安然自得于事实上的主体缺席，而缺乏对过分实用主义的警惕。在“客观性”的说辞下，实用沦丧为机会主义，功能蜕变为功利主义，设计工作更多成为一种商业目标的视觉翻译，而无须信赖内容。大量生产出来的现代风格的视觉形式，只是单向地宣传着现代生活，成为媒介操控受众的工具。因此不论设计看起来多么复杂，其实质却可能是为了掩盖思想的平庸与匮乏。这种类似雇佣军般的态度使今天绝大多数的平面设计日益远离真正的文化生产，而沦落为图形图像的软件编排。当然，设计的生产关系需要设计师在项目中保有某种客观性。但这种客观性绝非一般的职业化设计服务，而是坚持本体的客观性。平等对话的设计姿态，让设计除保持介入的热情之外，拥有某种旁观者的冷静。

书籍与平面设计，不管设计师是否愿意，严格来讲都可算是一种公共性媒介，即一种“公器”，过度必然成为设计的滥用或误用。实际上今天平面设计的复杂性与深刻性依然不为人知。平面设计隐含在物质形式背后的思想性与精神性，以及其中所蕴含的综合性与开放性，如同早期现代主义那样，先锋的设计师与艺术家、建筑师、作家一起，共同去面对那个工业化的新时代。

在现在这个时代，何浩的实践更是一种对于书籍媒介的反思。在大规模机械生产之后，书籍在今天越来越远离消费主义的生产模式，而趋向于成为纯粹的文化形式和精神堡垒，其中则蕴含从容优雅的力量与含蓄节制的坚决。经由亲手摩挲时对质感的体验，书籍媒介所具有的通透与虚空、空间性与时间性兼顾的特性，以一种更为复杂的认知传递给阅读者，影响其感官与情绪。显然，支撑书籍的，是结实的内容而非海量的信息，是“之间”的物质性而非形式感。因此，今天的书籍媒介反而需要设计师保留足够的空间与余地，将注视的重心加以转移，超越之前执着于风格层面的设计观，远离设计形式主义的极端与偏执。设计愈多，意义愈少。在数字复制时代的今天，书籍可能比任何时候都依赖于草船借箭式的智慧。

设计是一种介入，设计师通过介入文化结构而参与历史，而设计作品成为介入的证据。因此设计的重要性，显然就在于介入和协作的状况、水平、层次与意义。今天的实践，更应放下对设计本身的执着，警惕不思进取的实用主义，建立一种相对客观的主体性，即去除主体表现的主体性。设计需要技术复制时代的高超本领，更需要被内化为个人能力的思想与行动。十年以来，何浩的设计实践一直发生在他与这些重要的当代艺术家之间，设计师的社会交往直接决定了设计项目的深度与广

在朝鲜平壤街头。从左至右：耿建翌、张培力、王蓬、王功新、何浩、林天苗、黄晓云（何浩妻子），2005

度。何浩的重要意义正是在于，其大量的独立出版设计项目宣示了今天的平面设计依然存在着对当代文化进程的介入姿态。

同时，设计是设计师与这个世界沉默的对话，十年间，这种对话方式也在悄然改变，与我们所处的社会一样，不经意间已发生了非常深刻的变化。今天的艺术与设计多陷于名闻利养，不管是体制内，还是市场中，已越来越带有名利场的味道。事实上，体制也好市场也罢，既可以成为施展才华、实现价值的平台，也可能成为吞噬人性与创造力的漩涡。身处体制与市场之间，何浩的设计保持了一份难得的淡定与从容、自在与安心。

何浩与人交往有古风，秉持一种率真随意的平等关系，不迎客来，不送客去，有酒且酌，无酒且止。设计即是君子之交，是同道之间的同心妙契；设计的过程，也是寻古探幽的同道旅程。设计师和艺术家之间近距离的交往、对话，其意义在于，既给予项目以结实的支点和准确的内容，也发展出一种试图搭建起设计与艺术之间精确关系的持久努力。何浩的实践也确实构建了这种平面设计与当代艺术之间的奇妙关系。那些曾与何浩一起工作的艺术家，大多珍视他的建议，将他的编辑与设计视为个人专业历程中一种特殊的独立批评，并将他引为知己。

对话，不仅仅意味着与同时代人的对话，也意味着与历史的对话。实际上，中国艺术从来就不满足于纯粹客观地再现世界，其创造力的源泉多半来自于士人传统的玄妙文心。“余事作诗人”[3]的那种山林气息与民间精神一直是中国艺术的厚重文脉，这种业余的专业性，常常构成中国艺术根深叶茂的民间生态，暗示了隐士与逸民的传统精神。这种精神，在今天这个时代，就是独立之精神、自由之思想、批判之态度。

古拙：设计实践的东方路径

四百年前的傅山箴言——“宁拙毋巧，宁丑毋媚，宁支离毋轻滑，宁真率毋安排”，由张进书赠何浩，被挂在工作室的显要位置。这种古拙的机锋，源于士人的生活态度，正是何浩设计创作潜藏的方法论。

古拙，既是中国传统重要的审美理想、东方艺术隐秘的创作方法，也是一种历经岁月打磨的生命体验，用以描述器物的气息意韵、创造主体的精神格调，一直深深地栖息在东方人的灵魂中。拙是一种开放的包容，是“无与有”的转换，是放下狭隘形式的执着，是照见本性之后的活脱与通透。所谓不求工而自工，于无设计中求设计，是完全放弃塑造个人风格企图的无我与无为。拙也是一种“少与多”的选择，是放弃更多更新更大的贪欲，而退到极少与专注。如筷子之于刀叉般的东方极少主义，以少胜多，以慢胜快，以柔胜刚，以小胜大。保持直面人生的智慧，一种持盈保泰、大直若屈的生命态度，并一一迹化于须弥。这种带有哲学意味的创作观，超越了西方艺术设计的形式观，也成为对东方主义伤感趣味的棒喝与粉碎。

古拙是一种对机巧的抵抗，是一种形式的责任。不是样式与风格，而是气象与格局。所谓大巧若拙，正是拙与巧在美学上的转换。设计最需要警惕的就是设计本身，即过于依赖“巧”而变成过度设计。在太多设计太多机巧的今天，平处呈奇，拙以寓巧，不但是一种美学体验，更是一种带有哲学意味的凝思与认知，

3. “多情怀酒伴，余事作诗人”出自韩愈《和席八十二韵》，作于元和十一年（公元816年）。

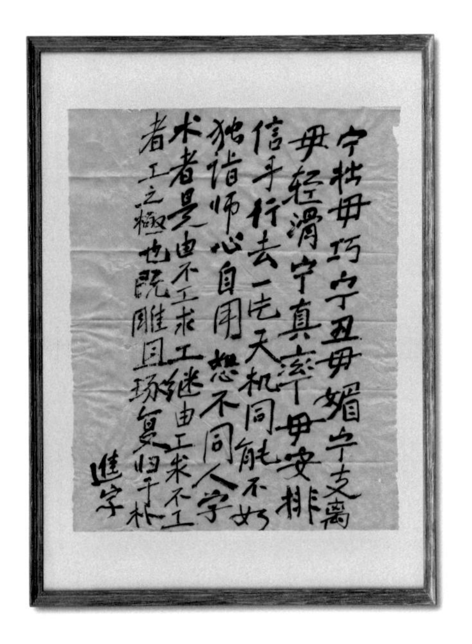

何浩工作室　张进书《宁拙毋巧》

是一种整体的认识论上的高古格调。何浩的设计，与设计师本人一样，平实而不炫耀，饱满厚重。这种古拙格调，来自那些伟大先行者的影响，也潜移默化地获益于当代艺术领域的同行者。对于这种节制之美，或可引传统器物之美的四种评判查之：沉溺样式，只存趣味；保持节制，则得清秀；阔绰开放，可抵大度；而古拙，则必须来自相由心生的浩然之气。何浩的设计，平实有力，直面真实，有一种来自坚定内心的持敬与郑重，或尚未抵古拙之境，但其古拙理想与周正大度，早已超越所谓“设计感”之小清新与小趣味。

古拙是一种对于古意的追寻，是一种当代的传统。中国在过去几十年间，往往用移植的现代性方案来改造平面设计，形成一种以“新”代“旧”的样式革新。用这种一味求新的话语装置来实现一种美学或风格的现代性，将传统视为策略，将现代视为工具，往往切断了与内容和文脉的关系。传统绝非历史，亦非古典，而是日常与当下。传统是一种活在当下的郑重态度，是直面每一个个体自身蕴藏的内在理路，一种在日常生活中生长的般若智慧。传统之重要，在于确认了“古”之于“今”的美学意味，一种关注当下的、有质量的时空观，意味着用这一时空文脉中生长的当代性代替移植的现代主义，用“古—今”共生的时空观来取代“新—旧”的话语装置，用“善”代替“新”。中国的当代性应是一种当下的传统，包含着传统自身蕴含的开放性。因此，借古开今，绝非将东方视为身份、将传统当作风格的“新东方主义”，而是将传统视为独立精神，一如陈师曾说的旧瓶新酒，一如白石翁谈的古风今雨。从这个意义上来说，八大、白石、宾虹，都是当代艺术。艾未未、徐冰、陈文骥，都是传统设计。这种在自身文脉下的当代实践，何浩同样追之思之，心向往之。

何浩坚持不雇用任何助手，所有工作由自己一个人独立完成，这种“以一当

何浩工作室　西周绳纹陶鬲

何浩工作室　陈文骥作品《函》

十”的工作模式，使得他必须以物尽其用的方式，逼迫自己退守到一种形式语言的隐匿状态，而成为有限条件下的极少主义，这正暗合士人文脉隐藏的高古格调。

何浩的项目，前期沟通准备细致绵长，案头操作过程则迅猛短暂，在这种“缓进速战”之后，是与工厂共同深入研究每个项目不同的工艺与材料，如同塑造之后的打磨抛光。这种工作方式在批量的工业化生产与精微的手工艺术之间取得了某种平衡，如同一个机械复制时代的手工艺匠人。这既是现代主义文脉下的书籍媒介实验，同时也蕴含了一种内在的东方精神。所谓“道—理—法—技”，如止于“技”的层面，也许仁者见仁、智者见智；而在“道”的层面，其内在理路必定是不约而同。比如齐白石，早年雕花木工的经历训练了他的缜密心思，实际上成就了白石山翁为人为艺的基本功夫。如同深入素描中悟道，如同经年临帖中追索。在一种工匠般的初心中，抵达大道的彼岸。

有意味的是，何浩的代表作其设计过程几乎都有与内容的反复砥砺，往往是深入问题之后的轻盈解答。几乎在每一个重要项目中，都需要“用最大的功力打进去，用最大勇气打出来”。平实有力，来自于饱满的情感；厚重结实，来自于重要的内容。所谓先能通透，方可出离，用一种直指人心的方式，将一个宏大复杂的问题变得清晰透彻，迅捷地完成在技术层面的设计工作，而将那种装饰层面的信息过滤掉。例如前文提及的《新摄影》项目在概念与形式上的内紧外松，放弃表现而保持真实，平衡了内在概念的结实与形式层面的放松，原始的粗糙与生硬被小心翼翼地保存下来。当代艺术书籍，其实验性往往在于观念而非技巧，形式的淡定与节制至关重要，当内容观念足够重要，那么任何形式层面对于艺术的简单挪用都是一种伪装成概念的"观念性"，沦为一种奇技淫巧与搔首弄姿。功力精湛的设计师总是能够漫不经心地将情绪转译为一种技术语言。

设计是一种劳动，是寂寞之道，如砚田劳作、书法日课一般。创作基于日常劳作，介入日常生活。作为一种学习手段，读书写字、师友交往都成为设计日课。这种方法层面的古拙，在理法与技术上是一种精纯的圆融，是熟后生，以看似愚笨的脚踏实地，放弃捷径与聪明劲儿，如同孙过庭所说的："初学分布，但求平正；既知平正，务追险绝；既能险绝，复归平正。"这种匠心是对于技术的圆融与通透，对于内容的尊重与对话，对于工匠系统的敬畏与传承。与紧张、自我、偏执的"匠气"完全不同，匠心是关注当下、心不二用的东方创作方法，对于此时此地心无旁骛的沉入，是格物致知、知行合一的体验，是自问与渡过。

或许荷兰书籍设计师伊玛·布[4]可以提供一个有趣的参照，与大多数西方的职业设计师不同，伊玛与何浩都是以书籍为媒介的艺术家，保持着在媒体艺术家与印刷工匠之间的设计姿态。初看起来，何浩的书籍与西方主流设计风格并无不同，都处于同一国际标准的高质量与工业体系中；但如同明式家具与Droog[5]产品之间有清晰的区别，伊玛的项目多属富豪的私人定制，不像中国，荷兰相对而言缺乏更为重要的项目。这些就私人主题展开的媒介试验，具有显性的、喧嚣的书籍媒介自身的特点，一看就是"伊玛"的书。伊玛所呈现的书籍，也有对新形式的依赖。在内容相对不够重要的时候，设计师并不需要对内容过分尊重，形式成为唯一的驱动理由，实际上每一本书籍都制造了一个小规模的技术难题。在西方常有的那些高质量的复数性工作之中，人们更为向往的是一种显而易见的形式感，这种显而易见的设计趣味为大众所期待与惊叹。这是典型的荷兰设计，有荷

4. 伊玛·布（Irma Boom），荷兰女设计师，以书籍设计闻名。

5. Droog 设计，荷兰产品设计机构。

兰那种肆无忌惮的媒介感与造型观，设计所透露的那种傲慢与任性，表现出西方现代主义以来的平面设计所陷入的某种困境。何浩熟悉现代主义设计史，早年也曾在这一模式下从事设计工作。但他对于传统的兴趣以及尊崇古拙的理想，使其内观诸己，有足够的自信完全拒绝平面设计的样式主义，试图在方法与格调层面承接起一种内在生长的中国传统文化精神，并重新确认一种“勇于不敢”的东方设计观。

平面设计的当代转向

设计就是对设计本身的定义，每个设计都是自身逻辑、日常生活与具体目标的语境化。作为思想的仪式、内容的装置，书籍聚合思想、保存内容、共享知识、传播文明，帮助我们触摸与感知世界。书籍几乎全程伴随人类文明产生至今的整个历史，在媒介与沟通方式转型的今天，纸质书籍已成为奢侈品，那么书籍与设计究竟意味着什么？

美国设计师保罗·兰德[6]在一个讲稿中谈到，“设计是一切艺术的基础”。如果艺术指向的是诗意，那么设计就是实现这种诗意的方法与路径。现代平面设计的出现，则依赖工业化印刷复制产生的知识传播能量。在这一过程中，平面设计成功地将手工艺术传统带入一个更为宏大的世界。对于平面设计而言，一切图像、语言或文字，都是符号学意义上的文本（text），作为设计的词语而存在。设计师就是在内容与形式之间建立某种仪式感，并努力提升形式的质量，重新设定事物的秩序。平面设计通过文本的复数性媒介生产，形成事件的现场（site of event）。在这个过程中，设计实际上就是在工业化的时代创造一种观点性的、物质感的、媒介性的“物”（thing/object）。归根到底，平面设计的核心就是为事与物注入观念，而设计的结果实际上是观念产生意义的证据。

从机械复制到数字交互，复数性的媒介与材料，实际上已经成为我们无法逃离的母体。无论是图像来源抑或是复制工艺，设计和艺术都与媒介复制脱不了干

---

6. 保罗·兰德（Paul Rand），美国现代主义平面设计师，耶鲁大学艺术学院教授。他设计了包括 IBM 在内的众多著名企业和机构的标志。

系。设计看似是指向目标的工作方法与实践路径，其背后真正重要的，则是思想。设计即知识形构，更准确地说，是思想赋形。事实上你可以认为，安迪·沃霍尔与博伊斯就是我们时代最重要的设计师。当代艺术，摆脱描摹与美化，深入内容与思想，所展现出来的实践主体与客观世界的那种复杂关系，引人深思。那些基于工业印刷品或版画等复数性媒介的艺术，不正是设计实践的场域吗？其中一个典型的例子，就是何浩在中央美术学院大受欢迎的完全用复印机来制作完成的书籍设计课程，鼓励学生用手工绘制拼贴原稿，然后用复印的手段反复试验，反复修改、不断深入，以实现书籍的媒介试验。课程的有趣之处在于，复印这一看似并不高端但实则在设计史中相当重要的技术又被重新注视，再现了媒介复制技术早期民主化阶段的景观，从而成功地激发了这些年轻设计师的匠心。设计的批判性实践同样基于材料与观念的表达，与当代艺术没什么区别——平面设计是创作主体缺席的当代艺术，而当代艺术则是主体在场的平面设计。当代艺术对于设计的影响是直接的，而设计对于当代艺术的影响则是隐秘的，只有去除了实用指向之后，这种复数性媒介时代的艺术观才会显得如此明确。

在媒介与技术转向的这几十年间，平面设计的水准大幅下降，这实际上与工匠体系的崩溃息息相关。之前，集设计制作于一身的工匠们能够完成几乎所有的出版工作。在工业化之后的早期现代主义时代，知识分子与艺术家对复数性工业化媒介的占领成为现代平面设计的开端。在数字媒介转型之后，真正的断裂终于到来，之前的工匠体系灰飞烟灭，那些“画墨稿”“照相制版”等工作在20世纪90年代之后迅速消失。正是在这一背景下，平面设计在迎来媒介民主化的同时，也迅速进入一个低潮时代。今天，当我们回顾这一历程的时候，我们能轻易发现，今天消亡了的工匠体系，实际上已被那些由大量工人密集劳动的大型印刷厂与各种不思进取的广告公司所分解取代。因此，何浩的实践成为了一种传承高超技艺与伟大匠心的可能。这种工匠般精神所呼唤的设计师的觉醒，才使书籍继续具有成为人类精神价值载体的仪式感，保存复数性媒介时代造就艺术品的可能性，让设计师成为工匠精神在今天的延续。

平面设计是一个学科，经历着持续不断的自我修正：基础的清理、边界的拓展、本体的反思、历史的重写、实践的档案、知识的重构……从80年代至今，无论是全球范围还是当代中国，平面设计实践的主体与疆界正在经历重大改变。与其说是对平面设计的重新定义，不如说是对设计师的重新定义。设计师对设计目的与意义的不断质询，是今天平面设计最紧迫的问题。事实上任何一个行业与学科，都有其自身的困难，也有其力量。而平面设计真正的危险在于，实践上过分受困于实用主义，研究中过分关注学科内部，正是这些偏执掩盖了真正的反思，闭塞了开放性与开拓性。

显然，平面设计就是这个技术复制时代的核心艺术。过去一个世纪的现代主义平面设计史，基本等于样式主义革新史，无论设计本身不产生存在感的“透明高脚杯”说，还是视觉传达的“邮递员”说，均游离于内容之外。当忽略了真正的内容，剩下的只是语法的分析与炫耀。今天，这样的设计观已然受到质疑。平面设计的当代转向，意味着重寻缺失已久的思想驱动力，正是这种思想性，使得平面设计成为一种真正的内容贡献，并由此重获其先锋性。今天的设计实践，需要设计师有更为坚实的主体觉醒，也需要重返平面设计发轫时代的那种综合性与主动性。设计师的角色从视觉翻译转变为内容的贡献者与媒介的创作者，这一趋势越来越与艺术生产相似。正是从这种实践的意义上来说，今天的平面设计可能依然是一个年轻的、充满可能性的、不断发展的学科。何浩的实践证明了这一点，正可视为平面设计范式当代转向的例证。事实上，这种当代转向意味着，这些新的实践者，与之前现代主义时期的同行，实际上是完全不同类型的人，他们已摒弃之前的职业设计模式，而变成行动者，一个特殊角度的作家、编辑、出版人与批评家。今天的设计实践主体用各自独立的方式演绎这个世界。也许做法各不相同，但实质上，他们在做着同样的事情——之前的平面设计更多地被视为一种“语法”，而今天的平面设计必须是一种“写作”，即：贡献内容，生产思想。

平面设计的危险一直都在，而设计师自己的看法，则是一种选择。今天，基于全新支点的平面设计正在重生。

2013 年 12 月初稿，2014 年 6 月完稿

蒋　华　设计实践与研究者，宁波国际设计双年展创始人，国际平面设计联盟（AGI）会员，中央美术学院设计学院副教授

2003—2018

戴汉志 **戴汉志的宝丽来照片**

中国艺术文件仓库，北京，2003

210mm×210mm×5mm，40 页

平装

1000 册

编辑、前言：艾未未

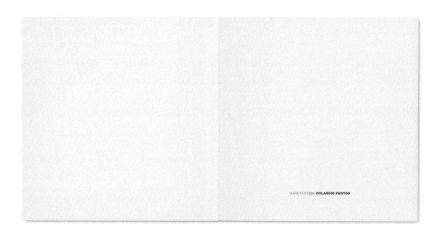

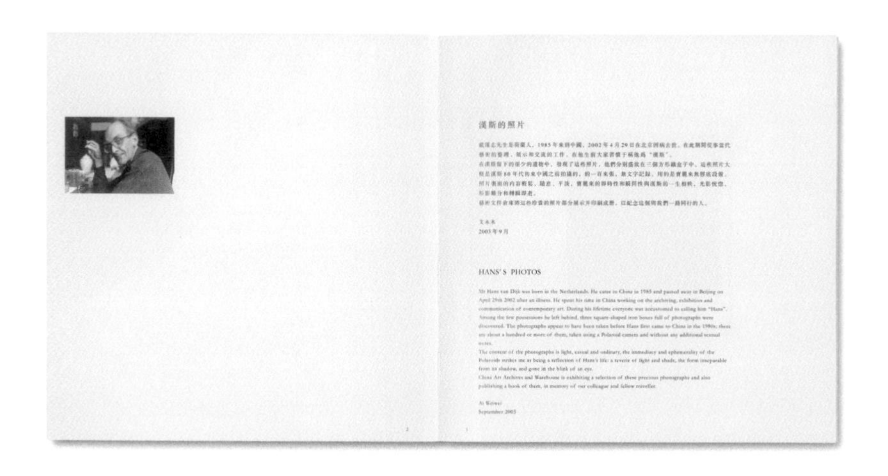

漢斯的照片

艾未未
2003 年 9 月

HANS' S PHOTOS

Mr Hans van Dijk was born in the Netherlands. He came to China in 1985 and passed away in Beijing on April 29th 2002 after an illness. He spent his time in China working on the archiving, exhibition and communication of contemporary art. During his lifetime everyone was accustomed to calling him "Hans". Among the few possessions he left behind, three square-shaped iron boxes full of photographs were discovered. The photographs appear to have been taken before Hans first came to China in the 1990s; there are about a hundred or more of them, taken using a Polaroid camera and without any additional textual notes.

The content of the photographs is light, casual and ordinary, the immediacy and ephemerality of the Polaroids strikes me as being a reflection of Hans's life: a reverie of light and shade, the form inseparable from its shadow, and gone in the blink of an eye.

China Art Archives and Warehouse is exhibiting a selection of these precious photographs and also publishing a book of them, in memory of our colleague and fellow traveller.

Ai Weiwei
September 2003

《戴汉志的宝丽来照片》

2003年初夏，肆虐数月的“非典”逐渐平息，北京街头稍现生机，但此时工作还没有恢复，大家仍普遍处在赋闲的状态中。就在这百无聊赖又蠢蠢欲动之际，艾未未给我打来电话，约我为他逝去的朋友戴汉志设计一本小书。他计划在戴汉志去世一周年之际举办一个戴汉志的摄影遗作展，同时出版部分作品以寄哀思。

戴汉志是荷兰人，1986年来中国，一直从事中国当代艺术的整理、展示和交流工作，直至2002年在北京去世。在戴汉志十分有限的遗物中，有百来张他生前拍摄的宝丽来照片，分装在三个铁盒中，之前没有人知道这批照片的存在。当我和未未一起翻看这些小小的已经发脆的照片时，他似乎有些伤感，但我作为“普通读者”，却从这些光影恍惚如梦如幻的影像中，真切地感受到了“乐而不淫，哀而不伤”的从容与淡然。

面对这本小书，我始终非常小心地保留着照片最初在情绪上给我的感染，并以此作为设计的基调。与多数平装书不同，这本书的封面和内文为同一种纸，质厚而色白：厚使得书页虽少，但仍挺括、强韧；白则十分微妙，这是一种类似汉白玉的颜色，单纯、温暖而又平朴、坚硬。为了追求手感的温润，我选择的这款纸没有表面涂层。一般来讲，“非涂纸”并不适于图像印刷，因为纸张吸墨，照片色彩的饱满程度会因此多少有些损失。但这一劣势对戴汉志的照片而言却意味着新的可能——用书籍语言来对原作进行“转译”，并因此强化其内在的情绪与韵味——拍摄时随机产生的色彩层次并非这些宝丽来照片所特别需要具体强调的，“味道”的把握与诠释在这个项目中比亦步亦趋地描摹更重要。此外，全书的所有文字被统一敷以灰褐色，如同书纸的白，这个颜色一样微妙而至关重要。虽然出版的预算不多，但我还是坚持额外增加了这版专色——越是看似简单的设计，越是要在表达的细微处下足功夫，一色之差即是天壤之别了。

在做这本书的一个月当中，我差不多每隔三两天就会去未未家，就设计的进展跟他讨论一下。未未从来不是一个随便的人，他只会提出更尖锐的问题和更严格的要求。多数情况是他纠正我作为一个设计师天生的“问题”：设计师总想靠一点形式变化的出位来实现自己的身份认知——要没设计感，能叫设计吗？然而他那时对我说得最多的一句话就是：“你要考虑的是内在的逻辑关系，而不是看上去的变化。”在过去的十年中，总有人会凭直觉认为我跟其他设计师有些不太一样，我想或许是因为我设计第一本书的时候，就把背负在身上的“设计师”的包袱卸去了。

现在回想起那段时光，我有时会为当时投入了如此多的时间和精力感到不可思议——毕竟这是一本统共只有区区四十页的小册子。但我又不得不承认，就在这样一点点分析、梳理、反复打磨中，我真正厘清了设计中的很多问题。多年后，我在《七十年代》一书中读到徐冰曾经用了整整一寒假的时间画了一张素描，但他说这张素描比他之前画的几百张素描解决的问题都多。我太能理解这句话的含义了，虽然我没有画过如此深入的素描，但我经历过同样深入的设计训练，我相信其中所蕴含的“理”没有差别。

林天苗 不零

东八时区，香港，2004

282mm×221mm×15mm，96 页

精装，裱特殊定制真丝，

裹半透明纸护封

1500 册

编辑：何浩

文章：凯伦·史密斯、皮力

访谈：皮力 / 林天苗

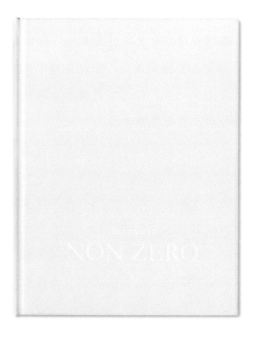

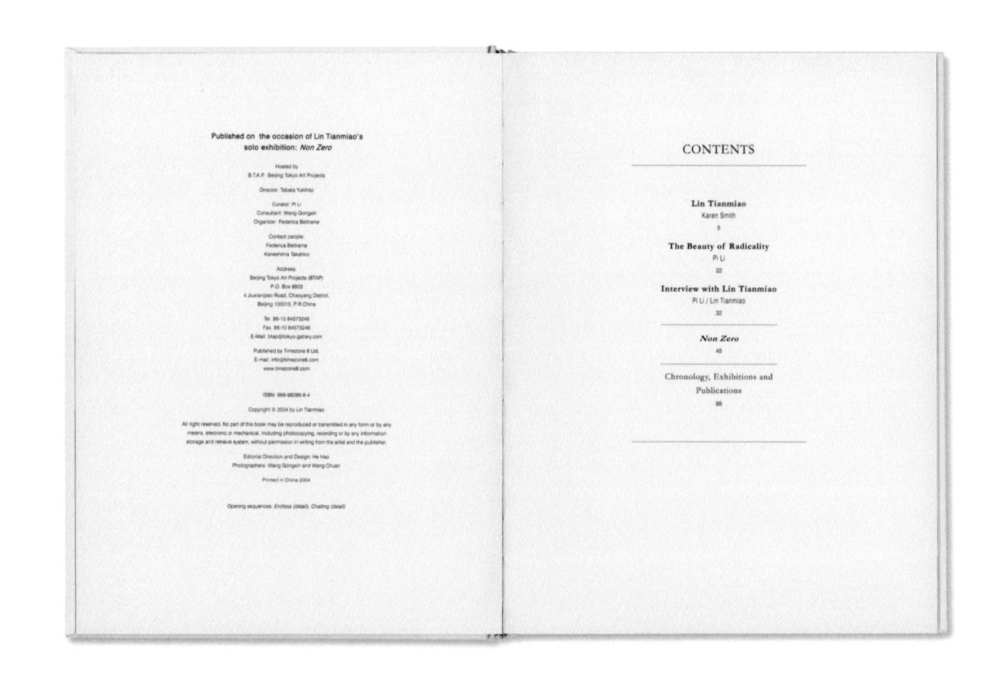

Published on the occasion of Lin Tianmiao's
solo exhibition: *Non Zero*

Hosted by
B.T.A.P. Beijing Tokyo Art Projects

Director: Tabata Yukihito

Curator: Pi Li
Consultant: Wang Gongxin
Organizer: Federica Beltrame

Contact people:
Federica Beltrame
Kaneshima Takahiro

Address:
Beijing Tokyo Art Projects (BTAP)
P.O. Box 8503
4 Jiuxianqiao Road, Chaoyang District,
Beijing 100015, P.R.China

Tel. 86-10 84573246
Fax. 86-10 84573246
E-Mail: btap@tokyo-gallery.com

Published by Timezone 8 Ltd.
E-mail: info@timezone8.com
www.timezone8.com

ISBN: 988-98086-8-4

Copyright © 2004 by Lin Tianmiao

All right reserved. No part of this book may be reproduced or transmitted in any form or by any means, electronic or mechanical, including photocopying, recording or by any information storage and retrieval system, without permission in writing from the artist and the publisher.

Editorial Direction and Design: He Hao
Photographers: Wang Gongxin and Wang Chuan

Printed in China 2004

Opening sequences: *Endless (detail)*, *Chatting (detail)*

CONTENTS

Lin Tianmiao
Karen Smith
5

The Beauty of Radicality
Pi Li
22

Interview with Lin Tianmiao
Pi Li / Lin Tianmiao
32

Non Zero
45

Chronology, Exhibitions and Publications
88

Artist Lin Tianmiao

Lin Tianmiao

Karen Smith

Compared to her contemporaries in China, Lin Tianmiao became a practising artist relatively late in years. Not fresh out of an academy, she was in her mid-thirties and had almost eight years of experience developing and sustaining a successful business in the field of textile design. I say that she was late in becoming a full-time artist – distinct from her life as a practising designer - because the fine arts had been an enormous preoccupation during an eight-year sojourn in New York with her artist husband Wang Gongxin between 1986 and 1994. Together, they were voraciously active in visiting the wide-ranging exhibitions that are held in New York each year, and were fortunate in securing a loft space in an artists' complex, which allowed for intimate interaction with the burgeoning community surrounding them.

Art had, in fact, also been the subject of Lin Tianmiao's studies at the Capital Normal University in Beijing in the early 1980s. But where the role of 'Normal' universities in China is that of a teacher-training college, the institutional approach to teaching 'art' – especially in the era that Lin Tianmiao pursued her studies - was as a set of principles and steps that could be learned by the students, remembered, and then followed in a classroom when the students graduated to take their place on teaching faculties nation-wide. Even in her childhood Lin Tianmiao demonstrated herself to be anything but a conformist. Yet, she did possess a natural talent for following established patterns, as she would subsequently reveal in New York, astounding those who witnessed the struggle that preceded her recognition as a successful textile designer.

"I was so hopeless at school that my parents almost gave up on me. I had absolutely no power of concentration. Nothing the teachers said sank in. I wasn't deliberately awkward, I just had a problem conforming - today it is what people describe as 'attention deficiency'. No matter how hard I tried, I was incapable of concentrating. Yet, I remember at the age of 14 forming a sudden interest in geometry. Within a week, I had gone through all the text books for an entire year!" Most subjects, however, continued to present challenges. "I was an extreme character - when I liked things I became obsessed with them. Those that aroused no interest went straight out of my head."

From the first, as the daughter of a recognised ink-painter father and a mother who was a respected authority on indigenous dance forms, Lin Tianmiao was artistic. In fact, the eccentricities she displayed are similar to those ascribed to child prodigies in environments that are ill equipped to recognise them. China of

The Proliferation of Thread Winding

Material: White cotton thread, 20000 needles, 12-15cm long, TV screen, video, bed, rice paper

Dimension: An area of approx. 20-40m²

Description: On a white bed covered with rice paper are 20000 needles closely stuck together in the central part of the mattress with the needlepoint upward. Each needle is connected to one thread, which leads to a ball of cotton the size of a ping-pong ball. The large numbers of thread balls are draped from the mattress on the bed and spread over the floor. Occupying the position of a pillow is a TV screen which shows a video image of a hand wrapping white cotton thread balls.

Production Time: 1995, 1997

Exhibition Place: 12# Studio, Beijing

Beijing Art Museum, Beijing

The Fifth International Istanbul Biennale, Istanbul, Turkey

from applying to their precious skin. En masse, the vapour from the perfume it contained stung the fragile membranes of eyeballs and sinuses. Yet, it was so tempting to draw near.

The reference to St. Teresa was only in part related to her sainthood, more to the perpetual conflict she endured between secular and convent life in her struggle to attain enlightenment and peace. For Lin Tianmiao, as a nun and a saint, Teresa embodied purity, and this allowed the artist to open up a cavern between the simplicity of surface appearance and what the work actually contained and imposed upon the viewer. Here, I refer to the onslaught of the chemical odour. This was beauty at a price, but that was not entirely Lin Tianmiao's point: "When I came across these worn toolboxes I was immediately fascinated - the builders thought I was nuts when I asked how much they wanted for them. To me, they looked like objects that had been used for centuries, yet they didn't have the fragility of antiques. I saw them as symbolising male crudity and dominance. Yet alone that wasn't enough. I knew they ought to contain something. This 'something' had to possess the quality of softness to overcome that hardness."

That turned out to be several litres of moisturising cream.

This concern to elicit the starkest contract between hard and soft, that paradox of the physical, as well as the abstract, world was to become a recurrent presence in Lin Tianmiao's work. Here, in "The Temptation of St. Teresa", she offered an initial, tentative – if literal – exploration of the qualities of *yin* and *yang* that govern everything within human existence and consciousness. In this regard, the work – and the 'proliferation' - was more than an invocation of 'male' and 'female', more than a question of the issues arising from being a woman in the contemporary world, or in China - by virtue of her experience living in the US, Lin Tianmiao was acutely aware of the problems women face, anywhere.

The Lin Tianmiao who returned from seven years in the United States was a different character to the girl who left Beijing to join her husband Wang Gongxin in 1986, although even then she owned the distinction of being one of the first licensed one-person companies - *ge ti hu* - in the capital. She would return permanently from New York in the autumn of 1995. At this point she was physically different too: she was six months pregnant.

Aside from the need to rest herself in preparation for motherhood, after years of working twelve-hour days to keep afloat in New York, Lin Tianmiao

extremely difficult to make rational judgements and choices about them. I too, found it confusing, so I simply bound all these things up."

Whilst this continued to be an approach to which Lin Tianmiao would return, she had no hesitation in combining her organic-flavoured thread with mediums born of the modern world that she claimed confused her. As previously described, she had incorporated an element of video art into her first 'thread' work in Beijing. Next, in 1997, as part of a huge installation piece for a solo show titled "Bound: Unbound", she projected an extraordinary videoed sequence of scissors cutting threads of cotton, this time onto a screen that she painstakingly formed from thousands of threads of the cotton vertically suspended in the space. The cutting of a thread is symbolic. It is the final part of Lin Tianmiao's process: the act of severing the umbilical cord that ties the artwork to its creator, or exacting a metaphoric break with the ties of, perhaps, family, friends or society, such that might effect a personal space for an individual. Juxtaposed with the large volume of bound objects strewn on the floor, which were, by dint of being household objects, dwarfed by the scale of the exhibition space, the monumental proportions of the screen magnified the impact of the scissor blades in motion. In a subtle way, here again was the hardness that Lin Tianmiao had perceived in the builders' hods, brought to heighten the softness of all the delicately wrapped objects. But here was that paradox of being bound and yearning to be free: the projected image of the scissors created the illusion of cutting, but where the blades could never touch the actual threads, it was but the projection of an untenable desire.

Within the space of a few years, Lin Tianmiao had bound everything imaginable from chopsticks to crockery, buckets to stoves and all else conceivably household between there and the kitchen sink, in a tight layer of thread. She was driven, obsessed, yet as her work progressed, her intensity brought forth clarity and led to tighter, more coherent forms.

"In the beginning it was a kind of torture. Eventually I became accustomed to it. Each day I would bind a little, every day the same. I became calm, composed. In this way, my concept of 'torture' changed. Was the external world 'torturing' me, or was I deliberately 'torturing' myself?"

She had even tackled entire trees. "Despite the tranquil birdsong that played in the space and the sprinkling of feathers on the floor, people felt that the inverted tree suggested a sense of loss and a confused world turned upside down, useless or frustrated. This surprised me because it was not. For me it was an

Tree

Material: Tape (birdsong), white cotton thread, white feathers, dead tree
Description: A dead tree 6-7m high, with a diameter of 5.5m is wrapped up in white cotton thread and hung upside down below the skylight of the exhibition place. Across the white floor below the tree are scattered white feathers, which move slightly when the audience walks nearby. A recording of birdsong plays quietly.
Production Time: 1997, 1998
Exhibition Place: The Modern Art Center, Breda, Holland; Women's Art Museum in Bonn, Germany

expression of an innate Chinese philosophy or sensibility in my blood. I felt it was truly a piece of contemporary art replete with a traditional Chinese outlook. Critics labelled me feminist. I looked in many books and catalogues of female artists to see if this was true but came to the conclusion that it is not the case. I had never thought I judged life, or presented a view of life from the perspective of being a woman, it was always from my own experience as a person who just happened to be a woman."

This perforce prompted Lin Tianmiao to give thought to the essential nature of 'woman'. Retaining the inalienable sensation of thread in relation to her body, to the fingers, the skin and the measure of the proportions of these to each other, she subsequently took her work in a slightly new direction in a large piece she titled "Spawn". It was to be the first of a series of works that began from a flat picture plane and which employed thread to create a three-dimensional space around the two-dimensional image. This new process resulted in a number of arresting installations that re-affirmed her refined understanding of physical space. Viewers would be confronted with the image of a human body cocooned within an arrangement of threads that mapped out their being as a delicate, ghostly aura.

She began with her own being, her face, her features, yet by removing all the hair from her head as if shaved clean, she managed to 'remove' her character and mitigate the feminine softness that hair generally provides. The result was ambiguity, further enhanced when the image was transferred from her computer to its printed form on canvas. She deliberately reduced the density of ink to the softest shades of grey whereby her face became pale as a shadow under cloudy skies. It becomes clearer at a distance, dissolving into abstraction when one draws near. To draw near is to discover myriad threads stitched across the already pale features, which serve to obscure them further and impart something of the softness that was razed with her hair.

"Many people asked me if this androgynous image was intended to negate the state of being female. I hadn't thought of it like this at all." This she had thought of in terms of *yin* and *yang*; watching an irrepressible maleness assert itself in her child as he gravitated towards guns and boys' games even as she attempted to create a neutral environment around him.

One senses that here, the impetus driving Lin Tianmiao's investigation of 'being' was much more specifically about the parameters for exerting individual-

The Beauty of Radicality

Pi Li

Background I. The Concealing of Female Awareness and the Removal of Female Identity

As early as the 1920s, Xiang Jingyu, a famous woman pioneer of the Communist Party of China (CPC), led a naked parade of women in Wuhan. It was the earliest and largest recorded Women's Movement in China, and was contemporary with the western Women's Movement. Yet, this event has been ignored, intentionally and unintentionally, in almost every kind of historical record in new China. As an important part of the modernist cultural movement, in China, the Women's Movement was totally different in form from other modern movements. In the 1920s, the land revolution led by the CPC solved the problem of a woman's right to vote and to possess land. After 1949, a woman's right to vote was further written into the Constitution. The result of mixing traditional asceticism and modern Feminism is that the Women's Movement in China has been limited to a focus on political and economic rights. 'Gender equality' as a political and economic slogan has in fact imperceptibly become a cultural standard, making 'women are as good as men' a basic tenet of women's creative expression ever since the founding of the People's Republic of China; whereas the gender identity and experience of women have been ignored. For this reason, in a country with a radical women's movement, women's art is neglected. The 'removal' of the female identity in political and economic fields, and the 'concealing' of female identity in the cultural field have become the target of cultural references in women's art in China, and distinguish it from that of the West.

Background II. 'Chinese Women's Art' as Determined by Subject Matters

The awareness of Feminist art in China began in the 1990s, and was initially characterized by an emphasis on gender and female life experience. Up to the mid-1990s the women's art movement was led by theory rather than practice. The translation and research of theories on female culture, as a part of the study of post-modernism that began in the 1990s in China, first took place within the literary community before shifting into the field of visual art. Therefore, the so-called early-stage research of Feminism in China was nothing but "attaching" Feminist theories to creative visual expression via traditional subject matters. Such examples include a dictionary-like classification of women artists from ancient times to today. Among them the most interesting case is the theoretical interpretation of the "Sick Lotus Series" by Zhou Sicong in her later years, or the female

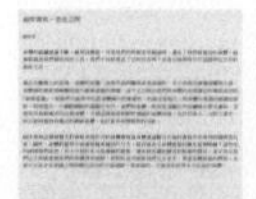

Shut Up!

Material: Paper, cotton thread
Description: Covering a curatorial concept using cotton embroidery until it disappears.
Production Time: 1999
Exhibition Place: It Park, Taipei, China

interpretation of the abstract painting of Pan Ying. Feminist art theories have also 'created' new 'women artists.' Depictions, metamorphosis and abstraction of sexual organs and women were the first topics to find release through new languages and symbols. Then expressions of babies, pregnancy and breast-feeding became a fad. The classical 'theory of determination by subject matter' in fine arts in the PRC was also an important feature of so-called Chinese women's art. Women's art became a field fought for by critics who lacked a basic ability to read the original theoretical texts and a personal power of reasoning. But no new possibility was seen from the 'art with women as the subject matter' (not Feminist art), which lacked new forms of language to support it. What one saw were confusing theories and works of art that could not withstand close scrutiny. In the mid-1990s, in regard of Feminism, theories were booming but the language at their disposal lagged behind. Compared with the trends towards exploring the language of art at that time, such as new media art and performance art, the Feminist art trumpeted by so-called 'Feminist critics' appeared as very weak. Excessive theories and unrestrained interpretations even concealed and killed the significance of exploring language and methodology, as shown in the works of artists with definite feminine consciousness, such as Yu Hong. For some time it seemed that even highly praised 'Feminist' artists felt ashamed to appear in the context of so-called Feminism, preferring to stress their identity as an artist, which in fact demonstrated that, to true artists, the changes in forms of language seemed to be much more important than 'subject matters' promoted by theories.

Dualistic Vacillation

Lin Tianmiao is one of the artists who emerged at this turning point. From the beginning her works represented the dawn of an age of art creation that probed into language and methods of creation. Her first work to attract people's attention was 'The Proliferation of Thread Winding', which was displayed at the Capital Art Museum (Beijing) in 1995. The first reason that it aroused interest was that the symbols in the work were transformed from real images, containing tortuous, diversified implications, which formed a sharp contrast with the other works of art of that time with realism as their main method. Secondly, in this work Lin Tianmiao demonstrated her special sensitivity to, and ability to transform, materials. For example, where she transformed steel needles into 'fur' and paper into cloth. The transformation of materials gave the audience an unexpected

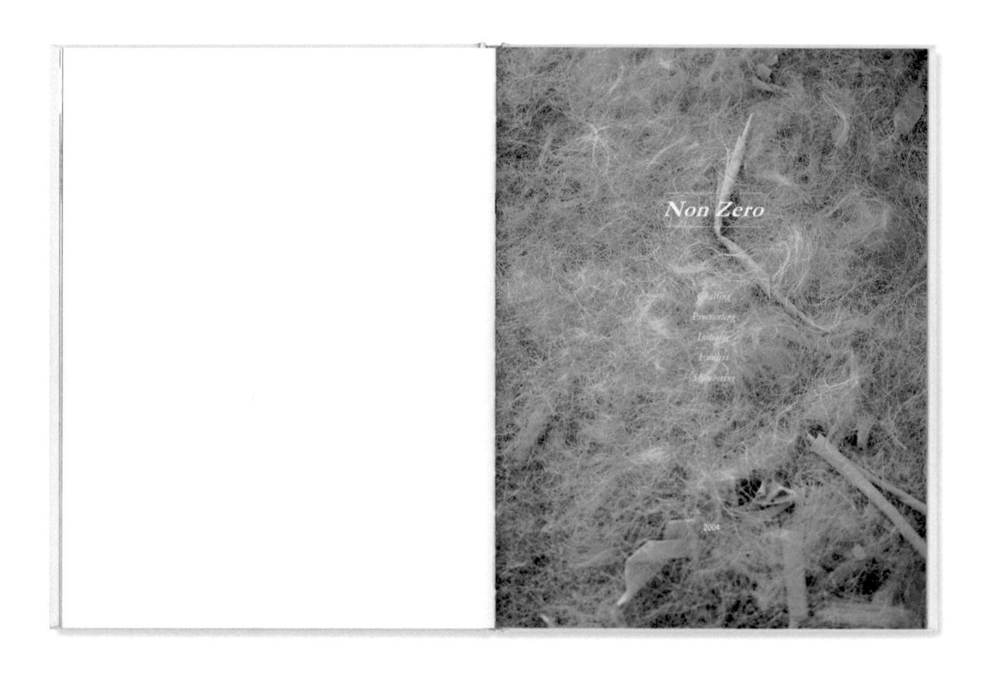
Non Zero
2004

Chatting
Procreating
Endless
Slumbering
2004

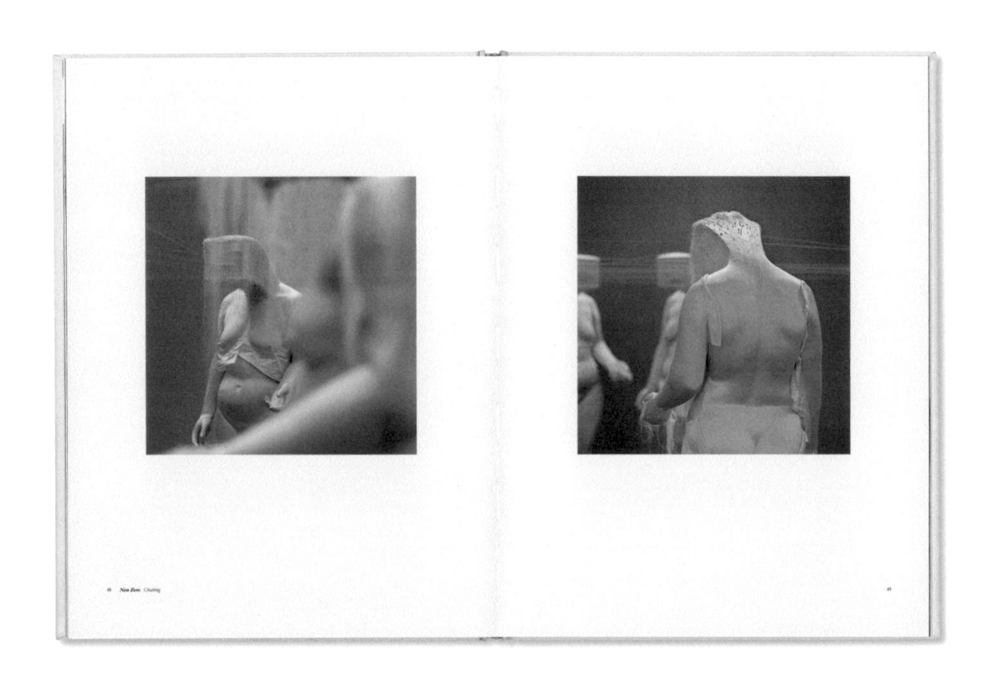

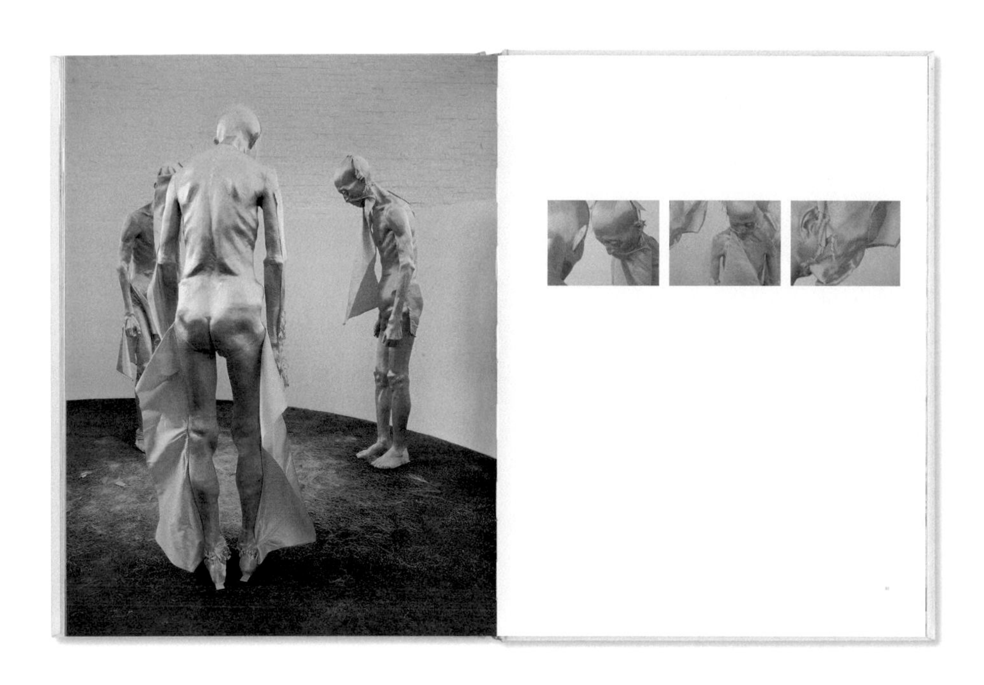

荣荣和映里 **蜕**

东八时区，香港，2004

291mm×291mm×35mm，286 页

布面精装，封面裱照片

2000 册

编辑：巫鸿、荣荣和映里

文本写作：巫鸿

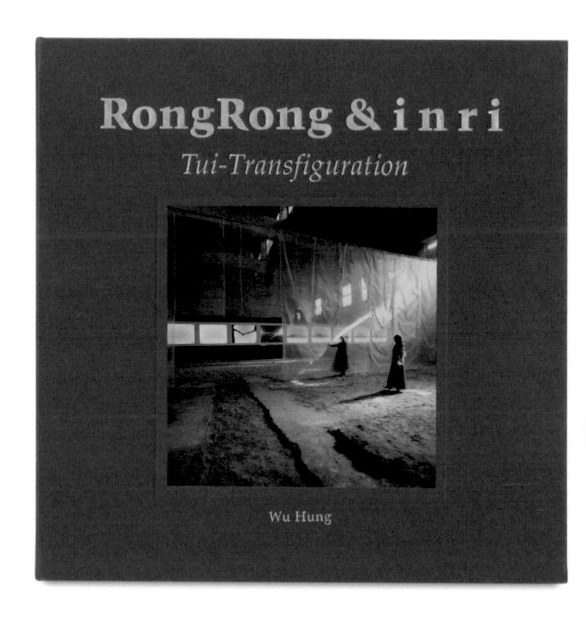

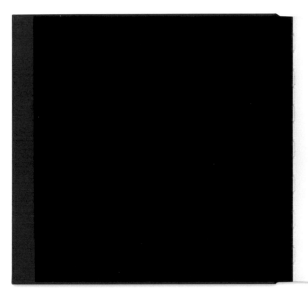

Tui-Transfiguration

A Site-Specific Exhibition at Factory 798

蜕

798 廠的一次特定場地展出

Tui-Transfiguration is a site-specific exhibition I curated at Beijing's Factory 798 in September 2003.[1] It displayed twelve series of photographs that RongRong and inri had created over the past decade, and can be considered a retrospective of these two artists. But I also conceived and planned the show as an experimental exhibition that posed questions about the language, site, audience, and function of a contemporary art exhibition itself. These two purposes are not separate. In fact, a principal goal of this project was to create a symbiosis between the exhibition's content and location: whereas RongRong and inri's photographs largely represent their intimate interactions with urban and natural environments, the exhibition site — an abandoned factory workshop — tells a story about Beijing and China. Combining these two stories in a coherent visual display, the exhibition reinforced their messages and created an artistic space with which the audience could interact.

This essay defines the exhibition as a site-specific event and introduces its social and artistic goals. I will first discuss the site: its origin, history, transformation, and current condition. I will then contextualize *Tui-Transfiguration* with a brief review of experimental exhibitions of contemporary Chinese art since the 1990s. Finally, I will characterize this exhibition as a series of interactions between the curator, the artists, the audience, and the place. In this way it represents a new kind of curatorial project that integrates art practice, museum installation, historiography, and social history in a single undertaking.[2]

This essay provides a background for the rest of the catalogue, which records and analyzes the exhibition itself. Part Two, the main component of the catalogue, describes the exhibition as a visual narrative consisting of four spatial/thematic sections. Mirroring the exhibition's structure, this part has four sections, each starting from a general discussion of the works it contains, followed by reproductions of

action with the artists. I arrived in Beijing five days prior to the exhibition's opening. RongRong and inri were waiting for me in the airport, and we took a taxi directly to Factory 798. RongRong could not wait to show me the place, which he had found after looking at several more conventional exhibition venues. I was stunned by the factory workshop when I entered it: it was over ten times bigger than the Beijing Tokyo Art Project gallery I had visited. For a while I had trouble comprehending what I was seeing. The space's immensity was nearly brutal, and its emptiness seemed to have a suffocating power. When RongRong talked to me, his voice vanished in the shadows surrounding those lifeless machines.

At the same time, a familiar feeling of excitement arose in me: the exhibition that had only existed in my mind would have to make this incredible, brooding place its home. I felt an urge to amplify RongRong's voice to fill the space, and to make the machines and their shadows speak. I knew that I was not the only one to feel this way: I could sense the same excitement from RongRong and inri next to me. Huang Rui then magically emerged, like the guardian deity of any locality in traditional Chinese folklore. With him were Zhan Wang, Zhang Dali, and Sui Jianguo – artists who had all worked with me in previous shows and who were now preparing their works next door for the exhibition *Left Hand and Right Hand (Zuoshou yu youshou)*, a group show featuring works by forty-eight Chinese and German artists. I went over to say hello to Feng Boyi, an old friend and my partner in organizing the Guangzhou Triennial. He told me that he was co-organizing the exhibition with Zhu Jinshi, who was also there. I had gotten to know Zhu well during the Guangzhou exhibition, for which he made a version of his astonishing *Impermanence* – a maze constructed of bamboo and white *xuan* paper. Both he and his wife Qin Yulen would make installations for my forthcoming exhibition on "*Negotiating Beauty*" in Berlin's Haus Kulturen der Welt.

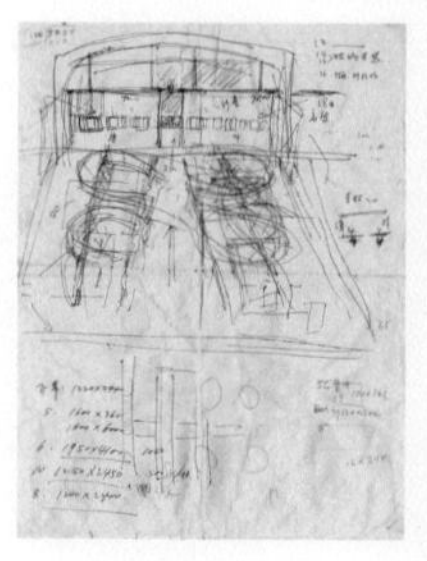

Fig. 28a,b Workers installing *Tui-Transfiguration*
工人布置《蜕》的情况

We greeted each other briefly as though we had just seen each other the day before —everyone in this business is constantly on the move. Besides, there was no time to waste: like our exhibition, their show was nowhere to be seen yet.

I mention these mundane social encounters because in my experience, such informal, "on-site" relationships between independent artists and curators constitute a particular kind of comradeship, which is crucial to this type of spontaneous, site-specific exhibition. A common situation is that all the participants in such a project decide to work together to realize a shared goal, but have to find actual solutions – even ideas – through interactions with the place and with each other. The success of a site-specific exhibition can never be secured in advance, and additional external factors can further heighten the sense of uncertainty. In the case of *Tui-Transfiguration*, both this exhibition and *Left Hand and Right Hand* were organized without official permission. The factory workshop which these two shows shared was not a licensed exhibition space, and for this reason alone they could be closed down by the authorities at any time. Such worries were not false. In fact, it is reported in an important art magazine that during an official news conference held prior to the Beijing Biennale, a government lawyer read a document, declaring that all "satellite shows" during the biennale period would be illegal. Some independent curators thus worked only with licensed exhibition venues for security. In other words, all our effort put into *Tui-Transfiguration* and *Left Hand and Right Hand* could be wasted. The forty-plus participants nevertheless set out to realize the exhibitions because they were willing to take the risk.

Making round after round in the factory workshop (including checking out all unlocked rooms and climbing onto the boiler room), I felt that I began to know the place. I also began to mentally superimpose the two artists' works with the remain-

[illegible]

The Image World of RongRong and inri:
Tui-Transfiguration as a Visual Narrative

榮榮和映里的影像世界：
《蜕》的視覺敘事

Tui-Transfiguration used one half of the largest workshop in 798, which originally housed the factory's central kiln and had a total footage of about 5,000 square meters (fig. 1). Having rented the entire space, RongRong divided it up with Feng Boyi, who used the other half for his exhibition *Left Hand and Right Hand*. Because no art events had been staged here before, these two exhibitions in effect transformed the factory workshop into a public art space, however temporarily.

After partitioning the shop room in the middle, the space available for *Tui-Transfiguration* was approximately 80 meters long, 32 meters wide, and 14 meters high. To realize the space's enormity one should not think of it as a single space, but as a "container" of groups of architectural structures and machines. Originally, a 50-meter long kiln connected to a huge chimney at one end of the room lay in the middle of the space like a giant serpent (fig. 2-3). The kiln had been destroyed before the exhibition (fig. 4-7), but the chimney still stood amidst rusty machines and industrial waste. Other remains from the past included some free-standing buildings with boarded-up windows and a row of two-level houses standing opposite the chimney (fig. 8). Most of its rooms were filled with junk and disused furniture (fig. 9); the factory's security guards used the two small rooms near the main gate as a hangout.

It is difficult to imagine a place more unsuitable for displaying photographs. But since my purpose was to create an interactive, site-specific exhibition, not a standard photography show, the challenge became how to merge the photographs and the architecture into a flow of images and spaces around a central concept. This concept– a temporal progression from death to rebirth–underlay both the photographic works and the transformation of the factory, and united the exhibition's content and site. In preparing for the show, I interviewed RongRong and inri and carefully studied their art–its stylistic and thematic development, and its relationship

[illegible]

RongRong: *East Village,1993-1998*

榮榮：《東村》

The official name of East Village is Dashan Zhuang -- literally the Manor on a Big Hill. After some wandering artists from the provinces made this place their home they called it the East Village, obviously inspired by its namesake on the other side of the globe. Here, "waste accumulates at the edges of the small ponds. This pollutes the water, generating noxious fumes in the summer. Raw sewage flows directly into the water. Slothful, threadbare dogs roam the narrow lanes between houses. People stare with the blankness of the illiterate and benighted."[1] It was, indeed, a place of death, the kind of space I have termed a wasteland: poisonous spaces filled with garbage; graveyards of dead objects that reject disintegration; "black holes" in an urban landscape that absorb time and escape change. Differing from ruins lamented in romantic poetry and painting, a wasteland never inspires sentiment or stirs up memory. Instead it is a contagious corpse, suffocating the living with its deadly excrement.

Viewed from this perspective, RongRong's moving into the Village did not just satisfy his need for cheap housing; instead his renewed creativity there must be considered the consequence of a voluntary self-exile. He and other Village artists were fully conscious of the "hellish" qualities of the Village, and the ambivalence toward making hell a home was best articulated by a series of performances by the Village artist Zhang Huan, which was designed to reduce the artist to the surrounding waste. RongRong's photographs, on the other hand, best capture Zhang's masochism; the absence of the photographer in these images attests to his existence as an invisible gaze. When Zhang Huan performed *12 Square Meters* — a project in which he stayed in a filthy public toilet for an hour with flies covering his naked body, RongRong was in the same toilet with his camera. The next day he wrote excitedly to his younger sister in the south:

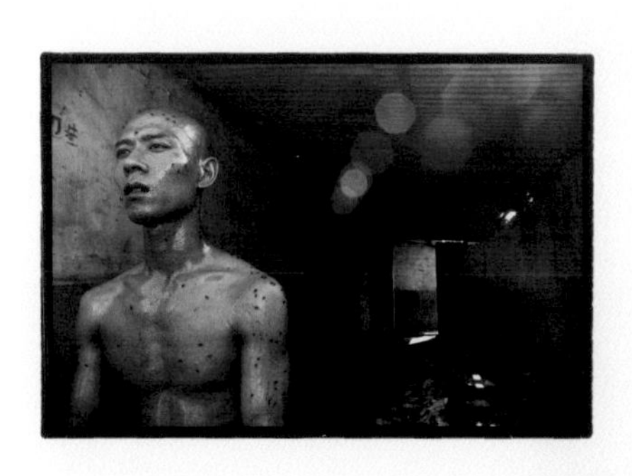

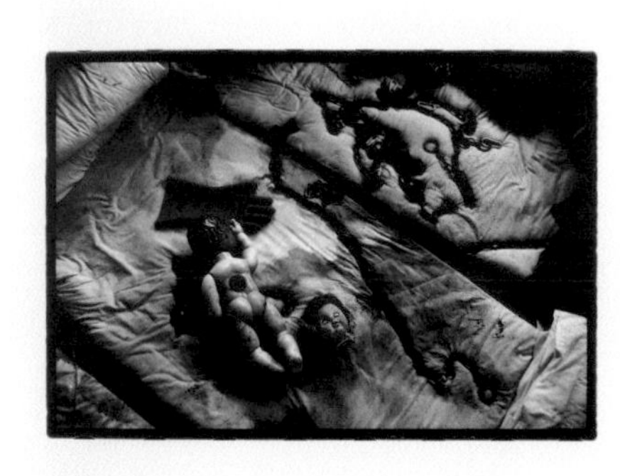

RongRong and inri : *In Nature*

荣荣和映里：《大自然》

inri:Traveling to Shikoku

My spirit and my body were split, scattered about. I thought the only way to unify them, to experience the reality I couldn't escape, was to make a pilgrimage. Ever since my youth I've had a desire to do this before I die, but I had never thought I'd do it so soon.

I had devoted myself to photography, but it was precisely because of this single-mindedness that I had become blind. I was constrained by the overload of things I saw. Actually I saw nothing. Though this was the case, all I could do was live by photography. At first I didn't want to take my camera, but then I changed my mind and brought it along. I could take pictures or not, but if something happened that I wanted to shoot I could.

Even if I went on a pilgrimage maybe nothing would change. Maybe I just wanted to experience the process, experience the unchangeable self. But though this was perhaps the case, I still had to do this in order to truly realize that I couldn't change. I was already 27 years old, but I was still so confused. I didn't know why, but I was surrounded by my own metamorphoses; my mind got heavier and heavier, more and more crowded. I felt I was stupid. It was only through action that I could think clearly, that I could experience understanding.

Upon returning from my trip to Shikoku, my spirit and body actually were unified. The excess fat and flesh was dispelled, emaciated to skin and bones. I became empty, utterly devoid of content. Pitifully watching the trash filling up my life, I thought, I didn't want anything.

Then, how would I live from now on? It was then, when I hazily looked upon a dimming cityscape, in that emptiness, my empty self, that I met RongRong.

There was an exhibit of his work in Tokyo. I was moved by his photography. Other

RongRong and inri:

荣荣和映里：

When, through the lens, the two scenes we each see become one, then the scenes have surpassed "scenery before us" and formed a single world.

People who go to visit those places cannot perceive the world of our experience.

That world does not exist anywhere in the real world. You can only possess that world with us through these photographs.

People often insert themselves into the work they are viewing, as if the subject of the photo has become themselves.

This truly can happen.

We cannot say for certain that the bodies in the photograph are ours; they just reflect something that has transcended the observed world.

Opening the senses, disappearing into vision, and then emerging on the other shore of sensation, this behavior of the observer is very similar to our own. Going through this process, an observer has used his mind to uncover the secret we have created.

The two of us have tried to enter a wordless world (nucleus).

In this world lacking material spirit, in which there is no chanting to spirits and gods, using prayer-like repetitive performance, our time shackled, an unbreakable prison.

This performance hints at faith.

The first three parts of the exhibition followed one other along the north-south axis of the factory workshop. The fourth and last group of photographs — self-portraits of RongRong and inri — occupied the two side walls (fig.14). In this position, these works played a double role to constitute an independent sequence of self-representations and to frame the other works with the artists' autobiographical references. The photographs on the west wall (i.e. on the partition wall that separated *Tui-Transfiguration* from *Left Hand and Right Hand*) consisted of self-portraits that RongRong and inri created before they met each other. The images are surprisingly similar in mood but markedly different in life experience: most of RongRong's self-portraits refer to the particular places in which he has lived: his old home in Fujian, the East Village, and a small farmhouse in Liulitun Village where he lived until 2002 (the last two locations have been demolished for new urban development). inri's self-portraits, on the other hand, construct an interior space detached from her physical surroundings. All images in the pictures belong to this space. One group of photographs, for example, juxtaposes her with a flowering cactus. Later in this volume I will quote her description about this series: through "discovering" the cactus she actually discovered her own individuality and sexuality.

On the opposite, east wall, three mural-size photographs show the two photographers inside a deserted, chaotic space — a factory workshop at 798 before it was turned into an art gallery for Beijing Tokyo Art Projects (fig.15). In the first of the three images, the couple stands motionlessly, confronting the space. Their naked bodies contrast with the harsh, dilapidated environment. But in the second and third images, mist rises around them and transforms the room into a dream world. The two artists embrace each other in the mist, absorbed in their own intimate relationship. From the west wall to the east wall, therefore, these self-portraits constitute a progress

Fig. 14 (page 230) Views of Section Four of *Tui-Transfiguration*: West wall

Fig. 15 Views of Section Four of *Tui-Transfiguration*: East Wall

parallel to the progression from deprivation to transcendence narrated in the first three parts. But the last three images in this sequence of self-representation have an additional significance: because these photographs were first shown in *Beijing Afloat* – the 2002 exhibition that formally started the transformation of Factory 798 into an art space – their reappearance in *Tui-Transfiguration* mark the enormous changes that took place here in just one year.

王蓬 **王蓬**

东八时区，香港，2005

48 页

布面精装，100 册，

286mm×221mm×10mm

平装，400 册，

280mm×215mm×5mm

编辑：何浩

文章：卢迎华、皮力

访谈：陈韦纯 / 王蓬

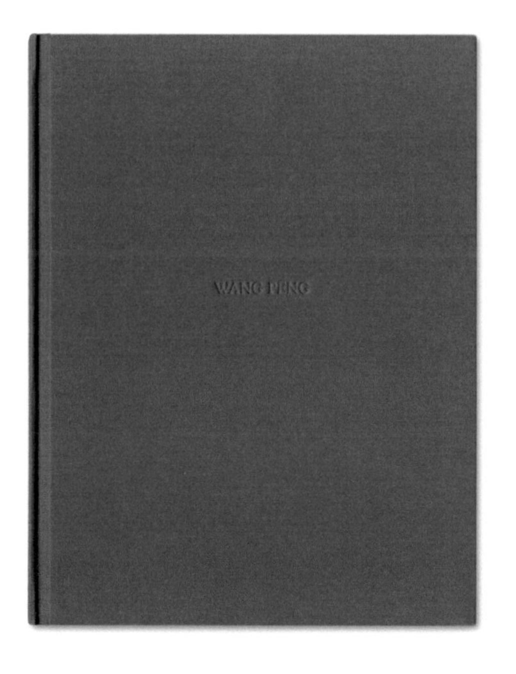

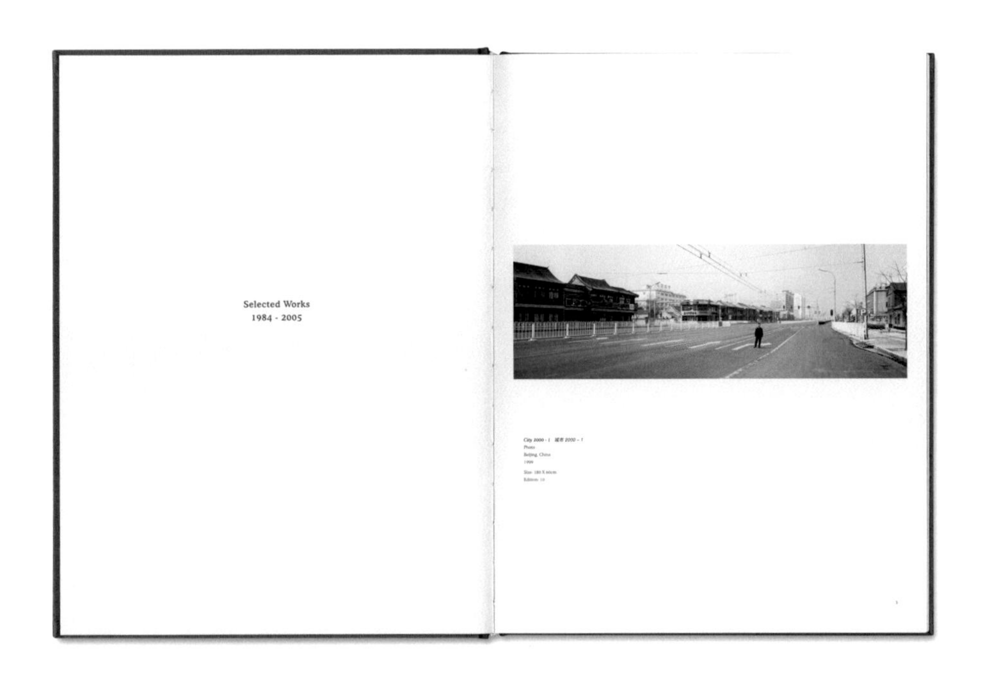
Selected Works
1984 - 2005

有預謀的無意義

——王蓬個展

盧迎華

Wall
Installation
Gallery of Contemporary Art, Beijing, China
1993
The artist bricked up the entrance of the contemporary art gallery in Beijing. The brick wall built by the artist was scheduled to be shown for three days but was forced down on the second day.

ate areas. This was also a kind of displacement between the "virtual" and the "real," the "weak individual" and the "powerful city." In the grand urban space, the insignificant traces of the individual were strengthened and the heartless urban space also began to be subtly divided by the individual. Once again they implied the confrontational relationship between the individual behavior and the urban space. Supposedly in "Three Days," the human action was solid and the city was changing. In "Passing Through," the city was solid and the human behavior was fragile. Of these two works, one was carried out in Beijing and the other in New York. Both of them are international cities but the artist's experiences of living in them indeed varied. In comparison, these two works revealed the consistency of the artist's methodology in his art practice as well as his sensitive perception of his own surroundings.

The third way of displacing: Where is the human body?

For Wang Peng, the urban space might as well be the background for highlighting the individual or a form of constraint on the individual. Be it Beijing or New York, when cities and the ways of living and thinking cultivated in cities develop to a certain extent, they would certainly go back to becoming a kind of restraint on the individual. It's exactly this kind of reflection of the city's restraint on the individual that made the reflection on the existence of the individual the inevitable outcome of his work. In this kind of view, the individual and the city become diverse and microscopic at the same time. The existence of the individual is represented as the connection between the space and the body. The city was translated into a set formula for the audience's thinking and viewing. In these works, Wang Peng abandoned the apparent moral judgment and implication in his early works and instead transferred the responsibility and conclusion of making judgments to the audience by producing paradoxical and provocative imagery.

This shift was fully present in some of Wang Peng's photography, which he began to make from 1998. In "Memory" (1998), he took many photographs of people in the city and then those of cities and people's clothes only and placed them together. The human flesh was either abandoned or existed as an unnecessary hint. What the artist presented to us were the clothes in the filthy corners of the city. They were the displacement of the human body as well as suggestions of traces. The annulment of the human body was further taken to an extreme state in "City 2000" (2000). The artist completely wiped out the human figures in the bustling urban landscape through digital technology. He nearly disposed of the existence of all the people in the urban space. The urban space shaped by human behaviors thus appeared to be cold-hearted and merciless. Although the artist himself was kept in the picture, the "body" now had been replaced by a "material" being. In both "Memory" and "City 2000," human and their bodies appeared to be all fragile and annullable in contrast to urban landscape and scenery. The key of his work lied in the dead end that his audience was forced into after the city has removed the flesh, as well as the relative line of thoughts that is aroused.

The disappearance of the human body has been an outstanding visual feature of Wang Peng's work after 2000. If we only analyze these works in comparison with his previous pieces, they lack the revolutionary nature of his pervious work ("Wall") and the heroic spirit of another work ("3 Days"), but they don't lead to a pessimistic conclusion. They have no less of a relatively objective and steady attitude. In his recent works, Wang Peng has removed the existence of "me" as an artist the way he removed the existence of the body. In these works, he selectively preserves the objectivity of things and landscapes and reveals the meaning of artwork by changing the viewing habit of the audience. In his photography work "Scan" (2002), a slice of the body is cut out so that the audience can only view a human body the way they look at an unintelligible object. In his photograph "Landscape," the artist altered the proportion of landscape photographs so that normal landscape acquires a certain bizarre look ("Landscape," 2002). There is a certain humor and intelligence in these works. The key of this kind of wisdom and humor lies in the artist's stressing the subject's (audience) subjectivity through removing the object's objectivity. By removing the artist himself and the human body, the artist has gained the freedom and pleasure of thinking in the solid urban space.

Actually, when we study Wang Peng's latest media works in detail, we would discover that the human body has not really disappeared. It has only disappeared visually and in contrary, has been replaced by a new existence in the urban and social space. This kind of existence is indicated through the action, behavior and thoughts of human beings in the city. In "Breathing" (2004), clothes fully covered by microphones turned the movements of human bodies into sound so that the movement of the body is replaced by sound. In "Call 13641041729" (2003), the communications among people were translated into the digital signal of mobiles to generate the obvious change of light and shadow of the work itself. As Wang Peng ventures into the field of multimedia, he's abandoned the exclusiveness of the visual and instead replaced the existence of the human body with physical existence. This is a process of the alienation of the individual and the flesh as well as the squeezing of manual creation by technological development, it is however, still presented as romantic and poetic as the nature by the artist.

Passing Through
Performance
New York
1996-1997
The artist hid a roll of tiny thread inside his clothes, with one end of the thread going through the back of his clothes and tied to an architecture in the city. The artist then walked freely on the streets of the city.

[illegible]

行走在内外之間
——王蓬訪談

陳韋純

Inside & Outside
Installation
Vermont Studio Center, Johnson, U.S.A
1996
The artist pasted the character "inside" on the outside of the gallery window and pasted the character "outside" on the inside of the gallery window.

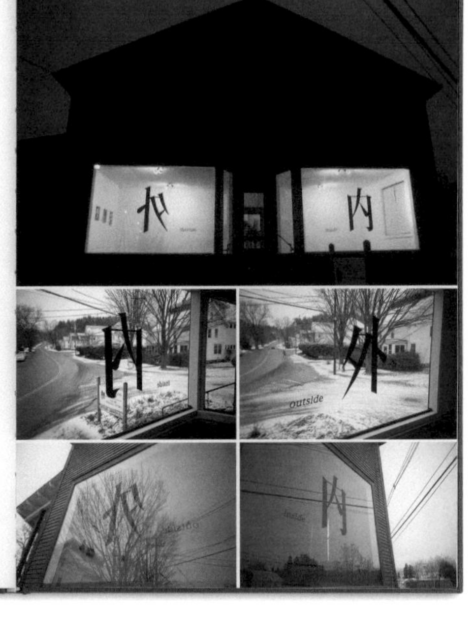

[illegible]

Gate
Performance

[illegible]

《王蓬》

我已经想不起来跟王蓬是在什么场合认识的了，但我们真正熟起来，是2005年的朝鲜之行。那年夏天，我和王功新、林天苗、王蓬、张培力、耿建翌几个人在丹东报了个旅行团，去朝鲜玩了一圈。当时这个团里二十来人，主要是去“三千里江山”追寻“卖花姑娘”的老头老太太。所以开始时朝鲜那边对我们看得也不严，但没想到我们一个劲乱拍乱照，一没看住就自己往外溜，弄得他们防不胜防。行程快结束时，朝鲜方面一定要检查我们的照片和录像，我们则扬言要找使馆寻求保护。王功新笑谈说朝鲜人怎么也想不到，这个团里混进了一帮中国的Avant-Garde。这话说得一点没错，尤其对王蓬，如果只用一个词来定义他，那只能是Avant-Garde。

王蓬是个有着心不在焉的决绝和魂不守舍的清醒的人，这是一种非常特殊的气质。1990年，他是个在中央美院附中教授最严谨的学院写实素描的老师。他张贴出海报，说某月某日他将在位于北京最繁华地段的当代美术馆举办主题名为《墙》的个人作品展。展览前夜，他雇佣一辆马车，拉着一车砖来到展览现场（当代美术馆位于北京最繁华稠密的商业地段，稍大点的汽车根本开不进来），几个农民三下五除二，在王蓬的指使下把美术馆大门砌得严严实实。第二天，观众如约而至，情形可想而知。同样可以想见，原本展期三天的展览当天上午就被封掉了。那年，我还是个初中生。

像王蓬这种从骨子里特立独行的人，在体制里自然痛快不了，几年后他离开了学校。而他作品中的非物质属性，使得艺术市场在他面前也有些不知所措。他也许因此获得了始终如一的独立，但同时也失去了一些他本应得到的世俗利益。在结束了朝鲜之旅回北京的火车上，王蓬说他秋天有个个展，预算有限，问我能否帮他设计个折页之类的东西。当时我没置可否，只是说回来看看内容再说，但心里已决定，要尽力帮他做一个比折页像样的设计。如果说一开始做设计我心中就有一个理想的话，那这样的委托人和项目就是我的理想，而且直到今天也从未改变过。

2005年，我已经有了一些书籍设计的经验，而且在材料供应商和印刷加工商当中也积累了些许口碑和人脉。我使用了这些资源，用设计之道平衡调配。比如我说动一个纸行把他们的一批质量极好，但因为数量有限始终没能卖掉的尾货纸半卖半送地给了王蓬，我再根据纸张数量来编辑内容、制定开本、规划页数。最终用了无法想象的低价，制作出了一部高品质画册。今天回过头看，这本书其实在我做过的项目中并不是节点式的，无论从哪个角度讲都不能说特别或重要。但这是我偏爱的一本书，因为它从另一个角度证明了设计的力量。

当时在做王蓬这本小书的时候，我脑子里面还有一个“理想”的方案。如同前面提到的《墙》，其实王蓬的很多作品都有着这种“棒喝”的意味，只是比起黄檗禅师的严厉，王蓬更像是在用恶作剧般的玩笑使人得到启示。因此我的“理想”是做一本中国传统古籍样式的书，夹层、筒子页，但所有内容都印在夹层内侧。整本书翻起来就像一册用废纸背面装订成的白本，图文隐隐透出但又无法观看——想看也很简单，“破墙而入”——裁开就是了。

八年后的2013年，我跟王蓬有了再次合作的机会。王蓬还没忘我当初的那个方案，旧话重提，只是半开玩笑地说这个方案好是好，但自己的作品就够“观念”的了，书就别弄得更“观念”了。其实我们想到一起去了，我如今对于设计的认识大概可以套用纪录片《舌尖上的中国》的一句解说词：“高端的食材往往只需要采用最朴素的烹饪方式。”而另一方面，千帆过尽，我们确实也不在“卖萌”的人生阶段了。

荣荣和映里 **超越——荣荣和映里的摄影近作**

Walsh 画廊，芝加哥，2005

215mm×292mm×7mm，60 页

平装

1000 册

前言：Julie Walsh

文章：巫鸿

Beyond

Wu Hung

Opening the senses, disappearing into vision, and then emerging on the other shore of sensation.
— inri

When I first saw *Mt. Fuji* and *We Are Here* — Rong Rong and inri's two photographic series which constitute this exhibition — they reminded me of a line I had heard before – "To go beyond is to vanish from view." Like the irony captured in this line, the photographs represent disappearance as a visual spectacle. *We Are Here* takes place in a state-owned factory in Beijing, now abandoned. (A blurred slogan on the wall – "Chairman Mao is the red sun in our hearts" — is a remnant of the Cultural Revolution.) The two artists stand motionlessly in the deserted, chaotic factory shop, confronting the space. Their naked bodies contrast with the harsh, dilapidated environment. But slowly, mist rises around them and transforms the industrial ruin into a dream world. They embrace each other and vanish into it. Miraculously, we are able to follow them to the other side of time and space, witnessing their rebirth. There, as the frozen, frightening winter landscape of Mt. Fuji inspires only joy, it is reborn with them.

§

In 1997 inri quit her job at *The Asahi Shimbun* as the popular newspaper's portrait photographer: she could not stand taking any more pictures of celebrities, even though people had praised her portraits for their psychological depth. She was free now, but hers was an empty freedom that frightened her. Looking through the photographs she had made in the past, she was astonished by their meaninglessness: they did not evoke in her any pleasure nor even nostalgia. What did this mean? — She asked herself. She had worked so hard to create these images, and so they should have been her treasures, the proof of her artistic pursuit. But now they disgusted her with their polished look and pretty figures. She stopped looking at these pictures and attempted "to expel" them out of her system. This effort again proved difficult: in her words, "the shadow of the past" was too thick and deep. The freedom she had gained had no substance. She traveled and tried to be busy; but her questions about her own worth as an artist and a person remained unanswered. Her autobiographical piece, *Traveling to Shikoku*, documents this crisis:

> *My spirit and my body were split, scattered about. I thought the only way to unify them, to experience the reality I couldn't escape, was to make a pilgrimage. Ever since youth I've had a desire to do this before I die, but I had never thought I'd do it so soon.*
>
> *I had devoted myself to photography, but it was precisely because of this single-mindedness that I had become blind. I was constrained by the overload of things I saw. Actually I saw nothing. Though this was the case, all I could do was live by photography. At first I didn't want to take my camera, but then I changed my mind and brought it along. I could take pictures or not, but if something happened that I wanted to shoot I could.*
>
> *Even if I went on a pilgrimage nothing may change. Maybe I just wanted to experience the process, experience the unchangeable self. But though this was perhaps the case, I still had to do this in order to truly realize that I couldn't change. I was already 27 years old, but I was still so confused. I didn't know why; but I was surrounded by my own metamorphoses; my mind got heavier and heavier, more and more crowded. I felt I was stupid. It was only through action that I could think clearly, that I could experience understanding.*
>
> *Upon returning from my trip to Shikoku, my spirit and body actually were unified. The excess fat and flesh was dispelled, emaciated to skin and bones. I became empty, utterly devoid of content. Pitifully watching the trash filling up my life, I thought, I didn't want anything.*
>
> *Then, how would I live from now on? It was then, when I hazily looked upon a dimming cityscape, in that emptiness, my empty self, that I met Rong Rong.*

§

Rong Rong arrived in Beijing in 1992 with wide-open eyes and a newly purchased camera. A farm boy from China's southeastern province of Fujian, he had never left home before. He was skilled in the fields but had failed almost every course in elementary and junior high school except for studio art. This was followed by three failed attempts to enter a local art school (mainly because of his poor performance in math and other general exams). By chance he discovered photography and developed a passion for it. First he rented a double-lens camera to take his sisters' portraits and landscape shots. He then struck a bargain with his father, the manager of a local shop, to work as his employee for three years in exchange for the freedom to leave home and a sum of money to start his new life with. So this was how he got his own camera and went to Beijing, China's capital, which by then had also become a Mecca for young avant-garde artists throughout the country.

In Beijing, Rong Rong attended photography classes and became quite good at making the kind of arty, sentimental pictures favored by popular photo magazines. One of his works even found its way to entering a National Photography Exhibition. But life was hard and the occasional public exposure of his pictures had little financial return. When his savings were gone he tried various odd jobs, including taking passport photos in a commercial studio. He changed addresses frequently, often guided by the cheapest housing on the market. In early 1993 he moved into a tumble-down village on the city's east fringe. Later known as Beijing's "East Village" (Dong cun), it would become home to an artistic community that produced some of the most daring works (mainly performance and photography) in contemporary Chinese art, before it was closed down by the local police in June 1994.

Rong Rong abandoned the kind of popular photographic style that had earned him a place in magazines and official exhibitions. The Village attracted him with its ugliness and anonymity. He explored the secrets of its refuse-filled dirt roads and courtyards with his camera, his young face and bare torso occasionally reflected in broken windows

In Fujisan, Japan 2001

王功新、林天苗

这儿？或那儿？

东八时区，香港，2005

286mm×221mm×19mm，144 页

精装，裱特殊定制丝绸，裹纸质护封

2000 册

编辑：何浩

文章：皮力、林似竹

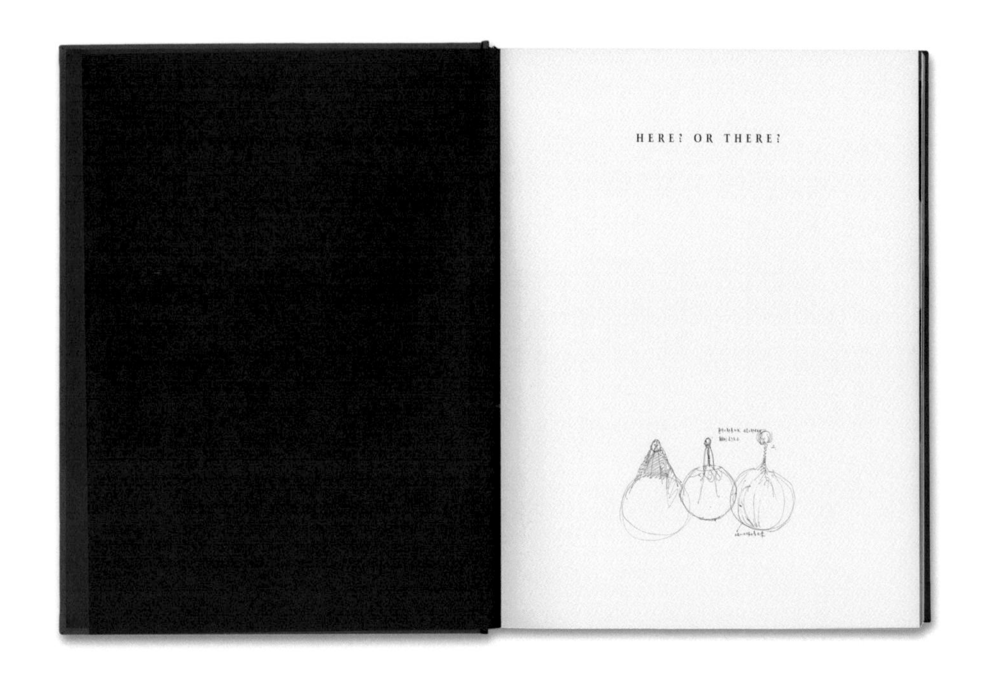

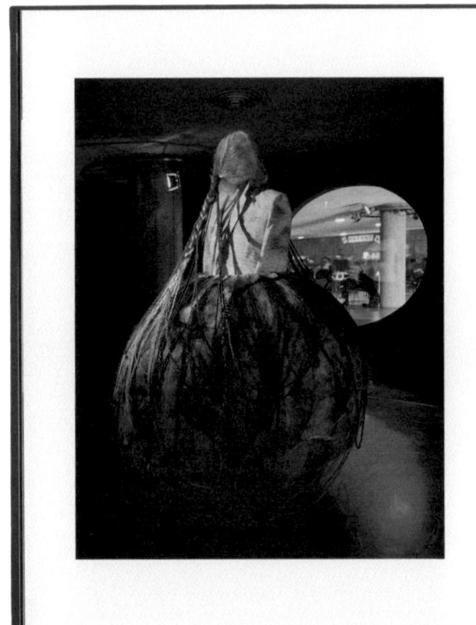

~ CONTENTS ~

Between Body and Landscape: Unscrambling 'Here? or There?'

Pi Li

Here? or There? was begun and completed in 2002. The spotlight on this work was the Shanghai Biennale 2002. But it was just a spotlight. In the fact, the work was the culmination of many important factors: factors that permit us to take this work as an example that illustrates certain structural changes in the development of contemporary art in China after 2000.

Wang Gongxin and Lin Tianmiao began working on *Here? or There?* almost ten years after their return to China from the US. Both can be said to exhibit a number of features typical to artists of their generation: they had each undergone artistic training pivoted on the techniques of academic, wholly realist drawing, before being initiated into an awareness of conceptual art aboard. Fortunately for them, the basic skills in the academic discipline of realism did not lead them to fall into the realm of commercialized galleries. Instead, they were inspired to focus their interests on learning new things. In the early 1990s, when they returned to China, they felt it imperative to integrate the aesthetic language they had acquired aboard with the vigorous mood of the "avant-garde" movement in China. In the 1990s, Chinese contemporary art was the subject of a series of exhibitions, which prompted interest from a handful of galleries abroad. The political and social climate, which had been catalyst for developments in the contemporary art scene in China, degenerated into a dispensable background. The primary reason for this was the sudden commercialisation of Chinese contemporary art. Commercialization typically required that an artist deverop a definite and recognizably personal style and iconography, and that there be repetition of the symbols employed. No doubt, for these two artists, who had lived aboard for nearly ten years, this had a

great impact upon the ideals they aspired to. They had been outside of the prevailing local ideology for ten years, and when they woke up to it, all became different.

Both adopted their own individual way to address their situation. In the beginning, Wang Gongxin explored the medium of video art. Scholars abroad could never understand the ways in which video art touched upon ideological issues when it began to emerge in China in the early 1990s. That we say it possesses some nature of an ideology implies that these video artists of earlier generations tried to realize a conceptualization of art through the medium of video, thereby challenging the tendency towards commercialization of art that permeated the art world. The works Wang Gongxin created through the 1990s explored video and installation. In these explorations, he highlights the contrast of life experience prompted by his return to China (like *Sky of Brooklyn*, No 1 *Shen Fen*). Another element was a strong desire to transform images through media technology (for example *Baby Cradle*). His works did not make ideology a trendy symbol for commercial thrust, something common to so many other works of the time.

In comparison with Wang Gongxin's slightly peripheral status, Lin Tianmiao seems to have been more fortunate during the 1990s. This was tied to her female identity, of which she had little interest. In order to survive in New York, Lin Tianmiao had worked as a textile designer. This measure of survival lent their life another aspect, awarding her skills in a certain artistic language, and alerting her to the need for a unique identity. These two aspects were a catalyst for her early works. In these works, she successfully transformed the process of weaving and twining employed for textiles into a visual language that corresponded with her own female identity. In the early 1990s, feminism became a new theme in China. Then like other areas of Chinese contemporary art, feminism became locked into superficial symbolization, so that at this time, Lin Tianmiao caught the attention of society as a result of the special "skills and techniques" she used. Audiences and critics were always surprised by her strong language and the visual effect of her works, but hardly gave mention to the concepts behind them.

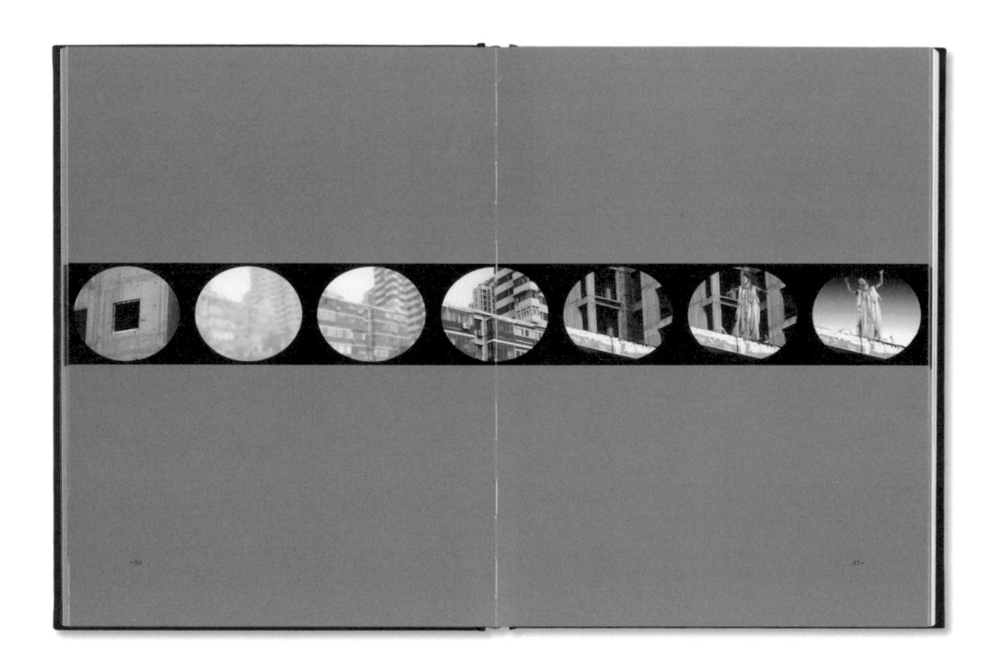

-98

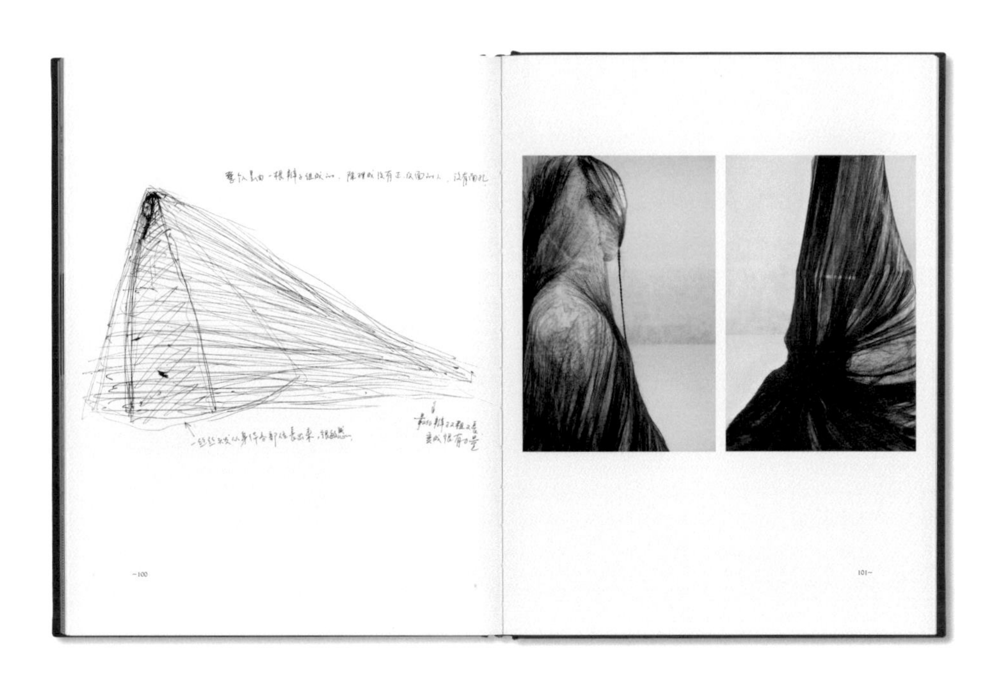
-100
101-

Here? Or There?: A History

Britta Erickson

Wang Gongxin and Lin Tianmiao's *Here? Or There?* was first exhibited at the 2002 *Shanghai Biennial*, where its physical beauty and conceptual power transfixed many in the audience: it was undoubtedly the stand-out work of the exhibition. Earlier in the year I had seen components of *Here? Or There?* in the studio and expected the final result to be impressive, but it went far beyond what I had anticipated. The work's extraordinary success is a tribute to the collaboration between the wife-and-husband team of Lin Tianmiao and Wang Gongxin. This was a rare instance of two strong artists with prominent individual careers melding their talents to produce something beyond their individual abilities, like two metals melted in proper proportion to create an alloy.

Concept

Last December I asked the artists to describe to me the process through which they had developed *Here? Or There?*. This is what they told me:

About a year before the 2002 *Shanghai Biennial*, the curator Pi Li informed Lin Tianmiao and Wang Gongxin that he wished to feature works by both of them in the biennial, but that he would only be able to select a very few Chinese artists. Since for him to choose between them presented a dilemma, they determined to develop a collaborative work. Lin Tianmiao had long been thinking of creating something relating to clothing, and Wang Gongxin suggested that this might be the right moment. It was very difficult to arrive at a specific approach, however. For about six months the two artists tried to find a meeting ground where they could collaborate, but were stymied. Their thinking was very different.

Finally, one evening they stumbled upon a common ground, in their per-

~110

ceptions regarding contemporary society. Before they had returned to Beijing from living in New York (from 1987 to 1995), they had felt they understood how people dressed and thought, how they went about their daily lives: things seemed predictable. When they returned to China, however, they were suddenly not so sure. Everything felt unstable. Although people now had money and could afford such important trappings of a stable life as a house, they conducted their lives as if they had no faith in society. Whenever they earned money they immediately spent it, as if the future was uncertain. They had returned to old superstitions, such as visiting the fortune-teller prior to making major decisions, or consulting the *fengshui* master. Why were people, particularly the rich, suddenly believing in these things? Lin and Wang talked about their shared observations for hours, honing in on the essentials: psychologically, people felt as if they had lost control of their living situation. It was as if a door had been opened between the human world and the spirit world, rendering the familiar suddenly strange, as happens in the famous Qing dynasty novel, *Liaozhai Zhiyi* (Strange Tales from a Chinese Studio).

Liaozhai Zhiyi by Pu Songling (1640-1715) provides a model of a world in which human and spirit realms are permeable. Humans enter into romantic relationships with figures who are later revealed to be ghosts or fox spirits; paintings take on life; humans morph into animals; and devils materialize to suck the life's blood from humans. Frequently, denizens of the spirit realm disguise themselves as humans, surprising their unsuspecting victims: nothing is certain, and in uncertainty lies danger.

The subject of the 2002 *Shanghai Biennial*, subtitled *Urban Creation*, was architecture and the city. Considering this topic, Wang Gongxin said, "When you talk about urban life, you aren't just talking about the buildings; you also have to talk about people's lives. That is absolutely essential! These days you get the sense that architecture is more important than the lives being lived inside of it. People are concerned with the architectural forms, and with the *fengshui*, but not with the life. That should be the starting point."[1]

The artists thus chose to focus on the life of the city, with an emphasis on uncertainty delivered by the supernatural.

111~

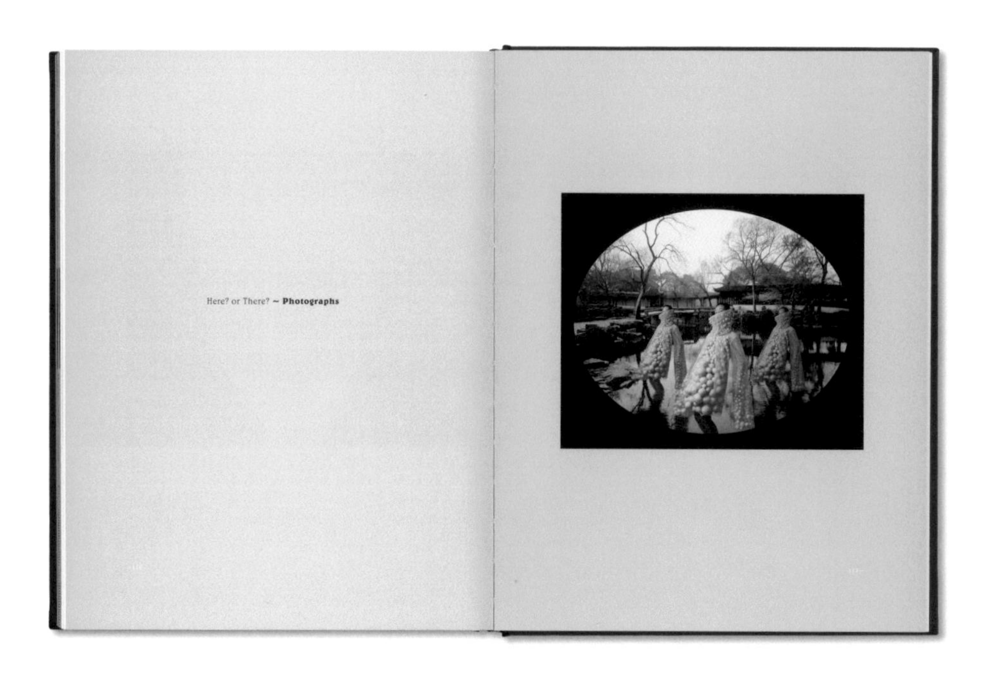

Here? or There? ~ Performance
April 21, 2004
The Gardens of Prince Gong's Mansion, Beijing
~150

郭凤怡 **谁是郭凤怡**

长征空间，北京，2005

420mm×282mm×18mm，96 页

布面精装，切口刷红

1500 册

编辑：卢杰

文章：张颂仁、高士明、卢杰

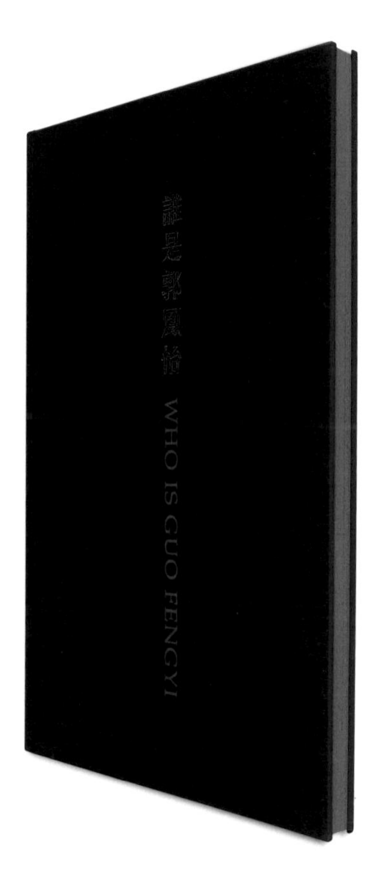

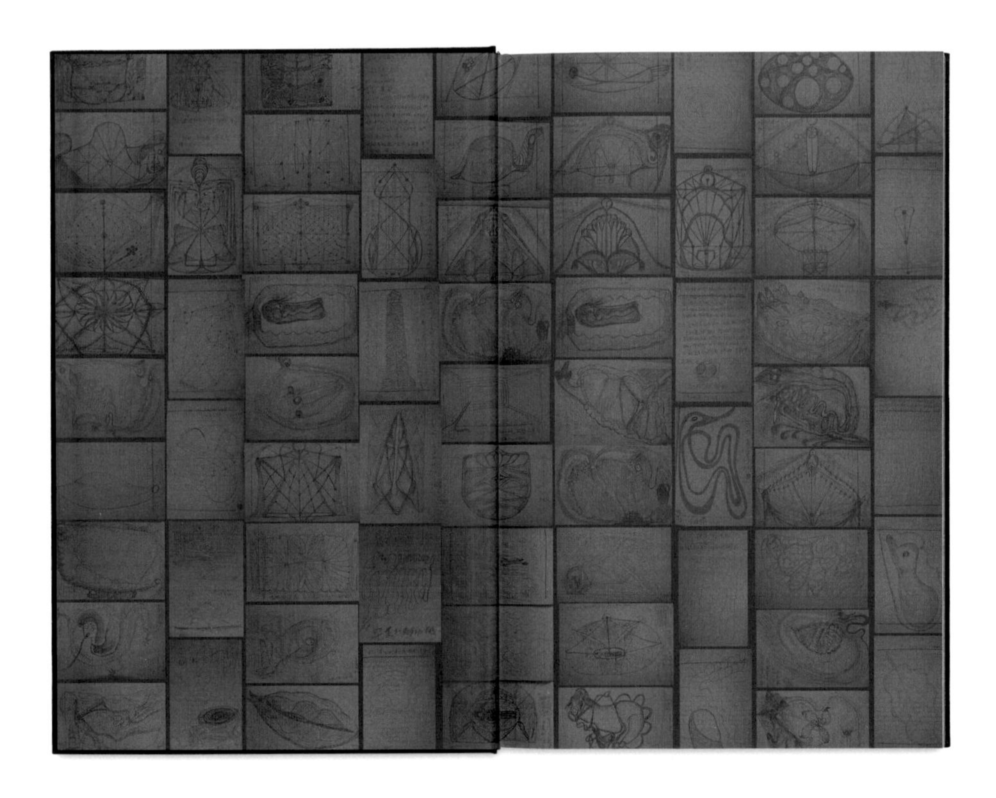

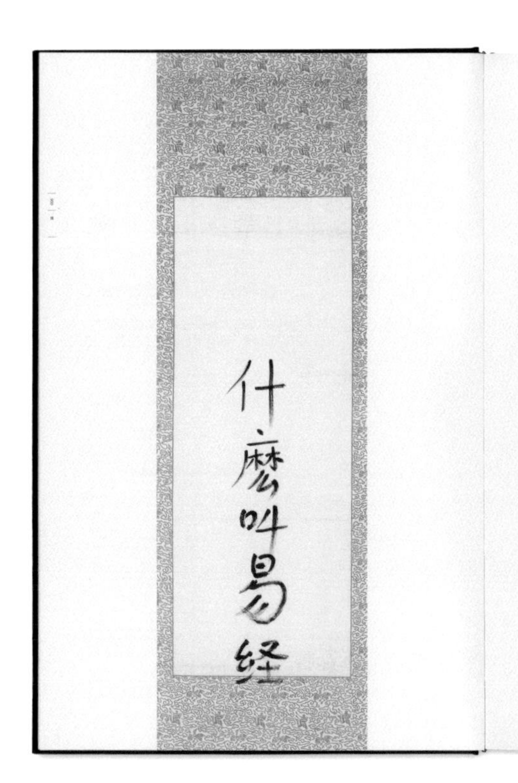

Another Universe: The art of Guo Fengyi

| Chang Tsong-zung

天外有天——思考郭鳳怡的繪畫 | 張頌仁

As a phenomenon of the art world, Guo Fengyi's significance is not just artistic, but also in the alternative worldview she has brought to that of the "modern". The fact that Guo is principally regarded as a spectacle reflects the limitation of cultural institutions; officially accepted forms of cultural expression are codified to such an extent that it is difficult to find a respectable position for worldviews originally indigenous and arising from older cultural memories. When we approach her works, it is not easy putting ourselves in the author's perspective: seeing them as cosmic diagrams, as divining apparatus of oracles. The best we can do is to place Guo in the category of the "creative" arts, and make her an "artist".

From the perspective of modern art, a seemingly acceptable role for Guo is that of "Outsider Art", or "Naive Art". What this means is acknowledging the difficulty of fitting certain "creative" products into the narrative of official art history; therefore, recognizing its genius as an alternative artistic production becomes a convenient solution. This approach does not come without problems. Surveying the major "Outsiders" canonized in the West, and taking as reference the collection put together by Jean Dubuffet and lodged in a museum of Naive art in Lausanne, it becomes evident that many of the representative geniuses are people medically diagnosed as mentally abnormal, some confined to psychiatric wards for life; and their works are read as coded texts emerging from the hidden recesses of man's psyche. The "Outsider" carries the stigma of the abnormal. Of course there are leagues of "Outsiders" not considered mad, but the defining feature of the "Outsider" still appears to be the strange, and not easily communicating, private world of the mind they live in. We believe these worlds communicate "truths" suppressed by daily life, secrets guarded only by the gifted or ones condemned, and delivered only in the language of symbols. Therefore, what "Outsiders" possess are more than alternative modes of art requiring the conventional art historian's reflection, but exceptional insight into unfathomable worlds. The more complex and rich the structure of their worlds, the more "scientifically" and artistically valuable they are as "Outsiders". This is the position in which we find Guo Fengyi's work, and it appears that this is also the best way to judge her work.

Judging from the social role of Guo and the type of work she produces, it is not difficult to see her as an example of the "Outsider". But Guo's paintings did not start out to be art; Guo had no ambitions, or intentions even, for the art world. She was recruited by the curatorial project of the "Long March", and now she is identified as an artist. Apart from this Guo does not seem to be much interested in art. She started to draw after practicing qi-gong for health; during these meditative states she experienced special visions that were put down on paper. Much of the subject matter of her drawings, as well as the concepts and physical structures she uses come from traditional Chinese philosophy, myths and medicine. Her claim to art is her coherent expressive style and consistent formal pictorial structure. The drawings are ghostly, uncanny to the degree of inspiring fear. The subjects she cover include mythical figures, historical anecdotes, astrology, geomancy and medical theory, which in themselves constitute separate cultural spheres each complete with sophisticated narratives. By mixing these aspects of cultural knowledge in her work, Guo's drawings become the meeting point of history and myth, knowledge and mysteries. These are the features that define Guo as an "Outsider" of the art world.

From the viewpoint of art, it is relatively easy to discuss Guo's work. Her drawings are impressive and emotionally touching, the pictorial character is distinct and strong, the expressive means are coherent even though limited. Both the mysteriousness suggested by the works themselves and the complexity of the cultural world they refer to are intriguing. All these are merits that justify the works as art. However, from the perspective of Guo, or from that of the "Outsider's" alternative world-view, the value of her works goes beyond that of art or has little to do with art. The alternative value should be that of their cultural and intellectual substance. If cultural content is the most significant aspect

[illegible]

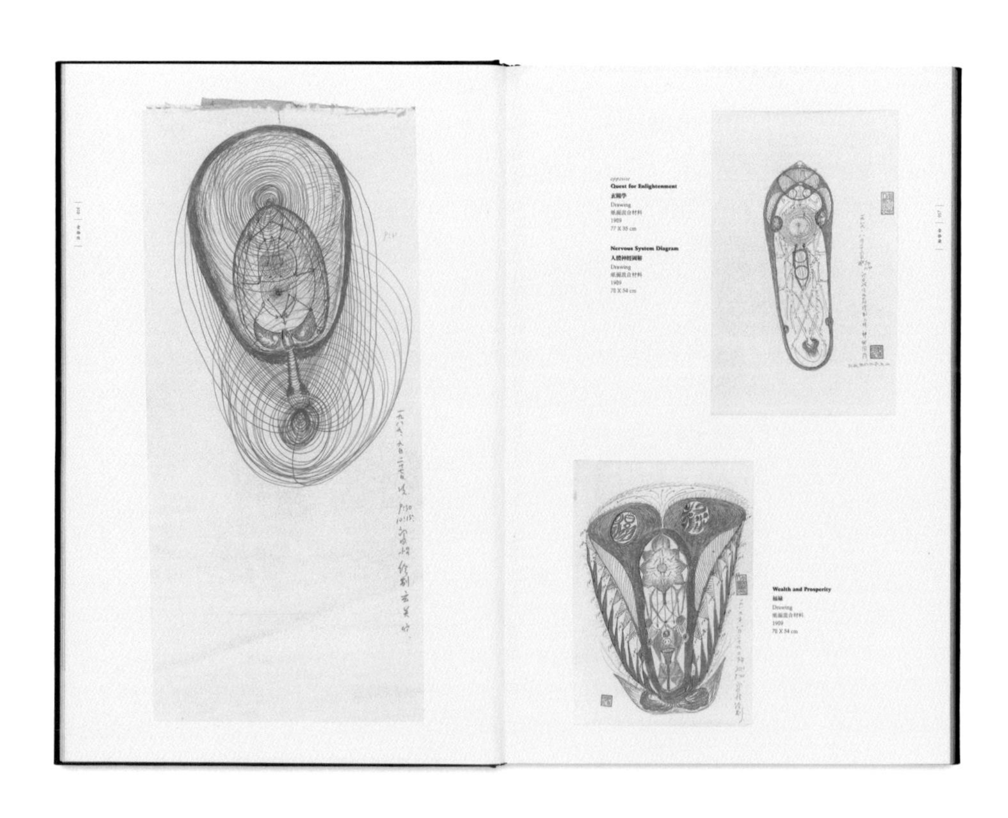
opposite
Quest for Enlightenment
Drawing
Nervous System Diagram
Drawing
Wealth and Prosperity
Drawing

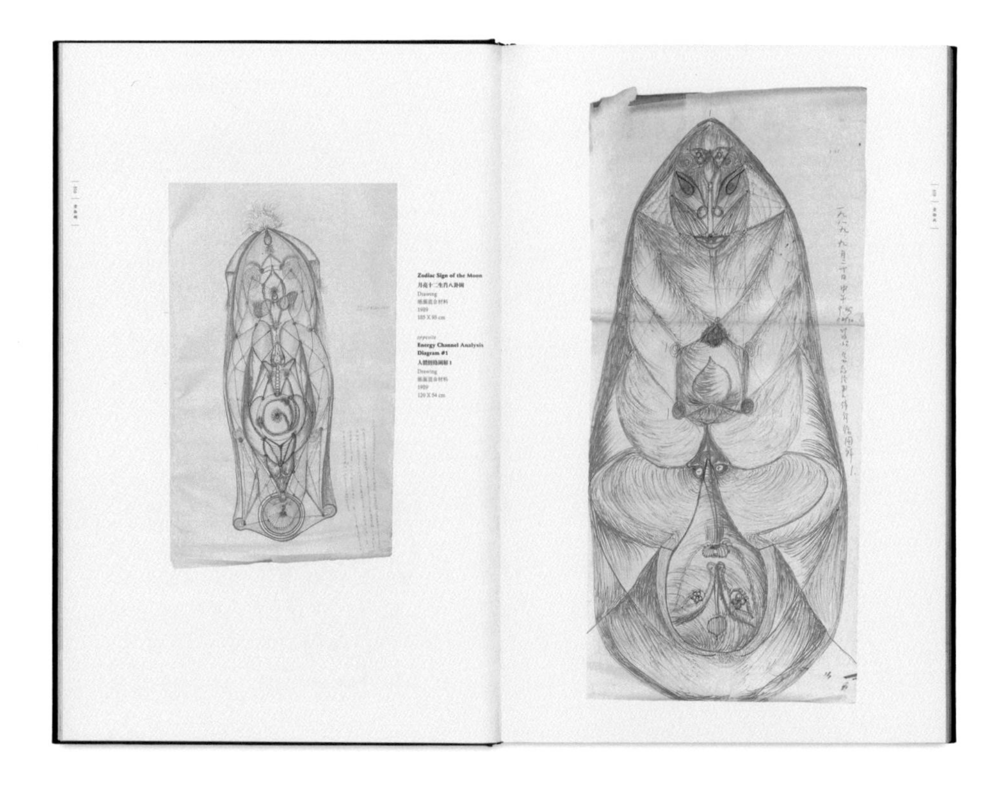
Zodiac Sign of the Moon
Drawing
opposite
Energy Channel Analysis Diagram #1
Drawing

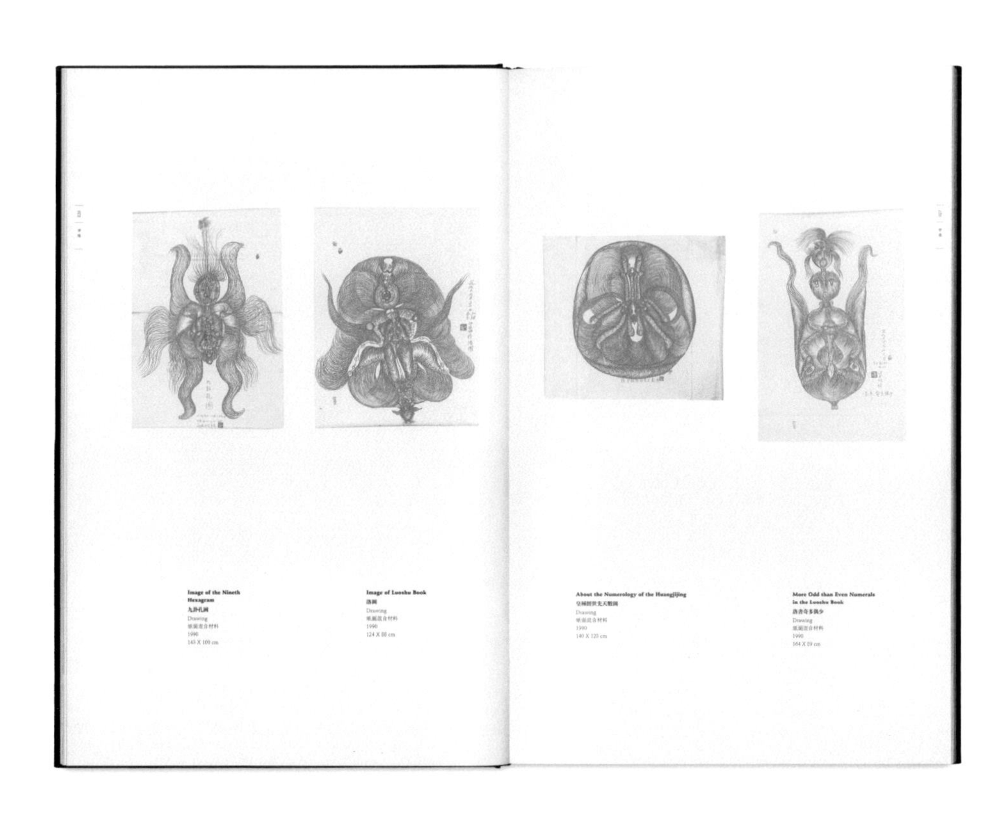
Image of the Nineth Hexagram
Drawing
纸面混合材料
1990
143 X 100 cm
Image of Luoshu Book
洛圖
Drawing
纸面混合材料
1990
124 X 88 cm
About the Numerology of the Huangjijing
皇極經世先天數圖
Drawing
纸面混合材料
1990
140 X 125 cm
More Odd than Even Numerals in the Luoshu Book
洛書奇多偶少
Drawing
纸面混合材料
1990
164 X 89 cm

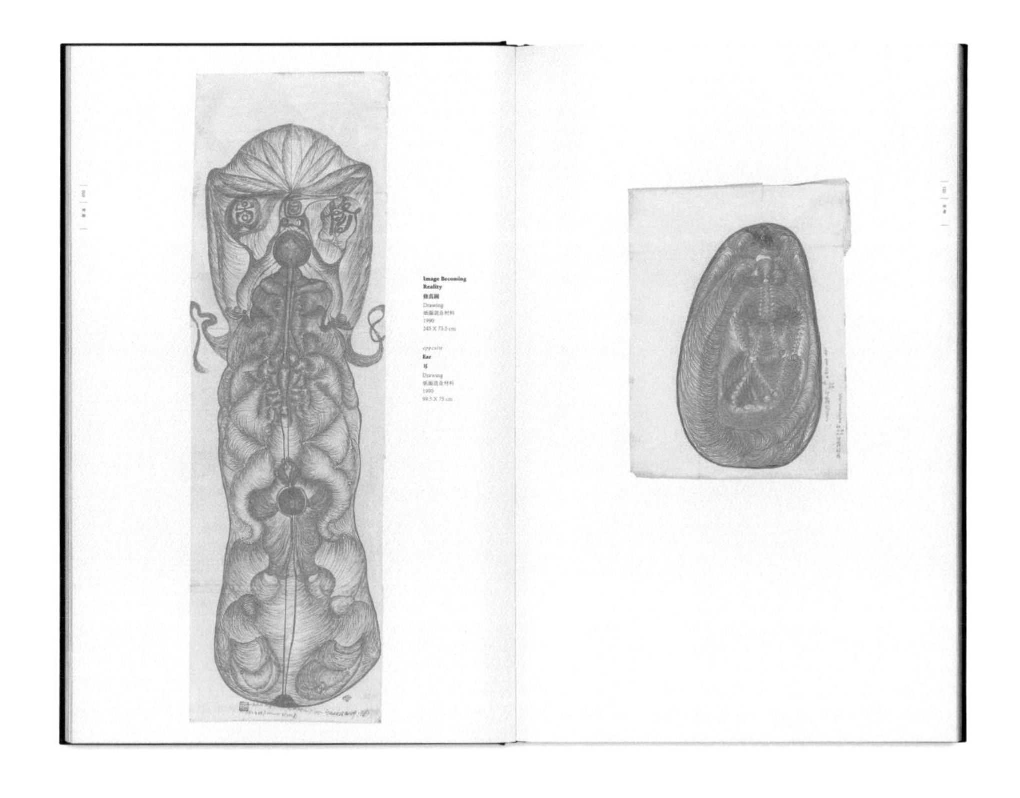
Image Becoming Reality
修真圖
Drawing
纸面混合材料
1990
245 X 73.5 cm
opposite
Ear
耳
Drawing
纸面混合材料
1990
99.5 X 75 cm

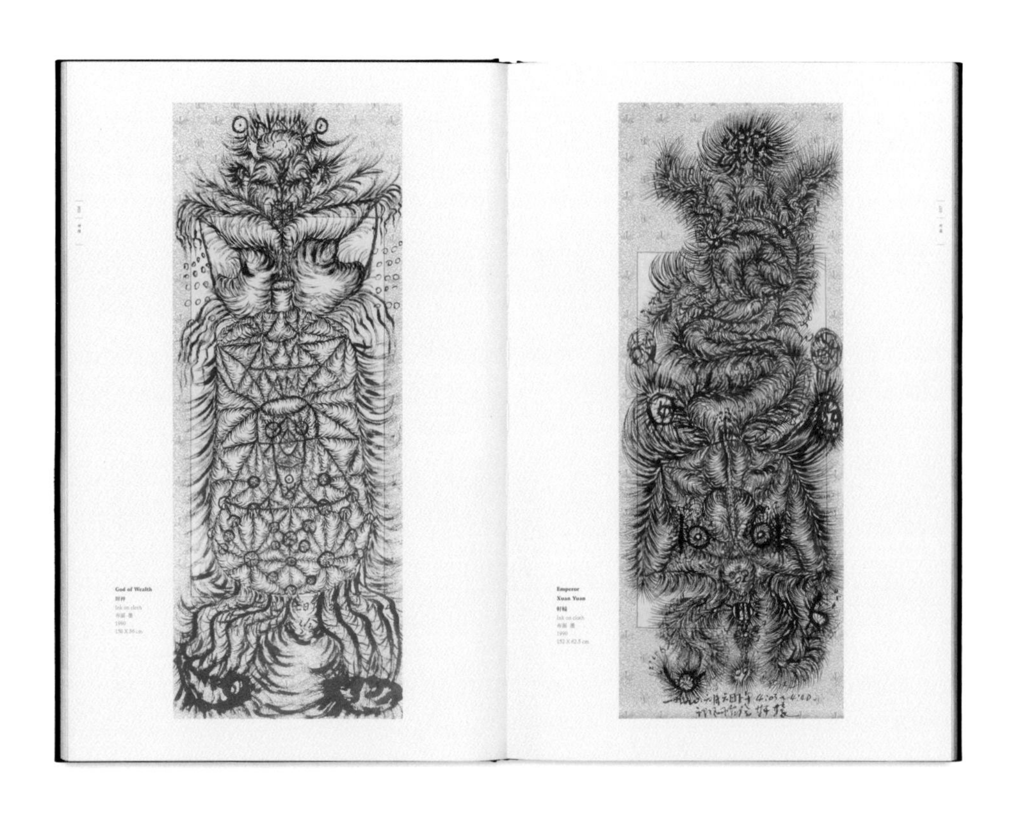

God of Wealth
Ink on cloth
Emperor
Xuan Yuan
Ink on cloth
1990

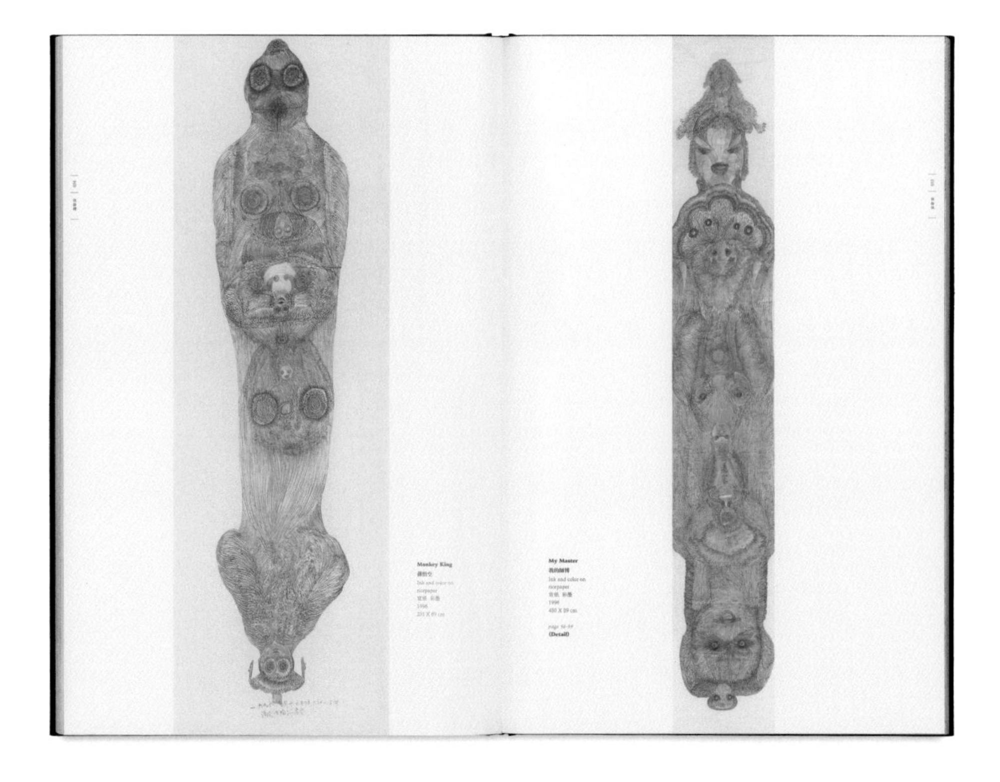

Monkey King
ricepaper
1996
My Master
Ink and color on
ricepaper
1996
(Detail)

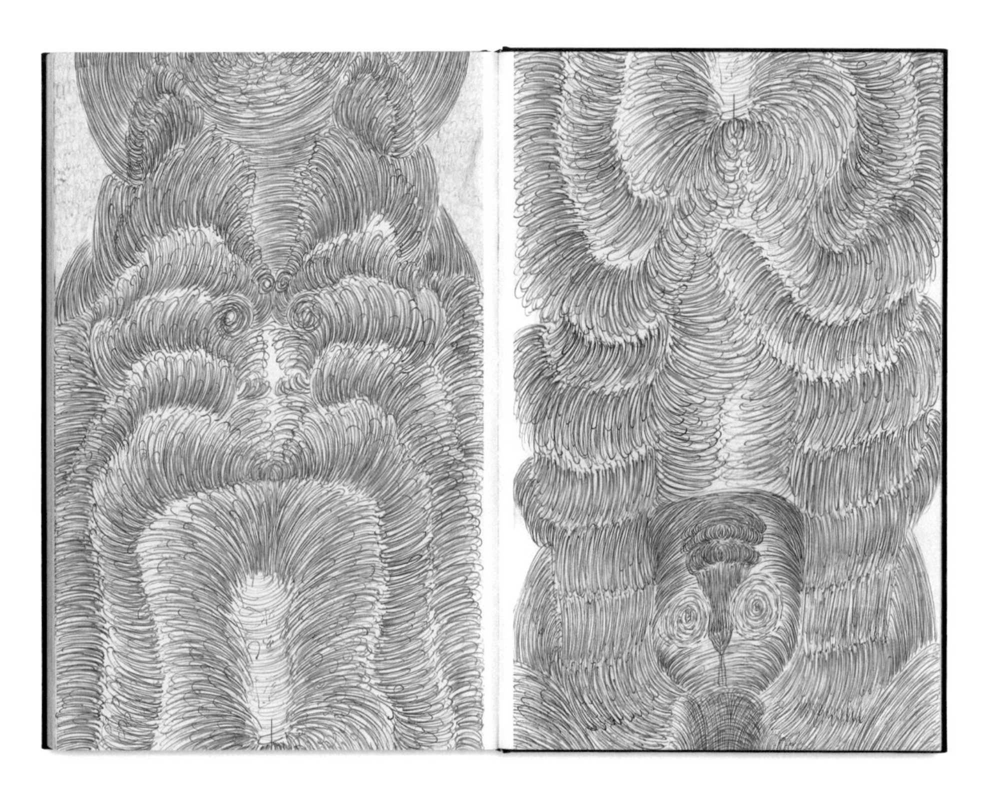

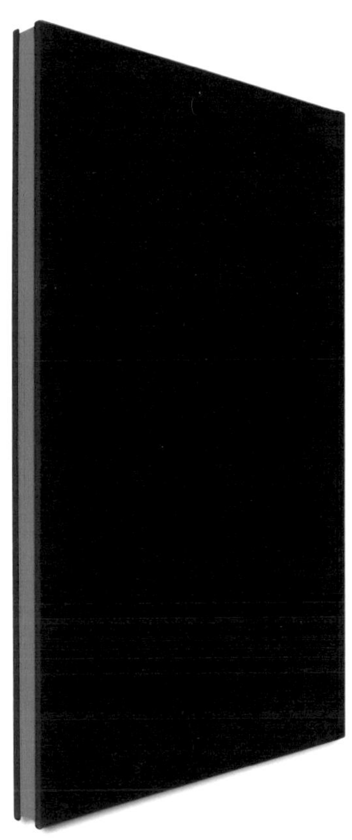

岳敏君 **复制的偶像：岳敏君作品 2004—2006**

何香凝美术馆，深圳，2006

286mm×221mm×12mm，128 页

纸面精装

5000 册

编辑：凯伦·史密斯

文章：冯博一

艺术家陈述：岳敏君

Reproduction Icons: Yue Minjun Works, 2004-2006

復制的偶像：岳敏君作品 2004–2006

何香凝美術館

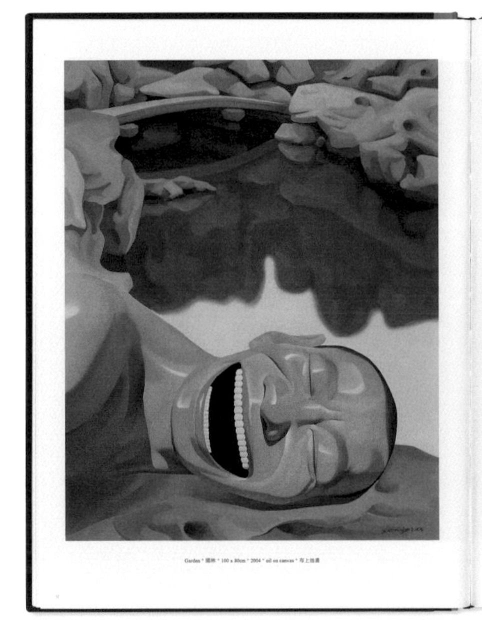

Garden * 園林 * 100 x 80cm * 2004 * oil on canvas * 布上油畫

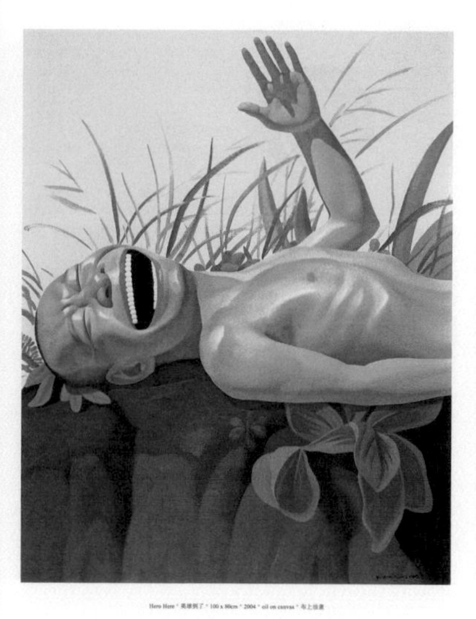

Hero Here * 英雄到了 * 100 x 80cm * 2004 * oil on canvas * 布上油畫

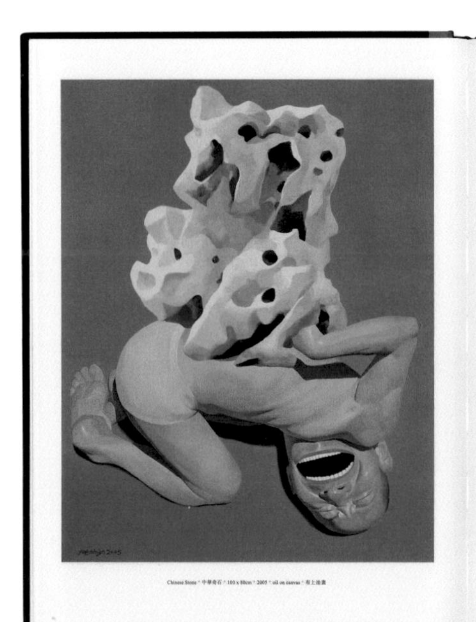

Chinese Stone * 中華奇石 * 100 x 80cm * 2005 * oil on canvas * 布上油畫

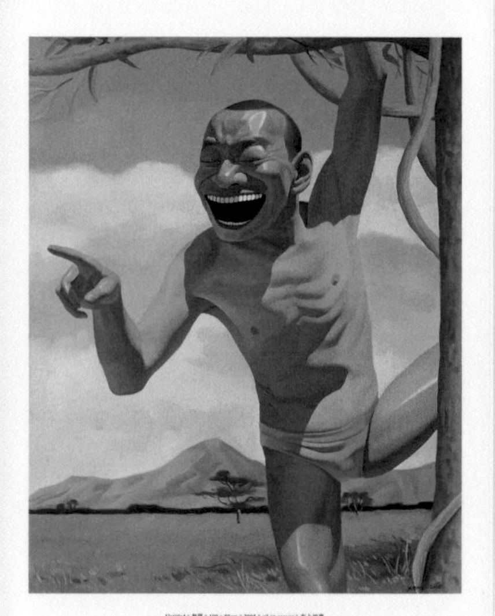

Untitled * 無題 * 100 x 80cm * 2005 * oil on canvas * 布上油畫

Hats * 帽子 * 82 x 82cm x 10pcs * 2004 * oil on canvas * 布上油畫

ISBN 0-9785764-0-3
9 780978 576400

缪晓春 **虚拟的最后审判**

Walsh 画廊，芝加哥，2006

275mm×230mm×6mm，64 页

平装

1200 册

编辑：缪晓春、何浩

前言：Julie Walsh

文章：巫鸿

艺术家陈述：缪晓春

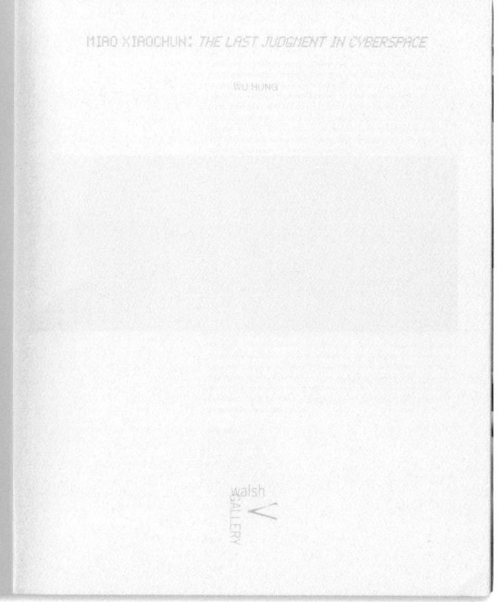

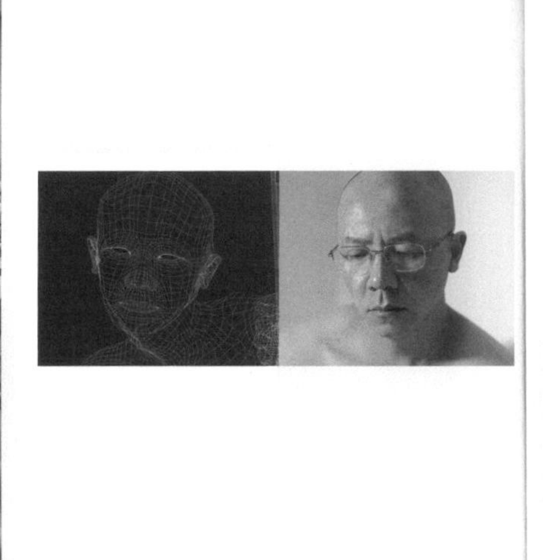

PREFACE

There are some artists who are not interested in conventional ways of sight and challenge and provoke new ways of seeing. In Miao Xiaochun's first solo at my gallery in 2004, he pioneered a technique of digital fusion to allow the viewer the illusion of actually looking around the virtual space he created. Wherever the viewer looked, everything was equally in focus. While looking at these photos you had the distinct feeling you were seeing something not quite right but it was difficult to figure out exactly what it was. The end result was that viewers in the gallery spent a lot of time just looking at Mr. Miao's photos as if to uncover what it was that was so entirely different. Few people suspected that it took the artist 70 photos compiled together to create the scene they were viewing, much less that it took him up to 3 months to create these epic depictions. With pieces such as "Celebration" where Mr Miao showed individuals at multiple points in time and space in the same picture frame, viewers would often sit for 10 minutes just staring at the piece as if they were watching a video. This reaction fascinated me because I felt that Xiaochun had in fact succeeded at creating a single-frame visual equivalent of the passing of time. In fact people were looking at these photographs in a new way... as if they were moving images.

No less stunning is Mr. Miao's attempt to reinterpret in cyberspace Michelangelo's "Last Judgment." The idea of not only creating a model based on the earlier artist's image but manipulating that model to replicate all 400 of the original figures beggars the imagination. Mr. Miao said that he and 3 assistants worked full-time for 6 months to achieve this technological feat.

When I saw this new work, I couldn't help imagining that the sense of awe I felt was similar to that experienced by medieval viewers beholding Michelangelo's painting for the first time. The sheer scale of Mr. Miao's project is mind-boggling. Maybe some artists would have been content with just this accomplishment, but I find it remarkable that Miao Xiaochun is only just beginning his investigation of "The Last Judgment." I picture him mentally body-surfing through cyberspace in search of meanings that may not have been fully visible in the fresco. It is as if via technology Mr. Miao was able to see and experience this ancient masterpiece in an entirely contemporary way. I remember the artist telling me a story: as a small boy in Wuxi, he had climbed a mountain and made his first painting from its summit; but it was only as a man, through the use of digital technology, that he was able to record and to communicate what he really saw. What could be more important to creating a masterpiece than to take something previously unseen and make it known in a new way? Mr. Miao's latest work does just that.

Julie Walsh
Director, Walsh Gallery

Fig. 1
Michelangelo, *The Last Judgment*

MIAO XIAOCHUN'S *LAST JUDGMENT*

Wu Hung

In praising Michelangelo's *Night* in the Medici Chapel, Giovanni Strozzi – a contemporary of the sculptor — imagined the stone statue being a "living image" of a sleeping woman conjured up by divine power. As if talking to a fellow visitor to the chapel, he promised that the statue would awake and speak upon a light touch:

> Night, which you see sleeping in such
> sweet attitudes, was carved in this stone
> by an angel; and because she sleeps,
> she has life. Wake her, if you don't
> believe it, and she will speak to you.

To these clever (but rather conventional) verses Michelangelo responded with a blunt epigram. He indeed let the statue speak, but only to reject any disturbance from an intrusive visitor:

> My sleep is dear to me, and more dear
> this being of stone, as long as the agony
> and shame last. Not to see, not to hear
> (or feel) is for me the best fortune. So
> do not wake me. Speak softly!

These lines, however, also beg the question — What would a painted or sculpted figure see, hear, and feel if he or she is *not* sleeping? This question becomes more tantalizing if the figure belongs to a large composition charged with strong emotion and dramatic intensity. The significance of the question lies in its redefinition of a painting from an external object of viewing to an organic body of internal visions, actions, and feelings. Once we accept the logic of this question, we begin to combine our seeing with active imagination. It is at this point we can turn to Michelangelo's *Last Judgment* (fig. 1), which has inspired Miao Xiaochun to create the multiple, large-format photographs and video work in this exhibition. Commonly upheld as the most ambitious and turbulent pictorial work from the Renaissance, Michelangelo's fresco has been skillfully described by S. J. Freedberg:

> In place of the actionless and hieratic scene of earlier *Last Judgment* illustrations, Michelangelo conceived an exalted drama, moved in every part, which is enacted by a multitude of beings in their essential nudity, still more superhuman in their breadth and muscularity of form than in the last stages of the ceiling (in the Sistine Chapel), and as exaggerated in their grandeur as the figures of the Medicean tombs. A youthful beardless Christ, compounded from antique conceptions of Hercules, Apollo, and Jupiter Fulminator, turns sinister towards the Damned, and makes the awesome gesture of their condemnation. Gathered tensile against His side beneath the gesturing arm, the Virgin averts her gaze and looks down on the Blessed. She cannot intercede for those whom Christ damns, nor can the surrounding agitated assembly of Saints. Christ's gesture generates their complex responses, which are those of giant powers here made powerless, bound by racking spiritual anxiety. The force and meaning of His gesture pass through the Saints and through the tangled Damned, who fall towards the crowded nightmare bark of Charon just below. Underneath the Christ, but in some immeasurable distance, angels summon the dead with trumpets, and they emerge and take on form as if from the very earth. Opposite the falling Damned, the Blessed levitate towards Heaven, most of them still numbed or half in sleep. In places wingless angels help them rise, and on the fringe of the ascending group one weightful, negroid pair are lifted by an angel on a rosary that denotes prayer. Christ's gesture sets in motion – not by its physical value but by its meaning – a gigantic slow rotation on the wall: descending, turning, and rising up to Him again. It is a motion subdivided almost endlessly into the convolutions of the densely grouped forms, but absolutely ineluctable: the great bodies are moved by and with it. The pattern of the whole movement and the way in which it functions in its parts appear to make a cosmic simile. Christ is seated in the heavens like a sun; the heavenly host around Him seem dense clouds made of human forms. Below, in a luminous aether, bodies fall to one side towards the water like clouds dissolving into rain; on the other they rise from the earth like moisture gathering again into clouds.

Interestingly, Freedberg describes the figures in the fresco as though they were living sentient beings and as though the whole painting were in motion. This passage from a standard art historical book thus legitimates the question – "What does a painted figure see?"

LooKing From Within

What do the figures in Michelangelo's *Last Judgment* – not only Christ and the Virgin but also the angels, the saints, the Damned, and the Blessed – see at this fatal moment? — What do they behold within the vast, mythical space in the fresco amidst a cosmic movement that is simultaneously orderly and chaotic? To Miao Xiaochun, to answer these questions means to enter the painting and to assume the varied gazes of the painted figures. Two pictures in his five-composition series result from this adventure, as they embody the *internal positions* of two figures and re-represent Michelangelo's masterpiece from their eyes. The picture entitled *The Last Judgment in Cyberspace: The Upward View* (fig. 2) is supposedly beheld by Figure I.35 (according to the numbering system in *The Last Judgment*, The Vatican Museum, Rizzoli) (fig. 3). In the original painting, the figure, a naked man lying on the ground, is looking up at the divinities while raising his left hand to cover his face. In Miao Xiaochun's composition, a large hand protrudes into the pictorial space from the lower edge, blocking the multitude of figures that recede into great distance. He explains:

This man seems to be tormented by great apprehension with the approaching final judgment, not knowing whether he will enter Heaven or be thrown into Hell. From his position he would first see his own hand and then a scene of salvation — a group of angels are rescuing suffering souls from the possession of hellish monsters. More angels appear further away, blowing their trumpets. Christ can be seen only at the edge of the sky in a greatly diminished size.

The other picture, entitled *The Last Judgment in Cyberspace: The Downward View*, is composed from a similar *internal position* (fig. 4):

> Figure C.1 shows an old woman holding open her hooded mantle with both hands (fig. 5). She occupies the upper left-hand corner of the picture. I suspect that Michelangelo placed her at this corner, holding open her hood, as a way of allowing her to witness the entire process of the last judgment as he imagined it. From the back, the old woman's position resembles a modern person holding up a camera and taking photographs. Thus, I would very much like to look down at the entire last judgment scene from her point of view, an angel from heaven looking down into Hell through a billowy human tide. In a split second of judgment, one could either fly into Heaven or crash into Hell. If a modern person came upon such a scene, he would certainly, either subconsciously or consciously, look for a tool with which to record it. This is like when the first moments of September 11 were subconsciously recorded by an amateur photographer and when the Gulf War was consciously recorded by a professional journalist. Such pictures have been presented over and over again before humanity.

These and other photographs in Miao Xiaochun's series resulted from a complex process of image translation and manipulation. The first step in this process was to create a 3D digital model of a figure based on his own image: he photographed himself from various angles, and assembled the fragmentary shots into a 3D-image on the computer. The next step was to use this model to copy all the figures in Michelangelo's *Last Judgment*, from Christ to a sinful soul. Employing the 3Dmax software, Miao Xiaochun was able to manipulate the model into different gestures and moments. The third step was to integrate these 3D figures into a virtual space according to the composition of *The Last Judgment*. Once Miao Xiaochun had achieved this 3D spatial construct, he could traverse it at will. (In his words, "I feel I can now move inside this space, selecting angles and taking pictures.") More specifically, he, or a built-in camera lens controlled by him, could not only view the composition from numerous *internal positions*, but could also assume vantage points outside the constructed pictorial space.

Fig. 2
Miao Xiaochun, *The Last Judgment in Cyberspace: The Upward View*

Fig. 3
Michelangelo, *The Last Judgment*, figure I.35 (*The Last Judgment*, The Vatican Museum, Rizzoli)

Fig. 4
Miao Xiaochun, *The Last Judgment in Cyberspace: The Downward View*

Fig. 5
Michelangelo, *The Last Judgment*, figure C.1

Looking From Without

Unlike *The Upward View* and *The Downward View* which are supposedly seen by specific figures in the painting, the remaining three pictures in the series are composed from positions outside the painting. We may compare these two types of viewing/representation to our relationship with the Milky Way: we are *inside* this enormous heavenly body but we can also see its objective existence in the sky. In Miao Xiaochun's case, the desire to see Michelangelo's painting from alternative *external positions* may have first motivated him to envision the project. Only later did he discover the potential of his 3D model in re-representing the painting from *internal positions*. His initial proposal thus started with a passage in which he imagines to seeing *The Last Judgment* from behind:

> A sculpture can be looked at from multiple sides, whereas a painting can only be viewed from the front. Imagine what would happen if we looked at a painting from the back?
>
> How would Michelangelo's *Last Judgment* appear from behind? I think the figures considered important in the original work would become less conspicuous, while the secondary figures situated on the edges of the picture plane would assume principal roles. The original meaning of the fresco would be dramatically transformed. Perhaps Michelangelo himself never imagined such a way of looking at his fresco.

The attempt to re-represent the painting from this and other *external positions* led to the creation of three of the five compositions in the series, including *The Last Judgment in Cyberspace–The Front View* (fig. 6, *see page 21*), *The Last Judgment in Cyberspace–The Rear View* (fig. 7, *see page 29*), and *The Last Judgment in Cyberspace–The Side View* (fig. 8, *see page 37*). Among these, *The Front View* is special because its composition largely coincides with the original fresco. It can thus be considered a *translation* of Michelangelo's work from a two-dimensional painting to a three-dimensional digital image. This image is particularly important because it provided the basis for composing other pictures in the series: as a *translation*, this digital image replaced the original fresco to become the subject of subsequent external and internal viewing.

The next composition Miao Xiaochun created was *The Side View* (fig. 8, *see page 37*). He made this composition before *The Rear View* because when he rotated the 3D image of *The Last Judgment* in the computer, a "side view" caught his fancy. He wrote to me that he was struck by similarities between what he saw and a traditional Chinese landscape painting. He imagined that if one were looking diagonally at Michelangelo's fresco from the balcony in the Sistine Chapel, it would be "like looking at landscape from an open

The Last Judgment in Cyberspace – The Side View

The Last Judgment in Cyberspace – The Below View
50
51

This Catalogue was published on the occasion of
Miao Xiaochun's solo exhibition: *The Last Judgment in Cyberspace*
at Walsh Gallery, Chicago, from April 21 to June 3, 2006
Curator: Wu Hung

Copyright © 2006 by Miao Xiaochun
Text Copyright © 2006 by Wu Hung and Miao Xiaochun

All right reserved. No part of this Catalogue may be reproduced or transmitted in any form or by any means, electronic or mechanical, including photocopying, recording or by any information storage and retrieval system, without permission in writing from the artist

Editor: Miao Xiaochun and He Hao
Designer: He Hao

Printed in China 2006

Cover: *The Last Judgment in Cyberspace - The Vertical View*

Walsh Gallery

118 N. Peoria Street, F2
Chicago, IL 60607, USA
P 312.829.3312 f 312.829.3316
T-Sa 10:30 – 5:30
www.walshgallery.com

缪晓春 **图像 + 想象**

奥沙当代艺术空间，香港，2006

300mm×230mm×8mm，72 页

平装

1500 册

文章：顾振清、唐忠信

艺术家陈述：缪晓春

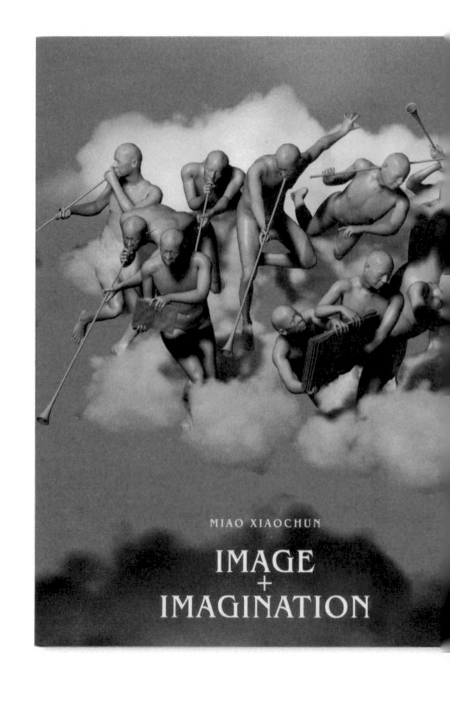

ao Xiaochun

nclined to think rationally. the superb effect of the anage his frames in the ny in a calm and collected can hardly be separated se the two are like the me thing. Therefore, he his works man's visual e wide spectrum, through nfolding great narrative nal images according to eful rearrangement and he boldly follows and

restores the panorama of the boundless world and human society. At the same time, he continues to raise the resolution ratio and the exhibition size of his works so as to bring them closer to reality. As a matter of fact, the reality Miao Xiaochun presses on towards is a kind of common visual experience and the life such experience represents. To be more exact, what he tries to present is a kind of field scene of dynamic reality, namely, the field of the actual life undergoing drastic changes in China's social transference. Facing Miao Xiaochun's works, the audience would feel as if they were personally on the scenes and experiencing everything. As a result, Miao Xiaochun's images give people a warm and intimate feeling and a familiar sense. Miao is good at arousing the audience's empathy in the recaptured reality, thus unostentatiously imbuing into them his own sense of form and concept.

Miao Xiaochun has always been concerned with the relation between seeing and being seen. He likes to watch his objects from various points of view and search for their interrelation. So he lays extreme importance to the interaction between his works and the audience. It is true that very often the audience has to view his works repeatedly before they can feel the elements the artists has meticulously melted into his works that are beyond their daily routine, thus comprehending the form, language, and concept suggested by these elements. Such seemingly random and ease elements are actually final touch Miao painstakingly designed in conceiving his ideas. There are in his works his self-sculpture, his self-taken pictures, and the images of three-dimensional constructed models, which all get involved in the common visual experience as well as the manifestation of the concept of time in the frames and the synchronic appearance of some diachronic figures. All these techniques lead the audience to enter a special visual experience different from their daily experience, switching their habitual way of learning visual things, even changing the logical relation between cause and effect in the audience's mind when they are watching his works.

Stumble 跌 2003

Miao Xiaochun has effectively touched and perceived the relation between human beings, social reality and the actual environment through his works, thus expressing his ideas behind this unique visual experience in a more transparent and unpretentious way.

Reviewed as a whole, Miao Xiaochun's experiences of creation not only have been influenced by but also have conscientiously inherited and carried forward Chinese traditional scholar landscapes and the realistic narrative mode, which has been flourishing in China's art institutes. Perhaps this has much to do with the cultural background in which he grew up and received his education. However, what determines his creation is the ever changing realistic context of existence.

Miao Xiaochun was born in southern China where talents have always abounded, and where he formed a solid foundation of traditional culture as well as a high cultural self-confidence. His study experiences in Germany offered him a realistic way of cross-cultural thinking, he thus became very much concerned about the implied obstacles and price of realizing cross-cultural communication. When he studied in Kassel in Germany, Miao Xiaochun began to make use of the image of an ancient Chinese scholar as his "self-sculpture", which served as a necessary object in his photo creation. His touch off point was aphasia in daily communication. Differences in cultivation, ideology, and traditions of value between China and Germany gave Miao Xiaochun enlightenment and he found there was an outstanding conflict in the linguistic context between his way of expression and modern German human environment. So Miao Xiaochun let that silent sculpture replace his own "himself" to appear on various social occasions, for examples in a class room, a family gathering, and so on. The incompatible dislocation of the sculpture and the living persons form the focus of visual conflict in his works. Such almost autobiographical photography expresses the artist's perplexity in morass as he lived in an alien land whose culture was greatly different from that of his native land. The replacement of the artist by the sculpture directly manifests certain surrealist nature of Miao Xiaochun's actual life.

11

Await 2005

linguistic context of modern art. In Miao's new works, conclusions in history of art have turned into questions, and the masters' works have become targets of artistic criticism, even material for experimental art recreation. The 3D digital technique enables Miao Xiaochun to look at Michelangelo's work through an unprecedented historical point of view in modern context. His recreated images express his modern explanation and criticism on the *Last Judgement*. His *Last Judgement in Cyberspace* allows the audience to re-examine the original scene of the Renaissance master from diverse points of view and even enter a virtual spatial relation to experience some "vivid" postures and atmosphere that are not in the original work.

Miao Xiaochun has been holding fast to his spiritual domain. Confronting modernity, the largest "mirage" of Chinese society, he maintains his attitude and position of raising questions. And such attitude and position form the inner motivation of Miao's artistic pursuance. He is, therefore, constantly renewing his experience, through which he links up with the common visual experience of the audience. He dexterously revises the way to view art works by taking up the dual role of the watcher and the watched, thus adjusting the interaction between the artist and the audience. We can detect the fact that his aspiration to communicate with the audience runs through almost all of his works. Such strong aspiration of communication will inevitably stimulate him to make his unique contributions to modern art.

Translated by Chen Dezhang

14

Eyewitness to History

Jonathan Thomson

For many art historians, writers and commentators, Michelangelo's *Last Judgement* is the crowning achievement of the Renaissance. This is no small claim as the Renaissance was an astoundingly fertile era, when in a staggering burst of creativity, humanist artists, sculptors and architects rediscovered and gave new meaning to painting, the depiction of the figure and the construction of buildings of timeless beauty.

In 1533 Michelangelo was invited by Pope Clement VII to paint a Last Judgement of the altar wall of the Sistine Chapel in Rome. However, Clement VII died before work could begin and it was his successor Paul III that confirmed the commission in 1535. The work was eventually unveiled in 1541. The Last Judgement is the Christian doctrine of the apocalypse and the Second Coming of Christ.

When Michelangelo's *Last Judgement* was unveiled in 1541, the crowds who came to view it were struck dumb by the "terribilita" of the work.[1] When Pope Paul III saw the painting, he is reported to have broken into prayer at the sight "Lord, charge me not with my sins when Thou shalt come on the Day of Judgement."[2] Michelangelo himself is reported to have attested to the painting the words of Dante: "The dead seemed dead, and the living seemed alive."[3]

How then could any later artist hope to make any work based on this sublime masterpiece that could add in any way to its power and its glory? Who would have the temerity to try?

In *The Last Judgement in Cyberspace*, Beijing artist Miao Xiaochun has created a three dimensional model of Michelangelo's painting and has populated it with digital versions of all four hundred figures that play a part in the unfolding drama. The technology that he employs has been used in architectural design and Hollywood film-making. But Miao's purpose is entirely different. He does not simply replicate the original, nor does he seek to animate the scene. Instead he deconstructs it and allows us to re-examine the work from within the picture itself. We are no longer tied to just one point of view but can now witness that sublime moment of truth, from within the painting and from the point of view of any one of the more than four hundred participants. Miao's use of technology has given us a new way of seeing. Our different perspectives, from within and without the painting, allow us to engage with the painting and its protagonists in precisely the same way as Michelangelo did himself. This is Miao's remarkable achievement.

In his classic novel about Michelangelo's life *The Agony and the Ecstasy*, Irving Stone imagines the way in which the artist was able to reconcile his Christian faith with his belief in humanism: "He did not believe he could paint the Last Judgement as something that had already happened, but only at the moment of inception. There he might portray man's agonising appraisal of himself."[4]

The iconographic features of the painting are drawn from many sources, including the Gospel of Saint Matthew. In Chapter 25 he describes what happens when the Last Judgement is upon us: "And when the Son of man shall come in his majesty, and all the angels with him, then shall he sit upon the seat of his majesty. And all nations shall be gathered together before him: and he shall separate them one from another, as the shepherd separates the sheep from the goats: Then shall the king say to them that shall be on his right hand: Come, ye blessed of my Father, possess you the kingdom prepared for you from the foundation of the world. Then he shall say to them also that shall be on his left hand: Depart from me, you cursed, into everlasting fire, which was prepared for the devil and his angels."[5]

Earlier depictions of the Last Judgement are mostly rather stiff hierarchical structures in which all of the figures are compartmentalised according to their standing. In Michelangelo's version, we observe the drama from outside the picture plane, as if the events are happening just beyond the end of the Sistine Chapel altar wall. This is the conventional perspective Miao has chosen for *The Last Judgement in Cyberspace: The Front View*. Figures rise from their graves, dazed and confused, some drawn upwards by angels, others held back by demons. The dead are summoned by a group of angels blowing on trumpets. The angel on the left (on Christ's right hand) reads the names of the saved from a small book. The angel on the right checks off the names of the reprobates in a rather larger book as they begin their descent into hell. All of the figures swirl up and around the figure of Christ in a maelstrom of humanity. Christ himself stands impassive at the centre of the whirlpool, his arm arrested in mid-swing as he drives the damned from him, perhaps more in sorrow than in anger. At his side, the Virgin averts her eyes, shielding her body from the damned in a self-protective gesture of denial. Looking on are the figures of the Blessed, bearing the signs of their martyrdom, powerless to intercede even if they wanted to. Some of the Elect, having been judged and found worthy, are being helped out of the tumult by angels, others greet one another with weary jubilation.

Michelangelo's figures come alive for us because he has imbued them with souls. They think and feel; they know anguish, dread, hopelessness, joy and compassion. For Michelangelo's contemporary biographer Giorgio Vasari: "the multitude of the figures and the magnificence and grandeur of the work are indescribable, for it is full of all the possible human emotions, all of which have been wonderfully expressed: the proud, the envious, the avaricious, the lustful, and all the other sinners can easily be distinguished from every blessed spirit, since Michelangelo observed every rule of decorum in portraying their expressions, poses and every other natural detail...Thus any person who has good judgement and an understanding of painting will see in this work the awesome power of the art of painting, for Michelangelo's figures reveal thoughts and emotions which were never depicted by anybody else; such a person will also see how he varies with diverse and strange gesture the many poses of the young and old, male and female: to whom do these figures not display the awesome power of his art along with the sense of grace with which Nature endowed him?"[6]

Detail of the Side View

22

23

5. The Last Judgement in Cyberspace - The Below View

40

41

Detail of the Vertical View
52
53

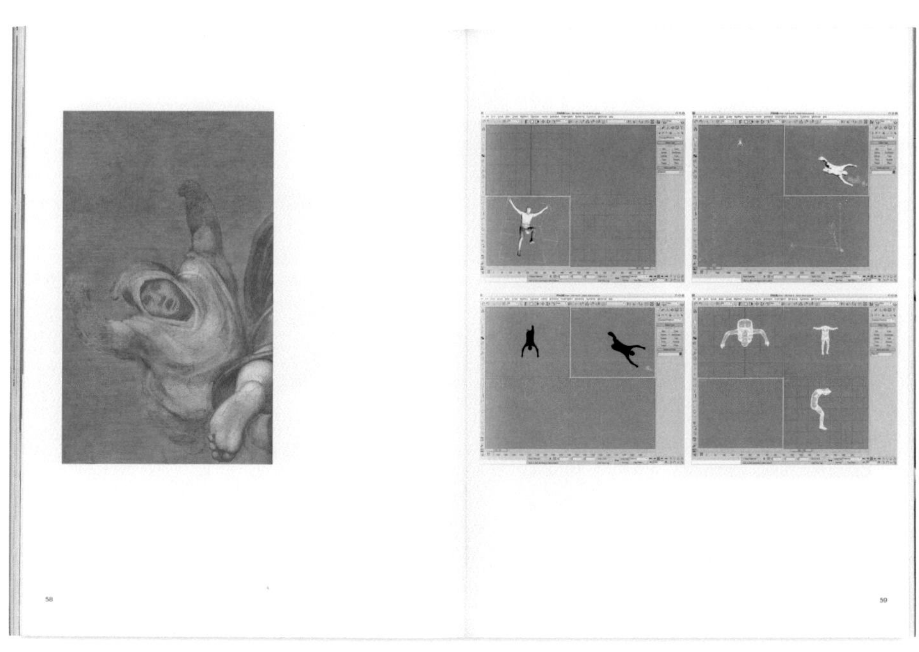

缪晓春 “水”的艺术史研究

Walsh 画廊，芝加哥，2007

285mm×195mm×4mm，

52 页

平装

1200 册

前言：Julie Walsh

文章：巫鸿

对话：巫鸿 / 缪晓春

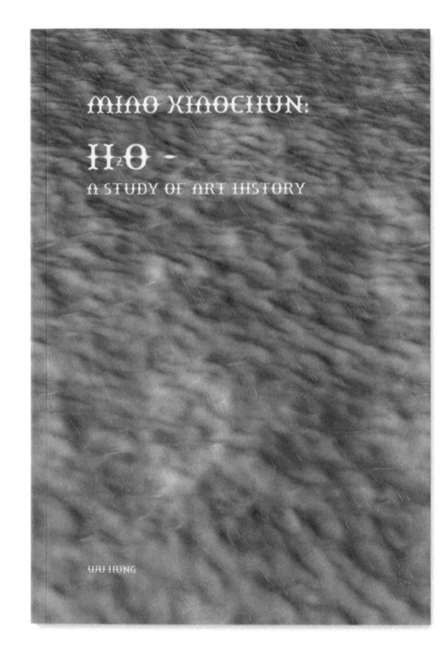

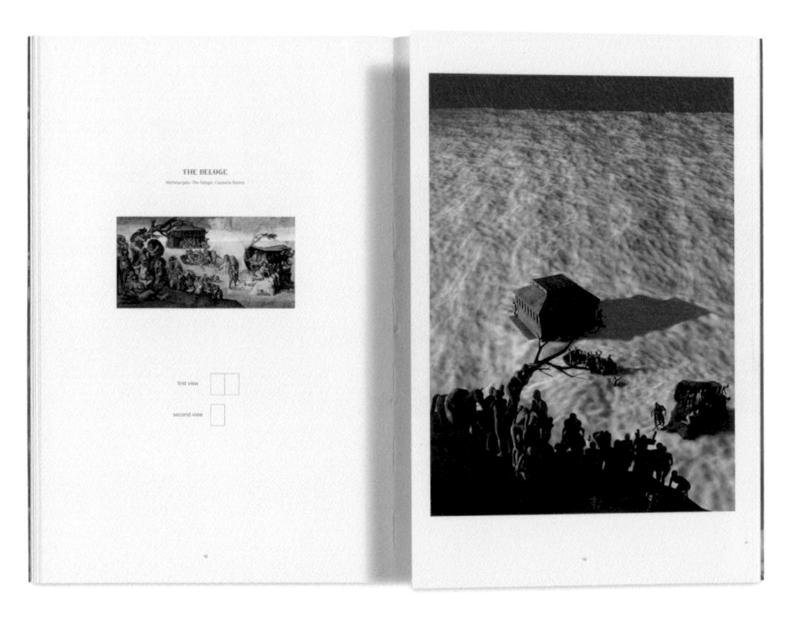

PREFACE

I am very proud to have another wonderful show of Miao Xiaochun's latest work. Also, special thanks to Wu Hung for his continued collaborations.

Julie Walsh
Director, Walsh Gallery

MIAO XIAOCHUN AND HIS H_2O

Wu Hung

Miao Xiaochun's newest art project, *H_2O: A Study of Art History*, contemplates on the continuity and metamorphoses of life and art. Philosophically it responds to a question he posed in his previous project, *The Last Judgment in Cyberspace*. As indicated by the title of a video in that project, the question is "Where will we go?" after human history. Using the technique of computer animation, the video features a person – a digital reconstruction of the artist himself – traversing the vast cosmic space inside Michelangelo's *Last Judgment*. Like a lost shooting star, he emerges from the depths of cyberspace and transforms into an infinite number of identical figures, both the divinities and the mortal beings who once lived. In Christian eschatology, on the day of the Last Judgment, every man and woman will present him or herself before Jesus Christ to have their conduct reviewed by their Lord. The meritorious ones will ascend to Heaven; the sinful ones will descend to Hell. But in Miao Xiaochun's computer animation, all the figures finally disperse, vanishing into the shapeless space from which they originally came. Their origins are beyond our knowledge and their future destinations are totally unknown. The video thus does not offer an answer to the question "Where will we go?" – but only raises it. On the other hand, by erasing any difference between Heaven and Hell and making all divine and human images identical, the artist effectively rejects the traditional Christian solution. *Constancy* and *continuity*, not differentiation and hierarchy, underlie the narrative structure and visual presentation of the computer animation. In this way, this work bridges *The Last Judgment in Cyberspace* and *H_2O: A Study of Art History*, because constancy and continuity now appear as the explicit theme of the later project. This also explains why Miao Xiaochun focuses on H_2O, a natural element which he believes best embodies these two concepts. He explains this idea in a "self-statement" about this project:

> I really don't know where I was from or where I will go, but I know many substances go into and out of my body every day, among which is water-H_2O. Before entering my body, it has gone through numerous living things, plants, animals, and human beings, and it will again go through numerous plants, animals and human beings after being released from my body. I am just one of the containers holding it temporarily, or one of the points it flows by. Water has been recycling through oceans, the sky, and the land. The process has begun ever since remote ages, continues to the present, and will continue into future, never stopping, repeating forever in an endless way. Does it carry and deliver certain information about the source and destination of life? All forms of life and water vitally interrelate with each other. Do they thus show compassion and concern for each other? Is it [not clear what "it" refers to] constantly changing life, which is delicate and subtle like water, recycling endlessly as water does? Will life that has vanished condense again at some other place and return to earth, just like evaporated drops of water condensing into rain, snow, or frost?[2]

The project does not attempt to document the natural transmission and transformation of water, however. Rather, Miao Xiaochun has selected a group of famous paintings as his "materials" for re-presentation. The rationale is that these images created by some of the greatest painters in art history, rather than natural phenomena in the objective world, more profoundly reveal the essential roles that water plays in human life. Just as in his creation of *The Last Judgment in Cyberspace*, Miao Xiaochun has replaced all the figures in these paintings with a single 3-D digital image of himself. But unlike *The Last Judgment in Cyberspace*, the meaning of this image has changed from unifying individual characters in a single work to connecting various historical paintings into "a kind of metabolism" (Miao Xiaochun's words[3]). In his view, if water flows from one organism to another in the natural world, then this project makes him (or his image) a neutral element "flowing" through works of art created in different times – a process which generates new works based on old themes.

1

H_2O: A Study of Art History thus has a two-fold purpose. On the one hand, this body of works, including a series of digital photographs and a computer animation, can be comprehended as an elaborate metaphor for the organic process of life, in which water is indispensable. On the other hand, these images re-present existing images based on "a study of art history," and can be considered "meta-images" – "representations of representations" that articulate the artist's critical reflection on art making. It should be pointed out that although Miao Xiaochun formerly studied art history in Beijing's Central Academy of Fine Arts – he actually obtained a Master's degree from the Department of Art History in that school – his selection of the European masterpieces for *H_2O* is not based on a standard art historical approach. Rather, he is guided by an artist's instinct – in his words, the idea of water in these works "touches him to the heart."[4] Not all these

BACCHANAL
first view
second view

BACCHANAL
first view
second view

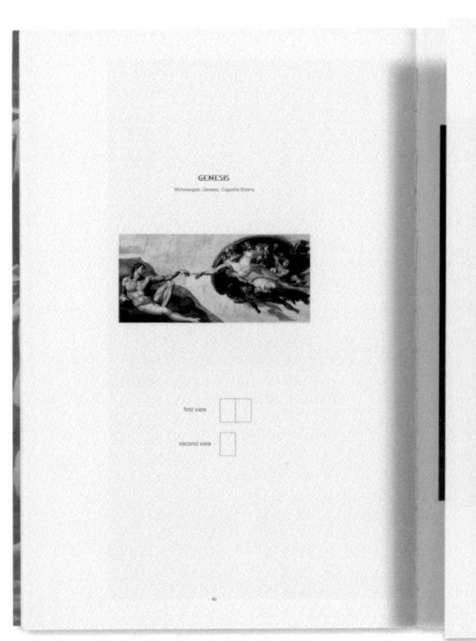
GENESIS
first view
second view

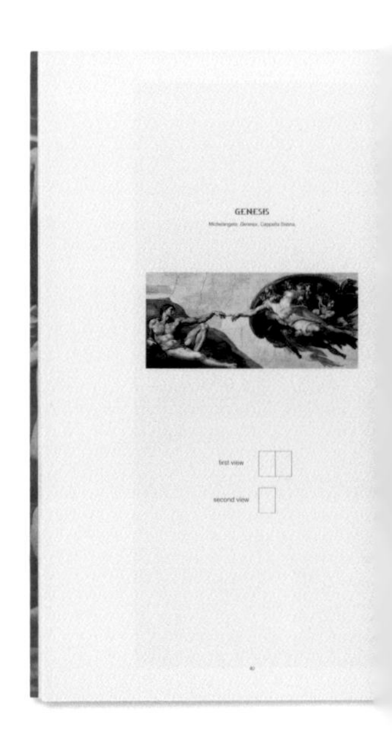
GENESIS
first view
second view

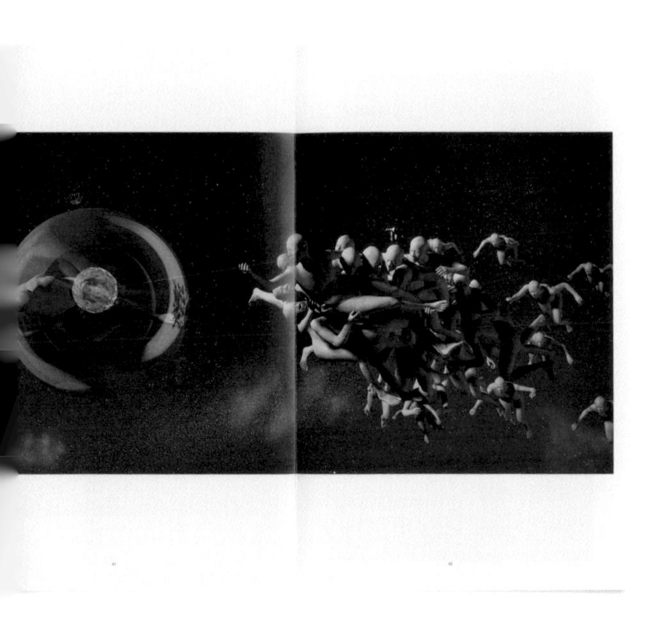

《虚拟的最后审判》《图像+想象》《“水”的艺术史研究》

之前因为荣荣和映里的项目，我跟芝加哥大学的巫鸿教授有过两次愉快的合作，2006年初，巫老师又给我发来邮件，说他最近在美国策划了一个中国艺术家缪晓春最新作品的个展，希望能由我来设计。如果可以，他即让缪晓春跟我联系。其实巫老师不知道，我跟老缪本就认识，我们同在美院任教，而且都在设计学院，虽然不熟，但平日里抬头不见低头见是少不了的。

很快老缪跟我约好，在他工作室给我展示了即将展览的新作。这是一组完全由电脑3D技术生成的虚拟影像，没有任何拍摄的成分。老缪选取米开朗琪罗名作《最后的审判》，用数字技术把画中近400个各色人等替换成了他自己的形象，并严格按照每个人在画中的姿态位置把他们置入到了一个宇宙般无边的立体空间中。观者可以从上下前后各种任意的角度来体察这群人，完全改变了原画的二维观看方式和主题性意味。老缪把这组作品取名《虚拟的最后审判》。说实话，我第一眼看到这组作品时有些惊讶，这种惊讶主要来自新作跟他以前作品的巨大差距——他之前那些大场景中所有细节都纤毫毕现的巨幅照片早已成为中国当代艺术的符号式作品，这种大跨度转变对于一个已被学术和市场双重认可的艺术家来讲绝非儿戏。对此老缪只是淡淡地讲，想到了，于是就做了——为此他和大批助手夜以继日地工作了整整一年。后来我跟老缪在工作中慢慢熟悉起来，发现他平日待人接物温良谦和与世无争，但在艺术上却无比坚忍强悍，永远在舍近求远不停歇的跋涉中。老缪的字典中绝无“捷径”二字，仅凭这一点，即可让我视他为最可敬重的师友之一。

小画册的设计很顺利，使用了荧光绿和数字感字体来强调作品的全新属性。随后，《虚拟的最后审判》在香港再次展出，我又为之设计了《缪晓春：图像+想象》。半年之中，为同一件作品设计两本书，并不太容易，而新的设计概念来自对作品不断触摸之后理解上的推进。在这本书中，我特别选取了正视和背视两个内容完全对应的局部，丝丝入扣地放入封面和封底，以此对作品的认知路径进行了最明确的提示，而封面封底所夹的内页部分，则成为了“可穿越”的观念容器。

2007年，老缪再次推出以相同方式创作的《“水”的艺术史研究》，5张中世纪与水有关的名画被数字化虚拟。与《虚拟的最后审判》不同，这次每张画作只对应生成两个角度的观看——两张影像，一横一竖。这种尺幅形态对展览无碍，但出现在书中却颇为棘手。首先开本无论横竖，都无法有效利用，必有一张很小。其次，虚拟生成的两张影像需要分别对照原画观看才好理解，但书籍打开只有左右两页，两张作品如何共用一张原画？如果原画前后接连出现两次，则看上去像个产品说明书，既繁复又迟滞，灵动尽失。当时我把5张原画10张影像作品全部缩小平铺在电脑屏幕上，对着它们一筹莫展，但看着看着我忽然发现，上述困境似用拉页的形式可以完美解决：首先，把开本设为竖长，左页放原画，右页放竖幅图像，彼此可对照观看；而右页拉开后，即变为两个页面大小的横幅页面，亦可对照左页原画——通过拉页，书籍的左右两页变成了左右三页，横竖图像的放置也不再受开本形状的局限。左右页、开本这些看起来根本无法触动的书籍恒定要素，在此生发出了新的可能。

转日，我约老缪来我工作室看方案，没开电脑，我只是用纸折了个模型示意。老缪看后跟我笑谈，他觉得这本书很不好做，以为会在我这儿待很久，为此推掉了另一个很重要的约会，要知道几分钟就结束了，还不如去赴约。其实，真正问题的解决大多就在一瞬间，只要你能找到开锁的那一把钥匙。

荣荣和映里 **六里屯**

三影堂摄影艺术中心，北京；前波画廊，纽约，2006

290mm×195mm×45mm

四册一函，外加函套

限量 1000 函，每函艺术家签名、编号

前言：茅为清

文章：张黎、艾未未、巫鸿

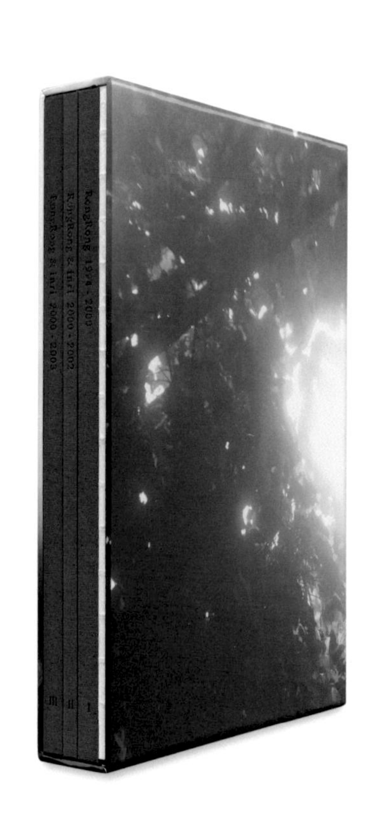

LIU LI TUN

六里屯

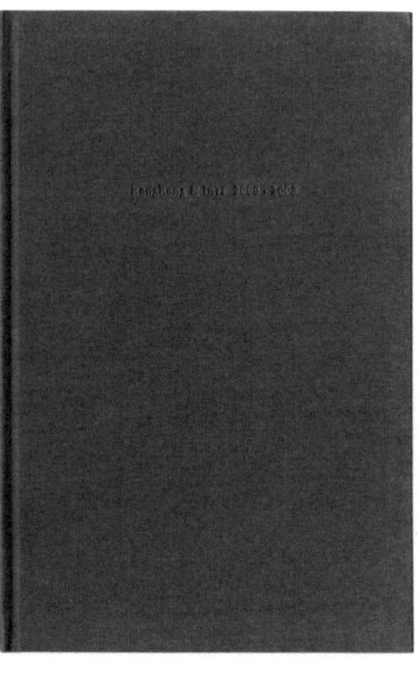

The Liu Li Tun Moment

By Wu Hung

A few years ago when I began writing *RongRong's East Village* (New York: Chambers Fine Art, 2003), I wanted to discuss three things. While each of these had its own distinct scope and theme, all were inseparable from understanding contemporary experimental art in China and the artists who practiced it. The first thing was to interpret works of art, particularly to unearth the best ones and explore their depths, as this is the basis of understanding any art or artist. The second task was to document and thus restore his living and working environment at that time, mainly by consolidating scattered pictures of his lived experience, emotion, and desire into a coherent artistic expression. Third was to place these works and artists into the larger framework of contemporary Chinese urban, social, and cultural space. What was "RongRong's East Village"? Why did he gather with other artists from far-flung reaches of the country in this tiny village on the periphery of Beijing, filled with garbage and industrial waste? What connection did they have to this place? What was the fate of the East Village? Why did it disappear without a trace, leaving not even a broken brick or shattered tile to mark its presence? When we look at the pictures of RongRong's East Village and ask these questions, we hope to understand—and indeed do gradually come to understand—not only the artistic nature of each photograph and the life experience of the photographer who made them, but the fate of a city, the intimacy and struggle with which people inhabit their environments, the uncontrollable tragedy of history.

The present compilation might be called a continuation of *RongRong's East Village*. The earliest works here were created shortly after the East Village artist community was disbanded, and RongRong moved to the nearby village of Liu Li Tun in 1995. Writing this brief article, I am still thinking of the three perspectives mentioned above: artworks, artist, city. As time passes by, Liu Li Tun leads us step-by-step through the last years of the Twentieth Century, and on into the Twenty-First. We see how RongRong's lens and life gradually evolve, and how Beijing's merciless expansion—that massive wheel of destruction and reconstruction—keeps on turning. At the dawn of the century, a Japanese woman who speaks with her eyes entered into this strange, small courtyard; she is inri. From this point on, she and RongRong and Liu Li Tun find meaning only in their soundless dialogue. Two years later, the wheel of destruction encroached, finally crushing Liu Li Tun. When their familiar courtyard was turned into a pile of bricks, RongRong and inri held a lonely funeral atop the ruins, holding fresh white flowers in their hands.

RongRong
1994 - 2000

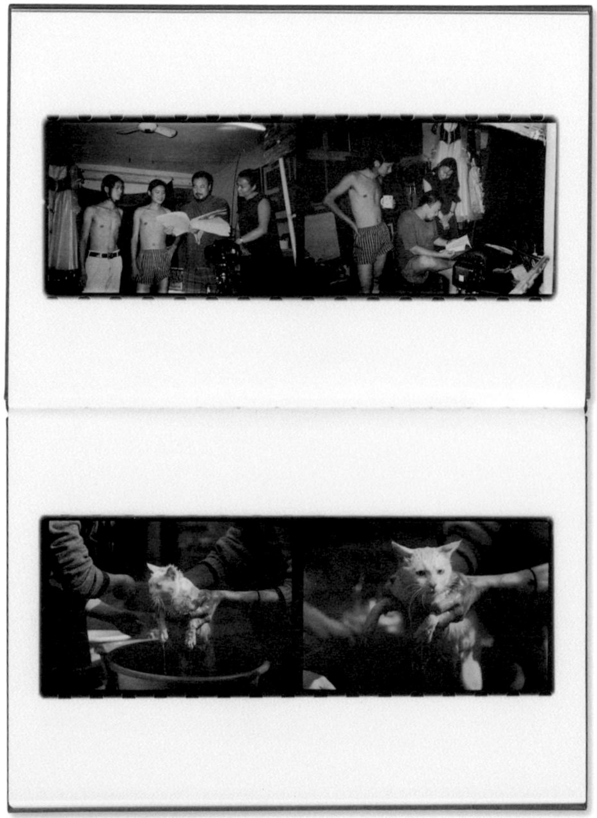

Rong Rong & inri
2000-2002

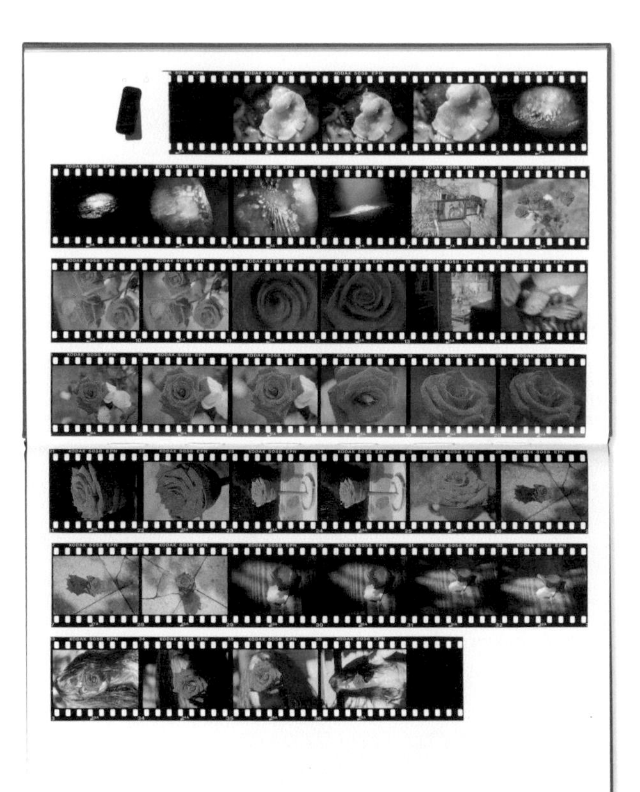

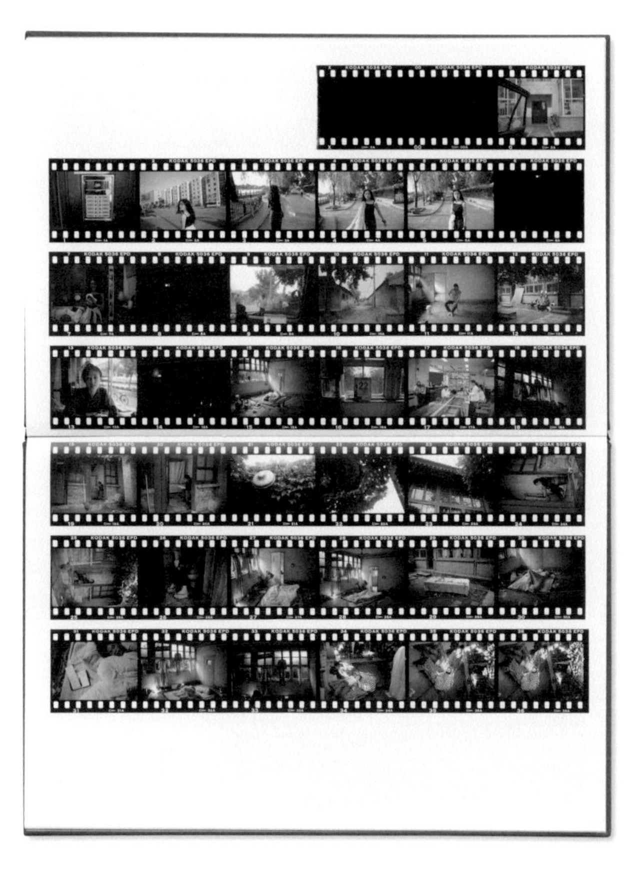

《六里屯》

荣荣最早是作为“东村”艺术家群体中的一员为人所知的。但事实上，“东村”这个著名的前卫艺术家聚落只存在了两年时间。1994年6月，荣荣再次来到北京，租住在靠近今天东四环的六里屯村。同“东村”一样，这是一个城市扩张进程中还没来得及被推倒的城中村，房租低廉，靠近市里的同时还毗邻使馆区。荣荣在这里冲洗底片、放大照片、装订自己编辑复印的杂志；和朋友们聚餐、喝酒、谈天说地……艰苦的日子与年轻的热情交织在一起，平实而激荡。2000年日本艺术家映里的加入，更为荣荣的这种波希米亚式的生活平添了绚烂颜色。虽然当时两人语言交流都还很困难，但他们从此将自己的生命和艺术结合到了一起。2003年，由于正对朝阳公园这样的绝佳地段，六里屯终究躲不过拆迁的命运，在荣荣和映里的注视下，他们的住所被铁锤和推土机夷为一片瓦砾。

三年后的六里屯已是高楼耸立。这一年，荣荣和映里获邀在纽约展出他们六里屯时期的照片并出版画册。作为他们的朋友和这部画册的设计者，我也因此有机会跟他们一同重新开启了这仅过数年却已恍如隔世的影像世界。

这是一批贯穿六里屯生活始终，数量巨大的影像日记。荣荣和映里通过镜头和胶片凝固了当初在那个小院里生活的点点滴滴。相对于一般意义上的系列摄影作品，这批照片的初始状态似乎用“影像碎片”来描述更为贴切。这种感觉，一方面来自于当初拍摄时的即兴和非“创作”意图，照片之间并无多少概念上的连接或上下文关系；而另一方面，照片的物质形态也是纷杂各异——黑白、彩色，正片、负片，120、135……不同质感，各种底片。荣荣和映里首先花大力气对这批素材进行了细致的挑选和大刀阔斧的删除，但即便这样，当我以工作者身份进入到这个项目中时，眼前的照片仍有数百张之巨，且无改碎片化属性。

按照我们开始的构想，这将是一本大而厚的图集，如同一个箱子，把数百张照片以目前的支离状态收纳其中。这看上去是个好想法，直接、真切、别具一格，也很符合照片所传递出的气息。但是，当我们真正开始将图片落实在纸面上时，却发现了“真实”表面之下潜伏的“危机”：由于照片之间缺乏明确的内容关联，因此在版面的排布上很难找到确实、有力的依据，只能从形式感出发。但没有内在逻辑的形式看上去总是似是而非，这使得费尽心力挑选出来的照片好像也变得不怎么靠谱了——如此大量的图片，似乎永远存在着进一步增减和换一种排列组合的可能。我们三人都为此感到十分挠头。当时映里正是有孕在身，怀着他们的第二个孩子，所以如果要一起商讨方案，往往是我到他们家里去。记得有一天我们又一起工作到很晚，但还是少有进展，映里可能在身体上有些难以支撑，不得不躺在旁边的沙发上稍作喘息。看着疲惫的映里和堆在桌子上成箱的底片，我忽然感觉不能再这样下去，必须要重新寻找出路。多年的设计实践让我相信一点：操作层面的持续障碍一定是来自概念认识上的偏差。我得先想想问题出在哪儿了。

从荣荣家出来，夜色如水。我步行了一段，脑子里盘算着书的事情：出版对于这批素材而言，有点像是要把一筐葡萄酿成葡萄酒，但目前我们的做法却只是把葡萄榨成了葡萄汁。看似挤出的都是精华，但因为缺少“酿造”这一化学反应的过程，所以怎么摆弄都去不掉素材的“生涩味”和“不确定感”，自然也无法升华为浑然的作品了。回到工作室，我把《六里屯》素材调出来再次梳理、分析，在细细的浏览中我逐渐发现看似东鳞西爪的照片其实线索暗含，而且一经发现便愈发清晰：135黑白负片都是荣荣所摄，而135彩色正片则全属映里，择出这两个部分，所余即是后期荣荣和映里的合作，这部分基本为120黑白负片，仅有个别几张大画幅底片。这三部分图片从时间、形态到作品内容上都有明确区分，只是因为之前呈单张散乱状态，又混杂在一起，所以不易分辨。

当发现这一规律后，困扰我们的问题随即迎刃而解。我跟荣荣和映里建议把原计划的“箱子”改为“抽屉”，把作品按照内容和形态由一厚册拆分为三薄册，每册以拍摄时间为标注，如《荣荣1994—2000》《映里2000—2002》《荣荣和映里2000—2003》，相关文章则单以一册轻薄的平装本收录。三册作品加一册文集合为一匣，同时缩小开本——影像日记的阶段性和私密感因此油然而生。

荣荣和映里欣然接受了这个建议，只是对个别信息做了细微纠正，比如彩色正片一册改为《荣荣和映里 2000—2002》——这里面的照片也有荣荣所摄，但之前被我所忽略。随后，荣荣和映里按照新的编辑概念重新制作了作品——既为展览，也为书籍。新作品强化了各自特征上的差别，而不再以支离形态示人：荣荣重新挑选 135 黑白负片，以两幅一联的形式曝光放大在等大尺寸的相纸上；彩色正片不再作任何挑选，整卷 36 张完全以印片（胶片贴着相纸曝光出与胶片等大的照片，一般放大之前看样用）方式集合在同一张相纸上，保留废片、齿孔及胶片上的所有痕迹，共 20 卷；两人合作部分则放大为巨幅照片，以传统的摄影作品样貌呈现。

一切做完，映里已是临产。我拿着粘好的书籍样册去医院看映里，在病房，他们孩子出生前的几个小时，我们最终确定了所有细节。另外值得顺便一提的，是我从 2005 年使用至今的这间工作室就在六里屯——我所在小区正是建在荣荣和映里当初居住的那片小院上。

确有其事——来自中国的当代艺术

泰特，利物浦，2007

285mm×190mm×20mm，208 页

平装，2000 册

序言、致谢：Christoph Grunenberg

文章：Simon Groom、凯伦 · 史密斯、徐震

文本写作：凯伦 · 史密斯

封面设计：Piccia Neri

艺术家：艾未未、曹斐、耿建翌、顾德鑫、何岸、李永斌、仇晓飞、邱志杰、王功新、王蓬、王卫、徐震、杨福东、阳江组、杨少斌、周铁海、周啸虎、庄辉

art@lastminute.cn

Karen Smith
Independent Curator

Real Problems

In China today, every imaginable aspect of modern life is unfolding simultaneously. One might cite numerous examples to illustrate the complex nature of this social change: the widening gap between rich and poor; the challenge of implementing a legal system; the surging force of materialism; the problems of regulating the professions and of heading off corruptive practices; as well as the positive growth in opportunities for individual members of society, and the economic benefits that are slowly filtering through to the general populace. All are the result of the unbelievable rate of redevelopment, and its kaleidoscopic impact upon existing ways of life, community values, and individual aspirations.

But that doesn't help us understand how individuals are dealing with the need to adapt their mindsets and daily habits to the new world that these complex social phenomenon are directing. Nor does it explain the ways in which they are affecting personal experience, or how they currently inform contemporary art. Artists, whose choice of career largely renders them independent of mainstream society in China, still provide a good example of the contradictory nature of this modern macrocosm, and, of course, demonstrate in their work the various directions in which art is being lead.

Whilst his working practice produces some of the most challenging and arresting pieces of art to emerge from contemporary art circles in China, in terms of managing his art-career affairs, Gu Dexin belongs to the old school. He chooses to live entirely in the present moment with little thought for tomorrow—which is why all works are titled by the date upon which they were executed, and once completed are forgotten. You can call him to secure a meeting for a visiting curator a week in advance and he'll tell you he's not sure ... that it's better to call him on the morning in question to confirm. This is not because he's busy, or because he needs to juggle his schedule, or because he likes to play hard to get. He is simply not used to thinking or planning that far ahead, and will certainly have forgotten. In this high-tech world of mobile phones—Gu Dexin is one of a tiny minority of urban Chinese people who does not own one—and email communication, this artist is perhaps an extreme case, but it was not so long ago

· 16 ·

that all Chinese people lived this way. Dinner invitations were regularly received at 5:30pm for the same evening, and, more often than not, in person as very few people had a phone line in their home. By 2000, technology would have changed that, but the majority of older generation artists—who emerged from the mid-1980s to the mid-1990s—still largely adheres to such practices. It is almost as if the act of committing to anything definitively, of regimenting their days with scheduled appointments, infringes upon the personal freedom they fought so hard to secure.

The younger generation, these children of the reform era, whose experience and knowledge is tied to the modern age, operates using equally modern methods and mindsets that have no kinship with 'traditional' socio-cultural protocols. In contrast to Gu Dexin, Shanghai-based Xu Zhen prefers to have his time managed by a team of assistants who oversee his hectic schedule and order his priorities, as he shifts back and forth between his dual roles of artist and curator. Call him to make an appointment for that same visiting curator, and the response is much more likely to be "Xu Zhen can squeeze you in a week on Friday..."

So here, in a simple, yet telling example, we have a broad sense of the chaotic contemporary reality that the works in *The Real Thing* attempt to address—the pace of life, personal ambitions and social pressure, demand, change, new navigating old. These are works created by artists who, in myriad ways, are successfully withstanding the tempestuous winds of change. Not least, the recent tide of commercialism which is currently peddling a wide array of highly attractive rose-tinted spectacles, all of which serve to distort reality to various degrees, but each of an allure that largely conceals the inevitable complications that accompany self-deception. Against the rise of such overwhelming temptations, *The Real Thing* is the vision of eighteen artists who thus far have refused or avoided the spectacles, preferring reality without enhancement no matter how coarse the view. As Shanghai-based filmmaker Yang Fudong revealed: "In our small circle, we are quick to throw cold water on anyone who looks like they might compromise their work."

The force of change, then, is a true test of character, and ultimately of individual artistic vision. As the rule of ideology concedes ground to the practical politics of economics so, too, the *Weltanschauung* of Chinese art finds itself being undermined by commerce. The dramatic impact of the New York auction to which Simon Groom refers, encouraged a mass proliferation of auctions across China, each attempting to adopt a similar model. With the limited amount of works actually available in the marketplace—the number of contemporary artists is still surprisingly small—individual pieces are flipped from one sale to another with

· 17 ·

A Few Words...

Xu Zhen
Artist and Curator

001 Elementary and all-pervasive information concerning Chinese contemporary art:

002 the situation does not encourage optimism.

003 It's rather chaotic...

004 **People don't have high expectations of exhibitions.**

005 Many things just begin to make headway and people lose interest in them...

006 In our hearts, each and every one of us hates those with too much power, no matter if they are good guys or bad guys...

007 *The greater majority of questions are assault and battery.*

008 *We all have a sense of crisis...*

009 **We've made a profession of doing things unprofessionally...**

010 Artists by now ought to be able to maintain their independence and get their own attitudes straight, not let their energy become diluted, nor be lax about their career. This is becoming a serious issue...

011 It is so crazy...here we are in the world spotlight...facing a non-stop stream of visitors...with no shortage of exhibition opportunities...of course this is a good thing.

012 Right now in China where it's almost impossible to make a good exhibition, the wait for someone to break the cycle gains increasing significance...

013 I think that there really hasn't been much advance in the last few years, other than an increasing number of individuals learning how to make their works fulfil the appearance of 'good artworks'.

014 You still think that this kind of approach is the way to go?

015 Comrades, you shouldn't do things you might regret...

016 'Chinese contemporary' should not be a Chinese edition of international contemporary art.

017 We'll make better exhibitions next year...

018 To a real extent, it says everything about the problems with professional curators: where exhibitions organised by the artists themselves are better than the ones put together by a curator.

· 24 ·

019 **Accepting the situation is tantamount to supporting it; we should learn to be proactive, instead of accepting things passively.**

020 The problem is that the artists shouldn't have to take the initiative. If all of us (artists) have options for exhibiting, then there would be less room for random selection, and the reality would change. It is necessary for curators to maintain their independence, to draw in their claws and to stop engaging in meaningless competition. We need solidarity.

021 Our artists spend a lot of time thinking about opportunities...

022 We should examine our path carefully...

023 Many Chinese people speak in a sophisticated fashion, but is it necessary to be sophisticated?

024 Contemporary Chinese art offers an alternative model of modernism. Its inscrutability, its non-rational and un-systematised force, and the graceless energy it abstracts from the experiences of daily life, are a crucial aspect of its character.

025 **'China' should not be just a term.**

026 We are afraid of moving out of Beijing or Shanghai, afraid of another long march, afraid to go to the countryside or overseas.

027 **Capitals will corrupt those artists who lack the force to resist them.**

028 Subtle power is surely more interesting than obviously powerful forces.

029 Why do we always kill each other? Chinese art is chaotic enough. We should be united. It's not good to be a laughing stock.

030 From my childhood, I have carried a feeling that everything my teacher told me would remain unchanged, so I always hope that as a teacher, everything I teach should be open to change no matter if it is understood or not.

031 It's so simple really: we should learn to make use of each other.

032 The work is extremely vibrant. Of course, western art is the result of a huge amount of experimentation, but comparatively speaking, these artists operated under a largely inflexible logic, and within their own small circles. However, Chinese culture sits amidst the upheaval of a society undergoing a dramatic transformation. This process ought to inspire some profound and

· 25 ·

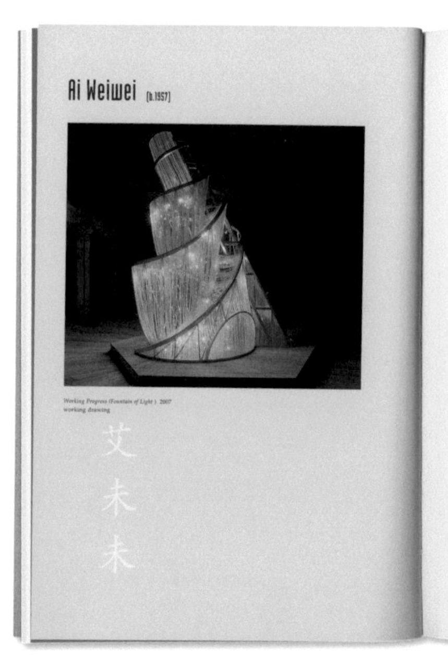

Ai Weiwei (b.1957)
Working Progress (Fountain of Light) 2007
working drawing
艾
未
未

Ai Weiwei came to art despising the political exigencies of a system that had sent his family into exile a year after he was born, and which condemned its society to a cultural impoverishment of immeasurable proportions. This intractable experience would later inform his work, but found initial expression in Ai Weiwei's discovery of Dada-inspired art when he left Beijing for New York in 1981, whose shock tactics and challenge to the socio-political status quo were to become distinctive features of his own work. His ability to deliver arresting statements in consistently elegant juxtapositions of material and form can be seen throughout his career, from the early works of 1986 such as *Violin with Spade Handle* and *Safe Sex* (a rubber raincoat with a condom for a pocket) through to more recent works, such as *Table and Beam* (2002) and *Fragments* (2006), where weighty philosophical abstractions assume the form of distinctly minimalist assemblages of materials. One of his most directly provocative activities involves his use of authentic, museum quality Han Dynasty (206BC-220AD) and Neolithic (4,000-3,000BC) vases, which he variously paints, whitewashes, or has been known to cover with commercial logos. His destructive approach to working with these pots is captured in photographs of Ai Weiwei dropping a particularly fine specimen from chest height causing it to smash spectacularly into a thousand pieces on the ground below. Reverence for antiquity clearly has no place in Ai Weiwei's world. His audacious affront to conservative traditionalists — which in addition to its reference to Duchamp, consciously invokes Mao's own command to the Red Guards to 'smash the old' — is a strategy calculated to provoke unsettling and uncomfortable questions about the role of culture, and its historical and ideological nature.

Painted Vases 2003
ceramic pots dipped in poster paint

· 31 ·

The Second State (detail) 1987
oil on canvas, four-panel work
Sigg Collection, Switzerland

with all the attendant aspirations for individual innovation, in its earliest incarnation in the 1980s, his generation has firsthand experience of the complexities of overcoming a profound absence of self-awareness and perception; Communism relieved individuals of all such burdens.

A further element that has played a role in shaping Geng Jianyi's aesthetic was his intuitive, reactive response to the tenets of art extolled by the State, which subjugated all expression to the socialist cause. This meant that Art was understood to present the ideal vision of social reality, but that ultimately the reality it depicted was defined by a political ideology to which all forms of expression, and subject matter had to conform. As a response, in all his work, Geng Jianyi deliberately subverts the idea that art can faithfully imitate or encapsulate something as subjective as reality: a direct challenge to those who would pretend otherwise.

· 56 ·

An Unapologetic Act of Sabotage 2007

Geng Jianyi likes to tell simple stories to illustrate the nature of the issues he explores, which serve as particularly useful aids in helping people from his own cultural background to engage with his works. In the context of describing the work he created for *The Real Thing*, the stories shed light on the philosophical nuances of ingrained Chinese attitudes such that point to the inspiration for the work he chose to produce, as well as the nature of the thought process that shape Geng Jianyi's particular approach.

'A grandfather possesses a magnificent beard of which he is extremely proud. One day, his grandson asks him, "Grandpa, when you sleep, do you put your beard inside the quilt or outside." Such a question had never occurred to the grandfather, but that night, when he retired to bed, the question perplexed him so much that he was unable to sleep. The next night was the same. Inside did not feel right, but outside was equally uncomfortable: the natural process of sleep was entirely disrupted by questioning something that he had had no previous need to consider. The issue was decided with the removal of the beard, which resolved the problem, allowing the grandfather to sleep soundly once again.'

This matter of consciousness is explored in *An Unapologetic Act of Sabotage*, as Geng Jianyi puts a similar question to a couple of unsuspecting bystanders. In a pilot programme, produced in the spring of 2006, the artist filmed a road sweeper at work in a street in Hangzhou. Geng Jianyi then approached him, and having assured the man of no mal intent, invited him into a studio at the China National Academy where Geng Jianyi teaches. Here, he showed the road sweeper the video recording of himself in action. Geng Jianyi asked if the man could repeat his actions, acting them out 'exactly as he had done in the street'. As the road sweeper contemplated his own image, the conscious mind attuned to the motion of the self in the most unexpected of ways, his earnest attempt to be natural demonstrates a marked degree of discomfort in common with the grandfather's attempts to sleep naturally with his beard.

An Unapologetic Act of Sabotage returns to this subject. Initially, the concept pivoted on a simple comparison as established with the road sweeper. But, as he thought about developing the idea, Geng Jianyi began to feel it might be too simple. He considered informing a chosen 'actor' of the concept before filming began, but realised that this would erase the element of

· 57 ·

For a brief period, when Gu Dexin married, the cramped room was both home and studio. The couple now lives in a slightly larger apartment but the walls are still hung from corner to corner with layer upon layer of the paintings that Gu Dexin made in his teens. Of greater interest here, though, are the shelves littered with abstract, gruesome forms, or with smooth-skinned, multiple-breasted beings, limbs and lips entwined, each of which appear to grow out of a larger fertility mother, like an inverse stack of Russian dolls, which Gu Dexin sculpts from children's plasticine.

Perhaps Gu Dexin does not like to talk about his work as the words really cannot do the work justice, beyond being obviously visceral, intoxicating as well as offensive, and always compelling. Often the works will undergo some physical change during the course of an exhibition, and the majority of the pieces are site-specific projects relevant to the time and place of the exhibitions in which he participates. Without precise knowledge of the exhibition environment, or the specific location and environs of a work, Gu Dexin is unable — and often unwilling — to conceive a plan for a work. The relationship between location, and the physical nature of the place, as well as the cultural framework in which it is sited, plays an enormous part in his choice of form, content and materials. Here, a site visit to Liverpool allowed him to encounter the red lighthouse boat that inspired the work created for *The Real Thing*.

opposite page
Meat 1996-2001
photographic record of
artist action

this page
2001.12.12 2001
site specific installation
using apples, Shenzhen

-62-

-63-

Qiu Zhijie (b.1969)

Railway from Lhasa to Kathmandu... 2006-7
multi-media installation and performance work

邱志杰

Qiu Zhijie is the intellectual dynamo of the contemporary art world in China. Engaging in parallel activities simultaneously, displaying all the multi-tasking mental gymnastics of a *wunderkind*, Qiu Zhijie is as much art critic, philosopher, social historian, and curator as visual artist. In all of these areas, he is never far from the frontline in thought, endeavour, intervention, and innovation. It is a cutting edge that has all the physical appeal of a razor-sharp precipice, but Qiu Zhijie doesn't seem to care. Curatorial positions, performances, critical theories, and opinions flouted on web-blogs have yet to reveal any chinks in his armour: even in the face of the outrage his views inevitably court, Qiu Zhijie parries controversy with evident flair, and takes on his critics with cocksure conviction of his position, his politics, and the premises he advances.

Artistically, Qiu Zhijie is also one of a kind. His works belie the paradox between traditional 'Chinese' aesthetics and contemporary values: he is as much at home in the philosophies and practices of classical Chinese culture, specifically the techniques and history of calligraphy, which he writes with consummate skill, as postmodern discourse. He simply mixes and matches as the mood takes him, which is demonstrated to greatest effect in a multimedia work titled *Rewriting the Lanting Xu* 1990-7. This involved the arduous physical act of writing a thousand times an inscription known as the *Lanting Xu* — the preface to *Poems Composed at the Orchid Pavilion*, written by Wang Xizhi in 353, an exemplary work of literature and calligraphic skill subsequently adopted as a model for mastery of the art — across the same stretch of paper until the surface was consumed by a solid black film of ink. The final work comprises the black expanse of paper, a video recording of the process, and a series of photographic stills taken at regular intervals during the process.

Qiu Zhijie's skills and interests collide in a body of work that encompasses canvas painting, calligraphy, video works, installations, performance, and most recently, a seam of sociological investigations into a diverse range of topics associated with the changing cultural climate across China. Evidence of his zealous engagement with socio-cultural phenomena is present in a number of earlier video works. The title of one such example of the approach is *Ping-pong* 1997, which references an actual historic circumstance that has come to be known as 'Ping-pong Diplomacy' — a term coined to describe a new era in Sino-US relations which was launched in 1971, when China invited the US table-tennis team to visit China for a friendly game with the local national team. The phrase 'ping-pong' was seen as an apt metaphor for relations between the two countries leading

-95-

informed of the dangers of spiritual pollution and modern — western — bourgeois thinking, was at times more than a little reluctant to respond to these novel interventions with too great a display of curiosity or fervour. Whilst this might have been disappointing to a truly devoted Maoist, Zhuang Hui was not blind to the pressures exerted upon the workers, or the necessity felt by most to side with the common, rather than independent, viewpoint. That did not mean, however, he had to give up.

Between 1990 and 1993, Zhuang Hui criss-crossed the country to realise a series of conceptual and site-specific projects under the name of *To Serve the People*. His attempts to get the workers involved would later emerge as a strategy for producing art. This strategy was implemented in the massive collaboration with various State organs and social groups required to produce a series of twelve group portraits, as well as a second photographic series titled *1+30*. The *Group Portraits* were conceived in 1995, and would be almost two years in the making. Each involved a protracted process of negotiation in order to get approval for gathering together what were enormous groups of people in one place at one time. Zhuang Hui displayed remarkable diplomatic skills, and considerable charm, in persuading the powers in charge to let him photograph the entire staff of a public hospital, for example, or the complete faculty and cadets of a police academy. The *1+30* series is divided into four sets of thirty images, each set representing a specific social group — workers, peasants, artists, and children — whilst the '1' referred to Zhuang Hui. This references the social reality of China during Mao's reign, when all were categorised as '*gong, nong, bing*' (workers, peasants, soldiers), and again required much persuasion to bring thirty individual representatives of each 'group' to stand next to Zhuang Hui in the photographs. Here, by attempting to focus on individuals, Zhuang Hui is interested in the collective dynamics of Chinese society. So, instead of forcing us to consider the individual, the process through which he realises his works and the final format of their presentation, points to the commonalities and sense of collectivism that bind the ordinary people in China in the course of doing an honest day's work.

Chashan County (detail) 2001
intallation

· 196 ·

Group Portrait Series: Officers and Cadets of the Fourth Artillery of 51410 Army Division, Handan, Hebei Province, July 23 1997
production still

From performance, photography, installation and more recently video and painting, Zhuang Hui's body of work spans a diverse range of materials and approach. The pace of work has slowed significantly in recent years as he concentrates his energies on a smaller number of monumental pieces, of which *Factory Floor* is perhaps the most ambitious. *Factory Floor* sits between an initial excursion into the contemporary human condition in 2001, *Chashan County*, and *Tunnel Building* 2004, which recreated the sparse living conditions afforded labourers and workers in the single-sex dormitories that were the bedrock of socialist housing.

Chashan County is a visual rendering of a real-life incident reported in a southern China newspaper, in which a girl was found mutilated — attacked whilst walking home at dusk, her eyeballs removed. The article was one of a series that appeared from 2000-2001 related to the gruesome 'profession' of organ stealing.

· 197 ·

LIST OF EXHIBITED WORKS

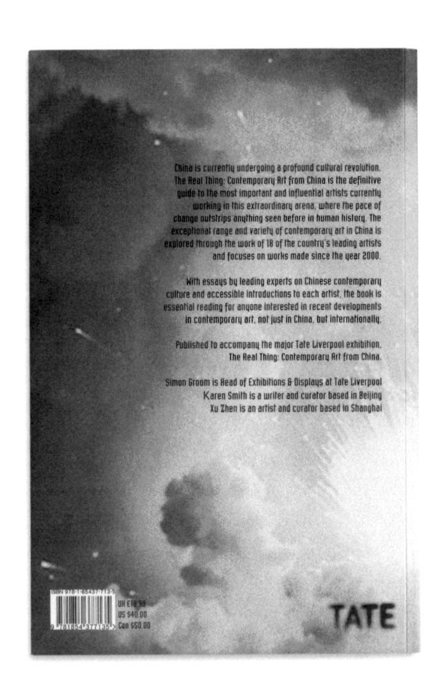

Dear He Hao

I hope you are well. You will be delighted to know that the catalogue you designed for "The Real Thing" was voted the 2nd best catalogue in the whole of the UK in 2007! Congratulations!

Many thanks,

*Simon Groom**

亲爱的何浩：

别来无恙。你一定会很高兴得知你设计的《确有其事》被评选为 2007 年全英国最佳画册第二名！祝贺！

十分感谢，

Simon Groom

* Simon Groom，苏格兰国立现代美术馆馆长，时任泰特利物浦美术馆展览负责人。

张鸥 **爸爸和我**

北京艺门画廊，北京，2007

275mm×215mm×5mm，48 页

平装

1000 册

编辑：马芝安

文章：Barbara Pollack

访谈：马芝安 / 张鸥

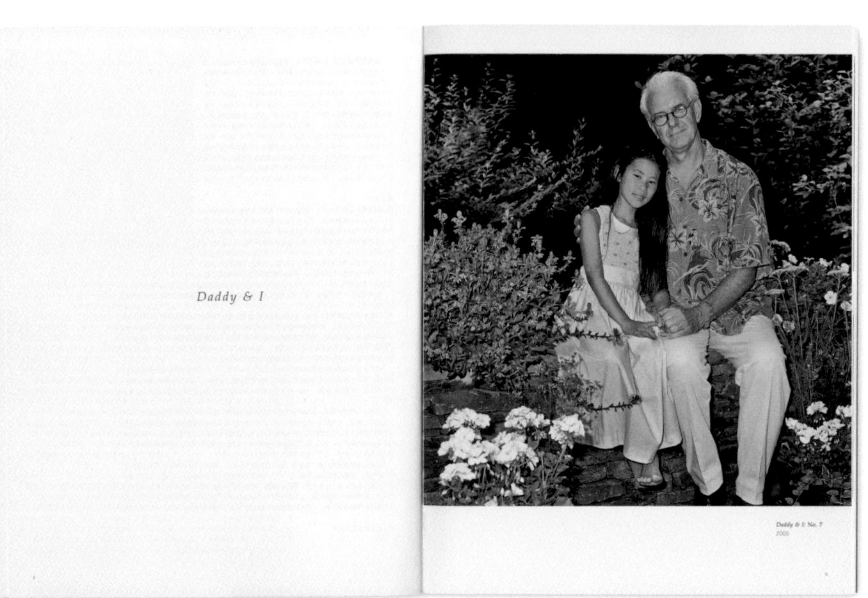

Daddy & I: No. 16
2005

Daddy & I: No. 5
2005

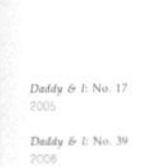

Daddy & I: No. 17
2005

Daddy & I: No. 39
2008

Daddy & I: No. 6
2005

Daddy & I: No. 33
2006

Daddy & I: No. 36
2006
Daddy & I: No. 31
2006
Daddy & I: No. 26
2006
Daddy & I: No. 4
2005
Daddy & I: No. 18
2006
Daddy & I: No. 38
2006
Daddy & I: No. 34
2006
Daddy & I: No. 11
2005

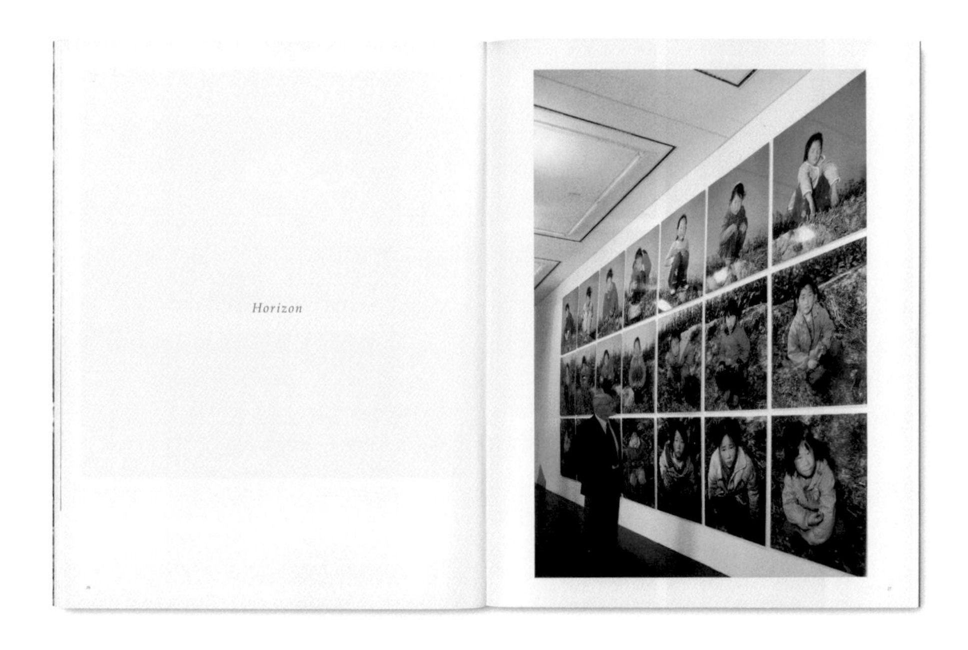
Horizon

Pékin Fine Arts Co. Ltd.
Printed in China 2007

莫毅　**我的街坊**

Walsh 画廊，芝加哥，2007

215mm×285mm×5mm，60 页

平装

1000 册

编辑、文章：巫鸿

文本写作：莫毅

Walsh Gallery is pleased to present the United States premier of Mo Yi's work
in collaboration with Wu Hung

Mo Yi: An Ethnographer in Contemporary Chinese Art

Wu Hung

The world of today's contemporary Chinese artists is full of glittering, self-congratulating celebrities, who vie with one another over auctions, and who find as much joy in their new cars as in their art. Should Mo Yi ever appear among them (a hypothetical situation which I have never witnessed), he would be seen as a total oddity. Looking older than his real age (he was born in 1958), he has a haggard face with an earnest expression. He is nearly bald, and his once long and dark beard has become sparse and grayish. He strikes one as a person who sleeps little and cares little about his appearance; and he could easily be mistaken for a migrant rural worker in Beijing or Tianjin (where he lives). I have never heard him praise his own work, although I have been seeing him almost every summer for the past seven years. Usually he travels to Beijing to pay me a visit. He calls from the Beijing Train Station before showing up at my door with rolls of photos tucked under his arm. (He only recently acquired a computer.) He unrolls the photos one by one, offering explanations now and then in a low, tentative voice. This may be his way of inviting suggestions and criticism—the purpose of his visits. As a result, I always find myself talking more than I usually do on such occasions, especially because I am equally interested in his ideas and projects, which differ considerably from what one sees in popular art magazines.

§

In an interview with me Mo Yi identified himself as "basically a Tibetan," although neither of his parents are Tibetan nor was he born in Tibet. When his mother was about to give birth to him, she was traveling in a truck that fell into a gorge. Most of the passengers died but she (and he) survived. During the Cultural Revolution his parents both worked in Lhasa. When he was fifteen he was recruited by the soccer team of the Tibetan Autonomous Region. He played professional soccer on and off for twelve years (from 1973 to 1985), first representing Tibet and then the city of Tianjin. In 1982 he became fascinated with photography and traveled around Tibet to take pictures with a "blockhead camera" (shagua xiangji— the Chinese term for a simple automatic camera). Around the same time he also tried to enter college, but this goal proved to be a fantasy for a person who never even finished junior high school. When he retired from professional soccer he asked to work "anywhere which needed a photographer." He was assigned such a job in one of Tianjin's children's hospitals.

Neither in the interview nor in his writing has Mo Yi shown a strong interest in politics. His involvement in politics was intuitive and emotional. To show his support for the June Fourth student demonstrations in 1989, for example, he designed and staged a "performance," marching on Tianjin's streets as a mourner holding a funerary banner (p. 49 in this catalogue).[1] On his white robe he wrote a couplet: "Laughing, I mourn for the death of the feudal system. / Crying, I am greeting the birth of democracy." He also wrote two large characters on the banner: "Qu ye" or "It's gone!" Mo Yi explains: "What is gone is my earlier blind faith [in the Party] and so I am laughing. I also cry because it is so difficult for democracy to arrive." A big crowd gathered around him, cheering and taking pictures. This "performance " made him a household name in Tianjin but also one of the "Ten Big Criminal Cases" (shi da an), handed over to the city's Police Department after the crackdown on the movement. Mo Yi lost his job, was under house arrest for several years, and in 1997—when I first met him—was still living in fear.

§

By Mo Yi's own account, his involvement in the June Fourth movement was largely an accident, but his interest in society has stayed at the center of his art. In particular he has been fascinated by the city— not a city's architecture but its "expression" (biaoqing), which he finds on people's faces and in their clothes, movements, and

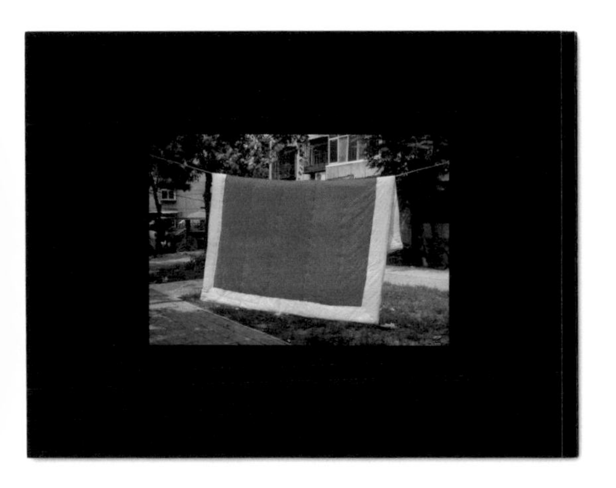

林天苗 **聚焦——纸上**

艺术家独立出版，北京，2007

400mm×336mm×15mm，96 页

布面精装，裹纸质护封，外加瓦楞纸书匣（未展示），

对页单幅作品之间装订薄隔页纸

限量 1200 册，每册艺术家签名、编号

编辑：何浩

文章：陆蓉之

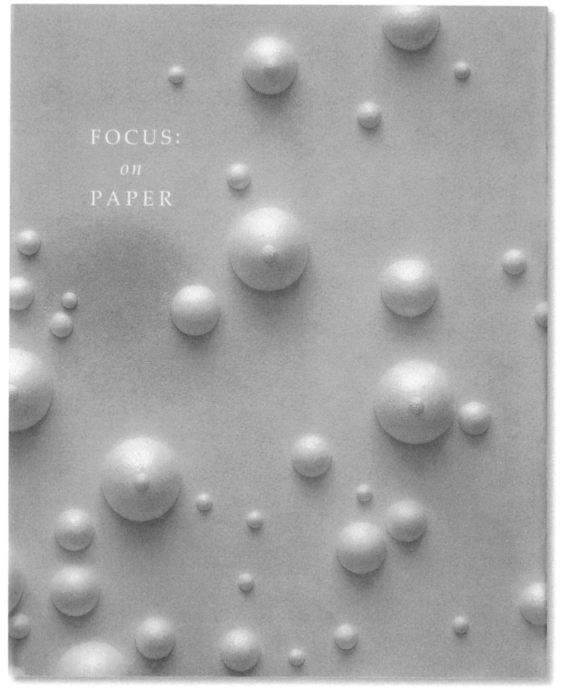

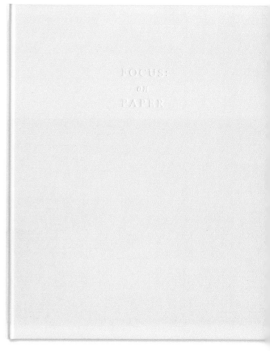

Mirror Images of God-Like Creatures:

A Glance at Lin Tianmiao's Work on Paper/Print-Making Series

In 2006, Lin Tianmiao was invited by the Singapore Tyler Print Institute (STPI) to make her first print series in a studio established by Ken Tyler, a world famous printmaker. These silk screen print images remain faithful to Lin's oeuvre despite the adaptation of a new medium, retaining common elements of her body of work that over-ride the various treatments of paper learned at STPI. For example, imprints and tangled lines interact with the paper like faint scars on one's skin. Lin Tianmiao's approach favors the gradual and colorless change of a monochrome palette of black, grey and white, used to express the most subtle changes of emotion and mood. Each variation on a subject portrait creates a new expression and focus, as reflected in each of these prints. Two dimensional print medium is transformed by Lin into multi-layered and three dimensional work with abundant surface texture. This richness of surface treatment is unprecedented in print making.

Lin Tianmiao's prints exalt human portraiture to a higher status of iconographic and god-like imagery. Her visages are similar to religious icons hung above the viewer at a reverential height, imbued with the gravity of portraiture that traditionally defies print-making. In other words, Lin's use of prints to capture these mystical portraits create mirror images of god-like subjects, and are a logical extension of her earlier body of work. However, the print portraits are naturally bound by the limits of the paper itself, so they appear as details of larger more monumental portraits. The artist deliberately blurs facial features and magnifies the human visage to the point of semi-abstraction; much like the fragments of memory that drift across human thought.

Lin's mono-prints are of the same stature as her earlier monumental photo and portrait works. The paper surface as new material allows Lin to multiply her earlier ideas with logical clarity befitting this new found medium. For twenty years, Lin Tianmiao's artistic pursuits have wandered adeptly from design to painting and photography; from sculpture to installation. And now reaches out to the field of print and paper-making. Her determination to conquer this new medium only confirms the force and vigor of her sensibilities as an artist. These new works bear witness to her journey, ever nearer to the temple of female artistic master.

Victoria Lu

Creative Director, Museum of Contemporary Art, Shanghai

《聚焦——纸上》

2006年夏天，天苗应新加坡泰勒版画院（STPI）的邀请，在他们的材料和技术的支持下进行了为期四周的版画创作。走之前天苗跟我约定好，回来要为她的版画新作出本书，虽然当时我们都还不确切知道最终出来的将是怎样的一批作品。

一个月后，天苗回到北京，电话中疲惫而兴奋。她说她这次一气完成了几十张作品，效果很好，由于之前从没做过版画，反而因此得以突破常规。在这组名为《聚焦——纸上》的系列作品中，天苗选取了一些她身边人的近距离肖像，模糊了他们的面目特征，使其身份难辨，然后在这些游离的、冥想着的头像上，层层叠置了交错的线、球或毛发，同时辅以造纸时独特工艺产生的纸张肌理，给人以羽毛划过皮肤一般的敏锐感觉。这种触碰，不仅是视觉的，更是心理的。

按照一般经验，版画用纸虽然种类繁多，但基本都是批量生产的成品纸，即便有的艺术家会针对某些作品特别订制，但也只是将成品纸拿来使用。版画的核心价值通常会被认为是印痕，而非承载物纸。天苗的作品却从另一个角度进入——STPI的技师在她的要求下研发出了一种直接在滴着水的半成品纸上印刷的技术，在印刷图像的同时继续堆积或破坏纸面，直至作品完成。在这种创作方式下，纸面肌理成为表达作品概念的重要元素，印痕反而退为背景。

从上面的描述可以看出，设计这本书有两个关键词，一是“聚焦”，二是“纸感”。首先，书籍翻开虽有左右两页，但并不能并置两幅作品，否则无法“聚焦”；其次，书籍开本要够大，纸张肌理的呈现有赖于图像的印刷尺幅，如若太小必是混沌一片。点破这两个关键词看似设计已定，其实不止于此。

书籍的阅读是平面观看，但作为拿在手里的物质，书籍又是一个六面体，对于这个六面体的形态认知与构建为书籍设计的首要问题。天苗这本书如果开本巨大且内文将近一半篇幅为空白页，书籍体量必然超大——这样的量感对于雕塑、壁画或史诗般的主题性创作都还合适，但罩在天苗这些敏感细腻的纸本作品上则厚重有余轻盈不足，有如用牛刀杀鸡，用着看着都很吃力。我想象的这本书要大而轻薄，有种往上升而非向下沉的力量。此外，为了不失手感同时又可在印刷中淋漓尽致地呈现作品的细微层次，我选择了一款价格昂贵的内文用纸，但如果书中一半空白，为此增加的成倍预算同样需要面对。

最终，我决定所有作品在书中连续放置，不留一页空白，但在每个左右对页之间插入一张极薄的隔页纸——所有问题都被这薄薄一页纸片瞬间化解。左右两页被薄纸隔开互不干扰又透过薄纸彼此隐约可见，对页图像经过透叠过滤更加不可名状，天苗“聚焦”在纸上，而此刻亦在书间。翻动薄纸时动作舒缓轻柔的“虚”，则在心理层面上凸显了作品敏感不可触碰的“实”，如果说翻动书页的这个行为亦是阅读体验的一个组成部分，那么在这里翻动即是感知。

为了使这个设计构想在技术上得以完美实现，我动了很多脑筋，最终印厂按照我设计的装订方案把薄厚两种纸张完全用套叠锁线的方式天衣无缝地装在了一起，无一页粘贴。印厂的技师为此不禁感叹我作为一个设计者对于装订原理的精熟。关于专业知识和技术，我觉得就像语言中的词汇和语法，决定了表达的边界。真正的言简意赅一语中的，只属于对大量词汇和语法规则精准掌握的人。说话写作如此，设计亦如此。

林天苗 看影

艺术家独立出版，北京，2007

185mm×335mm×13mm

24 张单页，装在布面精装函套内

1200 册

编辑：何浩

文字：林天苗

LIN TIANMIAO Seeing Shadows

My thoughts about thread:

Thread can change the value of things, turning the useful into futile, and futile into useful.

Thread can both collect and break up power.

Thread can represent gender and change identity.

Thread is both real and imaginary.

Thread is sensitive and sharp.

Thread is a process, something you go through.

Lin TianMiao

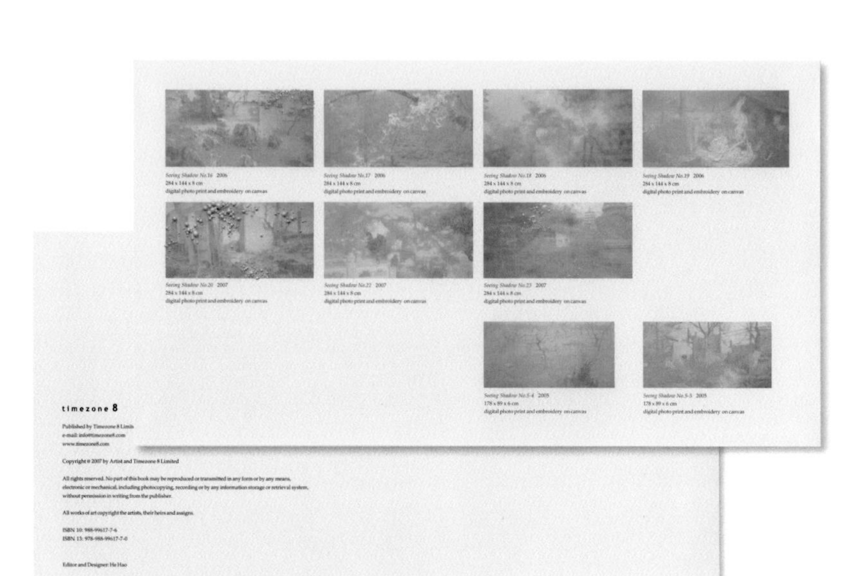

Seeing Shadow No.16 2006
284 x 144 x 8 cm
digital photo print and embroidery on canvas

Seeing Shadow No.17 2006
284 x 144 x 8 cm
digital photo print and embroidery on canvas

Seeing Shadow No.18 2006
284 x 144 x 8 cm
digital photo print and embroidery on canvas

Seeing Shadow No.19 2006
284 x 144 x 8 cm
digital photo print and embroidery on canvas

Seeing Shadow No.20 2007
284 x 144 x 8 cm
digital photo print and embroidery on canvas

Seeing Shadow No.22 2007
284 x 144 x 8 cm
digital photo print and embroidery on canvas

Seeing Shadow No.23 2007
284 x 144 x 8 cm
digital photo print and embroidery on canvas

Seeing Shadow No.5-4 2005
178 x 89 x 6 cm
digital photo print and embroidery on canvas

Seeing Shadow No.5-3 2005
178 x 89 x 6 cm
digital photo print and embroidery on canvas

timezone 8

Published by Timezone 8 Limi
e-mail: info@timezone8.com
www.timezone8.com

ISBN 10: 988-99617-7-6
ISBN 13: 978-988-99617-7-0

Editor and Designer: He Hao

Printed in China 2007

《新摄影》十年

前波画廊，纽约；卡罗琳娜·尼采当代艺术，纽约；三影堂摄影艺术中心，北京，2007

420mm×300mm×55mm

五册一函：四册作品，线装（原版设计：荣荣、刘铮）；一册文集，平装

亚麻布面函套，外加瓦楞纸书匣（未展示）

限量600函，每函编号；初始的70函内包含艺术家原作，由参与的十五位艺术家签名，收入在皮质书盒内，第41至70号为AP版（未展示）

艺术家：刘铮、晋永权、荣荣、王旭、刘安平、庄辉、高波、赵亮、邱志杰、蒋志、安宏、洪磊、管策、三毛（冯卫东）、郑国谷

NEWPHOTO 10 Years
WU HUNG
and
ZHANG LI
OTO
NO.1
OTO
OTO
NO. 3
OTO
NO. 4

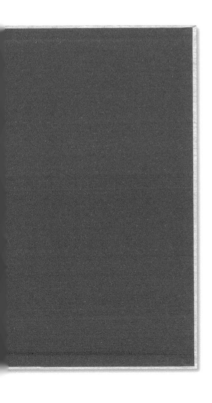

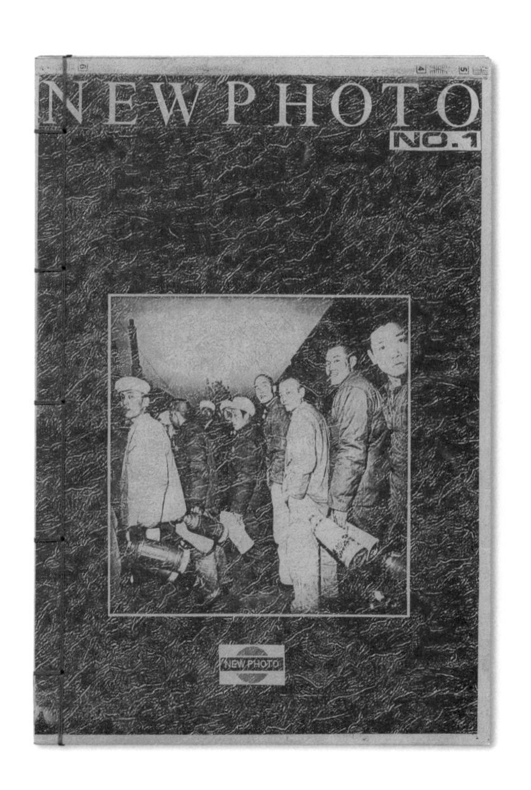
NEW PHOTO
NO.1
NEW PHOTO

NEW PHOTO
献给理想主义的朋友们！
- 2 -

NEW PHOTO
- 9 -

NEW PHOTO
- 30 -

寫真、攝影、照相、
PHOTOGRAHPY

指用光、放射綫、粒子綫等的能量形成、
記録能够用視覺 識别影像的方法和記
録下來的結果。

— 摘自《攝影術語詞典》
P155 － 156頁
1996.8.RR

1995. 西藏 原件 80 × 100 cm
明膠鹵化銀手工相紙

观念进入中国摄影，就如同封闭已久的房屋中豁然敞开了一扇窗户，呼吸舒畅多了，此时我们也明白了**新**摄影之**新**的涵义。

于《新摄影》创刊一年

洪　磊 HONG LEI

第　页

鸟的死亡
是身伪滥情虚假和浮华
饰掩脂粉来庸
失了生之向望
我喘息的思想
和喉咙哽咽的血液
高贵而优雅
轻盈地推梦
翱翔
去择取精神自杀
获得偷猝辉煌
于是我的紧闭
双眼
针刺心一刻
无意再看天下
——洪磊

20×20=400　江苏省哲学社会科学联合会

新摄影

1998.8

第四辑

邱志杰 QIU ZHIJIE
刘铮 LIU ZHENG
管策 GUAN CHE
荣荣 RONG RONG
三毛 SAN MAO
洪磊 HONG LEI
赵亮 ZHAO LIANG
郑国谷 ZHENG GUOGU
蒋志 JIANG ZHI

《新摄影》十年（文集）

420mm×295mm×12mm，152 页

编辑：巫鸿、张黎

前言：荣荣和映里

文章：巫鸿、凯伦·史密斯、清水敏男、张黎、克里斯托弗·菲利普斯

访谈 / 问卷：十五个艺术家

ESSAYS

NEW PHOTO AND ITS HISTORICAL CONDITIONS

WU HUNG

New Photo is a "tongren kanwu" compiled and published by the experimental photographers Liu Zheng and Rong Rong in Beijing from 1996 to 1998. The Chinese word "tongren kanwu" generally means an unofficial serial publication initiated, organized, and funded by a group of similar-minded writers or artists. The closest English term may be a "private collaborative journal." Viewed in the context of 1996-1998, however, the publication of *New Photo* had a significance that went far beyond a private endeavor within a small circle. Most importantly, it helped develop "experimental photography" into an independent contemporary art field through articulating its basic characteristics and forging ties between experimental photographers. This field did not yet exist in the first half of the 1990s, when the official Association of Chinese Photographers still largely dominated Chinese photography and controlled the exhibition and publication channels. Although some collective activities were organized around fields such as "artistic photography" and "photo journalism," most of these activities were affiliated with the Association or were entirely amateur.

In the field of contemporary art, Chinese "avant-garde" art (also known as "contemporary art" or "experimental art") developed rapidly in the 1980s and 90s. But until the mid-90s, photography played only a minor role in this movement. Nonetheless, it is worth noting that starting from the late 80s, some artists had already begun to explore alternative photographic styles. These artists had different backgrounds and motivations; some were experimental artists using other art mediums, while others were photographers critical of the mainstream dominance. Their initially separate experiments began to constitute a trend around the mid-90s. The compilation and publication of *New Photo* at this moment clearly marked this historical turning point. Moreover, through the introductions of mutual friends and increased contacts, a group of young artists who had similar ideas about art and life came together around this journal. That the first issue of *New Photo* (published on May 25, 1996) devoted an entire page to printing a single sentence — "To our idealistic friends" — expressed such comradeship, but also made it clear that in 1996, the concept of "new photography" still remained an "ideal."

A decade has passed since that moment. One can now say without hesitation that the "ideal" of *New Photo* has been fully realized. Internationally, the exhibition *Between Past and Future: New Photography and Video from China* traveled to different places from 2004-2007 and generated broad interest, making collecting and showing experimental Chinese photography fashionable. Domestically, the founding of the Three Shadows Photography Art Centre by Rong Rong and inri demonstrates the maturity of photography as a field of art creation as well as art historical research.

Considering the importance of *New Photo* in contemporary Chinese art, my introduction here takes an art historical approach, discussing the conditions of this journal's appearance and analyzing the basic features of "experimental photography" that it promoted. For these purposes, I will cover a broader period: a review of the period from the late 1970s to 1980s will establish the starting point of contemporary Chinese photography. The following discussion will then focus on two aspects of experimental photography since the 1990s: its relationship with China's social transformation and the artists' changing self-identity, and the interaction between experimental Chinese photography and postmodern theories, conceptual art, and other new forms of contemporary art such as performance and installation. In this way, this discussion will also outline the general development of contemporary Chinese photography over the past thirty years.

The Rise of Unofficial Photography (1976–1979)

Three consecutive events from 1977 to early 1979 together constituted a turning point in the history of contemporary Chinese photography. First, a group of amateur photographers formed an underground network, compiling their private records of a suppressed political movement into volumes for public circulation. This movement — the mass mourning for Premier Zhou Enlai in 1976 — was the first large-scale public demonstration in the capital of the People's Republic of China. Most of these photographs were taken by amateurs. Some of them, such as Wang Zhiping and Li Xiaobin, would later become the leaders of the Photographic New Wave in the 1980s. But in 1976 they were beginners with little knowledge of the art of photography; what they had were cheap cameras and a burning desire to record the mass demonstration for posterity. Working individually, each of them took hundreds of photos in the Square throughout the April Fifth Movement, and preserved the negatives during the subsequent political persecution. They became friends and comrades only later, when they saw each others' April Fifth photos and embarked on a collaborative effort to publicize these private records to represent public memory. Before this, some of them had compiled individual photo albums of the movement and presented them to Deng Xiaoping and Deng Yingchao.[1] Now they decided to compile the most powerful photos of the movement into a single volume and publish it nationwide. A seven-member editorial committee was established in 1977 to collect and select images.[2] Because the April Fifth Movement was still labeled an anti-government movement, they worked secretly and relied on very limited resources. It remains unclear how many negatives they actually collected, but according to Li Xiaobin, prints they enlarged from these negatives numbered between 20,000 and 30,000.[3] Some 500 images were selected for the volume, which eventually came out in January 1979 under the title *People's Mourning (Renmin de daonian)*.[4]

Published after the rehabilitation of the April Fifth Movement, however, this volume was no longer a private undertaking as the editors had planned, but became an official project endorsed by China's top leaders. While no authorship of the individual photos is explicitly identified, the volume's title page bears a dedication by Hua Guofeng, then the Chairman of the Chinese Communist Party. The volume thus helped Hua gain public support and legitimated his mandate. This official patronage also brought unexpected fame to the editors, whose "heroic deeds" were reported in newspapers, and who were invited to join the mainstream Association of Chinese Photographers. No longer considering themselves amateurs, they now took photography seriously as a lifelong pursuit; disillusioned by the official hijacking of their project, however, they turned away from political involvement to pursuing an artistic photography outside the government's agendas. Li Xiaobin recalled that right before the publication of *People's Mourning*, he and Wang Zhiping took a trip to the Old Summer Palace (Yuanming Yuan) to photograph the famous ruins there, while Wang suddenly turned to him and said: "Let's stop making a career in politics. Let's just create art and organize our own exhibitions!"[5] This idea soon spread to other members of the group, and in early 1979, they established the April Photo Society. The club's first exhibition, *Nature, Society and Man*, opened in April.

Two photographic groups or "salons" in Beijing formed the core of the April Photo Society. One group—most of whom had taken part in the *People's Mourning* project — met regularly in Wang Zhiping's small apartment in the eastern part of the city. The other group, formed as early as the winter of 1976, had 30 to 40 members who gathered every Friday evening in the dorm of the young photographer Chi Xiaoning in the western part of the city (the dorm belonged to the Northern Film Studio). The spiritual leader of this second group, known to its members as Xingqiwu shalongt — the Every Friday Salon — was Di Cangyuan, a special effect photographer working in Beijing's Science Film Studio. Di's younger followers admired him for his extensive knowledge in the history of photography, and he served as the main lecturer in the group's meetings. Other photographers, film directors, and artists were also invited as guest speakers, attracting young artists citywide, among whom Wang Keping, Huang Rui, Gu Cheng, and A Cheng would soon emerge as representative avant-garde artists and writers of their generation. In addition to these

8

9

《〈新摄影〉十年》

开幕首展对于任何一个艺术空间而言都是头件大事。2007年，荣荣和映里创办的三影堂摄影艺术中心开幕，首展推出《〈新摄影〉十年》。

《新摄影》是荣荣和艺术家刘铮在1996—1998年独立编辑出版的一本观念摄影民刊。选择在新空间开幕之际，把当年刊登过的所有原作拿出来再次展示，我想这既是荣荣作为双重创始人所看重的，更是三影堂成立初始的无声宣言——我相信荣荣讲的，在某种程度上，三影堂的创办，正是《新摄影》理想在十年后的另一种延续。

存在于1996—1998年之间的《新摄影》一共出版过四期，每期"发行"数量不确定，约几十册——具体印量要看荣荣和刘铮在当时能凑够的钱数。《新摄影》的印刷完全用复印方式完成，之所以这么做，同样源于资金的问题——荣荣他们无法支付在印刷机上批量印刷的费用。为了获得相对较好的印刷效果，荣荣他们试遍了街边打印店里的各种复印机，但偶尔出现的彩色照片更令人头疼——据说为了防止被用来制作假钞，当时所有的彩色复印机都被严格管控，私人很难拥有。为此荣荣他们不得不托一个在国营单位工作、办公室里正好有台彩色复印机的朋友，趁周末没人时，偷偷印个几十张。荣荣和刘铮将每期复印好的一摞摞单页抱回住处，自己动手配页、装订。我曾问荣荣为什么要采用古籍般的线装形式，他回答说当时租住的六里屯村口有个修鞋的，只有他的工具能够穿透几十张纸，因此每回配好页后都到这个修鞋摊打孔，然后回去自己穿线。现实条件就是这样，自然也没有什么更多的选择。

星星之火，可以燎原。短短十年，中国当代艺术的境遇已是天翻地覆。按照大家最初的设想，这部配合开幕展的同名画册《〈新摄影〉十年》毫无疑问要弥补当年的"遗憾"，所有原作将重新扫描、精美印刷，还之以作品的本来面目。如果沿着这个思路做下去，我们将会得到一本标准的群展画册，其"高大上"程度甚至可以不亚于世界上任何一个著名美术馆的出版物，但这真的是我们所需要的吗？带着一点习惯性的质疑，我再次仔细翻阅了荣荣手上仅存的一套《新摄影》孤本，看看其中是否还有暗含的"真相"。事实上，随着阅读的深入，我确实感到了另辟蹊径的必要：当年《新摄影》四册刊登了十五个艺术家的作品，这些人中的绝大多数在十年后的今天已是声名赫赫，而当时收录其中的那些作品，也在后来的日子里随着艺术家的崛起被反复出版、广泛传播，成为中国当代艺术最具代表性的一部分。在这样的背景下，此次出版无论书籍装帧如何考究、图片印制怎样精美，恐怕也仅是在出版履历上又增加了一项而已，其价值跟十年前相比已不可同日而语。与这种情形相对的，是随着时间的推移和环境的改变，当年那四册不得已而为之的复印件却在日趋冰冷与光鲜的今天显得愈发动人——那因复印而产生的如刀刻般凌厉的影像、业余但饱含真诚与信仰的排版、不合理却合情的装订方式……中国观念摄影发轫之初的呐喊与勃发尽在其中，粗粝而炙手。

转天，我跟荣荣和映里见面提出了我的意见——不再做任何“新”的设计，把《新摄影》四册“复刻”就好，那些与之相关的文献资料则单做一本文献集，五册一函，构成《〈新摄影〉十年》这一完整的出版项目。听过我的陈述，荣荣和映里也觉得这是一个好想法，只是当时旁边有人提议，说如果这样还不如做得更彻底一点——买一台复印机，完全还原当年《新摄影》的原貌。这个问题我也想过，但觉得并不妥当——虽然根据原版复刻，但我们出版的是一本完全的新书，而非原书的“赝品”。时过境迁，两部《新摄影》所要传递的信息完全不同，而实现这一不同的根本在于“转换”——我们要用最先进的设备和最高级的材料复制出当年的粗糙，全新的意味就在这两者的碰撞间生成，反差越大，撞击的力量就越大。

话说到这儿，看似所有的问题都已解决，但没想到应该最简单的印刷却遇到了难题——一般的印刷油墨无法复制出复印机碳粉特有的那种黑。为了解决这个问题，我跟印厂的技师一起反复试验，直到约定时间的最后一刻，才终于创造性地勾兑出了所要的油墨配方。在屡试屡败的试制过程中我曾断想：诸葛亮在决定“草船借箭”之后，一定也对每条船上的草靶数和船只的承载力做过十分精确的计算和测试，甚至还为了草靶跟船体到底应该怎样连接颇费苦心。其实，任何“谈笑间樯橹灰飞烟灭”的背后都一定要靠大量艰苦细致的工作来支撑，这才是真实不虚的计谋。

多年以来，我一直认为最好的设计就是“不设计”，起码要“少设计”，所谓“上兵伐谋”，“不战而屈人之兵”。“不设计”并非不作为，而是洞彻形势之后的顺势而为。《〈新摄影〉十年》大概可算一例。

王秉龙 1979—1989 中国戏曲年画摄影

三影堂摄影艺术中心，北京，2007

235mm×305mm×8mm，60 页

布面精装

限量 500 册，每册艺术家签名、编号

主编：荣荣、张黎

编委：荣荣、王卫、张黎、赵亮、何浩

文章：张黎

Chinese Opera Calendar Photography and Me

Wong Bingjong

I took these photographs in the ten years between 1979 and 1989. Essentially, I brought the theater troupe out into the open, and photographed opera stories for publication in calendars. I called it "Chinese Opera Calendar Photography." I was the first to take pictures of opera in this manner. There were a lot of people who used this style afterwards, but they didn't have the conditions that I had, nor did they have the volume or variety of works that I did. In 1990, I moved to South America, and because of market saturation and a lack of people, this style of photography withered. Not too long ago, the publishing company that I had worked for moved, and lost the original prints that I had published. Unfortunately, the only surviving documents of this opera photography are the films that I have in my hand.

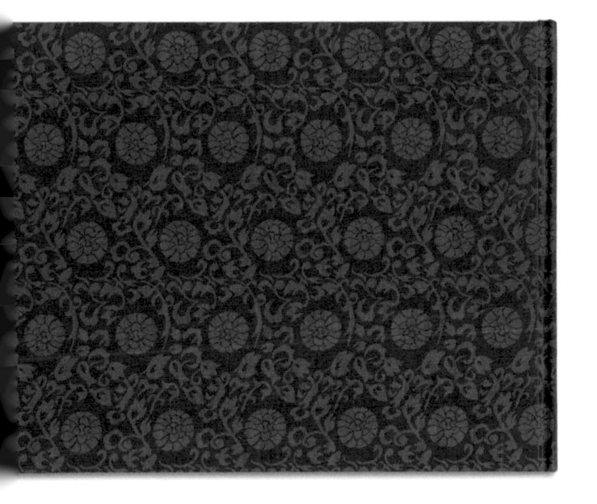

陆亮 **夜行者**

艺术家独立出版，北京，2007

293mm×382mm×7mm，64 页

纸面精装，书脊包布

1000 册

编辑：陆亮、何浩

文章：李旭

访谈：向莉 / 陆亮

艺术家陈述：陆亮

The Metropolitan Fable of A Classicist

Li Xu

Director of Z-Art Center, Shanghai

It has already been ten years since I met Lu Liang for the first time.

Then he was still studying in the mural painting department in Central academy of Fine Arts, and showed me the photos of his exercise works of sketch and oil painting when he visited his family in Shanghai during the summer vacation. I only felt that he was a very diligent student with good foundation of plastic art, though his creation hadn't started yet. He didn't choose the smart and ingratiating style, which was concerned with his steady and conscientious personality. Well-disciplined brushwork of classic academicism is low-yielding and quite difficult, and it really needs inexorable perseverance to form individuality in this.

Ten years later, when I saw Lu Liang again, he had gradated as a MA from the academy and taught there. Influenced by his tutors Chen Wenji and Cao Li, his individual style has initially formed, and the series of works after 2004 is unexpectedly refreshing.

Without any sense of gaudiness or extravagance, Lu's works are obviously opposite to the classic style popular in the art market. Almost all the models under his paintbrush are male, mostly the peasant workers common in life. The paintings are generally shrouded in the solemnity of dark hue, depicting the traditional Chinese fables unfamiliar to the contemporary people, and prevailing with the atmosphere of philosophic speculation. From the perspective of painting language, Lu directly benefits from the western classic masters, such as Rembrandt, Caravaggio, Latour and etc, however, he narrates extremely personal stories in such a language. Living in Beijing for a long time makes him suspicious and worried about the vicissitudes of contemporary metropolises. He always sighs and mediates about the endless expansion of city's scale, pollution and damage of ecological environment and the disseverment and separation in historical memory. The creation of series on the themes of fable, legend and fantasy expresses his meditation on these solemn topics in the way of surrealism.

Besides the figure, another important theme of Lu's works is landscape. Without daylight, these landscapes only have night, dawn and twilight. Like those figure paintings, all the scenes are unfolded under various lights: the unclear sun, dim fluorescent light, shocking lightning, aggressive flashlight, and etc. As the absolute leading character hidden in Lu's works, the light traverses air and floating dust and integrates with shade and color. Eventually, all the scenes are reflected on our retinae in the form of light. Influenced by Gerhard Richter, a German master, many domestic artists who create with the help of photography would usually emphasize the tendency of flatness, lightness and void, while Lu is quite different. To precisely depict the subtle reflection of light on different objects and numerous details of sense of reality, he often re-operates on his works again and again, and the process of depiction, polish and re-depiction would repeat for seven or eight times, sometimes even ten times. Therefore, no print would ever realistically reproduce the actual effect of these works, which is apparently a great pity for the audience unable to see the originals. It may be said that, Lu is a perseverant artist insisting on depicting the light, and his name Liang means light, which is probably not a mere coincidence.

In the times when various international exhibitions are filled with the forms of multi-media, painting has been quietly returning. In the circle of Chinese contemporary art, the trend of conceptual painting is gradually flourishing. Could the skill of classic painting revive in the current times? Could the concept of realism generate a whole-new significance? I believe, Lu Liang's continuous efforts would surely present an answer with extreme individual charm for us.

November 15, 2007

废墟与夜游人—— 陆亮访谈

向莉 · 陆亮

[illegible]

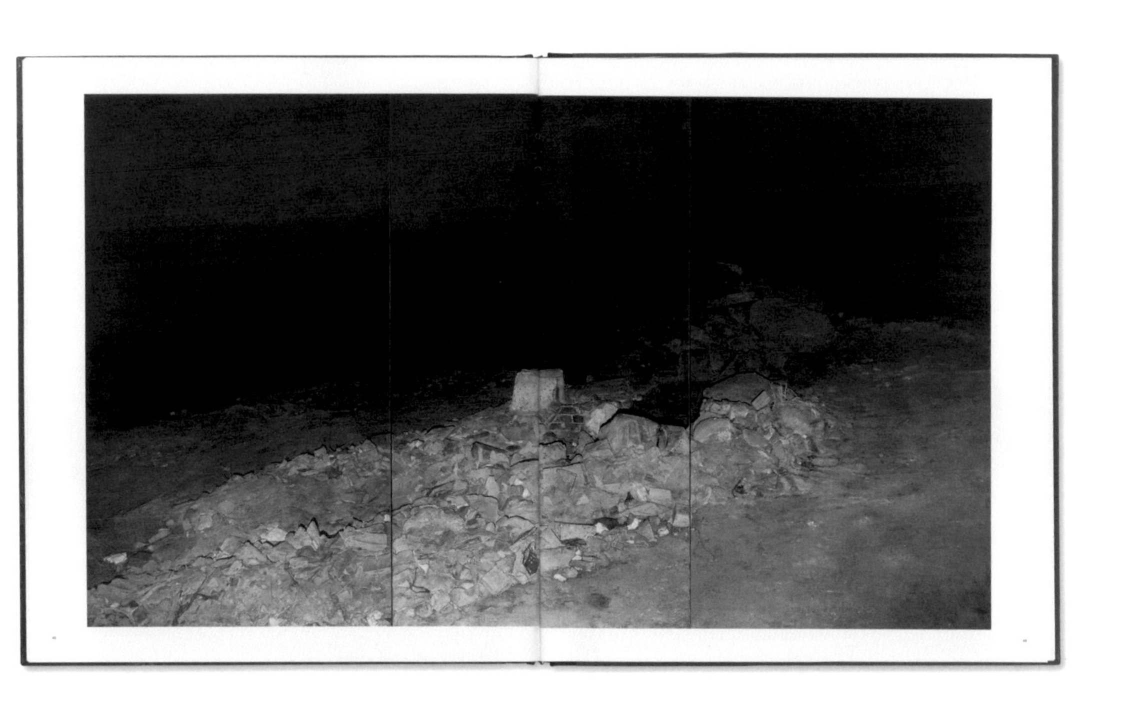

Jan Worst 绘画 1988—2008

Sperone Westwater，纽约，2008

210mm×291mm×30mm，238 页

缎面精装，外加缎面函套

1000 册

前言：Gian Enzo Sperone

文章：Adrian Dannatt

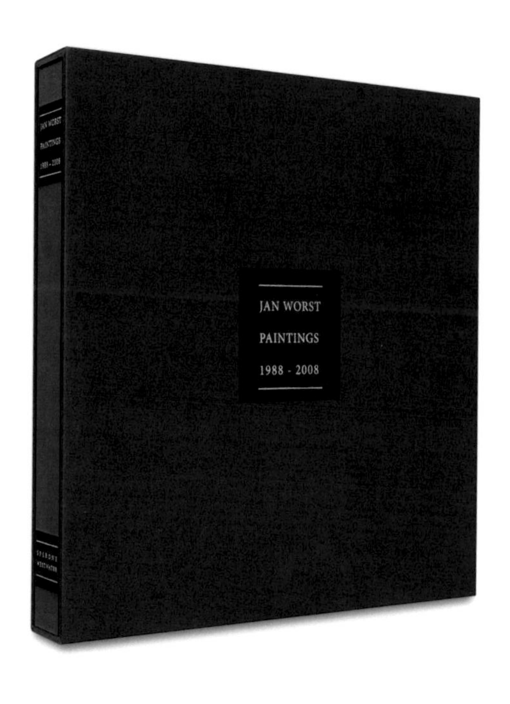

A Matter of Light

Gian Enzo Sperone

The mantra that was repeated in my high school days, "art is a lyrical perception of reality", was created by Benedetto Croce in such a simple way that it seemed as unbearable as a mathematical formula or a good syllogism for all situations, yet misleading; vaguely threatening. It's like this, or it's like this. And we thought: and if we took away "of reality"? or "lyrical perception"?

We'd be left with "art is". And we laughed heartily, as we did for almost all the things that allowed us to get through the endless school hours (which we'd later miss) with as little damage as possible.

When I first saw a work of Jan Worst – so formal, cold, calm – I couldn't help thinking it was strictly predetermined, with a rather philosophical precision and little of anything "lyrical". I was wrong. The paintings of this lanky Dutchman, with his measured ways and his beautiful bespoke English shoes of a considerable size, are an enchanted, treacherous world where, in a large complexity of elements, what lords it over all the rest is the light. And then there are the thousand pentimenti that take the natural slowness of this type of work to alarming levels.

I've never been to Heerenveen, the Dutch town not far from the sea where Worst was born, but from a map I discovered it's 169 km from Leyden, where Rembrandt was born.

As it's highly probable there are no mountains or forests with their inherent light and shade effects, the light dominates everything, evenly enveloping the entire visible world. Let's be clear about this: not that suffocating Mediterranean light, but the frugal yet clear-cut one of Vermeer interiors.

I can't resist the temptation to stray a bit more with idle considerations. In 400 years, the art of painting has changed a great deal, and especially in just a few years of the early 20th century, rather as the stones exposed to the weather have been corroded more during this century than over periods of hundreds or thousands of years. Painting, too, has been corroded by the ravenous intelligence of the painters – modern cannibals – and never before with such dis - inhibition and ferocity.

From this perspective, much contemporary painting is frankly impossible to look at, perhaps not at first glance, but probably the second time around.

Despite setting aside the standards of classicism, now dead and eaten by the worms (each one absorbing from the previous one, only to then swallow and expel without mercy), it's not clear how the splintering, dismembering, unravelling and mutilating can leave room for a form of representation which, however you look at it, is still the heart of painting. Even where the aim is just to hint.

Laying aside the problem of the light as a vehicle of expressiveness, and denying the relevance of the old rule of art being somehow decipherable, there is no desire to engage in new metaphors (because who notices anyway?). It would seem more practical to camouflage singsongs that have been picked up, recycling the rhythms deleted by the cinema, fashion, television or other various trash banalities. At the most, the aim is to arrive at the staging of a feeble toy theatre. The hope that a pubic hair, disinfected by the lens of a fashion photographer, can alone constitute a slightly reheated linguistic starting point is not enough to produce painting material.

Painting is a bewitching, unconventional glance, not mediated by anything else apart from the artist's interior mirrors; sometimes by the entrails.

It's also true that painting is just a system of signs that now have no order, more or less allusive, more or less penetrating. But the essential thing, yesterday as today, is to transpose one's own experience as a sensitive subject into the language; into one's own language, not that picked up from the dominating culture.

The art of Jan Worst is the representation of a silent, static life theatre, where a few humans are surrounded and dominated by objects. A world where nothing decisive happens and any tension promising energy and activity just wilts amid the formal perfection. A softened world with the good manners of classic elegance. Nothing personal, everything nearly calm: like an ideal scenario, so that the light becomes the real protagonist and anxiety can knock on the window of painting.

This filtering light of the inside is the only clue to an outside world. As if nothing existed apart from the domestic walls and everything, from birth to death, took place in that scenario of perfect furnishings.

This form of indifference and intransigence towards the viruses of nature makes the work of Jan Worst unique in its extreme, exaggerated fixedness.

New York, 5 November 2007

Mere Dreams, Mere Dreams!

Adrian Dannatt

What if the glory of escutcheoned doors,
And buildings that a haughtier age designed,
The pacing to and fro on polished floors
Amid great chambers and long galleries, lined
With famous portraits of our ancestors;
What if those things the greatest of mankind
Consider most to magnify, or to bless,
But take our greatness with our bitterness!

from Ancestral Houses by WB Yeats

Le lieu est lui-même en toi. Ce n'est pas toi qui est dans le lieu, le lieu est en toi.

The place is itself in you. It is not you that is in the place, but the place is within you.

Jacques Derrida - Sauf le Nom

As a precocious teenager Jan Worst was already involved in the street-fights of '68 as a "Provo", one of those avant-garde Dutch agitators, and ever since he has remained true to revolutionary anarchist beliefs, a radical punk, illegal squatter, even arrested for attacks upon the Queen of Holland's palace. Actually none of this is true, or maybe it is, for I know absolutely nothing of Worst's life other than dates of birth and art-school education. But the sumptuous oddity, the sheer singularity of his oeuvre is such that one reaches for any sort of biographical assistance, whatever clues his own history or character might grant to an interpretation of this hermetic domain. And surely our interpretation of Worst's work would shift if it was known that he had indeed been imprisoned for, say, spray painting upon the walls of Huis ten Bosch, the graffiti slogan " ALL PROPERTY IS THEFT!" Likewise, we somehow presume that Worst is not an actual aristocrat himself, has not literally grown up amongst and spent most of his life within just such rooms, and that these paintings are thus not to be understood as straightforward celebration of privilege. But it is not impossible, there is a strong art historical precedent of aristocratic (think Balthus) or even royal painters and there are certainly a sprinkling of contemporary artists, particularly of Italian and English origin, who come from social backgrounds almost as grand as those portrayed by Worst.

What would Worst's oeuvre "mean," how would we interpret it, if we knew that he was portraying his own home life, his family mansion, his relatives? Surely we would assume it was a non-critical, loving and sensuous homage to his own class, the exact opposite of the meaning we would ascribe were he the anarchist that we first imagined. Lured by Worst's world we immediately want to know about the artist himself, or more specifically about his social relationship to his aesthetic, his intentions, his own ideas, what he is here to tell us. Above all, what we fundamentally need to know is the artist's own relation to the very specific realm that he deliberately portrays again and again to the extraordinary exclusion of all other subject matter. What are his personal attachments or aspirations regarding this exceptional milieu? For however much we hope to try and judge art irrespective of any biographical information, just by the work itself, it is an entirely human failing to be swayed in our judgement by what we know about the artists themselves. We always need to construct stories, to resolve their plots, and the narrative that Worst has presented us with is so utterly mysterious, so seductively resonant, that we all naturally cast about for further clues.

For a true connoisseur the first thing that strikes one about Worst's work might be the dexterity and subtlety of his deployment of paint, the creation of such complex effects from such relatively simple brushstrokes, how his seemingly formal realism is built from near abstract pattern planes. But for most of us, naturally, what we immediately notice is Worst's subject matter. This is, to put it mildly, a most unusual theme for a contemporary artist, a strikingly unique topic, whose extremity blatantly challenges us. That challenge is inherently political because it goads us into examining and admitting our own relationship to wealth, privilege, luxury, and female beauty. Do we wholeheartedly approve? Do we want to have it, do we aspire toward these things, are these the secret sum of our hidden ambitions? Is this, let us for once be entirely honest with ourselves, the world that we want to inhabit and towards which all our everyday energies are directed? Do we approve of this world or do we disapprove, want it or reject it? Should such overt superiority, unashamed grandeur, be allowed to exist in our supposedly egalitarian times? Because he has posed this question to us, because he has forced us to consider such intimate ambitions, the extent of our complicity with this fantasy, we obviously would like to first establish Worst's own relationship to this world. Does he aspire and desire or analyze and despise? We will never know and that is precisely the deep potency, the magic of his oeuvre. As long as we do not know, cannot ever exactly establish Worst's own attitude to the milieu he so obsessively portrays then we are left in a state of ambiguity, moral and political, as to our own feelings. We may create fantastical hypotheses around him, invent these extreme biographical versions of a fictional Worst, but never divine his intention.

This biographical or even anecdotal imperative is partly inspired by a question of national identity, for the most unexpected, improbable dimension of Worst's work is that it should be made by a Dutch person. For nothing in Dutch society, politics or art history could prepare one for the emergence of Worst's oeuvre, and one can only imagine that its reception within that culture must be one of outrage, bewilderment or outright hostility. Indeed if some Dadaist provocateur was going to try and think up the one subject-matter, one style, that could actually rile and ruffle those famously tolerant, open-minded and liberal Dutch they might well arrive at Worst's world. For it is impossible to exaggerate the anti-elitism, the social liberalism and democratic conformity of the contemporary Dutch character, a country with the catchphrase "To be ordinary is extra-ordinary enough" and where any outward sign of

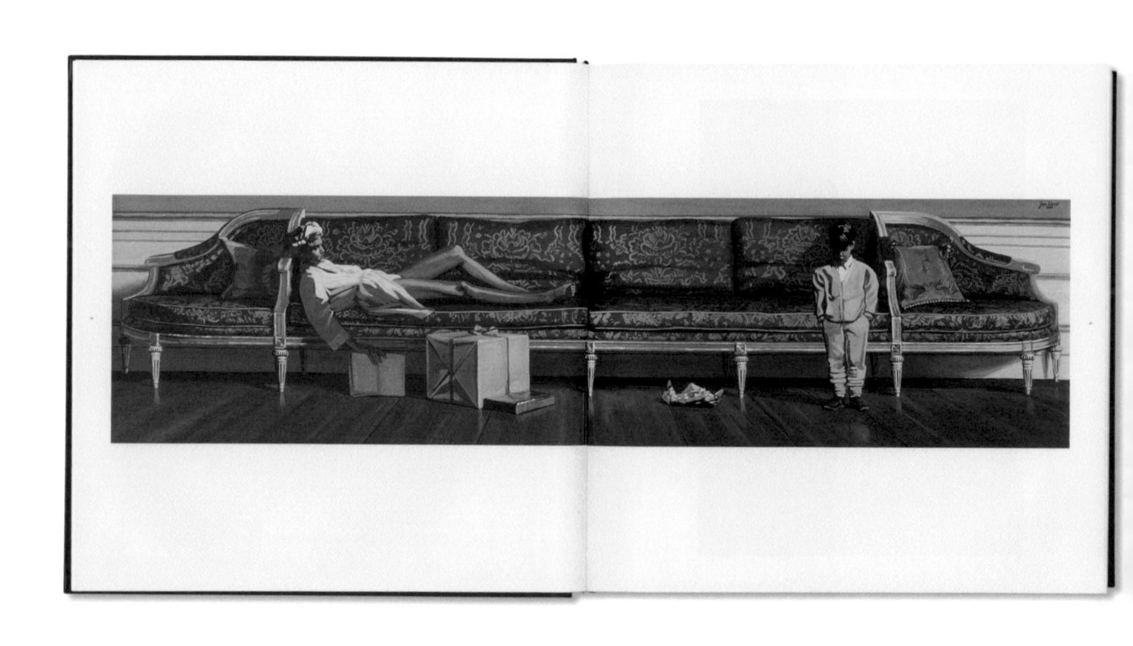

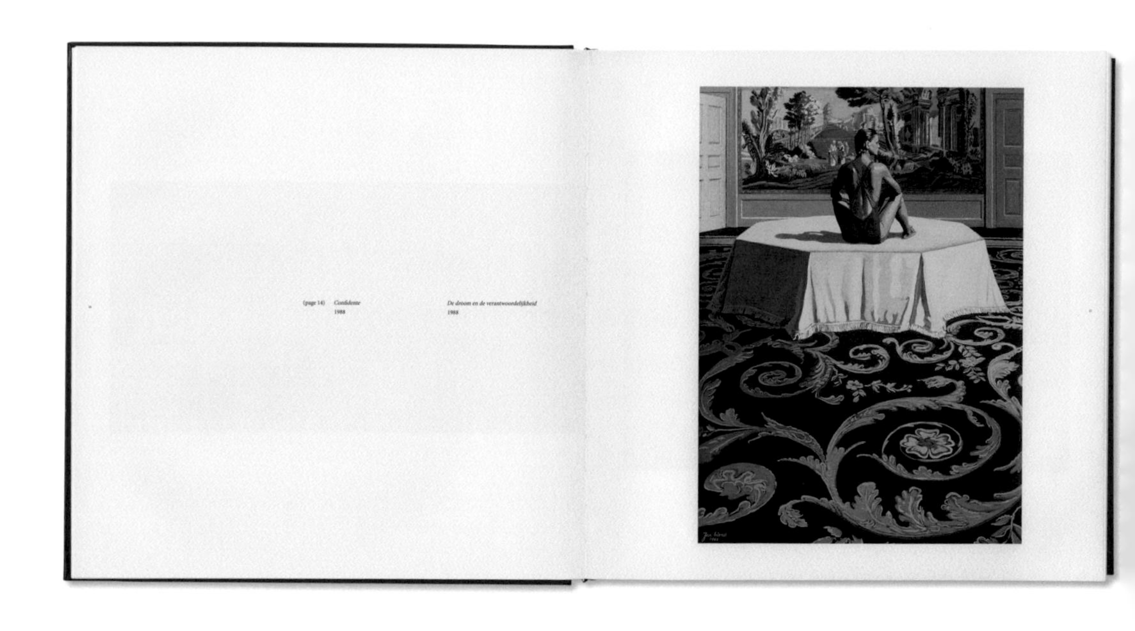
(page 14) *Confidente*
1988
De droom en de verantwoordelijkheid
1988

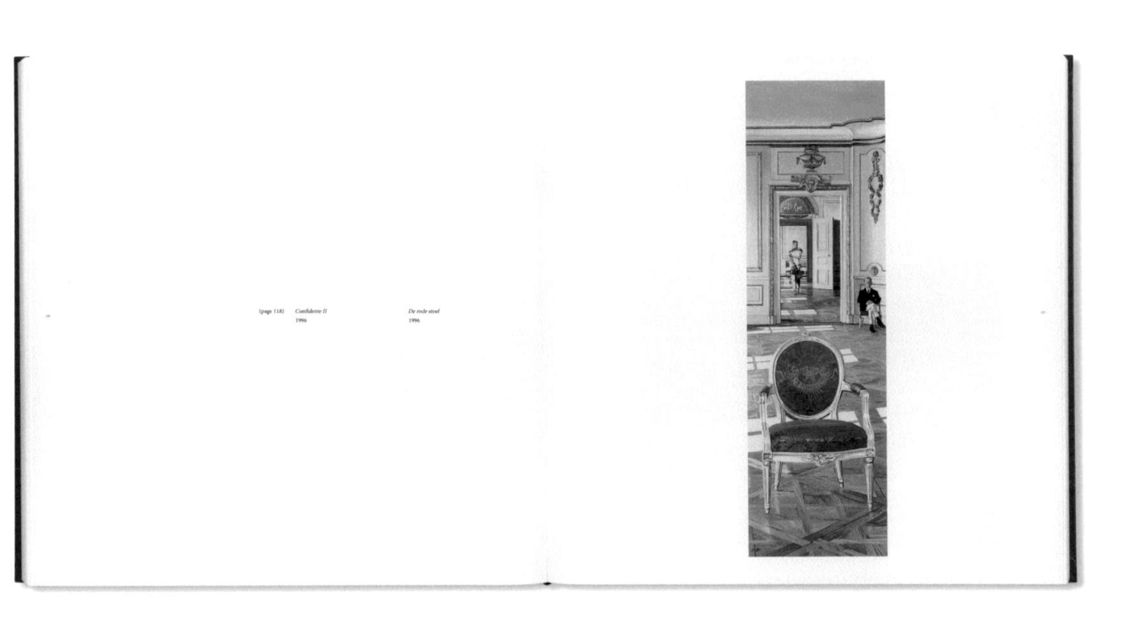
(page 118)
Confidente II
1996
De rode stoel
1996

De rode kamer
1998

*Hello Jenny *,*

After a couple of days turning over the pages again and again, putting the book down and picking it up, I am now extremely satisfied and happy with the result. The plates look really great in clear daylight. The format of the photographs, the feeling of the paper and the distinguished exterior combines brilliantly. The book is without exaggeration a milestone in my career.

Will you please tell all the persons who were involved in the project and especially He Hao, how satisfied I am with the result?

Best regards,

Jan Worst

Jenny 你好，

我这几天反反复复地翻开一页页内文，把书拿起又放下——我对这本书极其满意和欣喜。书中的画作在白天的光线下真的很漂亮。图片的形式、纸张的感觉和杰出的外观出色地结合在一起。毫不夸张地说，这本书是我事业的一个里程碑。

能否请你告诉所有参与这个项目的人，特别是何浩，我对这本书有多么满意？

祝好，

Jan Worst

* Jenny，雷宛萤（晚晚），木木美术馆联合创始人，时任这个出版项目的协调人，负责设计师和画廊、艺术家之间的沟通联络，贡献良多。

外象

三影堂摄影艺术中心，北京，2008

176mm×120mm×45mm

五册一函，平装，外加函套

1000 函

编辑：荣荣和映里、张黎、赵亮、董晓安

文章：张黎

对话：蔡卫东、丘、卢彦鹏、阿斗、荣荣、姜亦朋、董晓安、刘垣、张黎

艺术家：阿斗、蔡卫东、卢彦鹏、丘

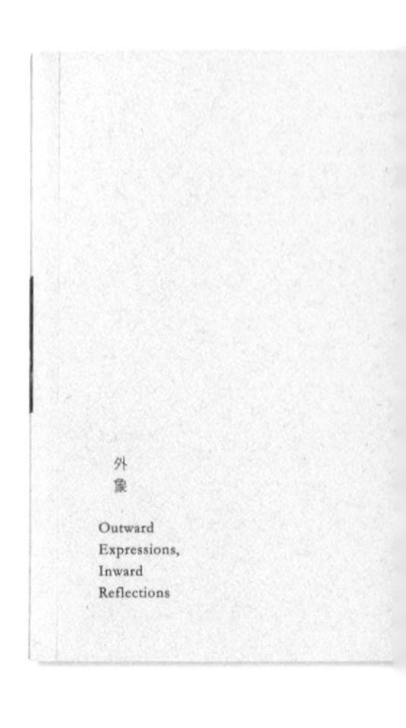

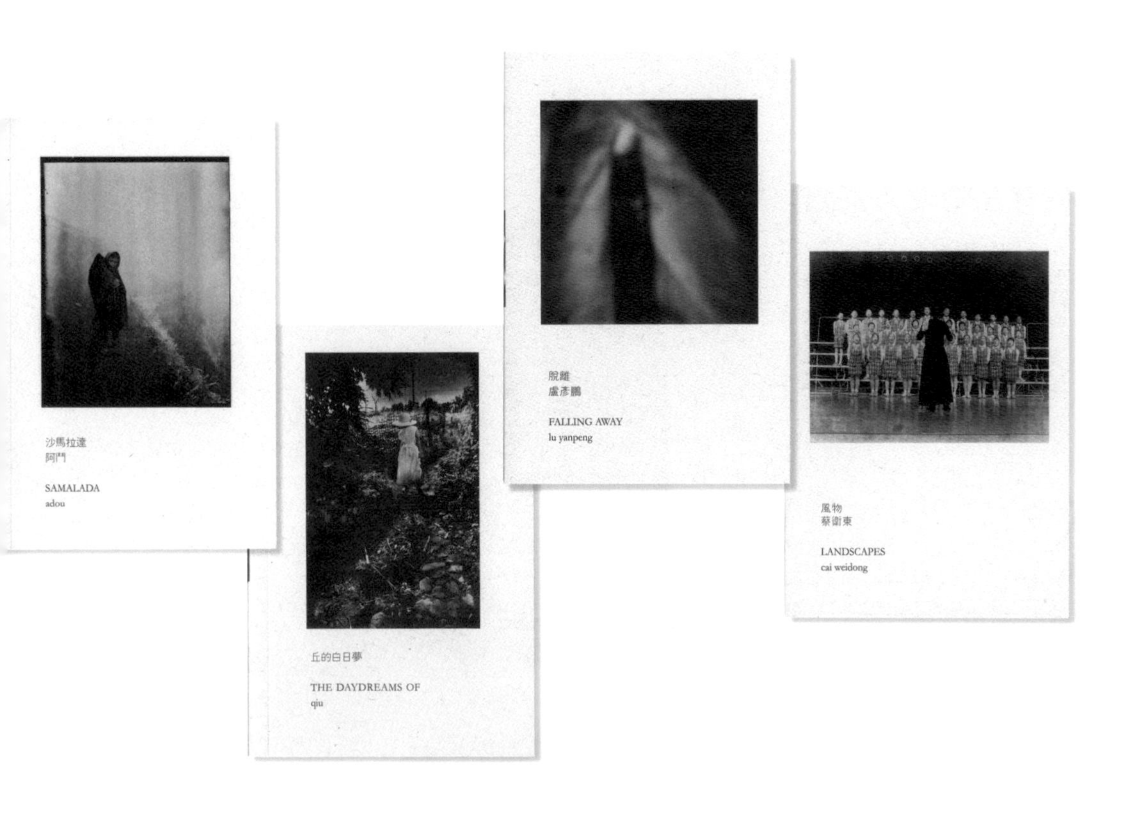
沙馬拉達
阿門
SAMALADA
adou
丘的白日夢
THE DAYDREAMS OF
qiu
脫離
盧彥鵬
FALLING AWAY
lu yanpeng
風物
蔡衛東
LANDSCAPES
cai weidong

Drip by drip drop
Tears
Which slowly turn into stone
The puppet collects the stones
Convinced that they are the sky's
Stars
Excerpted from "Puppets and Stones" 2005,
Lu Yanpeng

海报，1000mm×700mm，100 张

凯伦 · 史密斯 九条命——新中国前卫艺术的诞生

东八时区，香港，2008

（第一版：Scalo，苏黎世，2005）

275mm×230mm×36mm，472 页

软精装

2000 册

前言：Marianne Brouwer

文本写作：凯伦 · 史密斯

内文版式设计：杜蒙

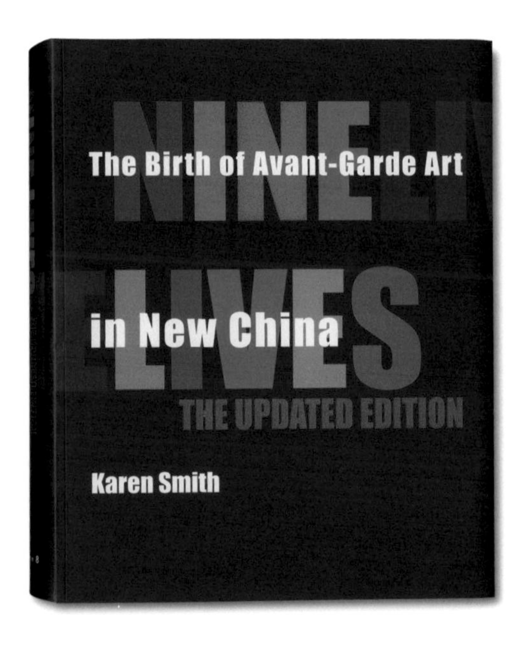

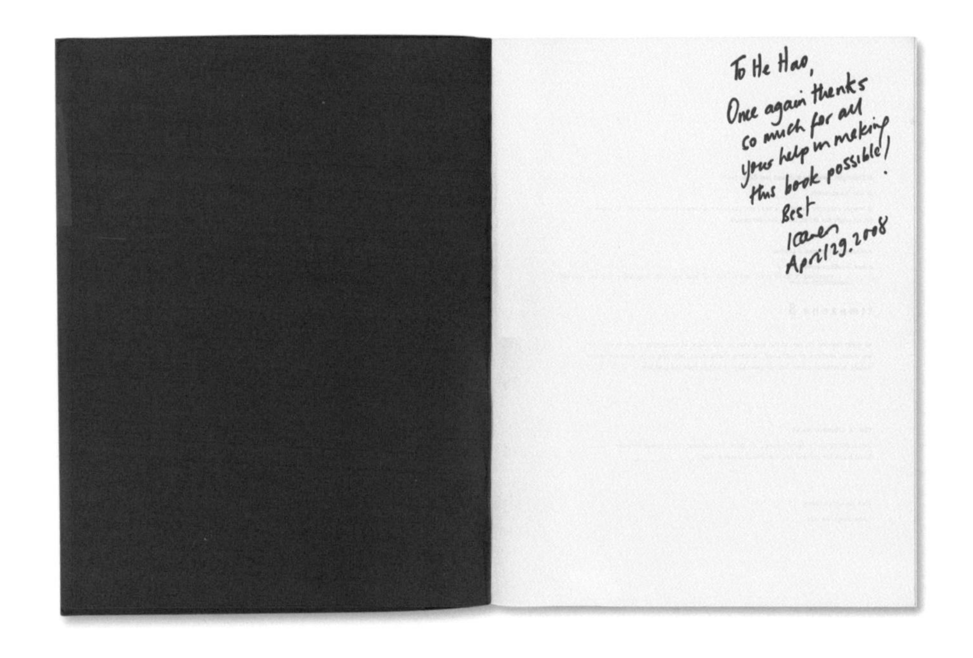

SECTION 1

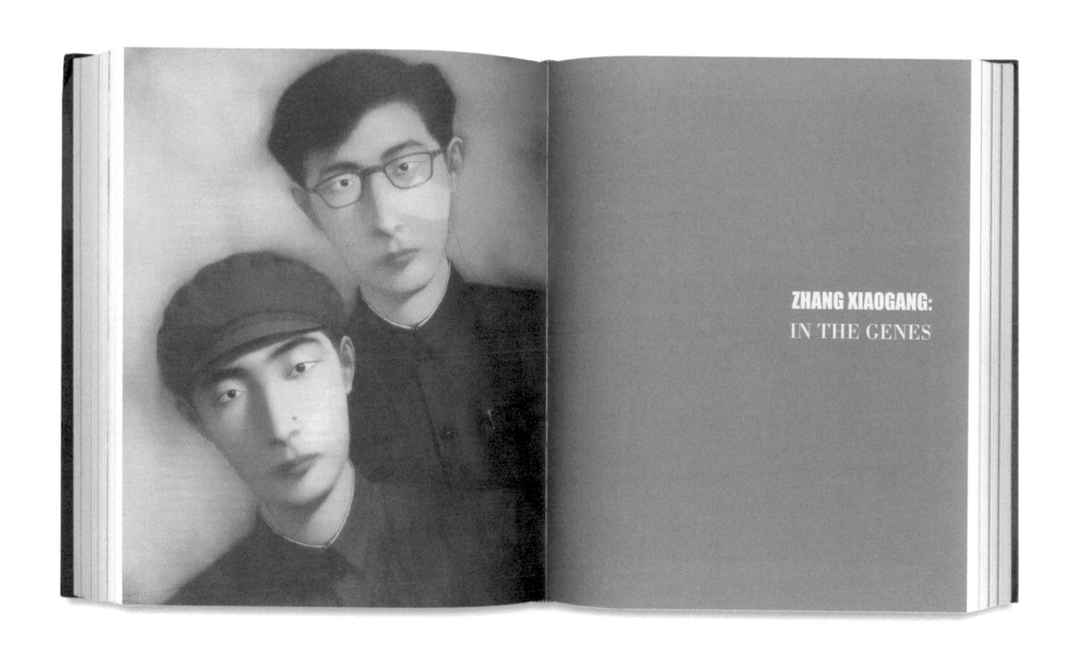
ZHANG XIAOGANG:
IN THE GENES

INTRODUCTION: "Hutong" is a kind of ancient city alley or lane typical in Beijing. Where hutongs may number several thousand. Many of them were built during the three dynasties of Yuan, Ming and Qing. Now they are mainly found around the Forbidden City. In the years of these dynasties, the emperors, in order to establish supreme power for themselves, planned the city and arranged residential areas according to the etiquette systems of the ZhouDynasty. The center of Beijing was the imperial palace—the Forbidden City—with the main streets laid out longitudinally and latitudinally. There are two kinds of hutongs. One kind is usually referred to as the regular hutongs, which are situated close to the east and west of the palace and orderly arranged alongside the streets. Most of the residents who used to live in these hutongs are imperial kinsmen and aristocrats. Another kind is the simple and crude hutongs, which are usually located far off the north and south of the palace. The residents there are merchants and other ordinary citizens. The main buildings in the hutongs are almost all known as "siheyuan" (quadrangles) —a compound with house around a square courtyard. The quadrangles vary in size and design with the social status of the residents. in former times the big quadrangles of high ranking officials and wealthy merchants were specially built with roof beams and pillars beautifully carved and painted, each with a front yard and a back yard. However ,the quadrangles for ordinary people were simply built with small gates and low houses. Hutongs, in fact, are passageways formed by many closely arranged quadrangles for ordinary people were simply built with small gates and low houses. Hutongs, in fact, are passageways formed by many closely arranged quadrangles of different sizes. The specially built quadrangles all face south for better lighting. That is why a lot of hutongs run from east to west. Between the big hutongs there are many small ones going from north to south for convenient passage. These lots, the city of Beijing appears like a magnified quadrangle, arranged symmetrically and neatly and surrounded rectangularly by high walls. This adds magnificence to the city.

By the end of the Qing Dynasty, China was in a backward state of natural economy. To feudal minds, to engage in trade was considered inferior. Therefore, Beijing was a consumer-city catering for pleasure of the emperors and aristocrats. People's life was mostly confined to hutongs. The quadrangle reflected the way of life and social culture at that time. Having experienced the changes of dynasties and vicissitudes of life, at the end of the Qing Dynasty, the ossified and closed China came under the influence from abroad. The stereotyped arrangement of hutongs was also affected. A lot of newly formed hutongs with irregular houses appeared outside the city while many old ones lost their former neat arrangement. The social status of the residents was also changed. Hutongs were no longer full of life. This was a reflection of the collapse of the feudal system. During the period of the Republic of China, Chinese society was unstable with frequent civil wars and repeated foreign invasions. Therefore, the conditions of Beijing suffered deterioration, and the conditions of the hutongs became worse. One quadrangle which used to be owned by owned by one family became a compound occupied by many households. After the founding of the People's Republic of China, the conditions of the hutongs were improved. But during the ten-year "Great Cultural Revolution", many historical and cultural relics in hutongs were destroyed. The reform and open policy has brought about great changes in Beijing. The houses in many buildings. Many residents have moved to new houses. Neither twisted hutongs nor close small quadrangles can be suited to the fast development of social productive forces and the rapid renewal of Chinese culture. Hutongs today seem to be fading into the shade in the eyes of tourists and inhabitants.

However, in the urban district of Beijing, houses along hutongs still occupy one third of the total area, providing residences to half of its population. So many Hutongs have survived. In this respect, we can see the old in the new in Beijing as an ancient yet modern city.

序：明代萬歷年間，有一對年輕的張氏夫婦，勤儉刻苦，在北京新街口東南小巷內，買下二十餘畝土地，以種菜為業。數年以後，又在園中闢地栽種牡丹，芍藥及滿塘的荷藕，每到春夏兩季，香飄風來，菊黃之秋，梅花映雪之日，這裏也別具風光，景色可謂四時得宜。當時許多文人墨客，士大夫紛紛前來賞花遊玩，從此這個地方即被人們稱為「百花深處」。張氏夫婦死後，花園無人經營，遂逐漸荒蕪。數百年過去了，遺跡已無處可尋，這個地方變成了一條小胡同，但「百花深處」這個美麗的名字，卻流傳到今天。北京有許許多多的胡同，也有許許多多這樣，那樣的故事……

徐樂　一九九零年清明

徐勇　**胡同十六图**

艺术家独立出版，北京，2008

275mm×230mm×36mm

18 张单页，装在布面精装函套内，

外加瓦楞纸书匣（未展示）

限量 800 册，每册艺术家签名、编号

前言：徐勇

杨福东 **竹林七贤**

Jarla Partilager，斯德哥尔摩，2008

286mm×223mm×25mm，224 页

布面精装，裹纸质护封

2000 册

编辑：田霏宇、贝安吉

文章：Mollly Nesbit

访谈：张亚璇 / 杨福东

Yang Fudong

Seven Intellectuals in Bamboo Forest

Jarla Partilager, Stockholm
Office for Discourse Engineering, Beijing

Yang Fudong: Seven Intellectuals in Bamboo Forest
Jarla Partilager, Stockholm
Office for Discourse Engineering, Beijing

Printed in China
ISBN 978-91-976351-3-4
First edition 2008

Editorial direction by Philip Tinari
Edited by Angie Baecker
Art direction by He Hao
Designed by He Hao and Du Meng

Published on the occasion of the exhibition *Yang Fudong: Seven Intellectuals in Bamboo Forest* at Jarla Partilager, Karlavägen 9, 114 24 Stockholm, Sweden

Special thanks to Marian Goodman Gallery, New York and ShanghART Gallery, Shanghai

Contents

MALE VOICEOVER: I went to Yellow Mountain eight or nine years ago. It is a well-known mountain, talked about by many people. When I was young, it was fashionable in many families to have pictures of Yellow Mountain hung on the walls of sitting rooms—an ocean of clouds, high and steep peaks, upright pine trees standing in the clouds, the red, round rising sun, as well as cranes flying above.

Seven Intellectuals in Bamboo Forest **Part One** 11

Wild Shanghai Grass

Molly Nesbit

In late April 1927, the writer Lu Xun sat in the southern city of Guangzhou and wrote a preface to his new group of prose poems, which he was calling *Wild Grass*. "When I am silent, I feel replete," he began. "As I open my mouth to speak, I am conscious of emptiness."

Words came. Past life had died. Dead life had decayed. From its clay, no trees grew, only wild grass. "I love my wild grass," he wrote, "but I detest the ground which decks itself with wild grass." He pointed to the fires underground that would one day erupt red and devour it.

At that point he would laugh out loud and sing, he claimed, and he repeated this laughing and singing, because this fierce turn of events was the fair proof that he had lived.[1] The poems that followed had been written to stand separately in the fray of the press. Pulled back and collected, they were a motley group—freeform meditations rife with inversions, vicious observations, visions exploding, cackling, tart, wafting, his dreams, he said—his wild grass. These pieces were not uniform. He shifted voices and cadences. He gave them titles that charted no path: "The Shadow's Leave-taking," "Snow," "The Passer-By," "Tremors of Degradation," "The Wise Man, the Fool, and the Slave."

At the end of the collection he put "The Awakening." That patch of wild grass began with Lu Xun at his writing desk at Peking University in April 1926; above him, bombers flew in to attack, and, mission accomplished, they departed; this was his baseline reality. He set himself to editing a pile of manuscripts by young writers, their words full of ambition, integrity, and anger; lovely words he felt: "Their spirits are roughened by the onslaught of wind and dust, for theirs is the spirit I love. I would gladly kiss this roughness dripping in blood but formless and colorless." He thought of Tolstoy and the thistle; he turned to the pages of the student journal, *The Sunken Bell*. He quoted the students:

> Some people say our society is desolate. If this were really the case, though rather desolate it should give you a sense of tranquility, though rather lonely it should give you a sense of infinity. It should not be so chaotic, gloomy, and above all so changeful as it is."

China was being torn apart by civil war, and Lu Xun's students were among those shot dead that year for demonstrating, but Lu Xun did not give those details. He wrote instead of the way events leave their mark. "The Awakening" had taken place in the depths of the night. It ended quietly:

34 Molly Nesbit

Fig. 1 Lu Xun photographed in 1933 in Shanghai. Image courtesy Beijing Lu Xun Museum.

Fig. 2 Cover for Lu Xun's Wild Grass. Image courtesy of Beijing Foreign Languages Press.

> While I have been editing the sun has set, and I carry on by lamplight. All kinds of youth flash past before my eyes, though around me is nothing but dusk. Tired, I take a cigarette, quietly close my eyes in indeterminate thought, and have a long, long dream. I wake with a start. All around is still nothing but dusk; cigarette smoke rises in the motionless air like tiny specks of cloud in the summer sky, to be slowly transformed into indefinable shapes.[2]

Lu Xun resigned his post in Beijing and escaped to the South. Ultimately, he would settle in Shanghai. *Wild Grass*, along with his stories like the "Diary of a Madman" and "The True Story of Ah Q" marks the arrival of a modern Chinese literature. Mao Zedong would be one of his greatest admirers. The wild grass ran together in the shadow of fire, with echoes of Nietzsche, and Turgenev, and with the ongoing Chinese revolution, which is also to say, he let it run with the shifts in the ground we call history.

How long does wild grass grow?

In 1991, Yang Fudong went south to Hangzhou, the old imperial city Marco Polo praised to the skies for the beauty of its freshwater lake, its pleasures, pavilions, and delights; centuries later, its reputation as a paradise remained intact. Yang Fudong went there to study oil painting at the Zhejiang Academy of Fine Arts, which would soon be renamed the China Academy of Art, although it would take many years for the name to catch on. The time was auspicious. A decade before, the Academy had become an inspired center of new experiment. Removed from the political imperatives of the Central Academy in Beijing, the Academy in Hangzhou was a place to think, to paint, to talk, to read. The state culture of realism was opening further; underground and overground, the avenues of translation and exchange that Lu Xun himself had done much to promote were being re-connected; at the Academy, they were listening to everything that came. Texts circulated. The circles of reference were increasingly unpredictable, unstable, electric.

They watched the video of Joseph Beuys' 1974 performance in New York where he arrived in an ambulance and then, wrapped in felt, lived in an art gallery, confined with a coyote and working on co-existence, proclaiming *I Like America and America Likes Me*. On another day at the Academy, the students were shown a video of an actual airplane accident played over and over again to gauge the numbness that came with repetition. Then there were the books.[3] There, for example, was Ludwig Wittgenstein's *Tractatus*, finished in the wake of the first World War, the book that began by drawing a

Wild Shanghai Grass 35

The First Intellectual

Fig. 8 The First Intellectual, photograph, 193 x 127 cm, 2000

> and I sat in front and had the warmest talk about the goodness and joy of life. Dean suddenly became tender. "Now dammit, look here, all of you, we all must admit that everything is fine and there's no need in the world to worry, and in fact we should realize what it would mean to us to UNDERSTAND that we're not REALLY worried about ANYTHING. Am I right?" We all agreed. "Here we go, we're all together... What did we do in New York? Let's forgive." We all had our spats back there. "That's behind us, merely by miles and inclinations. Now we're heading down to New Orleans to dig Old Bull Lee and ain't that going to be kicks and listen will you to this old tenorman blow his top"—he shot up the radio volume till the car shuddered—"and listen to him till the story and put down true relaxation and knowledge."
>
> We all jumped to the music and agreed. The purity of the road. The white line in the middle of the highway unrolled and hugged our left front tire as if glued to our groove. Dean hunched his muscular neck, T-shirted in the winter night, and blasted the car along. He insisted I drive through Baltimore for traffic practice; that was all right except he and Marylou insisted on steering while they kissed and fooled around. It was crazy; the radio was on full blast. Dean beat drums on the dashboard till a great sag developed in it; I did too. The poor Hudson—the slow boat to China—was receiving her beating.[10]

§

The *Seven Intellectuals* could only have been made by slowing the boat way down; there would need to be time, hours, years, for the drift. What does the intellectual do now? Where do the words go? The dreams? The life? These questions had haunted the work of Yang Fudong ever since he heard the trains in Hangzhou.

He had realized that his work would involve making movies and in 1996, wrote a script for the film that the next year became *An Estranged Paradise*. At the time, he told his friends that it was "a minor intellectual movie." He had invented the term (*xiao wenren dianying*) and so had to explain it. "Minor intellectual movies are about walking in the rain on a rainy day," he told them. "They are about your

38 Molly Nesbit

Figs. 9a-c Stills from *An Estranged Paradise*, 35mm black-and-white film transferred to DVD, 76 min., 2002.

emotions and moods; about the dreams that you cannot make true but cannot let go. They are about each detail of your life; they are what you think your life should be; they are the books you have read; they may also be a cliché."[11]

The film was shot as a pure experiment on expired black and white 35 mm film.[12] The actors were non-professionals; in other words, friends. In the film, they barely spoke; they were shown submerged in everyday life; they walked through it, disengaged, trying for more. Yang Fudong chose settings that he knew, seeking the kind found on outmoded wall calendars of beautiful Hangzhou in the rainy season. The film began with a lecture on Chinese landscape painting. It began, in other words, with lessons from a distant past. As a painter makes a branch, a leaning figure, a bridge, a mountain peak, a voice explains:

> Poetic mood is the life of painting and without it, painting just equals with a living dead. But how can the poetic mood be achieved? The great poet Su Shi once wrote: 'One fails to see what Lushan Mountain really looks like because one oneself is in the mountain,' which suggests that we need to get inspiration from our life.

After the lesson, a modern story of a man. He undergoes a breakdown, an illness crossed by three women, persistent dissatisfaction and a lethargy from which he finally recovers; he learns to compromise with life as it comes. He learns to enjoy life and love. The story ends inexplicably with a scene. A strange, half-mad man climbs up on a platform near the traintrack, does a dance, stripping, and lets out a string of yells toward infinity, or maybe the train. The landscape painting had provided a prologue, but not this conclusion. On this point, Yang Fudong remarked, "Actually, well-known adages, to a certain degree, should be the conclusions after compromise. They are all correct..."[13] These things are inherited and carried on. The end is the beginning. The beginning was the end. One is oneself in the mountain. One is oneself at the West Lake. Another one is beside himself by the train track.

What becomes of the mountain?

The funds to complete *An Estranged Paradise* came with Okwui Enwezor's invitation to exhibit at *documenta 11* in 2002. At *documenta*, the film received a great deal of attention from abroad, and a different conversation began. Invitations to exhibit outside China would now come one after the

Wild Shanghai Grass 39

(bedroom scene, MALE 2 is licking torso of FEMALE 1)
MALE 2: *(whispering)* What are you laughing at?
FEMALE 1: *(whispering)* What're you writing?
MALE 2: Nothing.
FEMALE 1: I see.
MALE 2: See what?
FEMALE 1: I see.

(GIRL ON STREET is applying make-up on a street corner. MALE ON MOTORCYCLE pulls up on his motorcycle.)
GIRL ON STREET: *(walking toward MALE ON MOTORCYCLE's motorcycle, speaking in Shanghainese)* Do you know what time it is? You only come over now. Where did you go last night?
(MALE ON MOTORCYCLE throws plastic bag full of food to GIRL ON STREET.)
GIRL ON STREET: You've been late so many times I've lost count. *(looking inside plastic bag)* This isn't what I wanted to eat. Do you even know what I want to eat? Eating this will make you fat.
(MALE ON MOTORCYCLE carries parcels off of motorcycle and drops a few, tripping over them)
GIRL ON STREET: *(smacking MALE ON MOTORCYCLE)* Hey, be careful, there's something really important in there. If you break it, you're going to have to pay for it. What are you, stupid or something? You can't even move anything without breaking things. What are you doing every night that you come back so late? All you know how to do is go out. *(smacking MALE ON MOTORCYCLE as he carries over more parcels)* You still haven't had enough, have you? Why don't you just go then?

Seven Intellectuals in Bamboo Forest **Part Two** 67

(MALE 1 and FEMALE 2 in the bath, FEMALE 2 splashing water over MALE 1)
MALE 1: After we caught the frog, we inserted a small bamboo stick into its ass, leaving half of it out. Then we put it on the ground and let it flee. When it jumped down onto the ground, the stick didn't hurt it. So it jumped up again, then hurt itself when it landed on the ground. In that way, the frog kept jumping on and on.
FEMALE 2: You are so cruel.
MALE 1: It is not cruel. We just wanted to help it exercise, get rid of its inertia.
FEMALE 2: That's an excuse for your evildoing.
MALE 1: Women's inertia is even bigger.
FEMALE 2: Hush! You really think so?
MALE 1: Of course. I'm helping you get rid of your inertia right now.
(FEMALE 2 alone in the bathtub, speaking directly into camera)
FEMALE 2: Now I have grown up and expect more from life. To be perfect and not to waste time, these are my goals for the future. That is aestheticism.

80 Yang Fudong

Seven Intellectuals in Bamboo Forest **Part Two** 81

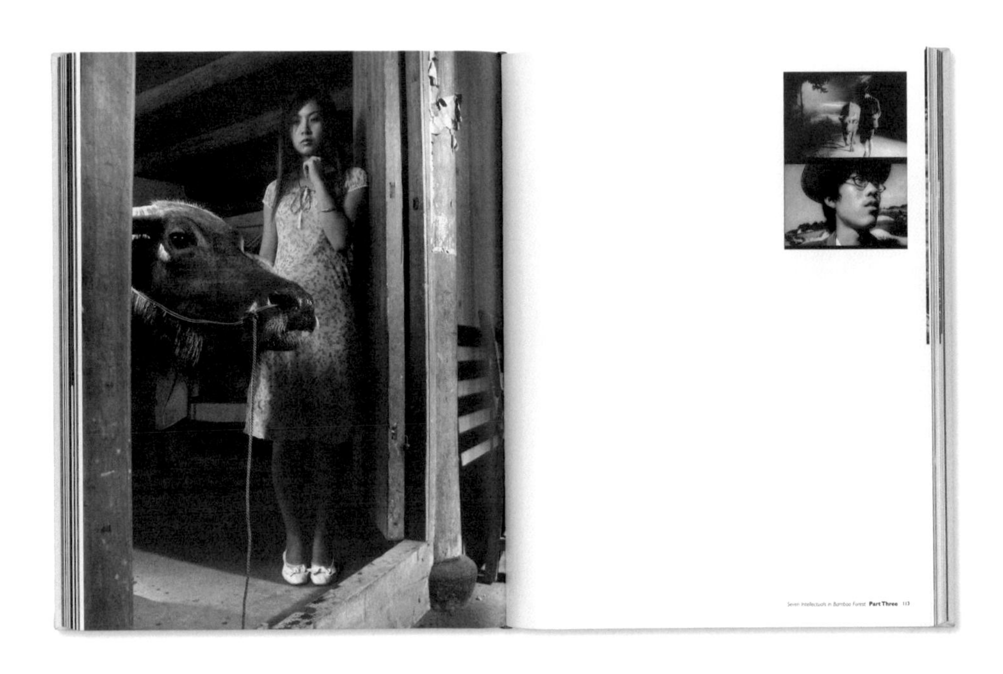

Interview: A Chill Spreading through the Air

Zhang Yaxuan, July 2005

Out of Context

Zhang Yaxuan: When we discussed *An Estranged Paradise* three years ago, you mentioned the idea of this cycle, *Seven Intellectuals in Bamboo Forest*. Now, you've really started working on it. Your present method, that is, filming one part every now and then, and spending several years to complete the whole work, does that stem from economic considerations or some other reason?

Yang Fudong: The original intention of *Seven Intellectuals in Bamboo Forest* was to slow down time. At the beginning, this was a relatively important principle to me. I think it's okay to make a movie all at once—you just pull it out of the air—but this is not quite the same as that kind of work situation. I want to make it so that work merges into life. It's a little like slowing down a work of art. This kind of slowing down allows your everyday thoughts and actions to slowly, intangibly seep into the work. When I shoot these films, money is not among my most pressing concerns. Because it's a low-cost production, money is not central. I want to quiet my mind, and allow myself to calmly go and try some things. Right now the plan is to finish the cycle in five years, and as for its temperament, its flavor, that will vary somewhat as my situation changes.

ZYX: Can you discuss the specifics of your working plan and your concept for each part?

YFD: Presently, three parts have been shot, and the work plan for the entire five-part cycle has become increasingly clear. Unless something really unexpected happens, I think it's a plan that's set in stone, and I will implement it.

I should first talk a bit about the overall structure of *Seven Intellectuals in Bamboo Forest*. The earliest original intention was to make this kind of film: five parts, each part established on its own, certainly with some different contrapositions and points of departure, and then borrowing a condition, a flavor, a sensibility from the Seven Sages of the Bamboo Grove to express some of my own feelings.

Initially, I thought these five stages would be—and actually this is also how I'm doing it now—about one part every year. *Part One* was filmed in April 2003, *Part Two* was filmed in April 2004, and in May this year, *Part Three*. Then, perhaps in the summer of 2006, I'll go and film *Part Four*. If timing works out, I plan to finish filming *Part Five* in the spring of 2007. Then there will be a half-year adjustment period, collecting the five parts to see if they should become an individual feature film, or five paragraph-style feature films. At this point, I haven't made a final decision.

Regarding the structure of these films, I'll blather on a little longer.

The Seven Sages of the Bamboo Grove are seven representative scholars of the Wei-Jin (220-420 A.D.) and Northern and Southern dynasties (420-589 A.D.) periods. I refer to them in my own film not in the spirit of a period piece. It's really more about contemporary young people. I think the films focus on ideals, beliefs and also life—these kinds of sensations. At that time, I decided to separate the cycle into five parts. The first part is travel notes from Huangshan. It is oriented as a postcard-style life, for everybody feels that beautiful places look like postcards. This kind of life is happy, a little bit like a wall calendar. *Part Two* is about closed-off life in a bustling city. A lot of times, the things in an individual's residence have nothing to do, in fact, with bustling. Everybody is living quietly in some corner of this city.

The facade of peace and prosperity, or nightlife—these kinds of things are completely different life situations from someone's evening time, or their time in their own place. Maybe if you put seven young people in a place, then some things will happen. *Part Three* is another kind of life. Seven young people from the city go to the countryside, and they plow and farm like the local peasants, living for a period in accordance with normal local lives, and again maybe some inexplicable things happen. And after that, *Part Four* is an island from our beliefs. Many people think of the island as a narrow, utopian idea. These seven youths spend some time living on an island during the summer, and perhaps there will be some special construction at that place, a harbor for taking shelter from storms, that sort of thing. And then *Part Five* returns to the city. It's kind of the closest thing to these seven people's real life situations, but perhaps, in terms of mentality, it's also disassociated. Basically, the movie will have this kind of structure.

Now that *Part Three* has been filmed, another idea has formed: maybe I'll also film a postscript type thing, ten or fifteen minutes long. Right now, I'm still not certain. It's an idea that emerged after *Part Three* was filmed. In the end, the entire cycle will be a film of three to three-and-a-half hours altogether. I don't want to fix the length yet, but probably it will be a slightly longer film.

ZYX: You've said before that using allusions is nothing more than taking something "out of context."

YFD: I feel that a good method for reading is to take things out of context, because sometimes the

YFD: Yes, I think the focus on real life in *Part Two* is a little greater, although in some places, it just happened that way. Actually, the state of life today is already totally different from before. For example, the volume of information is very large. People come into contact with so many things, and people are more and more open-minded. That is to say that life today is already changing, and the feelings between individuals in the face of society are more and more direct. This kind of incisiveness can turn into selfishness. I often think that *Part Two* has these focuses, but it's not really directly expressed, and a lot of these situations are transposed onto sex.

ZYX: Do you think that the perspective of sex can shed light more incisively on the relationships between people?

YFD: First of all, speaking of greater significance, no movie can resolve a revolution. Society's problems will not be settled by a movie. But I think that some of my own opinions have blended into this movie. It's a little like a pressure cooker slowly coming to a boil, and it's stuck in there. The pressure cooker doesn't burst open, and it just stays muffled inside. It's not easy for me to say real life is like this or like that. It's a tiny bit obscure, a bit cold, and perhaps I can also say that reality is brutal.

ZYX: This part also made me feel like there was an air of violence. Aside from the boy in the bathtub talking about mistreating animals in his childhood, there's also something you've told me about before, something that couldn't be used in the end because the actor's form was no good, from your own childhood experience—dressing up as a Japanese soldier and acting in a certain way to a girl. Hearing it makes one shudder.

YFD: Yes. The language in *Part Two* is actually relatively aggressive. It more or less brings a sense of violent speech. But it's the most commonly seen thing in life. In *Part Three*, people and nature are actually very balanced. We even chose some relatively beautiful scenery, with everyone doing farm work in a terraced field. But here, there's the issue of the cow's life, and there's the directly violent scene in which the cow is killed. Other than violent speech, some violent images must also directly appear, because I think that these things must be faced. In *Part Three*, there's the scene where the cow is killed, but in fact, you don't see blood in that scene. I think it has a lot of violent tension, and at the time, all the actors cried. Perhaps they lost control of their emotions. Here's where the problem comes out. If you grow up in the countryside, you see a lot of these real-life things.

People being born, getting old, getting sick and dying—that's an issue, and so is animals being born, getting old, getting sick and dying, and the animals don't get to decide that for themselves.

For example, if people want to eat the meat of an animal, then they resolve that beforehand. In this way, life in the countryside is very quotidian. You could call it a series of farming skills and tricks. It's very simple. For example, the village butcher has to kill a lot of cows and pigs every day. For him, it's a routine thing, and everyone is already used to it. They don't see this as violence. They take it as normal, just a job. But when the seven youths see this kind of violence, they come from the city, and sometimes they might make a mountain out of a molehill. And other people who don't live in the countryside might really pity this kind of life. People's understandings of violence differ. In the city, if even a little cat or dog dies, everybody turns it into a big mess. So *Part Three* has some feelings that address life and survival, and it has some brutality in it. It's quite weird; I am, at the present, slightly interested in survival and death.

ZYX: I was looking at the source material just now, and I saw the part where the two girls among the seven people perform sacrificial rites with a paper cow. Is it that women are more sensitive and empathetic?

YFD: In general, people think a woman is more neutral, more gentle, and they'll entrust their hopes to her because she's a carrier of life. It's subconscious. I think the feeling of the girls offering the sacrifices is appropriate.

ZYX: In your movies, people almost never speak to each other. *Part One*, in Huangshan, has only a voice-over. *Part Two* is set inside, and only the dinner table conversation can be considered public discussion. What are your opinions and attitudes toward communication between people?

YFD: Speaking about linguistic exchange often reflects attitudes about life, and attitudes are sometimes doubtful. People are going forward together, but you discover that everybody is located in parallel positions. These things are all unavoidable, but they appear everywhere; there are very few genuine points of human intersection, and they're ephemeral. Occasionally, you reach one of those flickering points, but everybody carelessly passes by it at maximum speed, and returns to that feeling of moving in parallel. It's hard to talk about these things. Sometimes, talking a lot leads to pessimistic

《竹林七贤》

杨福东《竹林七贤》这本书一直是我有些偏爱的。也许是喜爱《竹林七贤》这部作品本身，也许是觉得设计做得还不错。大概兼而有之吧。

2008年年初的一天，我接到田霏宇打来的电话。我跟霏宇本不认识，但在曾经的项目中略有交集，彼此知道。霏宇是美国人，在华多年，中文极佳。当时他主持一家名为“话坊”的独立出版机构，出版一些跟中国当代艺术有关的书，至于他做尤伦斯艺术中心（UCCA）的馆长，是后话。霏宇在电话中跟我说，有一家瑞典的私人展览空间收藏了杨福东的影片《竹林七贤》，想在秋天专门为这个作品搞个展览，同时出版一部画册。画册他来编辑，杨福东看过我去年给英国泰特美术馆设计的画册《确有其事——来自中国的当代艺术》，十分欣赏，故来询问我是否有时间和兴趣设计这本书。我当然有兴趣，杨福东的作品我虽谈不上特别熟悉但也不陌生。他在作品中用戏剧性的影像语言，意象地传达出了当代城市人内心最隐秘的彷徨，这种微妙的情绪我相信哪怕只看过几张剧照的人都会印象深刻，而《竹林七贤》无疑是杨福东最具代表性的作品。很快，霏宇把作品拷贝了一份给我，当我完整看过一遍之后，才真正了解传说中的《竹林七贤》到底是怎么一回事。《竹林七贤》是五部黑白电影的统称，每部电影皆七个演员，五男二女，五部片中成员有延续有更替。前两部有简单对白，后三部则完全为默片。说实话，长时间看无情节默片不是件轻松的事，所以后来杨福东跟我开玩笑说，大概全世界把这五部电影从头看到尾的只有他我二人。但更不轻松的，是我需要把这五部电影在纸面上静态地呈现出来。

6月，我为设计的事去上海跟杨福东见了一面。当时他在筹拍新片，杂事缠身，我则是一个人工作，来去无碍。杨福东江湖昵称“杨福”，长发，微胖，嗜烟，待人极亲切，与之交往如沐春风。我们在他铺着深色地板、摆满Art Deco风格家具的家里聊了整整一下午，其间并没怎么谈设计，多数只是闲聊，他说了不少拍片的趣事。比如说最早的演员都是周围朋友，现在基本上是用职业演员了；演员们都说杨导的电影最好演，只要迷茫地望着远方就可以了；要有裸戏得集中一次拍完，不能动不动就让人脱衣服，诸如此类云云。

如同看原作，跟作者的深入交谈亦是设计的关键，设计问题的解决之道，大多来自设计之外。正是这样的交谈，使得我对作者和作品的认知有了最真实的依托。一本书，看似由一件事引发，其实从更深层上讲，是来自事情背后的人。临别时，杨福东交给我一叠光盘，说是为画册准备的资料，包括剧照、影片静帧截图和拍摄现场工作照，我可以选择使用。

杨福东给我的资料是海量的，筛选很困难。而更困难的，是三类图片由于在同一现场拍摄，相似度很高，如何梳理编排才能传神达意而不致杂乱无章？最终，我决定以清晰的剧照作为框架图片，大图，出血，同一比例关系；静帧截图作为情节填充，连排且保留黑框；工作照则作为文章插图。依此原则，全书设计一周内一气呵成。“缓进速战”一直是我在工作方法上的主张，深入梳理，细致分析，但真正着手操作时，却是斩钉截铁，如庖丁解牛。照此说法，这本书该是典型。

另外值得一提的，是这本书的封面。精装压凹“竹林七贤”四个汉字，这个字体本是第一部电影的片头字，后来杨福东觉得设计形式感过强，后面就没有沿用。这四字在片中白底黑字出现时，确实略显突兀，但经过媒介转换，变为白色织物上压凹时，反而平添了书籍的质感与韵味。护封简单得无法再简单——由电影中一帧截图放大而成，粗糙，但准确提示了这是一本关于电影而非摄影的画册，所有要说的都有了：七个人，电影，“杨氏美学”。其实设计真正要面对的问题，永远非常简单，只是点破之前难以发现罢了。

唐晖 **唐晖 1991—2008**

伊森·科恩画廊，纽约，2008

291mm×221mm×20mm，192 页

丝面精装

1200 册

编辑：唐晖、何浩

文章：范迪安、皮力、杨小彦、

孙景波、唐晖、严善錞

访谈：张敢 / 唐晖

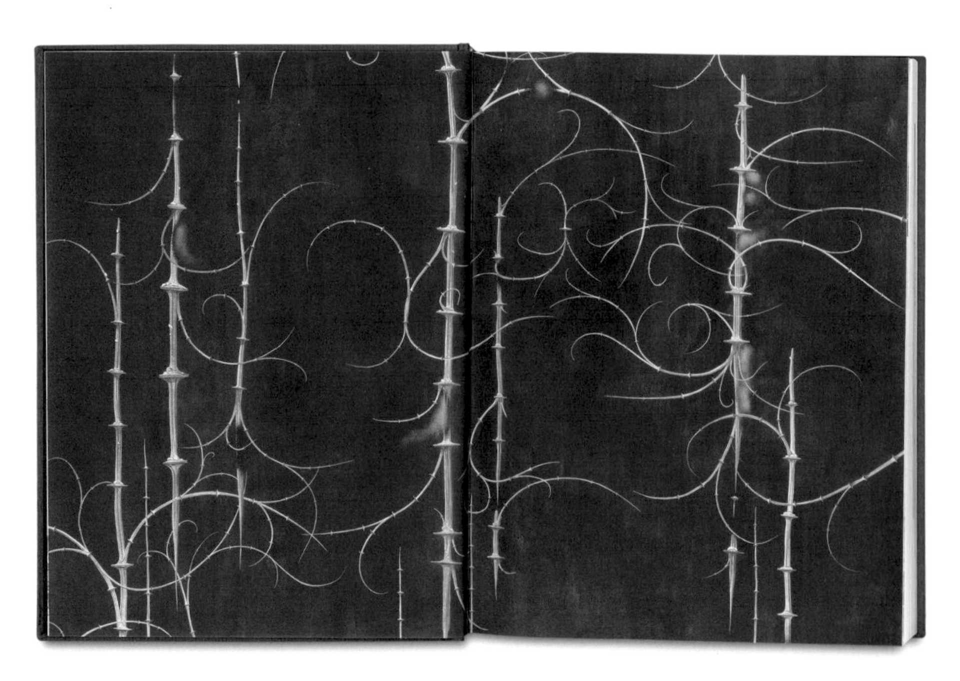

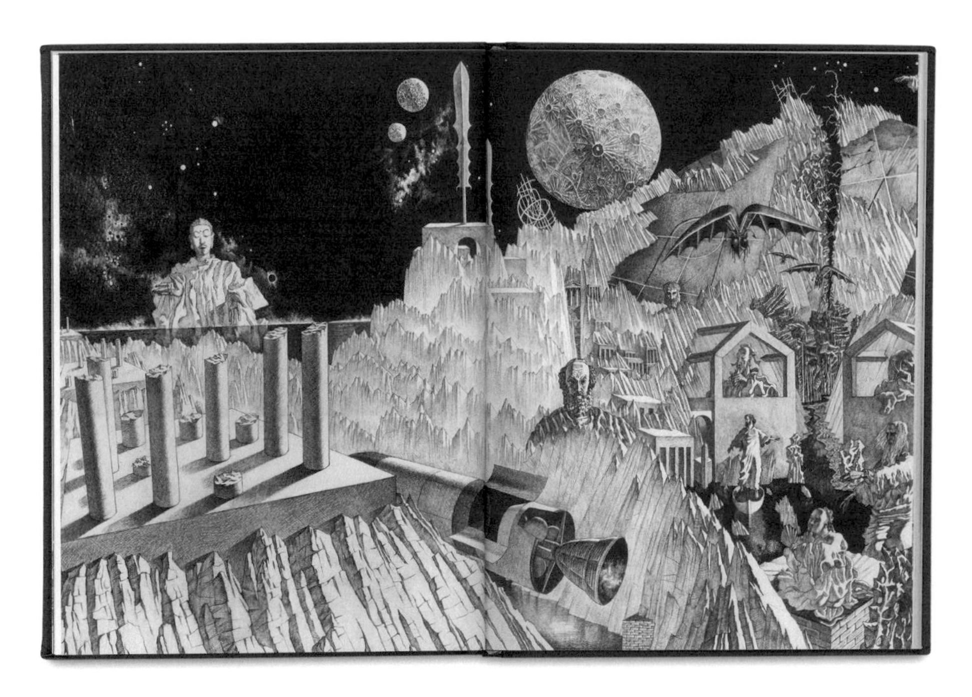

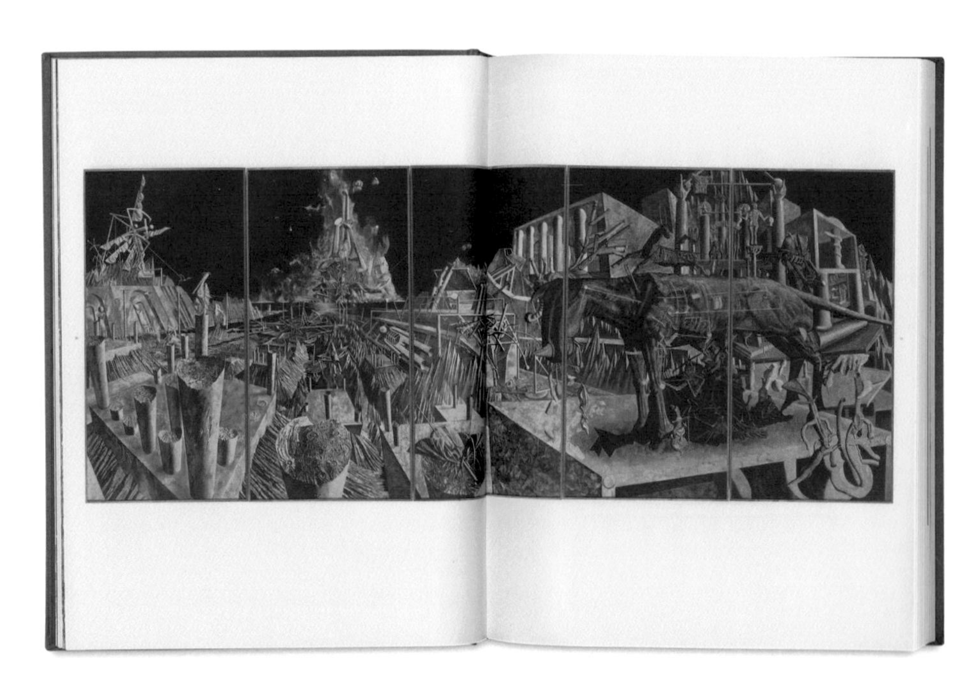

第14-15页 时空一击·素描，205×37cm，纸上铅笔，1991
第16-22页 时空一击·素描（局部）
第24-25页 时空一击，360×240cm，木板丙烯，1991
第27页 角落，50×70cm，布上丙烯，1992
第29页 悲情城市，55×55cm，布上丙烯，1994
第30-31页 时空一击·续，550×220cm，布上丙烯，1995
第33页 红色飞行器，110×220cm，布上丙烯，1995
第35页 蛋白质记忆体，55×55cm，布上丙烯，1995
第37页 飞行，55×55cm，布上丙烯，1995
第39页 时间机器，110×220cm，布上丙烯，1995
第41页 在DOS的轨道上，110×220cm，布上丙烯，1995
第42页 让我们来比比牙，60×50cm，布上丙烯，1995

page 14-15 **In Time · Sketch**, 205 × 37cm, pencil on paper, 1991
page 16-22 **In Time · Sketch** (detail)
page 24-25 **In Time**, 360 × 240cm, acrylic on board, 1991
page 27 **Corner**, 50 × 70cm, acrylic on canvas, 1992
page 29 **City**, 55 × 55cm, acrylic on canvas, 1994
page 30-31 **In Time · annex**, 550 × 220cm, acrylic on canvas, 1995
page 33 **Red Flyer**, 110 × 220cm, acrylic on canvas, 1995
page 35 **Protein Memory**, 55 × 55cm, acrylic on canvas, 1995
page 37 **Flight**, 55 × 55cm, acrylic on canvas, 1995
page 39 **Time Machine**, 110 × 220cm, acrylic on canvas, 1995
page 41 **On the DOS Track**, 110 × 220cm, acrylic on canvas, 1995
page 42 **Tooth for Tooth**, 60 × 50cm, acrylic on canvas, 1995

第74页 工人，32×32cm，木板丙烯，1997
第75页 红色生活，32×32cm，木板丙烯，1997
第76页 1968，32×32cm，木板丙烯，1997
第77页 红星酒，32×32cm，木板丙烯，1997
第78页 咪咪，32×32cm，木板丙烯，1997
第79页 山手线女孩，32×32cm，木板丙烯，1997
第80页 家，32×32cm，木板丙烯，1997
第81页 武藏丸，32×32cm，木板丙烯，1997
第82页 岛袋道浩，32×32cm，木板丙烯，1997
第83页 寿司，91×116cm，布上丙烯，1997
第84页 岛国，55×25cm，木板丙烯，1997
第85页 鞋的工厂，55×25cm，木板丙烯，1997
第86页 飞蛾，55×25cm，木板丙烯，1997
第87页 书写·三，25×55cm，木板丙烯，1997
第88页 书写·一，54×16cm，木板丙烯，1997
第89页 书写·二，54×16cm，木板丙烯，1997
第90页 富士山，112×146cm，布上丙烯，1997
第91页 新宿的雪，194×112cm，布上丙烯，1997
第92页 只是一声问候·一，91×116cm，布上丙烯，1997
第93页 只是一声问候·二，91×116cm，布上丙烯，1997
第94页 男人的宇宙·装置，46×30×27cm，1997

page 74 **Worker**, 32 × 32cm, acrylic on board, 1997
page 75 **Red Life**, 32 × 32cm, acrylic on board, 1997
page 76 **1968**, 32 × 32cm, acrylic on board, 1997
page 77 **Red Star Liquor**, 32 × 32cm, acrylic on board, 1997
page 78 **Cat**, 32 × 32cm, acrylic on board, 1997
page 79 **Girl on the Yamanote Line**, 32 × 32cm, acrylic on board, 1997
page 80 **Family**, 32 × 32cm, acrylic on board, 1997
page 81 **Pangya**, 32 × 32cm, acrylic on board, 1997
page 82 **SHIMABUKU**, 32 × 32cm, acrylic on board, 1997
page 83 **Sushi**, 91 × 116cm, acrylic on canvas, 1997
page 84 **Island Nation**, 55 × 25cm, acrylic on board, 1997
page 85 **Shoes Factory**, 55 × 25cm, acrylic on board, 1997
page 86 **Moth**, 55 × 25cm, acrylic on board, 1997
page 87 **Writing · III**, 25× 55cm, acrylic on board, 1997
page 88 **Writing · I**, 54 × 16cm, acrylic on board, 1997
page 89 **Writing · II**, 54 × 16cm, acrylic on board, 1997
page 90 **Fuji Mountain**, 112 × 146cm, acrylic on canvas, 1997
page 91 **Snow at Shinjuku**, 194 × 112cm, acrylic on canvas, 1997
page 92 **Just Say Hello · I**, 91 × 116cm, acrylic on canvas, 1997
page 93 **Just Say Hello · II**, 91 × 116cm, acrylic on canvas, 1997
page 94 **Men's Cosmos** · installation, 46 × 30 ×27cm, 1997

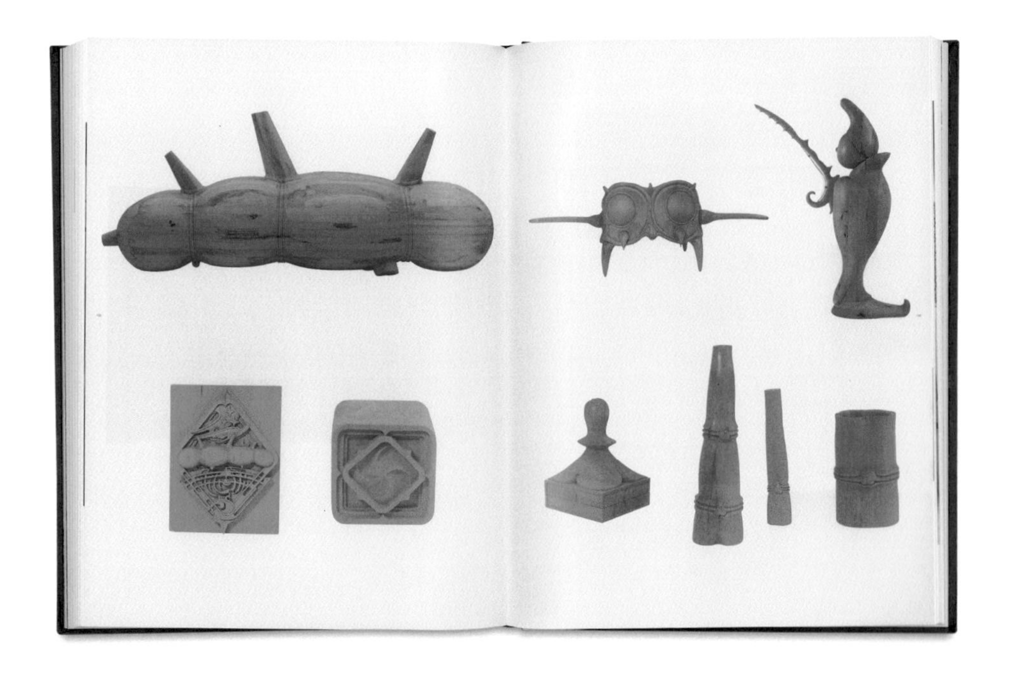

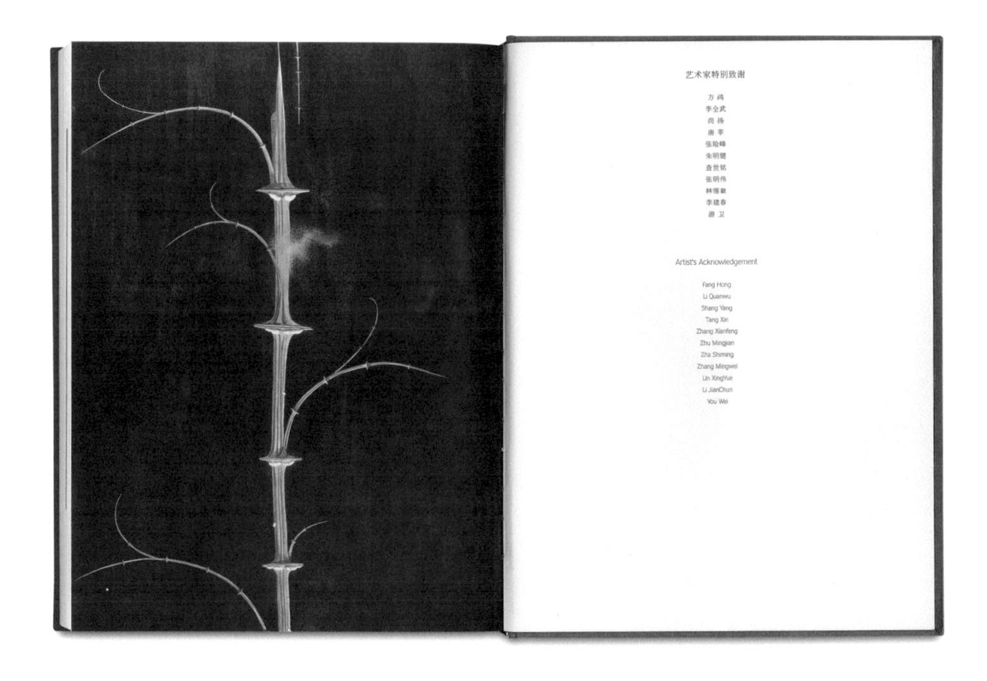

艺术家特别致谢
方 鸿
李全武
尚 扬
唐 莘
张险峰
朱明健
查世铭
张明伟
林惺嶽
李建春
游 卫
Artist's Acknowledgement
Fang Hong
Li Quanwu
Shang Yang
Tang Xin
Zhang Xianfeng
Zhu Mingjian
Zha Shiming
Zhang Mingwei
Lin XingYue
Li JianChun
You Wei

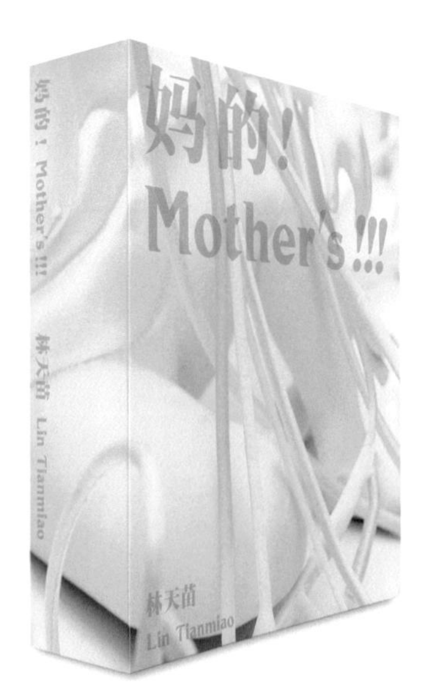
妈的！Mother's !!!
林天苗 Lin Tianmiao
妈的！
Mother's !!!
林天苗
Lin Tianmiao

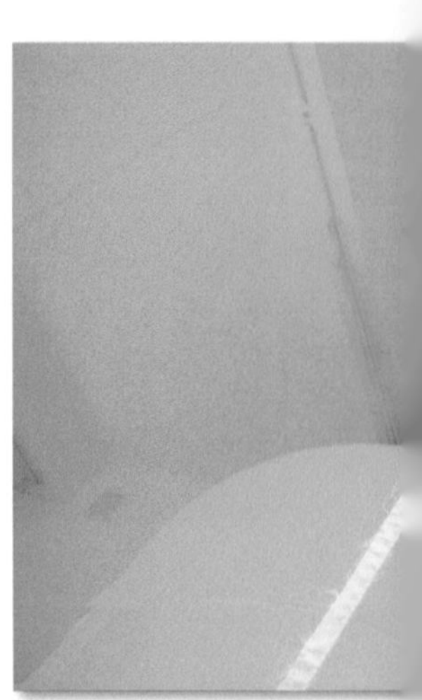

林天苗 妈的！

长征空间，北京，2008

210mm×174mm×17mm，144 页

对裱卡纸内页、封面

2000 册

文字：林天苗

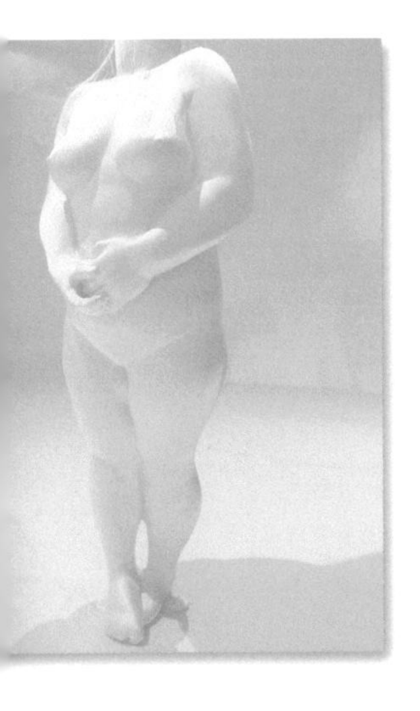

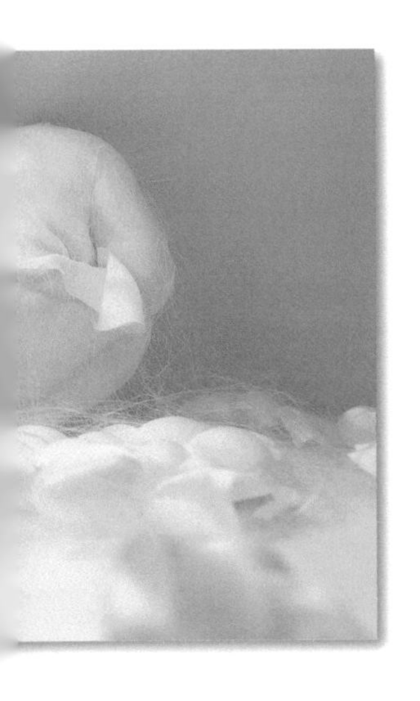

刘铮 **惊梦**

新北京画廊，北京，2008

345mm×305mm×11mm，60 页

布面精装，裹纸质护封

1000 册

文章：杨小彦

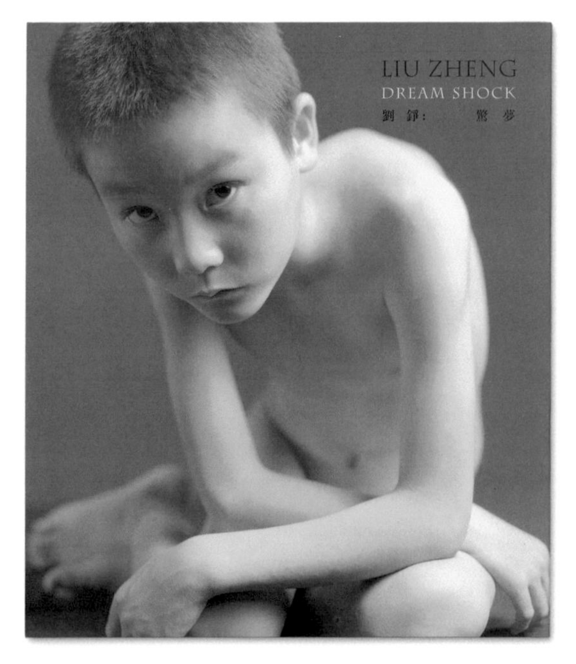

楊小彥

驚夢：作為微觀政治學的身體敘事

YANG XIAOYAN

DREAM SHOCK: THE NARRATIVES OF A STUDY IN MICROPOLITICS

It was over ten years ago in Beijing that I first crossed paths with Liu Zheng and listened to him talk about photography. He struck me as an extremely conscientious and sincere young man, a human being who was almost too harsh on himself. Liu Zheng talked about the quality of photography, the responsibilities of photographers and the cruelty of the lens. His methods of persuasion were soothing and convincing, engaging in discussion, but ultimately nailing down conclusions with little room for doubt or debate. His gaze fixed on you, words clearly spoken and eloquently organized, Liu Zheng cast a spell on his audience, under which they felt compelled to listen attentively and reply earnestly.

I learned that it was back then Liu Zheng engaged himself in his controversial "Fellow Countryman" series, which was quite a sensation. At that time, documentary photography had become the trophy of most younger experimental photographers. It had become so faddish that one's status as a photographer would be questioned if one didn't include the topic in a conversation or have some work of the sort in one's portfolio.

Liu Zheng and I never had in-depth discussion that covered the subject, but it is possible that when we first met, I fell for the habit of labeling him a "documentary" photographer. My direct impression was that through his lens, Liu Zheng tried to express a full set of visual concepts and a macro-existence of a reality that was Chinese. Liu Zheng hoped to create a presentation of a powerful, significant world of independence. Just like the German photographer Sander, Liu Zheng wanted to make a profile of a nation.

But in time, I realized this interpretation was flawed. First of all, throughout Fellow Countryman, Liu Zheng featured people with strangely distinguished facial structures. After carefully studying their visages, I realized that Liu Zheng had his own unique feeling towards images. And this dictated him in his choice of models. It had nothing to do with the commonly understood idea of "Documentary Photography". Liu Zheng wouldn't find satisfaction in responses stirred by ordinary images; neither would he linger on the shallowness of novelty. He was digging for the deeply hidden universal in all the irregular, irrelevant and irreplaceable images. Please pay attention to the word universal here. I need to emphasize here that in Liu Zheng's decision-making, universal doesn't even exist. This is a concept too fake and too posed. Universal exists on a heating body. The details have been molded over and over again by the repeating course of history. Once embodied into different bodies of flesh, these details radiate awe-inspiring composure and style.

In his search for this composure and style, Liu Zheng even zoomed in on those drawing their last breaths. Their flesh carried the syndromes of naturalism and chanted grief over losing its grip on life. He also examined bodies that were once exuberant and lively, but interrupted brutally by ugly accidents; corpses swollen and decomposed; the specimen of a deformed fetus. He never shunned away from inevitable cruelty when he was challenged to look. Instead, he wanted the concept of vision to return to the state of being challenged to look. To him, this is the true nature of vision.

Liu Zheng's lens stayed on the dead body of a young female: every trace of life drained. Still and silent, a naked body with all the aesthetics traits lay in the morgue. The delicate shadows of the image froze the rapid changes the body of the deceased was going through. Underneath the skin, there was an overly severe interrogation of watching. When my eyes fixed on this photo, I came to understand what Liu Zheng strived to achieve here. I sensed his inner tension, a stress of life that could not be calmed or comforted, and a magical power released from the tension and stress. I can imagine how Liu Zheng held his breath in his practice of the responsibilities of the onlookers. Throughout the process, he never lost his sincerity and conscientiousness most vision workers did not have. Through his lens, he stared at the image of his choice before depressing the shutter release, then he must have closed his eyes and dealt with the muted waves crashing from the darkness of his heart. After all this, he led the waves into his lonely dark room and allowed them to flow into the dark red narrow spaces.

At some point, Liu Zheng went through a transition. He no longer continued adding to Fellow Countryman. The superficial explanation was this task had been "finished". In actuality, he had discovered a new concept.

To those who never understood him, his transition seemed to be so rapid that it was even a bit unbelievable. At this stage, Liu Zheng was not merely looking for objects. He was creating

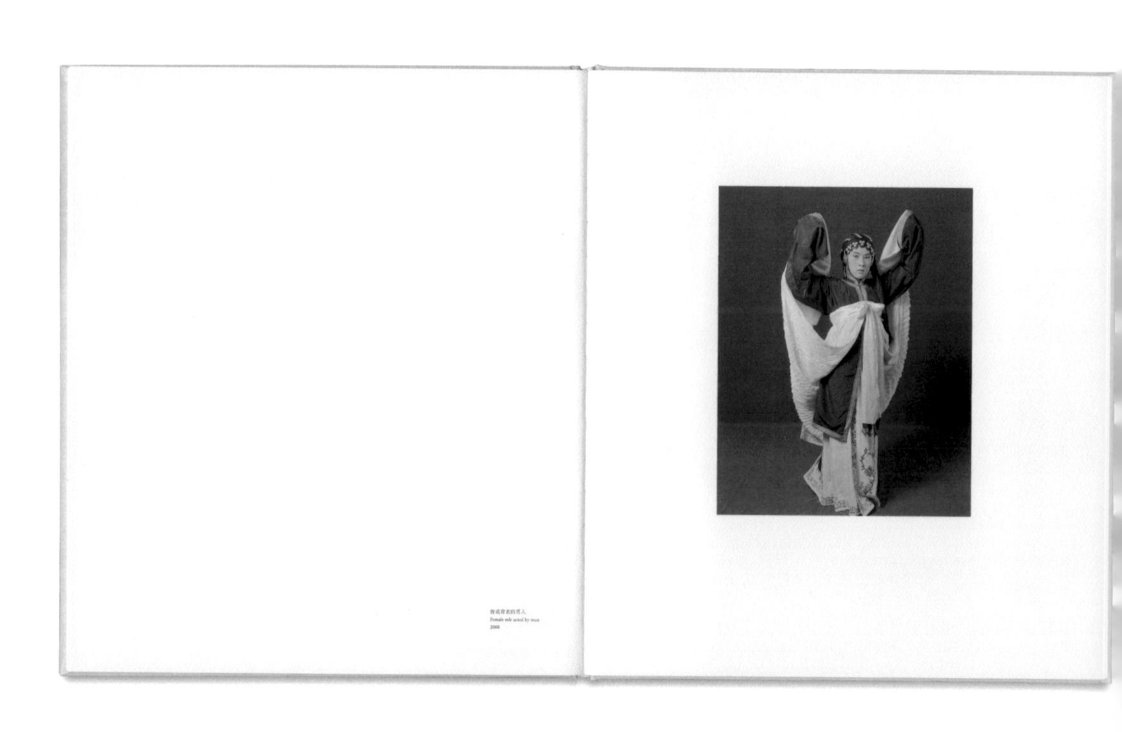
Female role acted by man
2008

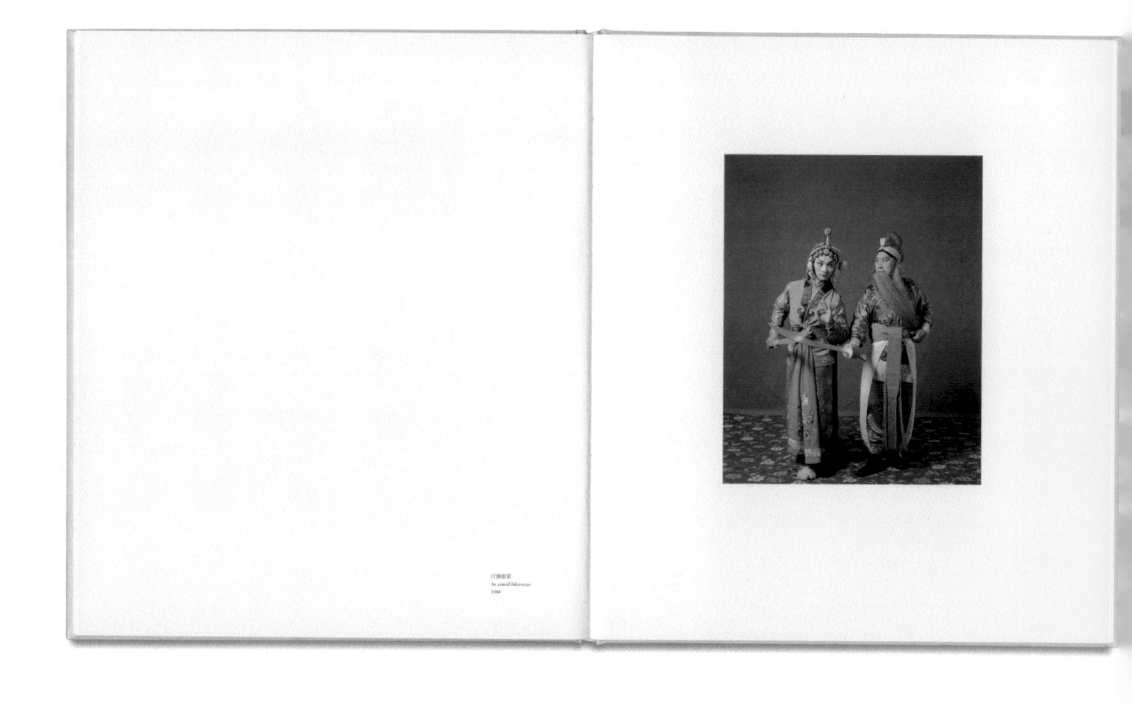
An armed fisherman

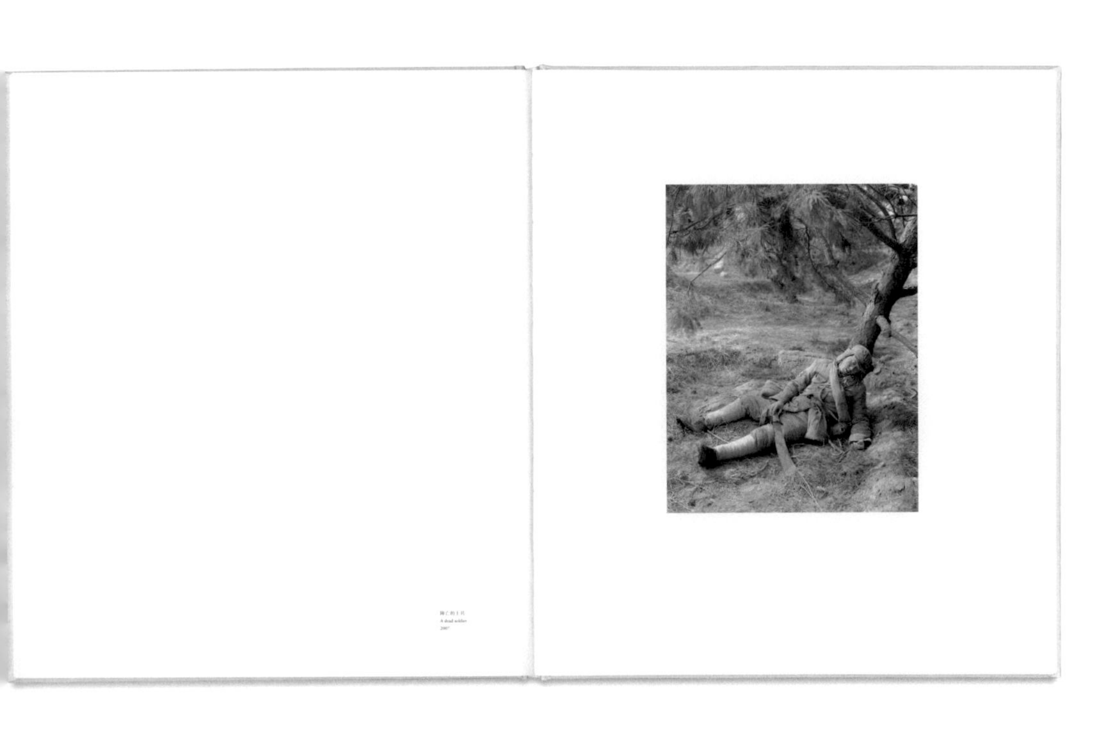

阵亡的士兵
A dead soldier
2007

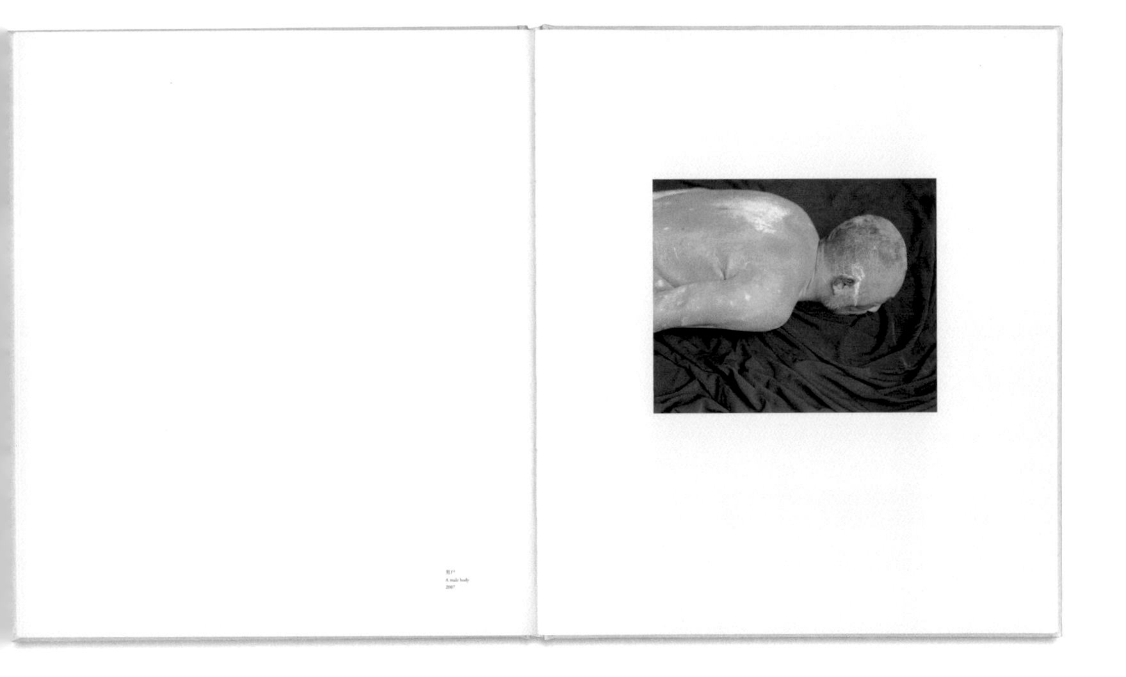

男尸
A male body
2007

谈话：三影堂摄影艺术中心 2007 年研讨会论文集

三影堂摄影艺术中心，北京，2008

210mm×142mm×15mm，192 页

平装

1000 册

编辑：荣荣和映里、张黎、毛卫东、姜亦朋、董晓安、韦春妍

談話：三影堂攝影藝術中心2007年研討會論文集

Conversations: Three Shadows Photography Art Centre's 2007 Symposium Series

Contemporary Photography Forum

都希望同樣的事情不會發生在榮榮身上。卡帕于1994年退休，之後ICP的展覽範圍戲劇性地大大拓寬了，展出內容包括當代攝影作品和歷史照片。

2000年，ICP搬離原址，遷往四十三街第六大道的路口旁邊，離時代廣場很近。

之前這座建築是一個汽車銷售展廳，ICP請了紐約非常優秀的建築設計師來徹底改造這個空間，把它變成一個近似于現代博物館的場所。

我們的展覽面積不算大，一共600平方米左右。所以作為策展人，我們在策展的時候要特別注意選擇舉辦哪些展覽以及如何辦。

ICP每年舉辦10到15個展覽，有的規模很大，把整個兩層空間都占滿，有的規模比較小，是一系列的小型展覽。

在紐約做策展人的挑戰，在于紐約人有很成熟的品位，他們每次來ICP都期望看到新東西。我們所做的就是，舉辦每次展覽的時候，把每個展覽空間做得很不一樣，我們會通過改變燈光、改變牆的顏色，請非常優秀的平面設計師、裝置設計師把展覽空間變得每次都與眾不同。

我們知道，假如我們不經常變化，來參觀的觀眾很快就會對我們的中心感到厭倦，進而轉向其他博物館。我們的觀眾是非常多元化的，主要的觀眾是學生和年輕攝影師。還有很多對攝影感興趣或對新媒體感興趣的游客，一到紐約就會立刻來我們的美術館。

因為我們所在的位置正好是在曼哈頓的中心，所以來我們那裏的觀眾很多，在過去的一、兩年裏，我們差不多一星期會有3000個來賓。有展覽的時候來的人會更多。比如說我們現在看到的是薩爾加多的展覽，整個展覽期間每天都有2000位來賓。

- 26 -

ICP的另一個優勢，就是我們周邊有很多辦公樓，是紐約著名報紙雜志的辦公所在地。這些媒體的工作人員隨時過來看我們的展覽，這樣一來，這些媒體會及時對展覽進行報道，比如說《紐約時報》。這樣，一方面提高了我們的知名度，一方面吸引更多的觀眾前來參觀。

同時，ICP吸引許多來賓和觀眾，他們對攝影史和曾被藝術家采用的攝影作品非常感興趣。比如說我們現在看到的這個展覽，展出的是1920至1930年代德國超現實主義藝術家漢斯·貝爾瑪（Hans Bellmar）的攝影作品。他的雕塑和繪畫也很出名，那麼我們在展示他的攝影作品的同時，也展示這些作品，這樣我們就可以讓大家了解到這位藝術家是如何采用不同視覺媒介來進行創作的。

在過去的五、六年間，ICP也和喬治·伊斯曼機構有過積極的合作，那就是今天上午演講的策展人艾利森所在的機構。現在看到的展覽，就是伊斯曼機構授權ICP展出的稀少的十九世紀末快照作品。這種展覽好處，就在于它允許ICP展出的自己的收藏品裏面沒有的珍貴歷史照片。攝影機構之間的此類合作，也許是三影堂以後可以借鑒的。

鑒于ICP和新聞攝影緊密聯系的歷史背景，我們一直在做一個系列，有關過去100年的不同時期攝影是如何被雜志和平面設計所運用。我們現在看到的展覽是1920年1930年間的早期雜志，那個時候新聞攝影作為一個新生力量出現。我們展出了雜志的樣本和平面設計，這些雜志來自法國、德國、俄國、日本和美國。可以看得出來，那時的發展已經相當國際化了。

科內爾·卡帕創建ICP的時候，他本不想讓它成為收藏作品的博物館。他所想的僅僅是建立一個展覽中心和供攝影家之間進行作品討論的場所。但是日積月累，ICP的收藏已經越來越多了。

現在我們擁有超過十萬張照片的收藏。大家在屏幕上可以看到，它

- 27 -

Because Bellmer was best known for his sculptures and drawings, we presented examples of works in those media in the exhibition along with photography, so the visitor would have a very clear sense of how Bellmer worked with different kinds of media.

For the last five or six years ICP has also had a very active collaboration with the George Eastman house, the museum where Alison works as a curator. In this exhibition, for example, the Eastman House allowed ICP to present examples of some very rare early snapshots from the late 19th century. This kind of collaboration allows ICP to present exhibitions of important historic works, which are not represented in our own collection. This idea of collaboration between different photographic institutions is an idea that Three Shadows, I hope, will actively explore in the years ahead.

Because ICP's history is so strongly connected to photojournalism, we have had a continuing series of exhibitions that examines the way photography has been used in magazines and in graphic design at different moments in the last one hundred years. In the exhibition that you see on the screen right now, we presented examples of early magazines from the 1920s and 1930s, the period when photojournalism first became a major force in the twentieth century. We showed examples of magazines and layouts from France, Germany, Russia, Japan, and the US. to show that this was really an international development.

Now when Cornell Capa established ICP, he did not want it to be a museum. He really wanted only an exhibition center, where photographers could be actively engaged in discussing their current projects. Nevertheless, over the years, a collection has grown up at ICP.

Today ICP has a collection that includes more than 100,000 photographs, all stored in the climate-controlled room you see on the screen. The

- 114 -

question of establishing a collection is a question that I know is being talked about at Three Shadows. I guess my best advice would be, think very carefully about the idea of starting a collection. It inevitably grows much bigger and much faster than you would ever imagine and requires enormous resources in terms of people's time and finances.

The collection at ICP has been assembled largely through donations from photographers themselves. For example, Henri Cartier-Bresson was a close friend of Cornell Capa and over the years donated more than 300 original prints to ICP's collection. Because of ICP's location in New York and because it's been in such close contact with magazine photographers over the years, ICP has a very large collection of fashion photographs as well. In the coming years, we expect to present a full year of fashion photography exhibitions, many of them drawn from ICP's collections.

ICP has also systematically collected magazines from the past one hundred years, in which the works of the great photographers appear. For example, this is a page from a French magazine from the 1930s with a picture story from the famous photographer André Kertész. For us, the interest in collecting this kind of magazine material is that it's in this form, on the magazine page, that these photographs really reached a wide audience at the time they were made. In the 1930s exhibitions of photography were very, very rare, but photographers found a way to use the picture magazines of the period so millions of people would see their work.

For those students and scholars who want to examine works from the collection, ICP has a study room. I think you may recognize some of the photo-

- 115 -

攝影論壇專題演講
博物館的作品收藏、保存、研究和推廣
・諾德斯托姆
影家》雜志對當代攝影的推廣

的當代攝影家
花木經惟、畠山直哉、奧諾黛拉・有機
太郎
國際攝影中心
托弗・菲利浦斯
攝影的制度化

當代攝影論壇現場發言

史密斯
美莎拉絲
・諾德斯托姆

鄭作品研討會

爾托・加西亞・阿利克斯作品研討會

人生的觀察者，變成一個說夢者，當然，真實的故事，要是講得不好，人人都覺得乏味。而一件虛構的事情，如果講得非常生動，就比什麼都感人。

在同一時期，我也創辦了一本以服務臺灣攝影家為主的雜志，叫做《影像》（Image Magazine）。《影像》雜志第四期介紹了一位臺灣攝影家李小鏡（Daniel Lee）的作品。到目前為止，我仍然認為，他是借助計算機完成肖像作品最成功的人之一。

我在訪問他時提出的一個問題，現在看起來，還是非常有意思的。他說：“有了計算機這個工具，我就想做些嘗試性的東西。而我的第一個反應是，可以把一些小時候聽過卻從沒看到過的神話人物做出來。我首先想到的是中國的十二生肖，所以就做了一系列這個東西。不過，我希望做出來的還是人，而不是變成怪物或動物，所以我的角度比較偏向人，祇是有一些動物的圖形或變體。我希望有那種幽昧出現。至於作品有些一般計算機攝影不一樣。我想，一般人用計算機做攝影嘗試時，多半是先知道計算機可以做哪些事，然後就計算機能做的事來找出一些東西，是比較被動的，等于是計算機在幫他們做那些東西。而我個人認為，計算機是一個工具，就像一根畫筆或一把刮刀，祇是計算機在記錄、儲存方面做得更好，更改也比較容易。我想我跟別人嘗試的方向唯一不同的地方就在這裏。我選擇計算機作為表達我個人想象的一個工具。”

他的優秀之處就在于沒有濫用科技，也沒有讓自己的想象力跨越以人為本的分寸。

在《攝影家》雜志裏，寓有性的攝影也不少。比如說，在“意大利”專號所介紹的保羅・治奧裏（Paolo Gioli），其作品既可以說是攝影，也可以說是繪畫。他把影像由底片直接轉印在畫圖紙上；影像滲透到纸張裏，紙張質料會強調出照片裏的事物。有時候，治奧裏會用刷子

- 17 -

關系。我覺得中國整個環境都在變化，所以對創作也產生了影響，也是挑戰性的。

阿：無論是技術還是交流方式的改變，最重要的是感覺一樣。每個藝術家應該表現內心的要求。大家都是攝影家，都記得自己拍過什麼作品，每一個作品都在自己以前的作品有什麼樣感覺？

就是一種記憶，是那段時間的經歷，這對我來說

了記憶的感覺？是不是在創作記憶。你和你拍是你透過了自己的角度，而不是記憶。

概念，那就是我用什麼方法看待攝影。這肯定存這種觀念的東西有可能對歷史、對其他人有意我自己來說，比如看我這本書，其中記錄的就

一樣，有很多東西需要記，需要知道你能不能對個燈光，你能不能用一個抖動的相機和一個閃光說，你如何拍攝是視角的問題，而記憶是另外

是我無法預料的，我做了我應該做的，對我來

角度。

我的照片，更多是想起角度，還是想起以前的

- 81 -

黎：比方說，今年在阿爾攝影節上，也有很多作品都用到攝影，有些是攝影家，有些是藝術家。

劉：你認為攝影的本質是什麼，你所堅信的是什麼？

阿：我沒有相信什麼，但開始問這個問題就會有很多回答。

劉：現在在西班牙，你從事的這類攝影占什麼地位？

阿：有段時間，年輕人開始把新技術運用攝影中，我的這種方式被看作傳統方式，人們對其它方式更感興趣。可最近可能很藝術經濟有關系，很多畫廊又開始關注我這類攝影，但還不是占重要地位。

劉：為什麼？

阿：因為那時這種照片不被認為是藝術照片，是記錄照片；可是現在人們認為也是藝術照片，這個概念是新的。

劉：我認為攝影與現實的關系永遠不能被割斷。

阿：我同意，而且這種關系也是很個人的。

劉：謝謝！

榮：王旭，你怎麼看？

王旭：佛朗哥時代可能很閉塞，你沒有讀也多書，現在你應該讀過很多書了，但是在看到這些書時，你是否還有一種激動？

阿：我讀的攝影書特別少，我更喜歡看小說。我喜歡花很長時間看一本書，否則記不住，所以看的少，看太多書對我來說有點可怕。

榮：我家裏本來有很多書，現在沒有了！都拿到圖書館給別人看了。（笑）

王寧德：千萬別相信藝術家說的話（笑）。

- 86 -

榮：這是你第一次來中國嗎？

阿：對，第一次，什麼都沒有看，沒有時間，因為明天就要走。

榮：第一次來中國有什麼感覺，和你想象中有什麼區別？

阿：我喜歡這裏所有的人，我在這兒誰都不認識，所有人我都想拍，因為我是個陌生人，你們也一樣，攝影家看生活好像是一個戲劇舞臺，這個舞臺好滿，舞臺演出我永遠在，因為生活的演出永遠不會停止，國旗給了我們一個新的生活。

榮：為什麼你不用數碼技術？

阿：我喜歡黑白照片，也喜歡這種工作方式，因為洗照片有個過程。

黎：無論數碼還是傳統，你都可以運用嗎？

阿：但我想還是有區別的，不一樣的兩條路。

阿：數碼最不好的方式就是會阻礙你自己的想象力，因為它立刻就可以看到結果。如果不是馬上可以看到結果，你還可以想一想該怎麼拍；數碼太直接，也沒有過程，我特別喜歡拍的時候不知道結果會是怎樣，之後再去發現結果的那種感覺。

榮：所以用數碼技術就沒有了衝動。

阿：我不是批評數碼不好。

榮：對，這是每個人的方式不同罷了。

黎：榮榮，有時候你用數碼嗎？

榮：拍我的小孩時用，因為小孩動得太快了（笑），但我的作品不用數碼。

張黎：我想討論一個問題，西班牙攝影和西班牙藝術有什麼特色？應該用什麼語言來概述？問這樣的問題是否有意義？

阿：我不知道有沒有一種西班牙攝影的特色，我喜歡攝影，喜歡創新。

比如說，今天看一些西班牙照片，拍攝的是一些傳統節日，你看這

- 87 -

we discuss less and less the actual values of pho-
ult of looking through a lens and going through this
process. I think the reason why it has become an
n the verge of an age where things are moving to-
id we have a certain numbness when we look at the
ody here uses the Internet and is used to seeing so
een that you don't even look at them. It's interest-
ntext. When you research and use the computer to
mpare a photograph to an artwork, the photograph
eadable and something people will look at because
oring to the computer screen.
e been looking at images in this way is because I'
partner Liu Heungshing on a book about Chinese
through the present. One of the most complex is-
e to the present time, the field of photography is
graphy and it becomes very difficult to find a way
utting these things together. It's very interesting to
ay about the cultural state of China in this moment.
graphy seem so far apart. Art photography seems
its time.
uotes about photography to clear my mind about
otography. You find that the idea of the document,
idea of saying something that cannot be captured
hat flow throughout the twentieth century.
it disjointed, it's because I've been finding it quite
od way to discuss photography in China. I'm very
just made his speech, because although it was

- 111 -

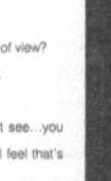

of an old friend or a certain point of view?

Alberto Garcia-Alix: All together.

Liu Zheng: For me the same.

Alberto Garcia-Alix: If you can't see...you are the director of your photos. I feel that's not natural.

Liu Zheng: Different people have different understandings about nature and reality.

Alberto Garcia-Alix: But the models are the same.

Liu Zheng:A model won't run to the photography studio by him or herself to meet you. I think the appearance of an entity in front of you and the model coming to your studio is the same. I think it's natural.

Today we mentioned taboo. Coincidentally, I am currently working on a series called "Taboo and Desire" or "Sex and Taboo." Alberto's works also have many taboo subjects. Did he have any trouble showing his works? In preparing my work for exhibition, some of my pieces have been pixilated. Only then can they be shown.

Alberto Garcia-Alix: Is the taboo you are talking about yours or others people's?

Liu Zheng: I don't have taboos, but other people do.

Alberto Garcia-Alix: But do you realize what is taboo when you are shooting?

Liu Zheng: Actually I keep the taboo on my mind. I'm trying to find out the taboo and then break it. My works always make efforts to solve this kind of problem. But the taboo will be something else when my works are on exhibition, such as morality and ethics.

Alberto Garcia-Alix: What is your taboo?

Liu Zheng: In people's mind there are always certain things that you can't

- 186 -

overcome. For example, if you put a chain on a baby elephant, when it grows big and strong it won't try to break free because it doesn't know any better. Today I am attempting to break through some taboos that we've had since we were young that are actually hidden by the process of growing up. I want to imagine what life would be like without taboos.

Alberto Garcia-Alix: How do you break your taboos?

Liu Zheng: This is complicated. When I saw Alberto's photos, I just passed through some, but others stayed in my mind. Why? What attracted me to those photos? Some broke through taboos. When you remember a photograph, you also remember the individual.

His style of photography moves me – it makes me feel nostalgic. In China today, this kind of work style is already disappearing. He, RongRong, and I are the same, paying close attention to photography's details. Many artists believe that what we've been talking about today – loving photographic details and cameras – is lowly. When RongRong and I created *New Photo*, we considered our position in traditional photography and how we should move forward. A wave of new photography appeared, called conceptual photography. Conceptual photography has developed to the point where the same problems appear. Twenty, thirty years later, which works will be remembered?

Alberto Garcia-Alix: I feel the same when I look at your works.

Liu Zheng: We are all very old, haha………

Berenice Angremy: But Alberto is very young

Alberto Garcia-Alix: Haha

Liu Zheng: Photography has already departed from the essence of photography in some ways. After some time people may go back and realize the essence of photography again. Maybe it will take a long time, but this is necessary. My understanding of the essence of photography is the relation-

- 187 -

天使的传说

红楼基金会，伦敦，2009

215mm×180mm×11mm，80 页

布面精装

500 册

文本写作：姜节泓

艺术家：高世强、蒋志、郭瑛、缪晓春、
史金淞、向京、萧昱

The Tale of Angels

Jiang Jiehong

Angels in the Bible

There are various views of angels in the world today. Some believe angels are human beings who have died, some see angels as impersonal sources of power, and others entirely deny the existence of angels. Although angels have been seen in classical myth and philosophy, and can be encountered in other religions including Zoroastrianism, Hinduism, Buddhism, Taoism and Islam, most people would think of Christianity when the word 'angel' appears. In both monotheistic and polytheistic traditions, angels are commonly believed to be messengers of God (or of the gods), or the inhabitants of an intermediate realm, sent by God to the world, and are entrusted with blessings to bestow on earth.

Angels are first described in the Bible, and a biblical understanding of Angelology tells us how angels relate to humanity and serve God's purposes. Sometime before the world, angels were created by God as an entirely different but a higher order of creatures in the universe than humans, with aspects of intelligence, emotions, and free will. Angels primarily praise, worship and serve God around His throne (Revelation 5:11-14). They seem to innately possess greater knowledge than humans by more thoroughly studying the Bible and the world (James 2:19; Revelation 12:12), observing human activities and knowing the past and future. Holy angels are sent by God to help believers, bring answers to prayer, and assist in converting people to Christ (Hebrews 1:14).

Most holy angels are not given names apart from only certain individuals, such as Michael the archangel, possibly the head of all the holy angels, and Gabriel, one of the principal messengers of God. Some angels are designated as Cherubim, which are described as creatures defending God's holiness from any defilement of sin (Genesis 3:24; Exodus 25:18, 20; Ezekiel 1:1-18), whilst Seraphim,

another high class of angels, are mentioned only once in Scripture (Isaiah 6:2-7). They hold the function of praising God, being God's messengers to Earth, and are especially concerned with the holiness of God. In contrast to the company of holy angels, are the fallen angels, led by Satan (also known as Lucifer), who have not maintained their holiness, defected and rebelled against God, and become sinful in their nature and work.

Time and space, as characteristics of our world, have limited our imagination greatly. As primarily physical beings but with spiritual elements ourselves, many of us would naturally expect an angel to take physical form as well. But, what on earth do these 'unseen' beings 'look' like? Unfortunately, the first-hand literature source – the Bible – does not actually offer us an explicit answer. In the Scripture, there is no specific visual depiction of God, and little of angels. The Bible does not even state that angels are created in the image and likeness of God, as humans are (Genesis 1:26). Angels are spirit beings (Hebrews 1:14), so they do not necessarily need any essentially physical form. However, angels have the ability to take on human forms to appear as normal males many times throughout the Bible, although never in the likeness of women. However, this visual presentation does not necessarily define the gender of angels. Instead, it coheres with a similar manner of referring to God especially in the patriarchal culture when the Bible was written. The same as the human form, the male look is the physical appearance that angels choose to employ. Possibly angels are sexless (Matthew 22:30), or at least not in the sense that we distinguish human gender. At other times, angels appeared not as humans, but as something otherworldly, whose appearance could be terrifying to those who encountered them (Luke 1:12, Matthew

284), whilst in other occasions, their appearance does not produce fear.

Saint Matthew describes an encounter with an angel as his 'appearance was like lightning, and his clothes were white as snow' (Matthew 28:3). Saint John records a mighty angel that he saw coming down from heaven, 'robed in a cloud, with a rainbow above his head; his face was like the sun, and his legs were like fiery pillars.' (Revelation 10:1) Whatever appearance angels take on, there is reason to believe they are incredibly beautiful in appearance, because they are continually in the presence of Almighty God, whose glory is reflected upon all that is around Him. The most commonly projected image of an angel is essentially a human being with wings, which actually has been described quite rarely in the Bible. For example, the Cherubim on the ark of the covenant have wings spread upward that cover the mercy seat (Exodus 25:20); and Isaiah also saw winged Seraphim in his vision of the throne of heaven, each one having six wings (Isaiah 6:2). However, as spirit beings, it is unclear as to why the angels would require wings. If a spirit being is not bound by the laws of the physical universe, then angels do not necessarily need wings in order to fly. However, our understanding of both angels and their wings described in the Bible remains narrow.

Angels in Art

In art, there have been numerous works representing people's imaginations of angels, who provide artists with an opportunity to portray heavenly things and the execution of God's will on earth. Angels are first and foremost painted in Christian art, particularly favoured during the Renaissance. Without this motif, the artistic tradition, Western in particular, could be quite different. But prior to depiction, one must develop the visual understanding of angels, simply whether they were male, female or child like. These disembodied beings were therefore visualised tangibly, and given concrete and artistic forms.

The limitation of our intellectual capacity seems to direct us first to think of angels with wings, which have been generally accepted, at least to distinguish themselves from human beings. In fact, in the first known image of the Annunciation, found in the Catacombs of Priscilla (Fig. 1), the angel has no wings. Similarly, in Dante Gabriel Rossetti's *Ecce Ancilla Domini* (Fig. 2), the angel Gabriel carries a flowering lily branch, symbol of purity, announcing to the Virgin Mary that she would give birth to the child of God. Noticeably, Gabriel is here depicted without wings, but his feet are engulfed in a supernatural fire, which instead generates the power for travelling between heaven and earth. In another example, *Madonna with Eight Angels*, Botticelli depicts eight wingless angels dressed in elegant tunics and set in two groups holding the branches of blossoming white lilies forming a crown around the Virgin and Child (Fig. 3).

Fig. 1, Annunciation, fresco, ca. 200-300, Rome. Catacombs of Priscilla

Fig. 2, Dante Gabriel Rossetti, *Ecce Ancilla Domini!*, 1850, London. Tate Gallery.

Fig. 3, Sandro Botticelli, *Madonna of the Pomegranate*, ca. 1487, Florence. Galleria degli Uffizi.

When the creatures moved, I heard the sound of their wings, like the roar of rushing waters, like the voice of the Almighty, like the tumult of an army. When they stood still, they lowered their wings. (Ezekiel 1:24)

诸物行走的时候，我听见翅膀的声响，像大水的声音，像全能者的声音，也像军队轰嚷的声音，活物站住的时候，便将翅膀垂下。（以西结书 1:24）

However, at a certain point in the history of Western art, most angels started to appear with wings, which imply the concept of God's heralds and flying creatures. Giotto's myriad angels in the Arena Chapel are swooping in from all directions (Fig. 4). Despite their wings and halos, all the angels express their profound sorrow and share in the mourning the death of Christ. Although these heavenly creatures were portrayed with human shapes, they were to be understood as having no sex or age. In the same way, Cimabue's angels surrounding the enthroned Mary and Child (Fig. 5) and Gabriel (Fig. 6) interpreted by Leonardo de Vinci in his *Annunciation* all appear genderless. As intermediaries between heaven and earth, wings were understood essential to carry angels swiftly from one world to the other. Relying on the scientific investigation of natural laws, artists started to describe the flight of angels with increasing literalness. As the Renaissance progressed, angel became more heavily lifelike and hence needed commensurately more substantial wings to lift them, while artists studied the movement of birds as believable models to master skills for conveying convincing illusions.[2] In *Saint Joseph's Dream* (Fig. 7), Rembrandt pictures an angel in a radiant robe speaking gently to Joseph in his sleep, with a pair of wings outspread to hold his ethereal body.

但是，在西方艺术史的某一个特定时期，大多数天使开始以有翼天使的形象出现，这似乎暗示了天使作为上帝军士的角色，以及他们的飞翔本领。乔托所描绘的成群的天使从四面八方飞来，一下子涌入阿雷那小礼拜堂（Fig. 4）。除了他们的翅膀和光晕，我们还可见所有的天使表达出的深深的哀伤，以及对于耶稣之死的悲恸。虽然这些来自天堂的生灵具有人类的形象，却没有性别与年龄。同样地，不论是由契马布埃创作的围绕在圣母和圣子周围的天使（Fig. 5），还是在达芬奇版的《天使来报》中的天使盲加列（Fig. 6），都很难辨别性别。对于天使来说，他们的翅膀至关重要，那是帮助他们轻盈地穿梭天堂与人间的工具。基于有关自然法则的科学研究与分析，艺术家将天使的飞行描绘得愈发精准。在文艺复兴时期的作品当中，天使的形象有所增大，因此便相应地需要更大的翅膀托起他们的身体，得以飞行。然而艺术家们将鸟类的飞行作为可信的依据，从而掌握高超的技法，最大化地将人类的幻想描写得让人信服。[2] 在作品《圣约瑟夫的梦》（Fig. 7）中，伦勃朗所描绘的天使身着发光的长袍，展开一对用以托起她轻盈身体的翅膀，正在向睡梦中的约瑟夫轻轻低语。

Angels are often depicted as children, as one way of looking at purity, perfection and the untainted being. In one of Raphael's best-known paintings, *the Sistine Madonna*, two naked infants with multicoloured birds' wings and tangled hair lean on a small, painted pedestal and glance mischievously at the figure above (Fig. 8). Other angelic children pictured in classical paintings include Rosso Fiorentino's *Angel Musician* (Fig. 9) and Joshua Reynolds' *Heads of Angels* (Fig. 10), which has become popular in contemporary print reproductions today.

天使还通常以孩子的形象出现，因为孩子代表着纯洁而又完美无瑕。在拉斐尔最著名的画作之一《西斯庭圣母》中，两个裸体小孩拥有鸟儿的五彩翅膀和凌乱的头发，俏皮地倚着彩色的柱脚，带着恶作剧般的眼神注视着上方（Fig. 8）。我们还可以在一些其他的作品中看到天使般的小孩，如罗索·菲伦蒂诺的《天使乐手》（Fig. 9）和雷诺兹的《天使的头部》（Fig. 10），这些都在当今的印刷复制工业中十分流行。

Closer to our own time, angels could be depicted less 'realistically'. James Tissot offers us an image of Gabriel with many large wings. The feminine face emanates a strong

从时间的角度来说，越是往后，天使越显得不那么“现实”。詹姆斯·提索为我们所现的天使拥有很多硕大的翅膀，女性化的脸

Fig. 4. Giotto, *The Lamentation*, fresco, ca. 1305-13. Padua: Arena Chapel.

Fig. 5. Cimabue, *Madonna and Child in Majesty Surrounded by Angels*, ca. 1270. Paris: Musée du Louvre.

Fig. 6. Leonardo de Vinci, *Annunciation*, ca. 1472. Florence: Galleria degli Uffizi.

Fig. 7. Rembrandt, *Saint Joseph's Dream*, 1645. Berlin: Gemäldegalerie.

Fig. 8. Raphael, *The Sistine Madonna* (detail), 1515-16. Dresden: Gemäldegalerie Alte Meister.

Fig. 9. Rosso Fiorentino, *Angel Musician*, 1520. Florence: Galleria degli Uffizi.

Fig. 10. Joshua Reynolds, *Heads of Angels*, 1787. London: Tate Gallery.

2 Grubb, Nancy. Angels in Art. New York: Abbeville Press, 1995, p. 85.

2. 南希·格拉布，《艺术中的天使》，纽约：阿贝维尔出版社，1995年，第85。

divine blue light, and the hands are raised in the greeting that accompanies the announcement (Fig. 11). Marc Chagall had a deep and sincere love for the Bible, which he found to be 'the greatest source of poetry'[3]. His angels with ambiguous body shapes are frequently appropriated to illustrate the Bible (Figs. 12), or create his dreamlike world. In *the Crossing of the Red Sea* (Fig. 13), an angel, who is depicted as a divine light in a peaceful movement, illuminates the composition, together with Moses, to lead and guard the people in the chaotic darkness. When angels started to depart from their original biblical description and become spiritual motifs and symbols, artists would have great latitude to portray them in a variety of ways. Paul Klee, in his *Angel Still Feminine* (Fig. 14), for instance, explores their significance as symbols for contemporary life. His angel is neither spirit nor human, shown as asexual and even unshaped, or in other words, in the process of transformation.

庞散发出蓝色的圣光，举起的双手表示致敬，更多的是传达着喜讯（Fig. 11）。马克·夏加尔一直都对《圣经》有着诚挚的热爱，他一直都认为《圣经》是“诗歌最伟大的素材”[3]。夏加尔所创造的天使形象没有清晰的肢体轮廓，却被人们广泛接受，认为其恰当地表现了《圣经》（Figs. 12），或者说是完美地再现了艺术家那如梦如幻的世界。在作品《穿越红海》（Fig. 13）中，一位天使安详沉稳，与摩西一起，用其神圣的光芒照亮了混沌的一切，引领并保守着人们走出可怖的黑暗。当天使的形象逐渐走出圣经起源而转化成某个精神的主题或象征时，艺术家们便有了更大的空间及自由度来对天使进行诠释。举例来说，保罗·克利在他的《天使仍女性》中（Fig. 14），探索了天使作为某种象征符号在当代生活的意义。他创造的天使既非精灵也非人类，无性无形；换句话说，呈现了一种转化的状态。

In sculptures, particularly in the contemporary context of public art, angels have been one of the traditional and ongoing themes. On the Bridge of Angels (*Ponte Sant'Angelo*) that spans the Tiber River in Rome, ten beautiful statues of angels are arranged, originally designed by Gian Lorenzo Bernini in the seventeenth century, (Fig. 15). And from Josep Llimona's *Guardian Angel* erected over the walls of an old ruined Gothic church in Comillas Spain (Fig. 16) to more recently, Antony Gormley's giant *Angel of the North* (Fig. 17) standing on a hill in Gateshead, with his wings measuring 54 metres wide across, angels live on in public

在雕塑艺术，尤其是在当代语境下的公共艺术中，天使是最具传统的延续不断的主题之一。在横跨罗马台伯河的圣天使桥上，由贝尼尼于十七世纪设计的十尊美丽的天使雕塑依次排开（Fig. 15）；在西班牙寇米拉斯，约瑟夫·里莫纳创造的作品《守护天使》矗立在一座废弃的哥特式教堂的围墙上（Fig. 16）；还有年代更近一些的，安东尼·高米利的大型雕塑《北方的天使》（Fig. 17）伫立

Fig. 11. James Tissot, *The Annunciation*, 1886-96. New York: Brooklyn Museum of Art.

3 Foray, Jean-Michel. in Marc Chagall. San Francisco: San Francisco Museum of Modern Art in association with Harry N. Abrams, Incorporated, New York, 2003, p. 204.

3. 让米歇尔·佛瑞，《马克·夏加尔》，旧金山：旧金山现代美术馆，2003年，第204。

Fig. 12. Marc Chagall, *The Rainbow, Sign of the Covenant between God and the Earth*, 1931. Nice: Musée National Message Biblique Marc Chagall.

Fig. 13. Marc Chagall, *The Crossing of the Red Sea*, 1955. Nice: Musée National Message Biblique Marc Chagall.

Fig. 14. Paul Klee, *Angel Still Feminine*, 1939. Bern: Paul Klee Foundation.

Fig. 15. Gian Lorenzo Bernini, *Angel with the Cross*, 1660-1668, Ponte Sant' Angelo, Rome, Italy

Fig. 16. Josep Llimona, *Guardian Angel*, 1894-1895, Comillas, Spain

Fig. 17. Antony Gormley, *Angel of the North*, 1995/98, Steel, 22 × 54 × 2.20 m
Commissioned by Gateshead Metropolitan Borough Council
© Courtesy of the artist and Jay Jopling/White Cube, London

Two Skins, 两张皮, silicon and wool, 2 pieces, 240 × 220 cm each, 2008

彭斯 **抱书独行**

世纪翰墨画廊，北京，2009

215mm×180mm×12mm，72 页

布面精装

1000 册

编辑：彭斯、何浩

文章：彭锋

2009 年度三影堂摄影奖作品展

三影堂摄影艺术中心，北京，2009

200mm×200mm×17mm，192 页

平装

1000 册

前言：荣荣和映里

文章：克里斯托弗 · 菲利普斯、巴斯 · 弗吉、顾铮、凯伦 · 史密斯

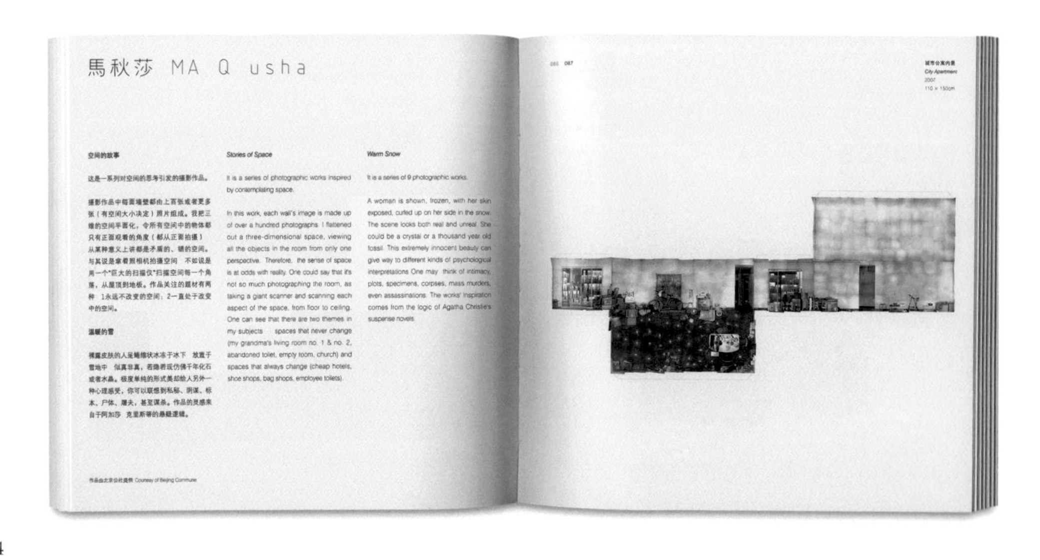

湯南南 TANG Nannan

我生活的三十多年里，中国持续发生着快速、激烈的变革。这组作品是不容喘息的变化背后遗留在内心之中的一面镜子。在这个镜子里,沉淀下来的影子象梦境一样，在今天的现实里穿行。我惊奇梦境里的色彩和情节 并试图以图像来捕捉那些似乎被遗忘的情绪，抓住似已逝去的纷乱世事和埋在深处的记忆，藉此了解今天的现实和周遭环境，即便混沌，也还可以在影像的影影绰绰中觉得不再孤独。

China has undergone a massive transformation during my lifetime. This series reflects the legacy of these rapid changes My photos are mirrors in which settled shadows resemble dreams seemingly floating towards reality. I am always surprised at the colors and the stories in these dreams. Through my images, I attempt to capture almost-forgotten emotions and grasp the deeply buried memories of a chaotic world before they are lost. I do this in order to comprehend the reality of what surrounds me today, and to peer into the vagueness of my images, not to feel lonely again.

Invisible Distance · Juko Road, Xiamen, 2006
2006
55 × 70cm

嚴明 YAN Ming

我的照片不是传统纪实 也不是当代艺术，而是用当代眼光去刻画当下。

中国是令人目眩的。我们生活在这里，满地都是意义、满地都是问题，有问题就有艺术。当下的问题与永恒价值，就分布在我迫切寻找的一个个码头，一个个意义的码头。发展变革中的中国，人和事时常处在集体的无奈与不确定中，我希望自己能在这变迁中与他们恰当地相遇相知，与我的所有拍摄对象一起感受平静、忐忑与尊严。

我的作品正是一种自我放逐的观察体验。在现实中把自己变成一个体力劳动者，哪怕是短期的、阶段性的。我坚持画面的纪实性、个人感受的直接性，在经历中关注人的状态，抓取事态的发展切片 不喜欢被题材束缚，将复杂问题简单化，而是在简单上纵深发展。

简单才能邀我同在、同看，把更多的空间留给我们善的情怀。

我爱这哭不出来的浪漫。

My pictures are neither traditional documentary nor contemporary art photographs, but rather depictions of our time by contemporary eyes.

China dazzles people. We live in a place full of problems and significance, and where there are problems, there is art. I discovered both present-day problems and eternal values at each pier that I sought out. The human and the inanimate co-exist in a state of collective impotence and uncertainty in the midst of China's rapid transformations. I hope that I can become accustomed to these changes, and experience peace, trouble, confusion and dignity with the subjects that I photograph.

My works are the product of self exile experience. I became a short term day labourer every once in a while. I concentrated on the documentary nature of my compositions and the immediacy of personal sense. I pay close attention to human experience and attempt to capture the way my surroundings evolve. I don't like feeling restrained by the subject matter. I simplify complex problems and then further develop the composition.

Only simplicity can allow us to exist and see, giving us benevolent feelings.

I love this romanticism that I cannot express by tears.

Post-Three Gorges · Child Picking up Garbage
2006
50 × 50cm

王川 **燕京八景**

艺术家独立出版，北京，2009

211mm×289mm×11mm

九帧印在折叠页里的照片，折叠页背面为此幅作品原大局部

布面精装

限量 500 册，每册艺术家签名、编号

艺术家陈述：王川

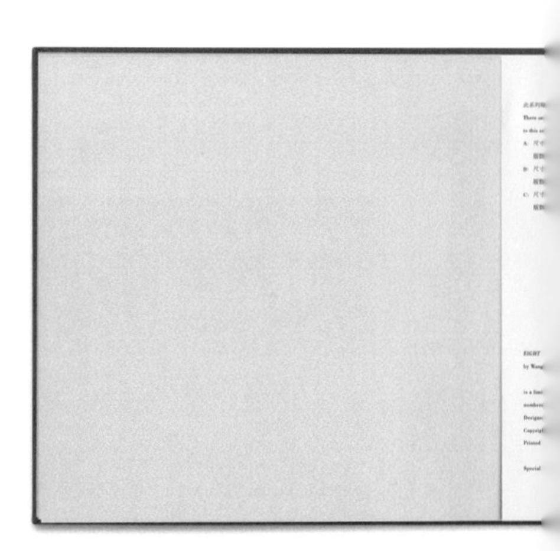

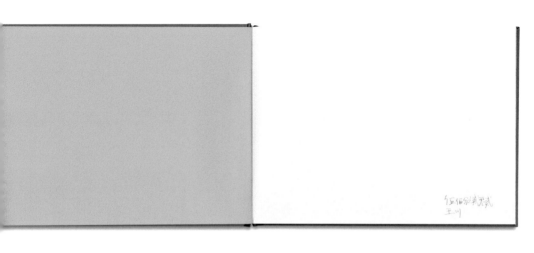

此系列每幅作品皆包括三种规格。

There are three dimensions and three editions to this series.

A. 尺寸 Size: 200×112cm

版数 Edition: 2

B. 尺寸 Size: 180×72cm

版数 Edition: 5

C. 尺寸 Size: 120×48cm

版数 Edition: 5

EIGHT GREAT SIGHTS OF BEIJING

by Wang Chuan

is a limited edition of 500 copies signed and numbered by Wang Chuan

Designed by He Hao

Copyright©2009 by Wang Chuan

Printed in China 2009

Special thanks: elinchrom

燕京八景

王川

Eight Great Sights of Beijing

Wang Chuan

About Eight Great Sights of Beijing

Wang Chuan

...an artist is to give thoughts to his time and responding in his own way with artistic means, ...esthetic experience of mankind or revealing some kind of positive meaning in life. Such artistic ...e time, an exploration as well as discovery of the media used by the artist.

...metropolis, for every change it experiences, Beijing can never break away from the links to its ...Such links are incorporated in all the aspects of politics, economy, ideology and social life. The ...es are, the easier such links are to be appreciated. As a result, I am frequently trapped in a rather ...hile recognizing those new things as well as changes brought about by the social development, ...onfused about the changes accompanying each development. Many things that used to help us ...the outside world, the past and ourselves are disappearing or being replaced. Though I live in ...I have not become more familiar with it as time passes by; instead, I am obsessed with a kind of ...gain. And the psychological recognition and emotions accumulated in years have emerged too. ...h history and specific meanings are either diminishing quickly or existing in a super realistic and ...rated from our life, or existing in real life without its original meaning. However, we still coexist ...y them.

...the Jin Dynasty (1168-1208A.D.) to Qianlong Emperor of the Qing Dynasty (1711-1799A.D.), ...modified and finally determined eight sights in Beijing as the best ones in their eyes to represent ...erve of the city and to meet agree with the regal aesthetic standards—the Eight Great Sights of ...familiar and strange to us. Qianlong Emperor even set up stone memorials with poetry for them, ...onfidence in politics and culture. All these can be found in the historical record of Beijing and ...of summarization of Beijing as well as clues to interpreting this ancient city. However, when I ...e sights, I am more attracted by the great discrepancies between their manner of existence in real ...escription of them than I am interested in the sceneries themselves, so I am impelled to find a

The camera can always separate personal choice from complicated realities, giving special meaning to these pictures. Pixel, the basic image element of today, is characterized by definite timeliness, turning it from a physical concept or technical term into a picture language, by means of which I am able to escape from the double temptation of realistic details and historical restoring towards photographing and turn to face my personal feelings directly. From the objects themselves to the language of picture elements constituting the works, our vision is wobbling between a kind of déjà vu and an uncertainty formed by the images deprived of details. I would like to know whether other people would experience the same familiar as well as strange observations and my helplessness and [illegible]. If I can receive the response in their hearts, then these personal feelings would belong to more people living in this city and to this age as well.

Wang Chuan, born in Beijing in 1957, graduated from the Department of Book Art of the Central Academy of Art and Design in 1990 and got a master's degree in the joint program Master of Arts in Visual Art (Photography) organized by China Central Academy of Fine Arts and the Queensland College of Art of Griffith University, Australia in 2005. He is head of the Department of Photography of the School of Design of China Central Academy of Fine Arts, and has held photographic exhibitions in China as well as in the U.S., Italy, UK, Australia, Austria, etc.

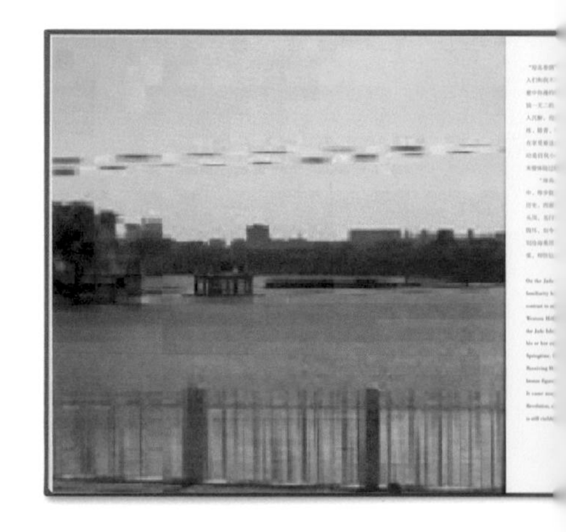

[illegible]

[illegible]

On the Jade Islet in shady springtime, I feel a familiarity beyond description, which is in sharp contrast to my [illegible] when appreciating the Western Hills Shimmering in Snow. Standing on the Jade Islet in spring wind, every one may have his or her own picture of the Jade Islet in Shady Springtime. Before us is the Platform of Immortal Receiving Heavenly Dew, featuring a north-facing bronze figure holding a plate high above his head. It came near to destruction during the Cultural Revolution, and the chasm on the white marble base is still visible.

滕菲、杰克·康宁翰

致陌生人

四大空间，上海，2009

250mm×180mm×10mm

48 页

精装

200 册

前言：姜节泓

访谈：姜节泓 / 滕菲、

姜节泓 / 杰克·康宁翰

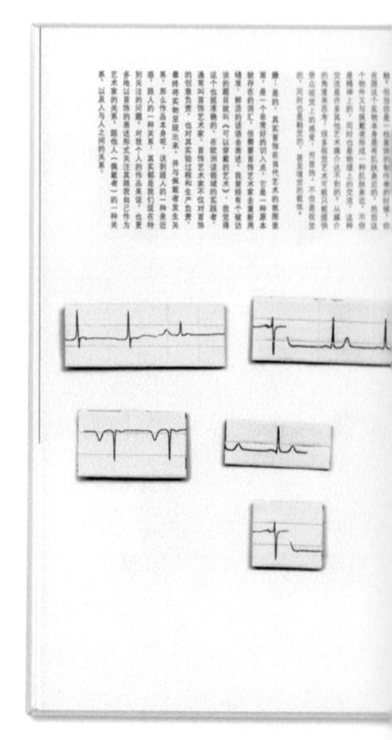

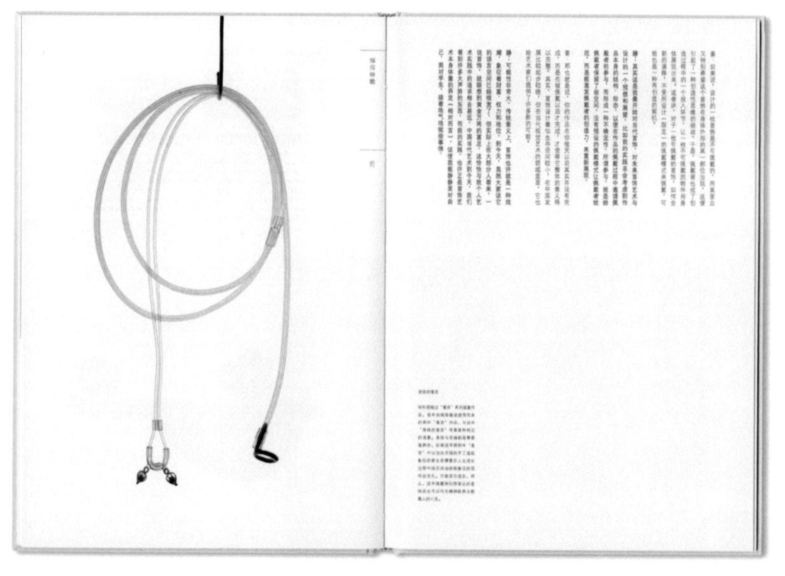

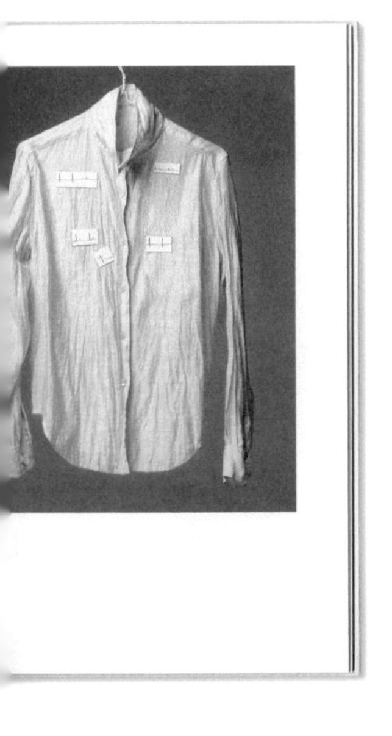

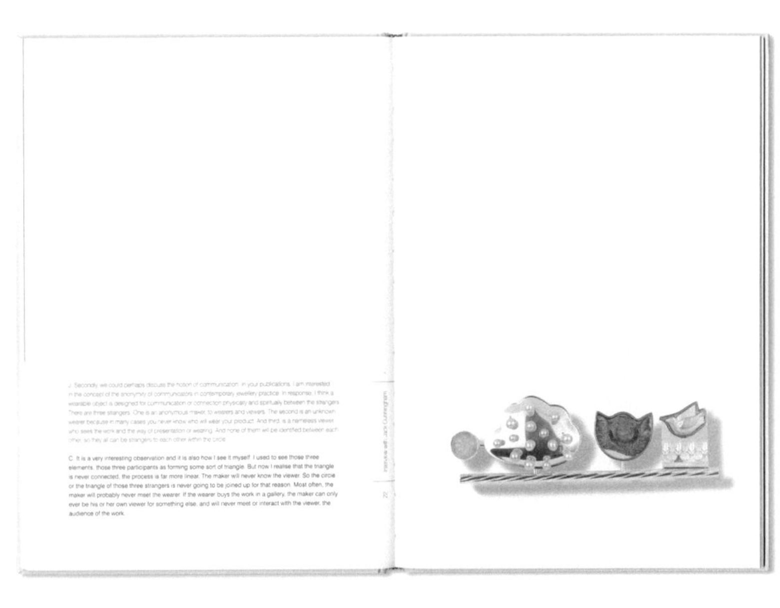
C: It is a very interesting observation and it is also how I see it myself. I used to see those three elements, those three participants as forming some sort of triangle. But now I realise that the triangle is never connected, the process is far more linear. The maker will never know the viewer. So the circle or the triangle of those three strangers is never going to be joined up for that reason. Most often, the maker will probably never meet the wearer. If the wearer buys the work in a gallery, the maker can only ever be his or her own viewer for something else, and will never meet or interact with the viewer, the audience of the work.

hree-way process, and the wearing stage is almost the start of the actual journey for
gh perhaps the scroll, which may physically change, as people are adding marks and
of jewellery can only change in the emotional sense by being worn, it doesn't generally

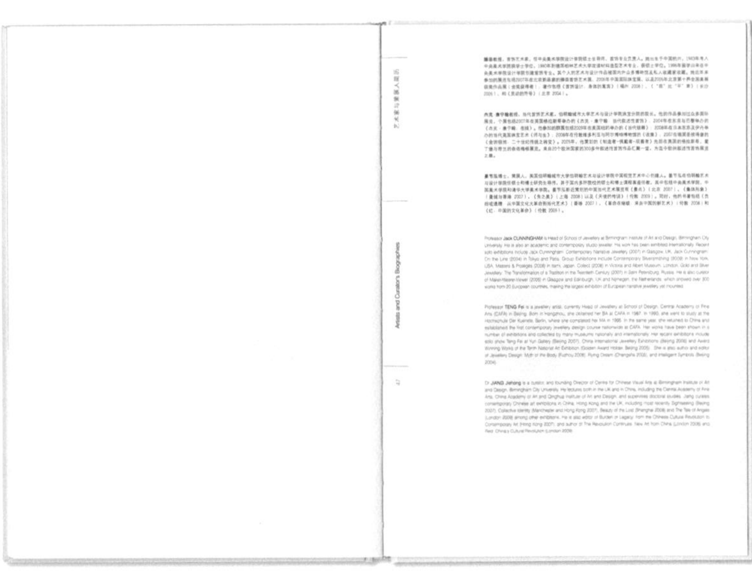

《致陌生人》

《致陌生人》是一本关于两位首饰艺术家——中国的滕菲和英国的杰克·康宁翰——的小书。书里除了两个人的作品之外，同时还包括学者姜节泓与他们分别的对话。所谓“致陌生人”，是指艺术家在制作首饰时，并不知道将来谁会佩戴，首饰是献给陌生人的艺术；对佩戴者而言，首饰的创作者往往也是匿名的，他（她）在接受一枚首饰的同时，也接受了一个陌生人的故事。

也许作品图片都是现成的，我记得图档收集异常顺利，这在以往的项目中并不多见。因为时间充裕，我甚至为康宁翰的作品在图片上不是完全镂空而带有淡淡的底色，还让他重新拍摄了照片。康宁翰回应之迅速，让我对英国人留下了好印象。但就当万事俱备，即可开工的时候，却出现了一个让人意想不到的情况：作为旅英华人学者，姜节泓用中文和英文分别与滕菲和康宁翰进行了对话并整理成文。按照惯例，这两组对话会同时翻译成另一种文字，在书中以双语形式对照出现，但此时姜节泓从英国打来电话，说自己翻译自己的写作实在勉为其难，但找到能让他满意的翻译一时也很困难，如果翻得不好，适得其反。他问我如果就目前的状态出书，是否可行。

按照以往我对图书的认识，要么单使用一种文字，要么双语或三语对照，但如此两种语言并置好像从没见过。节泓似乎也觉得让我为难，言语间颇有些踯躅。我说可以试试。事实上我当时就隐约有一点预感，也许一个更有意思的解决方案会因此产生。多年的设计实践使我相信，任何新的挑战其实即意味着新的契机。设计者的智慧就是把最严苛的限制转化为最关键的提示，在复杂的迷径当中借此来照亮出口。“纸里包不住火”，但灯笼可以，这就是设计。

再次反复阅读图文之后，我决定用间隔穿插的排序方式来呈现两位艺术家的作品，即一页滕菲，一页康宁翰，交替往复。而对话文本亦打散，紧贴作品，中文竖排，英文横排，借此强化交替的差异感。对话非连续性文章，话题间截断无碍，一唱一和，终止开始，反而更有言语间的现场感。设计至此，读者成为了艺术家、佩戴者之外的第三个陌生人——首饰的观看者，这种由于语言阻碍造成的不确定性，反而形成了陌生人之间才有的好奇与猜想，也造就了阅读交流中的变幻与惊喜。

设计最终落地时，我选用了一种长纤维、表面略微粗糙的纸张。虽然手感极佳，但这类纸张通常会被认为是器物印刷的天敌，因为纸张微观表面不平整，图片印上去颜色光泽会变得暗淡失真。但我发现正是这种“劣势”，如果控制得当，却能使冰冷的首饰在纸面上散发出完全有别于以往观看体验的柔和光泽，坚硬却温暖。书籍的真实如同一切艺术作品的真实，永远不是对现实丝毫不差的描摹，而是经过媒介转换之后在心理层面上对读者的提示与触摸。

陈文骥 **陈文骥**

东站画廊，北京，2009

221mm×303mm×16mm，114 页

布面精装

1500 册

文章：尹吉男、田恺

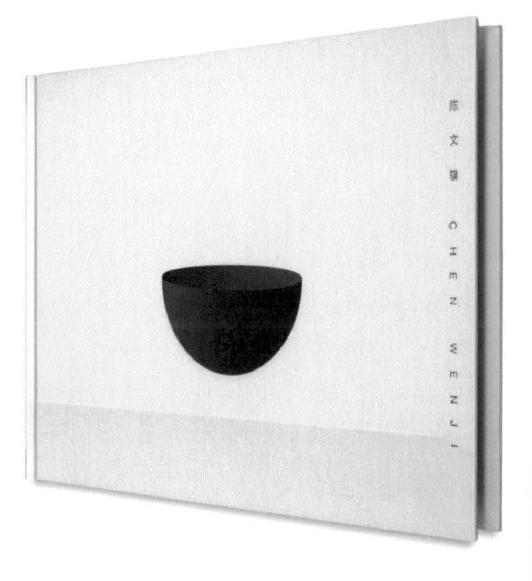

THE FREEDOM OF THE OBSERVER
THE UNIQUE ARTISTIC TRAITS OF
CHEN WENJI

Yin Jinan

Summarizing the artistic traits that distinguish Mr. Chen Wenji is incredibly difficult and I seem to have collided with a most demanding task.

His painting style is changing endlessly. From realistic still life painting to landscapes surrounding Beijing and later a fondness for industrial products and geometrical figures—his experimental process makes it abundantly clear change is his normal state. And this is merely discussing his paintings themselves. However, there is one unambiguous point: his art lacks any active relationship with the artistic trends that have unfolded in China over the previous thirty years. It's as if he's been living in his own private world all along, just being his same old self. His style of art is independent, his individualism vibrant. He hasn't necessarily voluntarily submit himself to seclusion, but in certain respects he is evading various ideological trends and movements.

Chen Wenji is a deeply efficient and beneficial professor at Central Academy of Art. As a painter hailing from Shanghai, and who has lived in Beijing for more than thirty years, the urban experience has become a primary resource in his art, as well as his primary object of close examination and criticism. In one respect he lives in the city, but in another he makes himself a sober onlooker onto it. These two gazes are always intermingled within his works, one is experiencing, the other is observing. What's more, the observer's gaze is more individualistic. Together, the helplessness of participation and the consciousness of observation create an unequal spiritual tension. This ambiguity of such an attitude baffles most critics (The passive acceptance of the reality of existence, yet giving expression to personal inclinations within an order one is unable to change.)

Contrary to the trend of a majority of Chinese painters who are fond of painting figures or social topics, Mr. Chen repeatedly manifests a material world vacant of life. More specifically, most of his paintings depict still life, architecture, industrial pipelines, man-made objects, geometric figures, lines or shadows. He works strictly within the formulas of painting to suspend human life and society before us, and his high mastery of painterly techniques endow these realistic materials with abstract results. The conceptual nature of his paintings isn't readily apparent, but is gradually realized.

With regard to China's cultural traditions, a similar participant observer mode exists in differing degrees in classical Chinese literati painting. Here we have life experience in the secular world, and the manifestation of individualistic ideals as a departure from the secular world. "Experiencing the world" and "renouncing the world" are positioned alongside one another. The literati lived under political authority and led worldly existences, but each one of them painted the natural world and its geographical phenomena (as in landscapes) or scenes from life (as in bird and flower painting). The oppositional relationship of nature and society is unequivocal, and this relationship becomes a metaphor symbolizing the tense relationship between personal freedom and the social order. In literati painting, the results of a painting most often indicate subjective abstraction. While they painted landscapes or plants of a certain type, these were not certain designated objects existing within time and space. Orchids, plum blossoms, chrysanthemums and bamboo were not specific things, but abstract classes of objects.

Casting aside living, natural objects (the landscapes, birds and flowers of classical literati painting), and replacing them with inanimate social objects (the city, industrially produced objects, etc.), the diametrical opposition between these lifeless social things and the living social things (people) is thus established. Different from the widespread expressions of the classical literati, and simultaneously independent from the customary tastes of contemporary painters, it is only in this sense can we interpret Mr. Chen's art as possessing personal traits. He persistently employs inanimate social objects in the same way that classical literati painters employed living objects in nature. It seems that Chen Wenji has glimpsed the power of a systematized human world, at the same time that he realizes that this power creates individual passivity. Thus he dreams of escaping this power's social inertia.

The occasion of this solo exhibition is similar to all of his exhibitions; it was realized at the energetic invitation of the host, the space is provided, not chosen. Mr. Chen enjoys reconstructing his artistic self through unfamiliar spaces, interacting with spaces and enjoying the experimental challenge to the creative process, expressing his personal freedom to the utmost through experimentation. As if within a limited reality, after natively gravitating between two sharply contrasting roles of "passive" and "active" he objectively completes a series of artistic transformation. Occasionally a childlike natural radiance shines from his frail silhouette; through the helpless face of an adult it beams out to meet you.

September 28, 2009
At Central Academy of Fine Arts

Prof. Yin Jinan
Dean of the School of Humanities, Central Academy of Fine Arts, Art Historian, Critic
Principal publications:
[illegible]

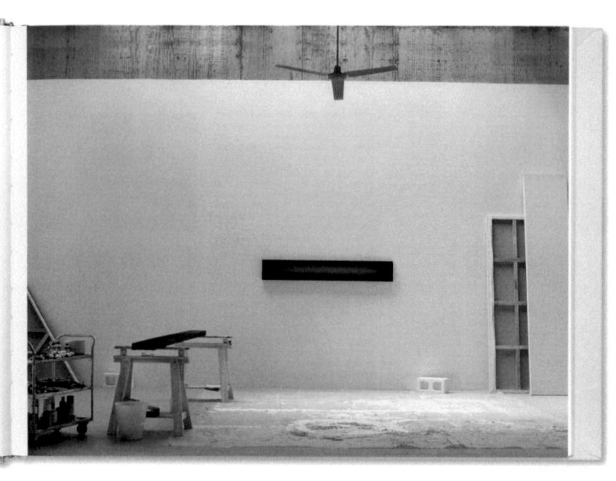

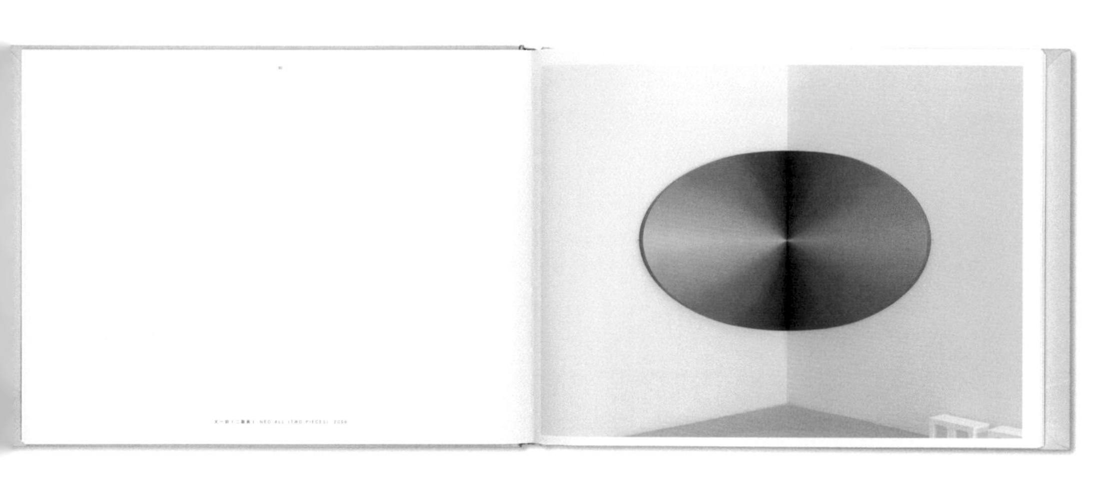

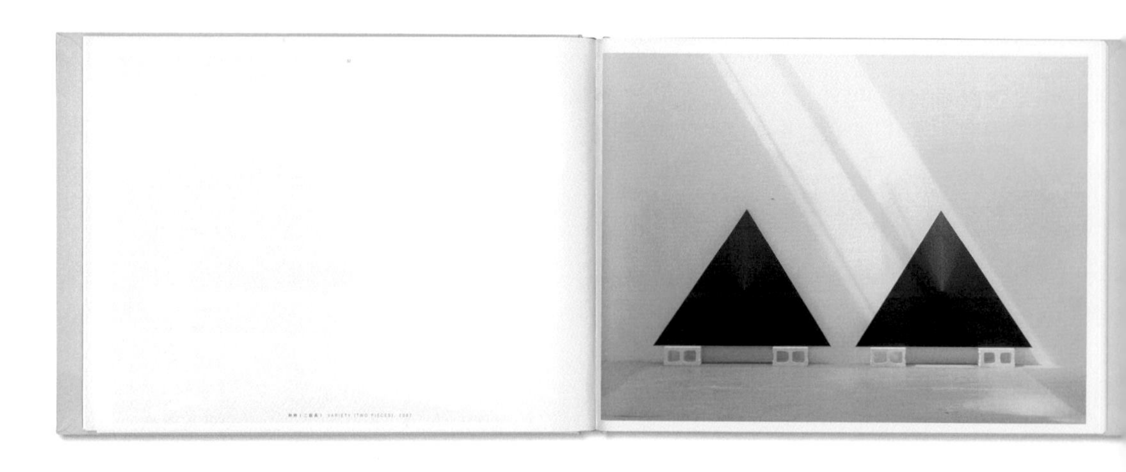

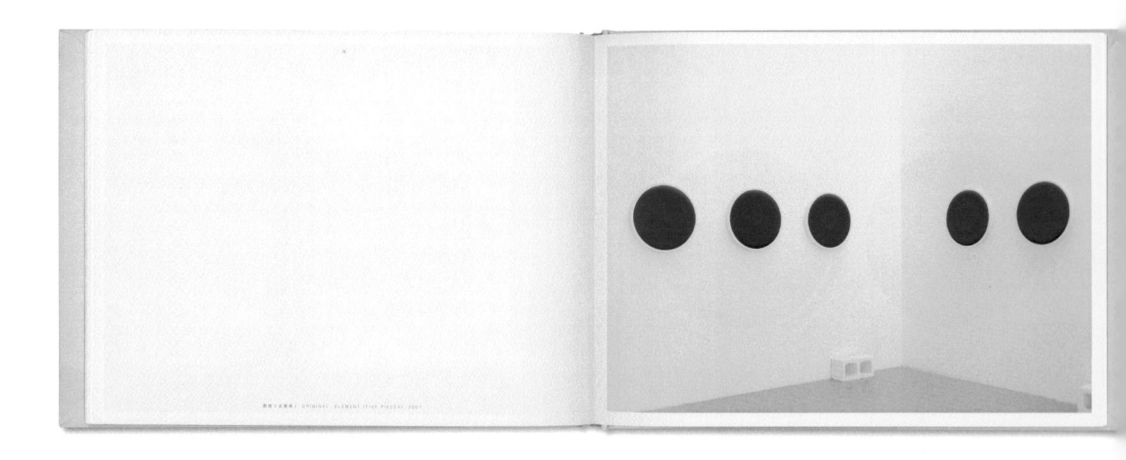

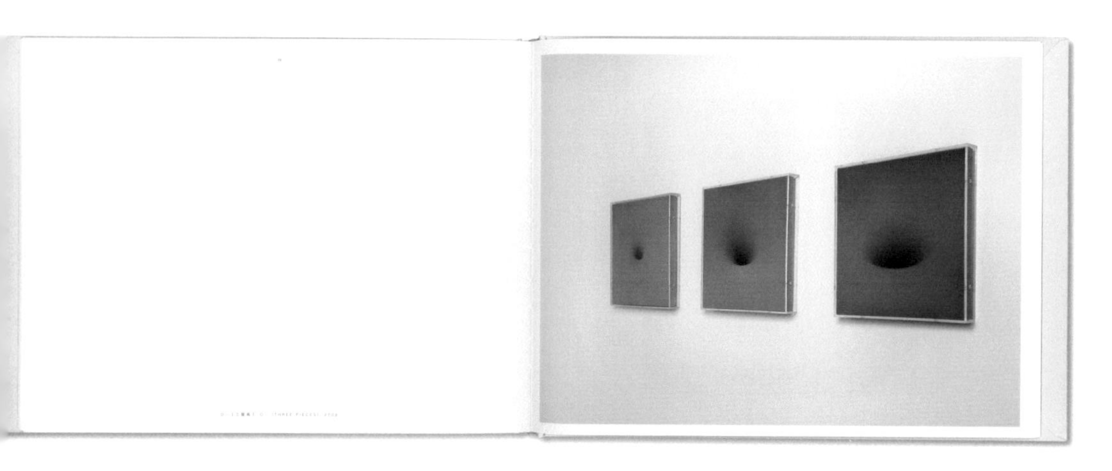

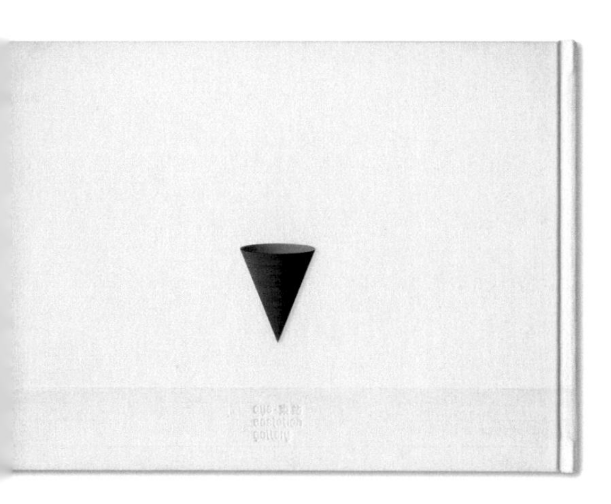

海报，700mm×1000mm，200 张

WORKS
Community,No.32

《陈文骥》

2008年底，持续了两年多的中国当代艺术的市场狂潮终于开始回落，作为裹挟在其中的一分子，我也因此得到喘息。当2009年暑假过后，艺术家陈文骥携新作来委托我进行书籍设计时，我已在外“云游”半年，之前因为被过度催逼而对设计工作产生的些许厌倦也在山水间逐渐褪去，合适的时机合适的项目，使我的设计历程有了新的开端。

陈文骥是我美院师长，虽不同系，但他的那些不怀旧、不愤世、不媚体制、不趋市场、始终独立于潮流之外的作品二十年前即为我所熟悉和喜爱。眼前这批始于2007年的新作，更将陈老师以往作品所含特质推向极致。这种极致，不只来自其远观的极简节制、超然出尘，更根植于其近看毫不虚妄的厚重坚实。画作的名字是解读这些“抽象”作品最好的钥匙——《一周》（205页，中）、《O——》（207页，中）……几十年的艺术修为，最终化作“提起放下”后的一笑，轻盈通透。

从物质形态上讲，陈老师的这批作品并非一般意义上的绘画，这种不一般来自其整体形态介于平面和立体之间的模糊属性——一方面，这些作品是不折不扣的平面油画；而另一方面，因为作品外轮廓根据画面随形而定，其飘忽变幻的外形又让画作具有了一种特殊的“立体感”，这种模糊性使得作品置于空间中时既不同于传统绘画也不同于雕塑装置，在画面和外形的双重作用下与空间产生了一种彼此存在、互为图底的奇妙幻觉。这种意味，在书中该如何呈现?

一般出现在书籍中的画作，往往被呈现为硬边图像，当一张在空间中拍摄的作品图片被沿着边缘裁切掉环境背景的同时，“物感”消退“图感”凸现——现实的画作成为了书中的“图版”。虽然这种转化对于绝大多数平面作品而言并无不妥，但针对陈老师的作品，这种约定俗成到几乎可以不假思索的呈现方式却需要被重新审视和讨论。

经过与陈老师的深入沟通，我们决定逆向而行，不仅不裁除环境，反而将之放大，完全在场景中呈现画作。

记得随后的那个有大阅兵的“十一”长假，我是完全对着电脑屏幕度过的，每日工作十小时以上，深入处理了每一张图片——调整画作与背景的比例、透视和明暗关系，去掉环境中的细微干扰，同时又十分小心地保留了具有真实质感的动人细节——墙上的一道裂缝，地面的几点斑驳……在极尽精微的调整、取舍之后，作品和环境之间形成了一种恰到好处的“理想的真实”。在这里，作品与空间的关系通过书籍得到最为清晰的提示，而作品更由此被揭示和延展出新的意味。此时，书籍已不再是对原作的被动复制和简单记录，而成为能够反作用于原作的、具有独立价值的另一重阐述。

尘埃落定之后，陈老师建议封面最好也能不动声色地“出格”一点，以对内页的设计概念有所回应。我把书籍翻开一侧的精装飘口（硬壳封面大出书芯的边缘部分）在常规基础上加长了5毫米，因此产生的虚空则由变幻的阴影来填充。如同那些置于空间中的作品，这本书也在真实中拥有了一丝虚幻。

徐冰 **徐冰版画**

文化艺术出版社，北京，2010

345mm×250mm×22mm，268 页

革面内衬海绵精装，

外加瓦楞纸书匣（未展示）

2000 册

文章：徐冰

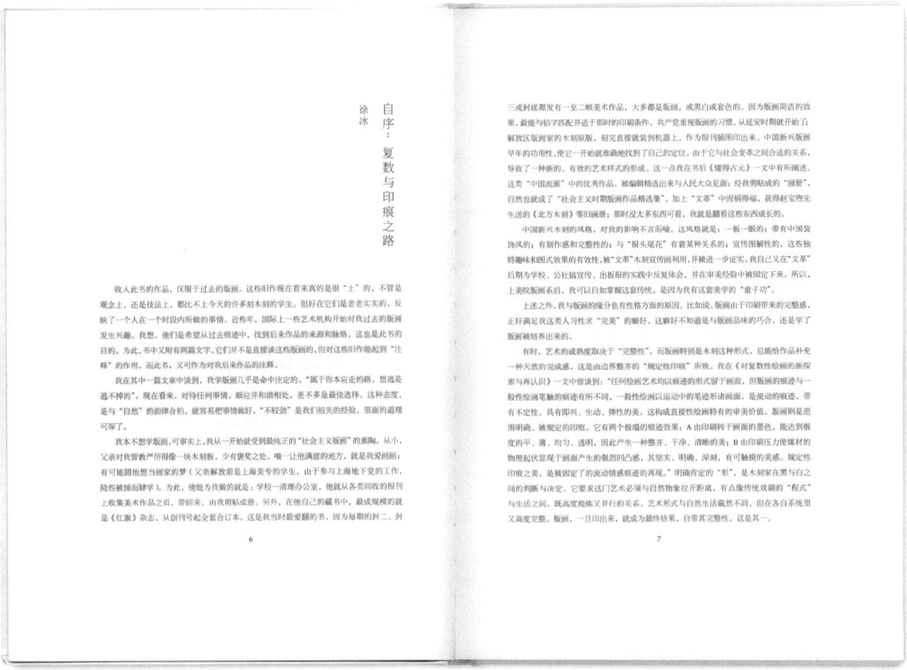

自序：复数与印痕之路

徐冰

收入此书的作品，仅限于过去的版画。这些旧作现在看来真的是很“土”的，不管是观念上，还是技法上，都比不上今天的许多刻木刻的学生。但好在它们是老老实实的，反映了一个人在一个时段内所做的事情。近些年，国际上一些艺术机构开始对我过去的版画发生兴趣。我想，他们是希望从过去痕迹中，找到后来作品的来源和脉络，这也是此书的目的。为此，书中又附有两篇文字，它们并不是直接谈这些版画的，但对这些旧作能起到“注释”的作用。而此书，又可作为对我后来作品的注释。

我在其中一篇文章中谈到，我学版画几乎是命中注定的。“属于你本应走的路，想逃是逃不掉的”。现在看来，对待任何事情，顺应并和谐相处，差不多是最佳选择。这种态度，是与“自然”的韵律合拍，就容易把事情做好。“不较劲”是我们祖先的经验，里面的道理可深了。

我本不想学版画，可事实上，我从一开始就受到最纯正的“社会主义版画”的熏陶。从小，父亲对我管教严厉得像一块木刻板，少有褒奖之处。唯一让他满意的地方，就是我爱画画：有可能圆他想当画家的梦（父亲解放前是上海美专的学生，由于参与上海地下党的工作，险些被捕而辍学）。为此，他能为我做的就是：学校一清理办公室，他就从各类回收的报刊上收集美术作品之页，带回来，由我剪贴成册。另外，在他自己的藏书中，最成规模的就是《红旗》杂志，从创刊号起全套合订本。这是我当时最爱翻的书，因为每期的封二、封

6

三或封底都发有一至二帧美术作品，大多都是版画，或黑白或套色的。因为版画简洁的效果，最能与铅字匹配并适于那时的印刷条件。共产党重视版画的习惯，从延安时期就开始了：解放区版画家的木刻原版，刻完直接就装到机器上，作为报刊插图印出来。中国新兴版画早年的功用性，使它一开始就准确地找到了自己的定位。由于它与社会变革之间合适的关系，导致了一种新的、有效的艺术样式的形成。这一点我在书后《懂得古元》一文中有所阐述。这类“中国流派”中的优秀作品，被编辑精选出来与人民大众见面；经我剪贴成的“画册”，自然也就成了“社会主义时期版画作品精选集”。加上“文革”中因祸得福，获得赵宝煦先生送的《北方木刻》等旧画册；那时没太多东西可看，我就是翻看这些东西成长的。

中国新兴木刻的风格，对我的影响不言而喻。这风格就是：一板一眼的；带有中国装饰风的；有制作感和完整性的；与“报头尾花”有着某种关系的；宣传图解性的，这些独特趣味和图式效果的有效性，被“文革”木刻宣传画利用，并被进一步证实。我自己又在“文革”后期为学校、公社搞宣传、出板报的实践中反复体会，并在审美经验中被固定下来。所以，上美院版画系后，我可以自如掌握这套传统，是因为我有这套美学的“童子功”。

上述之外，我与版画的缘分也有性格方面的原因。比如说，版画由于印刷带来的完整感，正好满足我这类人习性求“完美”的癖好，这癖好不知道是与版画品味的巧合，还是学了版画被培养出来的。

有时，艺术的成熟度取决于“完整性”。而版画特别是木刻这种形式，总能给作品补充一种天然的完成感，这是由边界整齐的“规定性印痕”所致。我在《对复数性绘画的新探索与再认识》一文中曾谈到：“任何绘画艺术均以痕迹的形式留于画面，但版画的痕迹与一般性绘画笔触的痕迹有所不同，一般性绘画以运动中的笔迹形诸画面，是流动的痕迹，带有不定性，具有即兴、生动、弹性的美。这构成直接性绘画特有的审美价值。版画则是范围明确、被规定的印痕。它有两个极端的痕迹效果：A 由印刷转于画面的墨色，能达到极度的平、薄、均匀、透明，因此产生一种整齐、干净、清晰的美；B 由印刷压力使媒材的物理起伏显现于画面产生的强烈凹凸感，其坚实、明确、深刻，有可触摸的美感。规定性印痕之美，是被固定了的流动情感痕迹的再现。”明确肯定的“形”，是木刻家在黑与白之间的判断与决定。它要求这门艺术必须与自然物象拉开距离，有点像传统戏剧的“程式”与生活之间，既高度提炼又并行的关系。艺术形式与自然生活截然不同，但在各自系统里又高度完整。版画，一旦印出来，就成为最终结果，自带其完整性。这是其一。

7

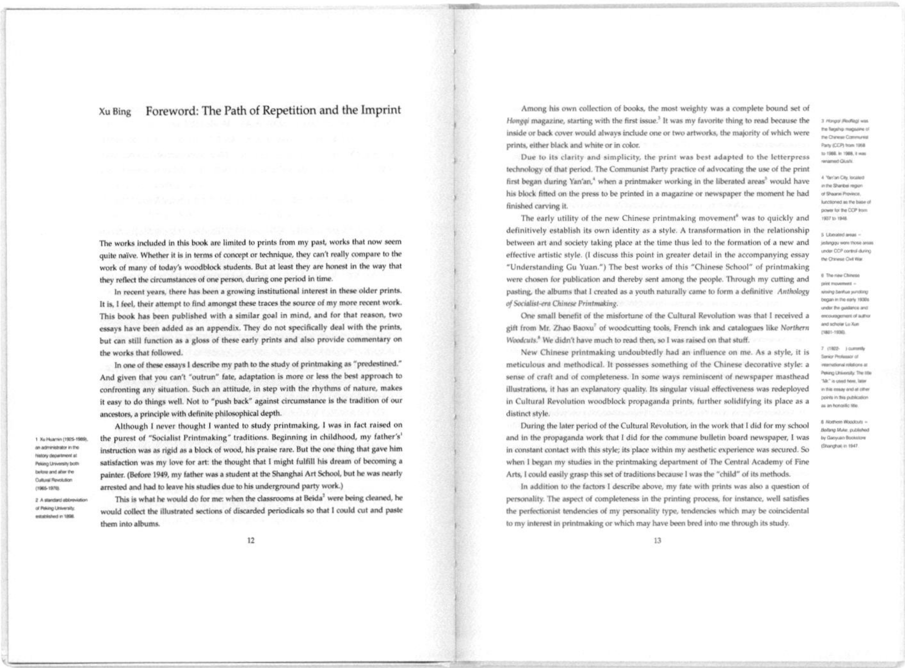

Xu Bing

Foreword: The Path of Repetition and the Imprint

The works included in this book are limited to prints from my past, works that now seem quite naïve. Whether it is in terms of concept or technique, they can't really compare to the work of many of today's woodblock students. But at least they are honest in the way that they reflect the circumstances of one person, during one period in time.

In recent years, there has been a growing institutional interest in these older prints. It is, I feel, their attempt to find amongst these traces the source of my more recent work. This book has been published with a similar goal in mind, and for that reason, two essays have been added as an appendix. They do not specifically deal with the prints, but can still function as a gloss of these early prints and also provide commentary on the works that followed.

In one of these essays I describe my path to the study of printmaking as "predestined." And given that you can't "outrun" fate, adaptation is more or less the best approach to confronting any situation. Such an attitude, in step with the rhythms of nature, makes it easy to do things well. Not to "push back" against circumstance is the tradition of our ancestors, a principle with definite philosophical depth.

Although I never thought I wanted to study printmaking, I was in fact raised on the purest of "Socialist Printmaking" traditions. Beginning in childhood, my father's[1] instruction was as rigid as a block of wood, his praise rare. But the one thing that gave him satisfaction was my love for art: the thought that I might fulfill his dream of becoming a painter. (Before 1949, my father was a student at the Shanghai Art School, but he was nearly arrested and had to leave his studies due to his underground party work.)

This is what he would do for me: when the classrooms at Beida[2] were being cleaned, he would collect the illustrated sections of discarded periodicals so that I could cut and paste them into albums.

1 Xu Huamin (1925-1989), an administrator in the history department at Peking University both before and after the Cultural Revolution (1965-1976).

2 A standard abbreviation of Peking University, established in 1898.

12

Among his own collection of books, the most weighty was a complete bound set of *Hongqi* magazine, starting with the first issue.[3] It was my favorite thing to read because the inside or back cover would always include one or two artworks, the majority of which were prints, either black and white or in color.

Due to its clarity and simplicity, the print was best adapted to the letterpress technology of that period. The Communist Party practice of advocating the use of the print first began during Yan'an,[4] when a printmaker working in the liberated areas[5] would have his block fitted on the press to be printed in a magazine or newspaper the moment he had finished carving it.

The early utility of the new Chinese printmaking movement[6] was to quickly and definitively establish its own identity as a style. A transformation in the relationship between art and society taking place at the time thus led to the formation of a new and effective artistic style. (I discuss this point in greater detail in the accompanying essay "Understanding Gu Yuan.") The best works of this "Chinese School" of printmaking were chosen for publication and thereby sent among the people. Through my cutting and pasting, the albums that I created as a youth naturally came to form a definitive *Anthology of Socialist-era Chinese Printmaking*.

One small benefit of the misfortune of the Cultural Revolution was that I received a gift from Mr. Zhao Baoxu[7] of woodcutting tools, French ink and catalogues like *Northern Woodcuts*.[8] We didn't have much to read then, so I was raised on that stuff.

New Chinese printmaking undoubtedly had an influence on me. As a style, it is meticulous and methodical. It possesses something of the Chinese decorative style: a sense of craft and of completeness. In some ways reminiscent of newspaper masthead illustrations, it has an explanatory quality. Its singular visual effectiveness was redeployed in Cultural Revolution woodblock propaganda prints, further solidifying its place as a distinct style.

During the later period of the Cultural Revolution, in the work that I did for my school and in the propaganda work that I did for the commune bulletin board newspaper, I was in constant contact with this style; its place within my aesthetic experience was secured. So when I began my studies in the printmaking department of The Central Academy of Fine Arts, I could easily grasp this set of traditions because I was the "child" of its methods.

In addition to the factors I describe above, my fate with prints was also a question of personality. The aspect of completeness in the printing process, for instance, well satisfies the perfectionist tendencies of my personality type, tendencies which may be coincidental to my interest in printmaking or which may have been bred into me through its study.

3 *Hongqi* (*Red Flag*) was the flagship magazine of the Chinese Communist Party (CCP) from 1958 to 1988. In 1988, it was renamed *Qiushi*.

4 Yan'an City, located in the Shanbei region of Shaanxi Province, functioned as the base of power for the CCP from 1937 to 1948.

5 Liberated areas – *jiefangqu* were those areas under CCP control during the Chinese Civil War.

6 The new Chinese print movement – *xinxing banhua yundong* began in the early 1930s under the guidance and encouragement of author and scholar Lu Xun (1881-1936).

7 (1922-) currently Senior Professor of international relations at Peking University. The title "Mr." is used here, later in this essay and at other points in this publication as an honorific title.

8 *Northern Woodcuts* – *Beifang Muke*, published by Ganyuan Bookstore (Shanghai) in 1947.

13

碎玉集
Shattered Jade
单幅作品
Individual Prints
复数系列
Series of Repetitions
开笔图・执笔图
Softening the Brush, Holding the Brush
铜版・石版・综合版画
Etchings, Lithographs and Mixed Prints

小麻雀 Sparrow
1981, 8.3 × 9.7 cm
田野的空间 Open Country Space
1985, 8.6 × 11.9 cm
鱼 Fish
1982, 8.5 × 7.4 cm
川北农家 Chuanbei Farmhouse
30
31

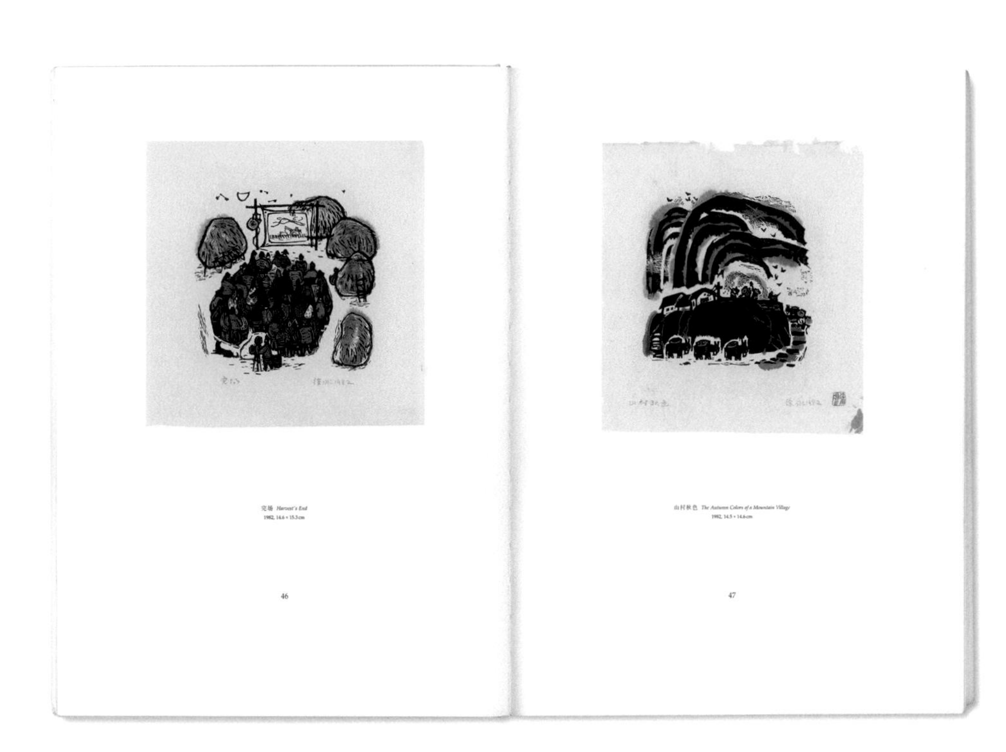
完场 Harvest's End
1982, 14.6 × 15.3 cm
46
山村秋色 The Autumn Colors of a Mountain Village
1982, 14.5 × 14.6 cm
47

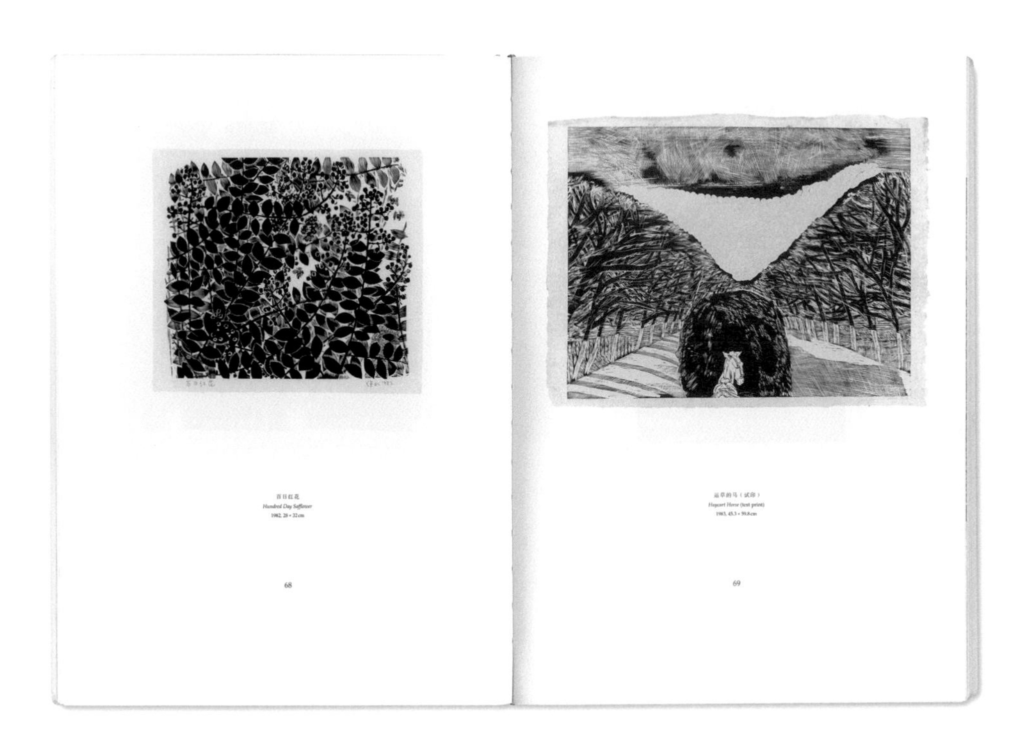
百日红花
Hundred Day Safflower
1982, 28 × 32 cm
68
运草的马（试印）
Haycart Horse (test print)
1983, 45.3 × 59.8 cm
69

复数试验（效果稿）
Reduction Test Print (installation mock-up)
1987, 189 × 300 cm

115

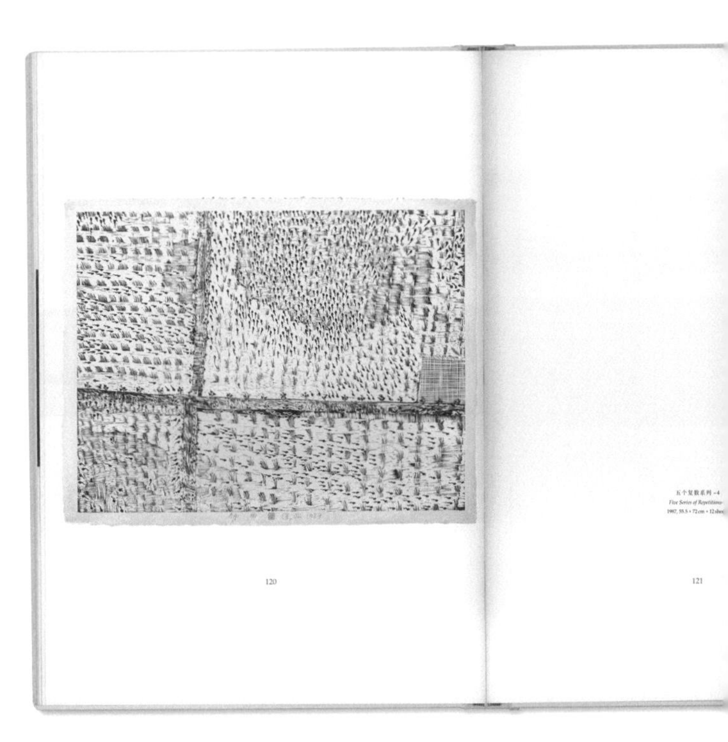
120
五个复数系列-4
Five Series of Repetitions-
1987, 55.5 × 72 cm × 12 shee
121

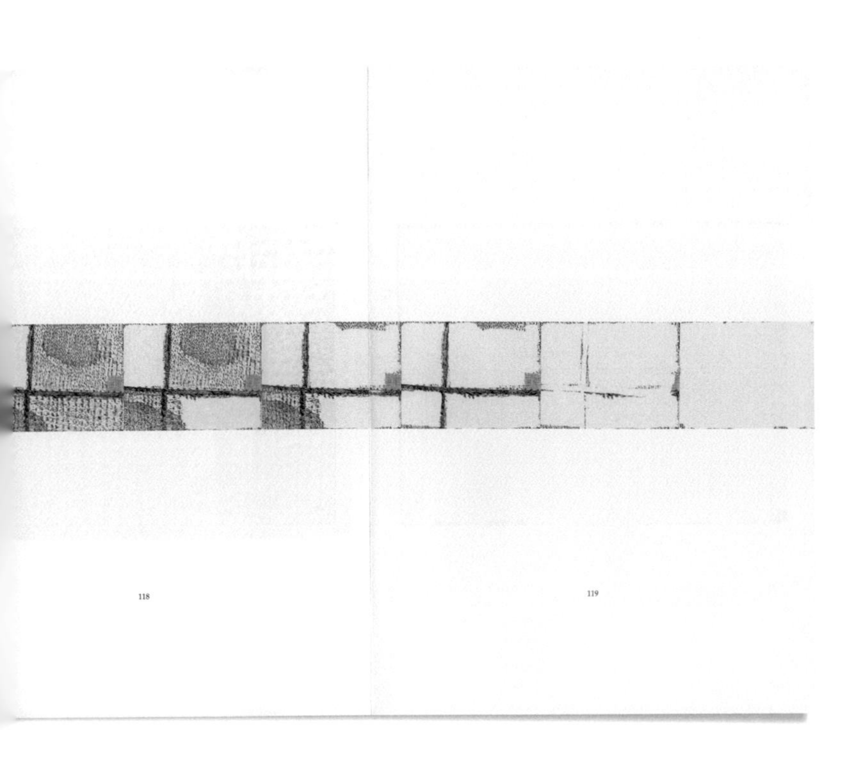
118
119

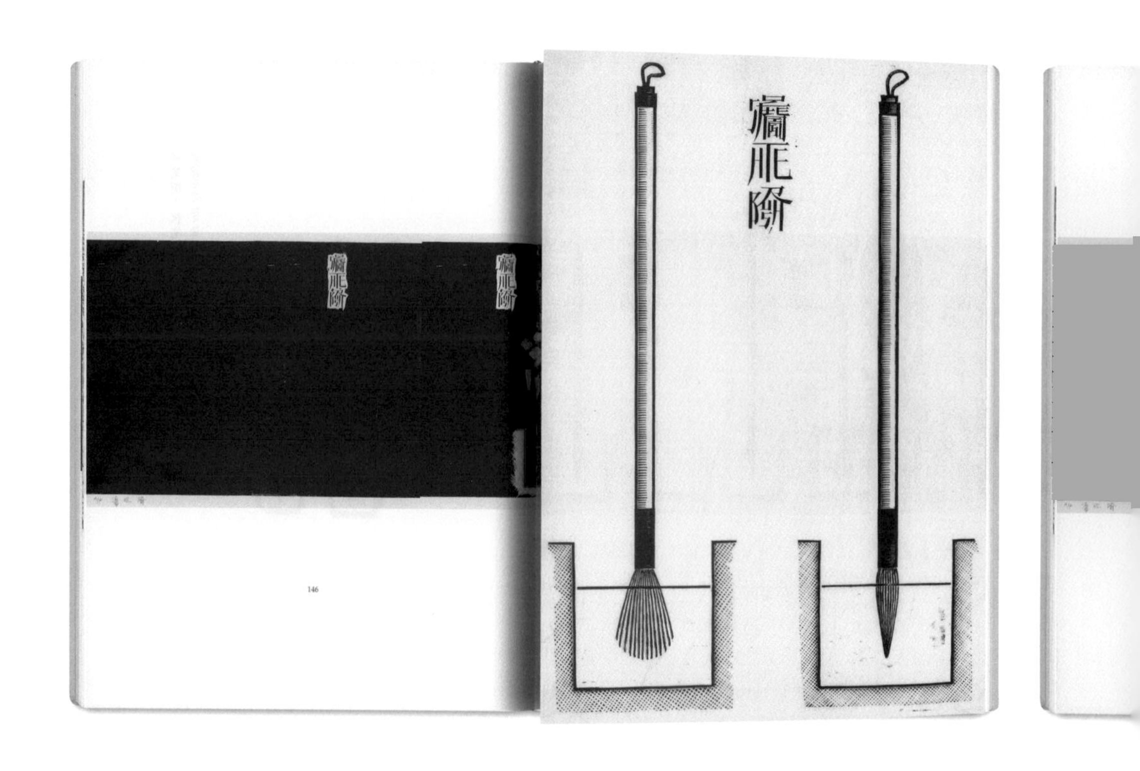

146

146
147

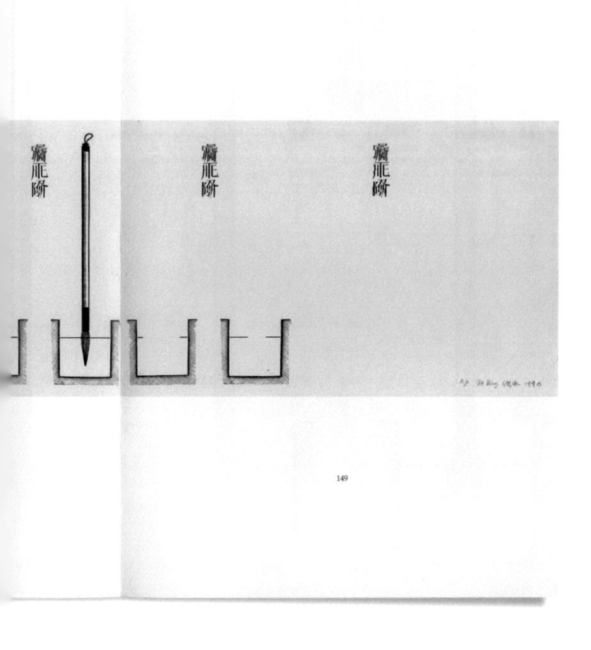

《天书》(1987-1991)，加拿大国家美术馆，1998 Book from the Sky (1987-1991), installation view at the National Gallery of Canada, 1998

徐冰年表
Xu Bing Timeline: Selected Events and Exhibitions
251

《徐冰版画》

我第一次跟徐冰见面是在他家，为这本书的设计。

同美院许多老师一样，徐老师也住在学校附近。但与众不同的是，他家面积虽大却有令人难以置信的朴素。后来在工作中我们慢慢熟络起来，我才意识到徐老师对于所谓世俗意义上的“享受”并无太大兴趣，他所有的注意力都在艺术上，孜孜以求，从无倦意。甚至我觉得他有点像文学作品中的早期共产党员——坚定的信仰使他能够屏蔽掉一切感知对于人的侵扰，无论享乐还是折磨——他曾亲口对我讲，当年在纽约时曾经为了要搭建一个展览，七天七夜没合过眼。

我们这次要做的书名为《徐冰版画》，是他20世纪70年代至90年代的版画作品合集。虽然徐老师以多媒介的观念作品闻名于世，但如他自己所说：“回头审视自己后来的这些综合材料的创作，与‘复数’和‘印痕’这些版画的核心概念之间，是否有一些关系。在行人一定会说，不但有，还很强。这是我的那些不好归类的艺术中的一条内在的线索，也成为我艺术语言得以延展的依据。”[1]作为研究徐冰艺术的“前传”，《徐冰版画》无疑是一本重要的书；而版画与书籍在媒介语言上天然的联系，也使得这本书有可能成为一本很“好看”的书。

出于对版画的兴趣，我之前看过不少版画的画册，但其中的绝大部分都让我觉得有些遗憾——版画作品一旦出现在书里就不太像画而是像图形设计了，尤其对于木刻，这种感觉更为明显。我想造成这一现象的根本原因，在于从作品到书籍的转化过程中画作质感的丢失——版画要是没有了纸张和印刷这两个质感因素，那留下的就只剩“图形”了。因此，如何最大限度地还原纸张与印痕之“美”，成为了《徐冰版画》这本书设计的关键所在。

我首先花了十来天的时间对所有将要放入书中的画作进行了非常仔细的修整，把其中多数已被裁切得规整的“图”通过挖补、挪移等手段还原成了边缘毛糙、实实在在印在纸上的“画”。“修图”这个环节一直被我视为设计工作

1. 徐冰：《复数与印痕之路》。

的一部分，哪怕只是去除扫描时留在画面上的脏点，或是调整一下拍摄时产生的细微透视畸变，我也会亲力亲为，不假他人之手。这里面既包含着一层“非人磨墨墨磨人”的功夫，有借此入定的意味，同时这个过程也是熟悉和理解作品的过程，其后的“设计”与之有着十分真切的关联。有一次我跟壁画系的同事叶剑青聊天，当时他刚画了一组每幅高达3.8米的观音像，我问他这么大的画他如何来控制造型，是否用了投影放大的方法，因为我知道很多人都是这么做的。但老叶说他没用投影，而是画了很细的等大素描稿，然后打上格誊到画布上。老叶讲，虽然画素描的方式看来吃力费时，但这个过程并不能简单地视为技术工序，他通过画素描把要塑造的对象非常仔细地“摸”了一遍，画过和没画过在理解上会非常不同，这种差异最终会反映到画作上，人物形象的生动程度自然也会很不一样。老叶的这番话正说到我心坎上，我太能理解这种感觉了。对设计而言，修图就是我的“素描”，虽然这两者看上去毫不相干。

“缓进速战”是我的工作方法，这本书也不例外。当我一点一滴修整图片的时候，如何排布也逐渐了然于心，落在版面上是水到渠成。书籍结构以创作阶段作为篇章的划分；相对于通常的书籍样式，我们适当地改窄了开本，同时加大字号、缩小版心。这是一种我和徐老师都非常喜欢的略微偏古典的品位，松弛的严谨和不刻意的考究相混杂。

在接下来的一个多月当中，我和徐老师在印厂对这本书进行了反复的“锤炼”。乍看“锤炼”这个词有些蹩脚，但也就是这两个字才能准确地概括出我们那段难忘的日子。调整图色，打样，再调整，再打样……中间夹杂着对于原稿的不断重新扫描和拍照——我们根据打样结果来调整扫描和拍摄的光源角度，以使纸张和油墨的质感得以更好地呈现。当整本书整整折返了五个来回之后，书籍页面最终发生了让人惊叹的质变——我们硬是把“图”又变回了“画”。

在我以往的阅读经验里，哪怕是印制极精美的画册，翻阅时也不会让你忘掉正在看的是一本书——人和作品总是隔着那么一层；要是设计得再过分点，就连书都读不进去了，满眼的喧嚣。但现在摆在面前的《徐冰版画》，却会让人不知不觉地忘掉“书”，而真正进入“画”境，仿佛触手可及。我想所谓“呼之欲出”，说的大概就是这种感觉吧。而我，也通过这个项目获得了对于书籍认识的新体验，甚至可以称之为“升华”。事实上，我的成就感与满足感也正是在这种不断拓展边界的过程中得以实现的。

徐冰 **凤凰的故事——徐冰《凤凰》项目**

徐冰工作室，北京，2010

263mm×354mm×7mm，100 页

平装，踩线钉，外加档案袋式封套

2000 册

文本写作：翟永明

访谈：胡赳赳 / 徐冰

徐冰
1. 北京CBD中心有许多财富大厦，它们大多位于东三环旁边，CCTV电视大楼是这里的象征符号。如果《凤凰》真的像一开始计划的那样挂于某个被认为“财富象征”的那座大楼，挂在连接两幢玻璃大厦的大堂中，我们有理由相信那是《凤凰》展出的最佳位置。
2. 2008年1月，台湾春美奥艺术集团总经理郭倩如小姐前来找到徐冰，请他为某财富大厦的大堂做一件作品。其中一个条件是只要徐冰答应将来有可能给美院学生一些资助。刚回美院的徐冰与春美奥的工作人员一起去看了现场。

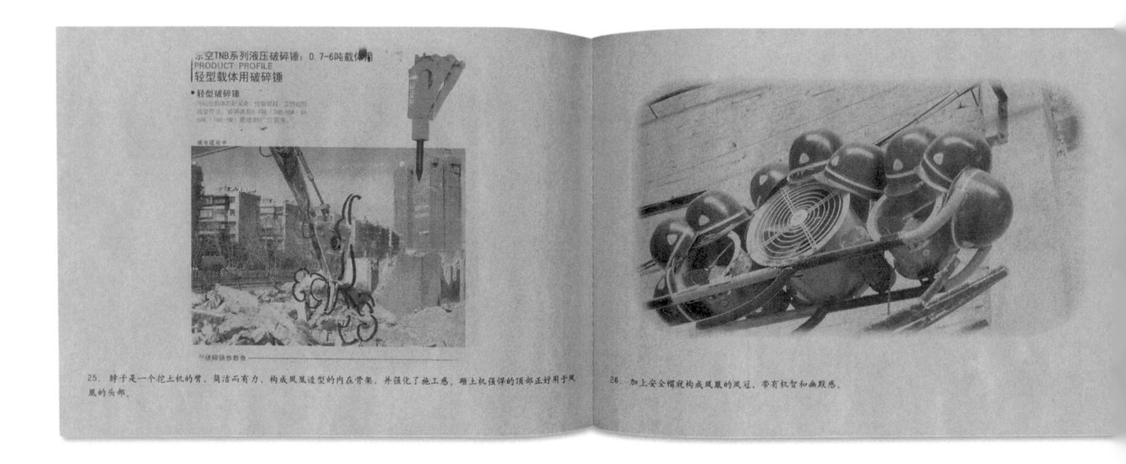

TNB系列液压破碎锤：0.7-6吨载体用
PRODUCT PROFILE
轻型载体用破碎锤
• 轻型破碎锤
25. 脖子是一个挖土机的臂，简洁而有力，构成凤凰造型的内在骨架，并强化了施工感。砸土机强悍的顶部正好用于凤凰的头部。
26. 加上安全帽就构成凤凰的凤冠，带有机智和幽默感。

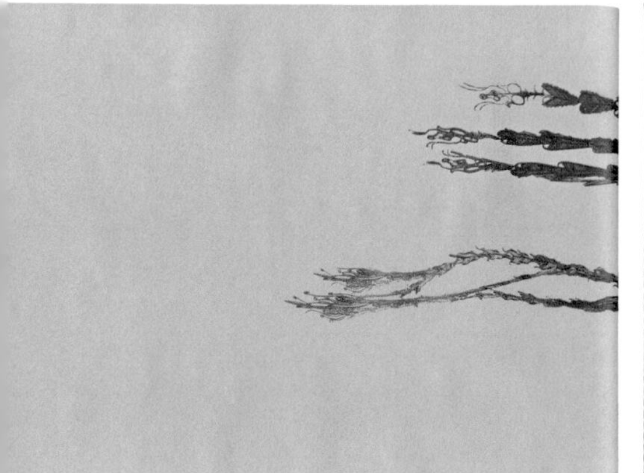

3、 工人的工作条件是非常原始的、低陋的，现代化的大楼和低级的施工方式给从未踏进过中国建筑工地的徐冰很大震动。

4、 当时施工现场全是建筑废料，徐冰就想用修建这幢大楼的建筑排泄物，以及民工的生产和生活用具来做一件作品，挂在金碧辉煌的财富大厦的大堂中。

王功新 **关联——王功新录像艺术**

东八时区，香港，2010

265mm×215mm×10mm，128 页

平装

2000 册

编辑：王功新、何浩

访谈：陈海涛 / 王功新

节选文章：Barbara Pollack、Barbara London

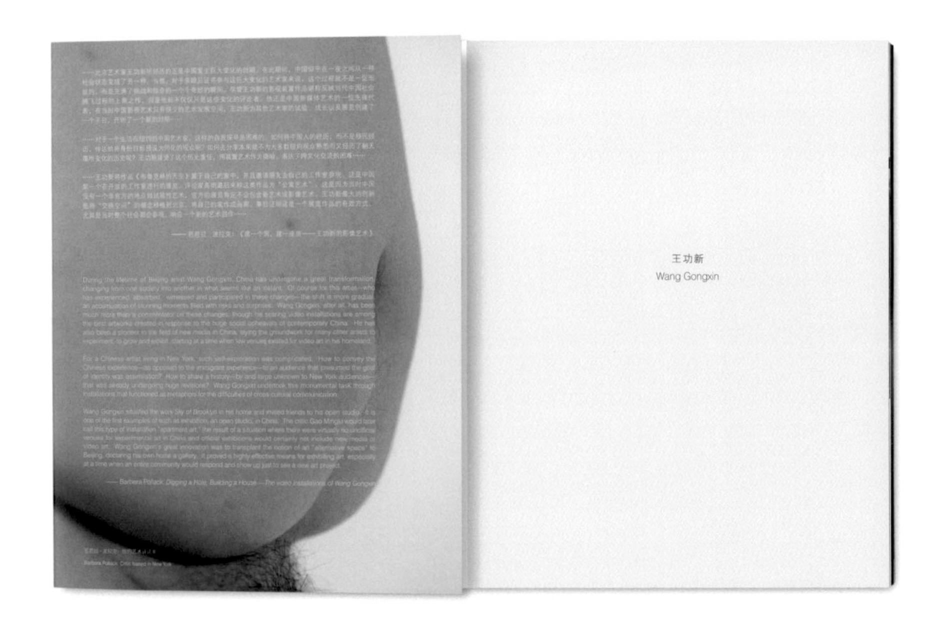

布鲁克林的天空 —— 在北京挖一个洞，1995

影像装置：单频录像，井
彩色，有声
3.5米深
北京东四报房胡同

The Sky of Brooklyn — dig a hole in Beijing, 1995

1-channel Video Installation: TV set and audiotapes, color, stereo
3.5m(depth)
Home of the Artist, Beijing

艺术家在自家的院子里挖了一个几米深的洞，在其中放置一台监视器，播放着自己在纽约布鲁克林的家拍摄的影像。屏幕上除了天空什么都没有。这个作品涉及了东西方文化交流过程中的单向审视问题。晴朗的天空似乎暗示着看客和被看的人的区别。有一个空洞的声音不断的重复着"看什么看？有什么好看的？"这个作品展示了西方看待东方的态度。

——第25届圣保罗双年展

The artist digs a hole several meters deep in his house, and in the hole he plays a videotape that he shot in Brooklyn, New York. On the screen there's nothing but the blue sky. The video work deals with the one-way watching and understanding of the west and the east. The clear sky seems to imply the indifference of those who are being watched towards those who are watching, while a persisting hollow voice seems to indicate the watchers' complicated psychology: "What are you looking at? What is there to look at?" This work expresses the west's watching attitude towards the east.

— the 25th Sao Paulo Biennial

我的太阳，2000
影像装置：3频录像
7分18秒，彩色，有声
该作品以三个数据投影仪制作呈现出一幅变幻的全景。中间的人物是一位处在空旷开阔土地上的老妇人，她不停的伸手去捕捉太阳的光线，得到的却只是抽象的空虚。
My sun, 2000
3-channel video installation
07'18, color, stereo
The light appears as the image of the sun rising and then entering the landscape as a visitation to this world leaving its traditional place of orbit. The light is an illogical form. Time and time again she reaches out to capture the energy, and embraces nothing but emptiness.

关联—与“丫”有关，2010
影像装置：9频录像
彩色，有声
Relating — It's about "Ya", 2010
9-channel Video Installation
color, stereo
9个投影仪播放着一系列以音频“ya”所引发变换的影像，通过电脑程序控制播放引导观众介入，按照艺术家设定的转换顺序依次观看
The 9-channel projectors broadcast a series of videos that start with the sound "Ya". All the videos are controlled by a computer in order to guide the audience to watch them according to the sequence set by the artist

红门，2002
影像装置：4频录像
5分30秒，彩色，有声
Red Door, 2002
4-channel video installation
05'30, color, stereo

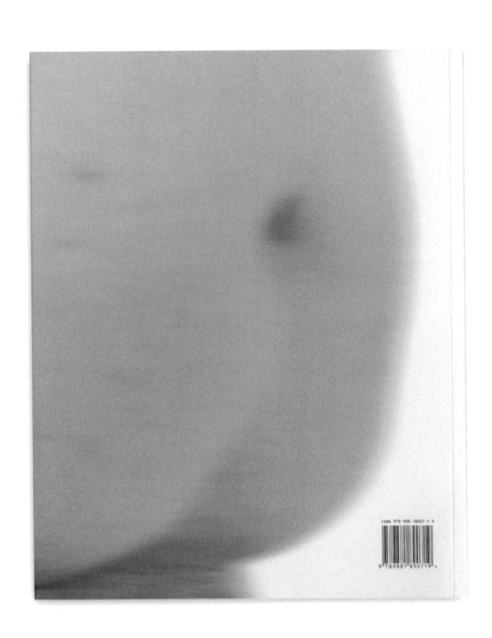
9 789881 890719 >

ZHAN WANG
The New Suyuan
Stone Catalogue
CHARTA

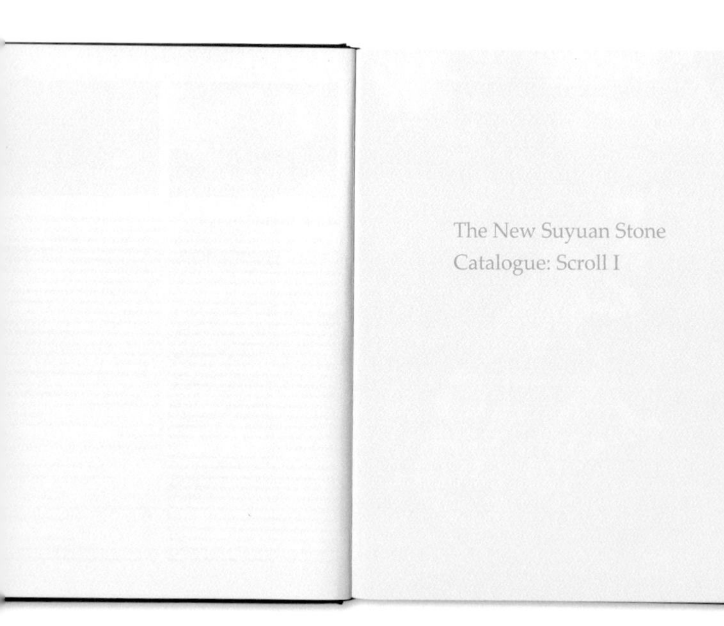
SCROLL I
The New Suyuan Stone
Catalogue: Scroll I
17

SCROLL I

陽氣之精上浮爲星散而隕墜當無形也然隕
時有聲金星之精墜于終南化爲白石彭蠡有
落星石又春秋五石隕越于宋近俞某早朝偶
覩星隕身側遽然有聲在地尚響視如硃砂小
石南都應天府學內有星隕石三塊

韓琮咏

的的墮芊芊蒼茫不見年幾逢疑虎將應逐犯
牛仙擇地依蘭畹題詩間錦箋何時成五色却

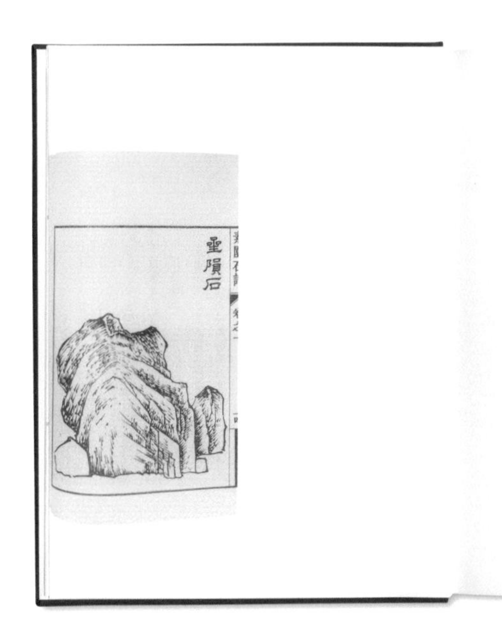

"Today, real life lies hidden in fantasies beneath dazzling surfaces, in stones now made hollow."

— Zhan Wang, 1995

)/1.22 *Artificial Rock No. 30*, 2001
s steel
3 x 138 cm
on of the Busan Museum of Modern Art, Korea
aph by Cho Sean-hyeong

Rocks

e Busan Museum of Modern Art.
t, we see that it was not put there
ir tradition, the Chinese are very
k is placed. At any rate, we need
. Actually, conceptually speaking,
use rocks do not have fixed direc-
rom Western sculpture as well as
Some ancient scholars' rocks have
ld be viewed appreciatively from
orks do not have clear front and
holars' rocks are abstract. Unlike
e of literati tradition, the Koreans'
ntrammeled.
aken inside the museum, we see
oor "at random." But because this
at it was "random." Indeed, how
andom? Therefore this "random"
nvincing! And so it is put there. A
ocess of an art object. This is both
my belabored choice of the front
a scholars' rock becoming "art"
eans were influenced by Chinese
um does not hold religiously to
ndoning traditional aesthetics to

1.24 (page 52) *Artificial Rock No. 19*, 2001
Stainless steel
53 x 43 x 42 cm

1.25 (page 53 left) *Artificial Rock No. 19* in the home of Britta Erickson

1.26 (page 53 right) *Artificial Rock No. 13*, 1999
Stainless steel
183 x 133 x 92 cm
Gallery Loft

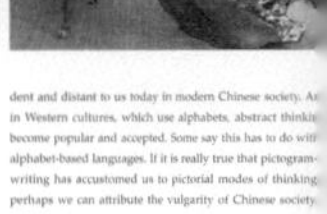

Puppy and Rock

This piece, at the San Francisco home of curator and scholar Britta Erickson, looks like a puppy, regardless of what direction it is displayed (Figure 1.26). We cannot escape from this kind of visual-metaphorical thinking. When thinking metaphorically, we relate everything we see to concrete objects like people, animals, tools, and agricultural products. This thinking negates the abstract thinking natural to humans, and it negates metaphysics. In China, many travel guides to natural sceneries relate ineffable beauties to animals. They can be anything from lions and tigers to rabbits, cats, and dogs, or even agricultural products like rice, potatoes, and bok choi. For a tourist with a bit of imagination, such images do not only drain all life out of the scenery, but also kill the appetite. True, this rock does make us think of animals, perhaps in a way beyond our conscious control, but most rocks are abstract in shape. Yet the popularity of this kind of visual-metaphorical thinking has turned the refined culture of literati rock appreciation into a most commonplace endeavor. The *Suyuan Stone Catalogue* hardly includes these rocks, which the literati probably found tasteless!

Sometimes, however, we have to resort to visual metaphors as mnemonic devices. When I learned to sketch in art school, my teacher taught a method that makes your sketching more accurate: when tackling a complicated area, you can relate it to an animal, and if you only draw this animal faithfully, then you also manage to draw the actual subject accurately and vividly. I myself experienced the magic of this method as a sketching student. This shows that visual-metaphorical thinking is actually quite practical. I suspect that the ancient Chinese pictograms, which "remember" the world by abstracting form, follow roughly the same principle. From this perspective, we see how popular culture promotes art generally, not to mention the dissemination of *jiashanshi* culture. The ancient literati's natural philosophy seems transcendent and distant to us today in modern Chinese society. A
in Western cultures, which use alphabets, abstract thinki
become popular and accepted. Some say this has to do wit
alphabet-based languages. If it is really true that pictogram-
writing has accustomed us to pictorial modes of thinking
perhaps we can attribute the vulgarity of Chinese society
writing system.

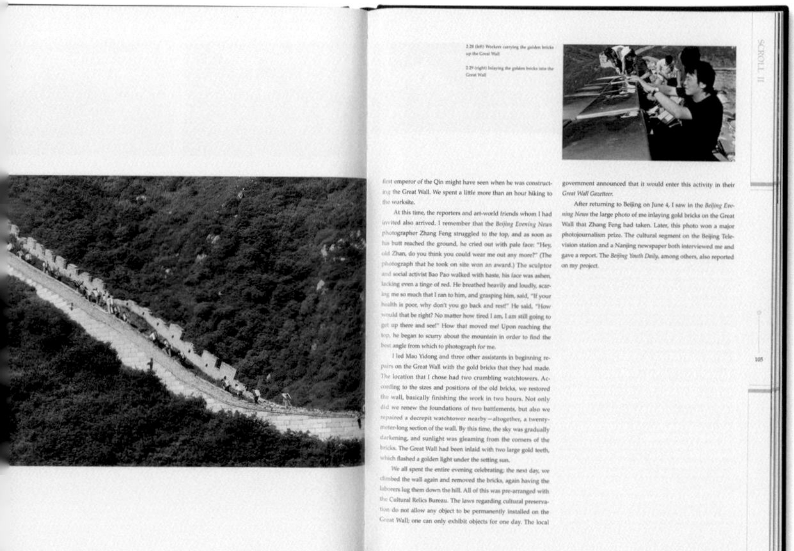

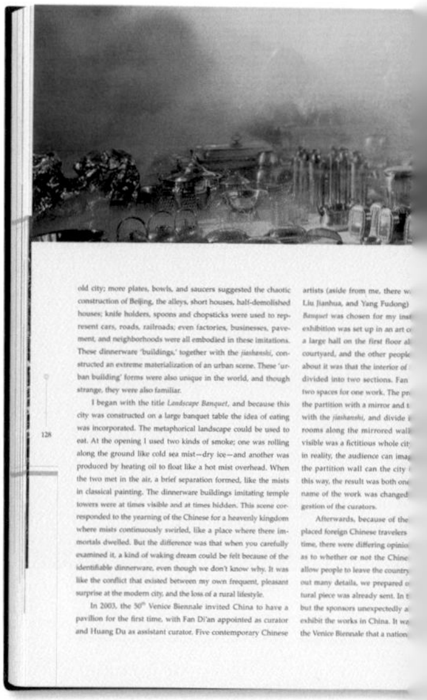

The New Suyuan Stone Catalogue: Scroll II

[illegible] (left) *Floating Rock and Fishy Play*, 1999
stainless steel, natural stone
120 × 250 × 260 cm (two pieces)
Collection of He Xiangning Art Museum, Shenzhen

Floating Rock and Fishy Play

In the traditional bird and flower paintings of the Song and Yuan dynasties, one often finds that landscape features or bird, flower, fish, and insect motifs were given elegant names. This cultural practice also became common in the gardens of the past as a means of establishing a visual source for the strange scenes encountered as one strolled. Such sites were assimilated to those paintings. In the modern city—for example, in some apartment complexes—people attempt to recreate such cultural practices, and they try to build ponds and create artificial scenery. Yet in the end, the times have changed; our eras are different. No matter what one does, that literati feeling is always difficult to summon, and thus, all of these artificial sites remain unintelligible.

Against this background, artists were invited to design works for an artificial lake then under construction for the Second Annual Exhibition of Contemporary Sculpture at the He Xiangning Art Museum in Shenzhen. The theme of the exhibition was defined by the critic Huang Zhuan: "Balanced Existence—Proposals for the Future of an Ecological City." I remember that we were in Beijing's Zhaolong Hotel and were specifically discussing works for this exhibition; finally, we decided on the project for the *Floating Rock and Fishy Play*. The project was based on the fact that this lake originally contained several rocks. I decided to have them find a large piece of rock as they were digging the lake. The assistant that I brought with me then created the work on site. In turn, the stainless steel stone became a float-rock drifting on the surface of the water, while the original stone [on which the stainless steel stone was modeled] was returned to the water. Thus, in the lake there were two pieces of rock, one genuine and one artificial. The real rock would not move. Meanwhile, although the artificial rock was fixed at a point on its bottom, on the water's surface it still moved back and forth, floating freely, creating a contrast of movement and calmness, brightness and darkness, shininess and coarseness, lightness and heaviness. "Fish" travel about bene[ath] the water, while the term "play" certainly does not refer to [the] original rock but instead to the floating stone. From these two [...]tions, the idea of *Floating Rock and Fishy Play* was created.

At that time, the technology for floating was far from perfe[ct]. The rock often took on water, and it even sank to the water's botto[m]. I had to ask divers to salvage it. It is now in the collection of [the] OCT Contemporary Art Terminal opened by the He Xiangni[ng] Art Museum.

[...] also formally published as the [...]bassy.
[...]se Embassy also held an open-[...]enice Biennale, and one at the [...]he scale of the exhibition was [...]he original plan that was to go [...]d in two exhibition halls with [...]eaning rhymed closely enough [...]scape scene was in the image [...]ely has mountains in the city [...]n, this exhibition became very [...]ewspapers reported this pecu-[...] also had differing views about [...]s of art and politics, giving rise [...]en I saw the famous Japanese [...]ught up this exhibition in par-[...]ut it.
[...]n exhibition ended, the exhibi-[...]ng of the curators to the Cen-[...]n in the capital. Here, I created [...]once more. The exhibition was [...]e fastest I ever installed *Urban* [...]articular exhibition space, the
[...]exhibited two more times in [...]new residential area in Beijing, [...]state Expo in a contemporary [...]cene of Shenzhen city and real estate advertisements, and models from the development business surrounded my piece, making it appear satirical.

Whenever I stood installing these dinnerware "cities," I always felt like a child standing among toys and idly playing with them, drunk off of the joy of controlling the world in my palm. I remember when I was between two and six years old, my favorite game was using the various things on a table, such as a matchbox, cigarette box, brush container, clock, pen, vase, and other objects, as well as a few building blocks, to set up complicated houses through which I pretended to drive the matchbox car. I could go an entire day without eating or drinking, kneeling on my stool designing my 'landscape' on the table. When it was perfectly set up, I would push it all over with a crash, then build a new 'scene,' and then knock it over again, over and over, as if I had forgotten that time existed at all. Could it be that my behavior now is a continuation of that hobby? Or is that just the form of life itself?

One day two years later, I suddenly received a letter from the Chicago Museum of Contemporary Art (MoCA) inviting me to participate in a large international art exhibition in 2005. The curator was the head curator of the 2003 Venice Biennale, Francesco Bonami. It seemed as if he wanted to solve the pitiable events of those works that could not leave the country. From that point on, *Urban Landscape* began an international tour. Luckily, I had just gotten married, and my newly wed wife accompanied me while traveling with *Urban Landscape*. Because this piece had to be installed anew on each site, my accompanying wife became my assistant as well as my translator for on-site installation. She

展望 新素园石谱

CHARTA，米兰；长征空间，北京，2011

280mm×220mm×13mm，144 页

布面精装，外裹纸质护封，2000 册

英文编辑：田霏宇、贝安吉

序言：巫鸿；前言：展望

文本写作：展望、黄笃、马钦忠、Maki Yoichi、高岭、孙振华、高名潞、张颂仁、邹跃进、张朝晖、栗宪庭、凯伦·史密斯、巫鸿、舒可文、黄燎原、冷林

徐勇 **十八度灰**

东八时区，香港，2010

267mm×352mm×22mm，178 页

布面精装，裹纸质护封、半透明纸腰封（未展示）

每幅作品之前装订 18 度灰色薄隔页纸，上印作品说明

1500 册

文章：舒阳

访谈：满宇 / 徐勇

腰封文字：艾未未

作品都属个人记忆与体验，均要对真实场景直接摄影完成。

These works, which are from purely personal memory and experiences, are shot and completed at once at the real sites.

徐勇访谈

时间：2010年6月22日

地点：望京新城

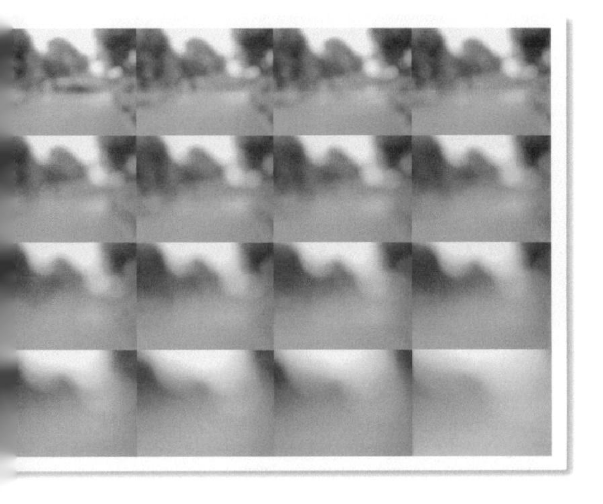

彭斯 **异世同流**

河北教育出版社，石家庄，2010

350mm×250mm×28mm，218 页

纸面精装，书脊包布

2000 册

前言：徐天进

文章：彭锋、理查德·舒斯特曼、库蒂斯·卡特、吴钊、简应隆

对话：徐天进、彭锋、齐东方、陈芳、彭斯

访谈：罗忠学 / 彭斯

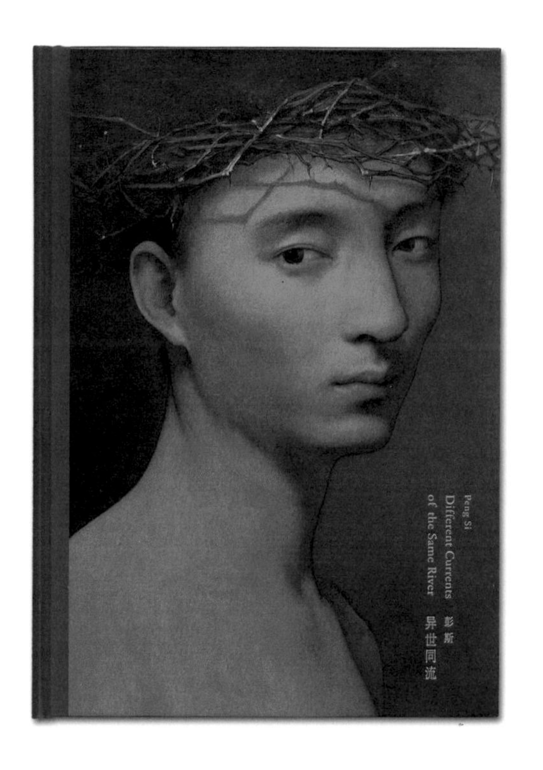

50
石倚 Leaning on Taihu Rock, 140 × 140cm, 2005, 布上油画, oil on canvas
51

曹子建来到洛水之畔，有感于宋玉的《神女赋》，思念自己心仪的美人，写下千古名篇《洛神赋》，亦真亦幻。

洛神宓妃在曹子建妙笔下「翩若惊鸿，婉若游龙。」「仿佛兮若轻云之蔽月，飘飖兮若流风之回雪。」洛神女子之风貌，引得后世无数文人雅士的「折腰」。

爱慕子建，也爱慕子建笔下的洛神。今岁去洛阳便不由得去了洛水之畔，流水无情，望着那水波荡漾的洛水，怀想着子建「背伊阙，越轘辕，经通谷，陵景山。」来到洛水之畔，做了那一场绮丽清梦，梦因着《洛神赋》还在，而那洛神女子的倩影似乎至今还袅袅不绝地飘忽在洛川的云水茫茫之间，山形大概依旧着，不曾记下了诗人的足迹，也无从觅得诗人的点点滴滴，诗人带着梦来，揣着梦离去，我却想把我的梦的种子撒播在这洛水之上，这山岗之间，水无情，毕竟东流去，唯有这山冈谷陇，似乎还存留着诗人一缕缕哀叹和那一场清丽的梦。

——彭斯

102
何处乡关 Where is My Hometown, 180 × 190cm, 2008, 布上油画, oil on canvas
103

118
落雪寒枝 Snow on Barren Branches, 80 × 80cm, 2009, 布上油画, oil on canvas
119

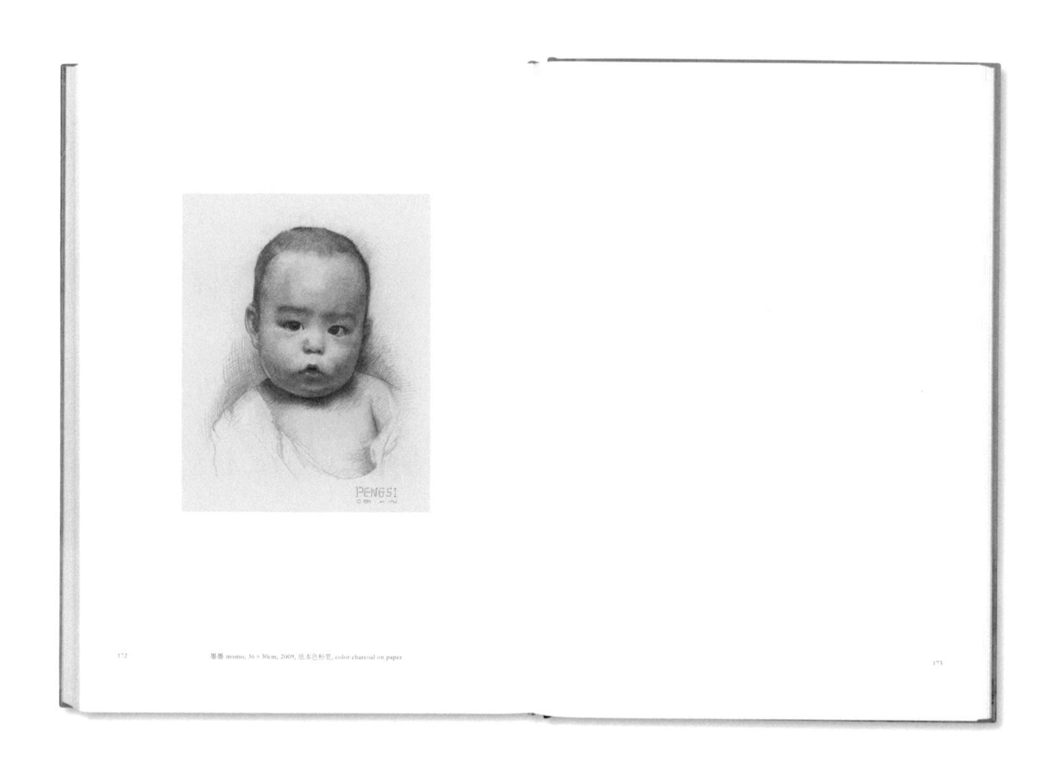

十年曝光——中央美术学院与中国当代影像

上海锦绣文章出版社，上海，2010

242mm×235mm×25mm，264 页

布面精装；2000 册

序言：潘公凯、谭平

文本写作：姜节泓

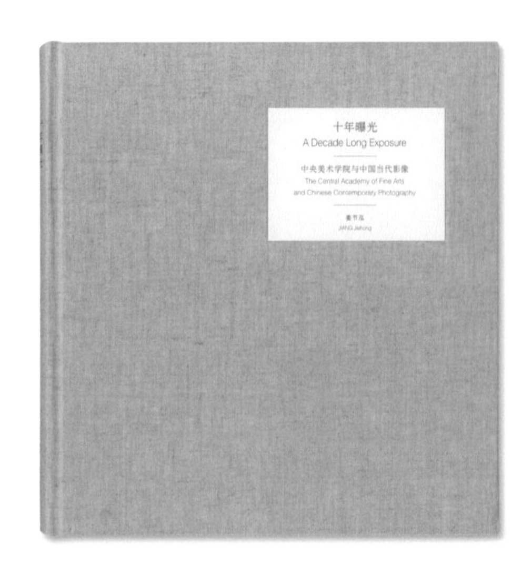

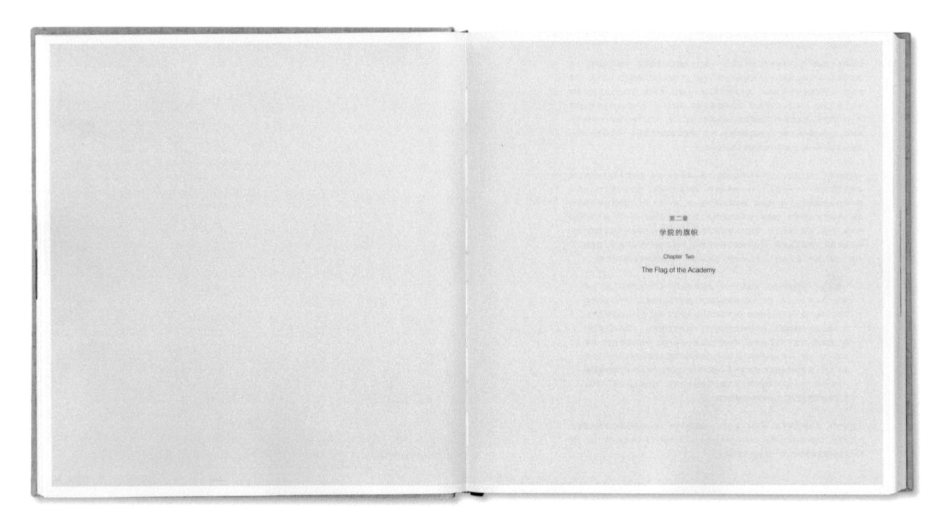

二零零七年开始，迟鹏开始了他自己的"西游记"，那个假借明代吴承恩的同名长篇小说为叙述框架的旅行。《西游记》的魅力不仅仅出于吴承恩，也出于它无数的读者，不仅仅出于写作，也出于阅读，出于中国人最浪漫最迷人的想象力。其实，真正在读者记忆深处的不是历史中唐玄奘的使命，也不是有没有取回真经的事实，而是那个美猴王孙悟空——他的神通广大，他的无畏和反叛精神。迟鹏在他的《西游记》系列中挪用了原作小说中的情节片段——从"五行山"、"大闹天宫"，到"三打白骨精"、"女儿国"——自己扮演了主角孙悟空，实施了一个双重演绎，一方面是那个浪漫神话故事，另一方面则是自己的成长经历。然而，正当传统故事中的玄幻世界与现实语境重叠的时候，当下社会岂不正是那个充满鬼魔、欺骗和陷阱的征程？

孙悟空对于迟鹏，意味着"一种记忆"，他眼中的现实呈现出"畸瘤的进步"，是一种"精神上后退"的现实。"社会的真实性在哪里？真假模糊，是非不辨，漠视恶行。社会需要一种正义的精神，而悟空就是那种精神。"[72]有意思的是，真正感染迟鹏的《西游记》并不是小说原著文本的《西游记》，而是电视剧、动画片里的《西游记》和孙悟空。换句话说，迟鹏的《西游记》不是完全出自吴承恩的一手文字，而是现代生活中某个导演、演员、摄影二手的视觉演绎：不是从文本到图像，而是从图像到图像的转换。但是，这不会给他带来任何限制，相反，许多更直接更自由的视觉想像。因为他影像中的孙悟空依然火眼金睛疾恶如仇，依然披荆斩棘驱妖除魔，却在一个没有紧箍咒的现实中。

王庆松许多早期影像的批评性也可以说是基于一种"模仿"，基于一种对美术史上的经典作品的再演绎——从中国美术史上的经典绘画如《韩熙载夜宴图》、《步辇图》，到西方名作《泉》《三女神》等，现实中的人物和模特替换了原作中的形象。在这种"模仿"的过程中，在历史与当下的比对之外，形成对于现实生活的批评。

> 中国一直在改变，从过去到现在，她真的改变了吗？摄影对于文学性、批评性来说是个很好用的工具。不管是摆拍还是抓拍，不管是编造还是纪实，摄影更能够抓住社会节奏的变化，很容易让人认为是真实的。大跃进和文革时期的视觉宣传中，浮夸造成了很多虚假，摆拍的现场被大多数老百姓当作事实相信了。而今天社会中的浮夸，通过镜头，同样造成了很多假像。"[73]

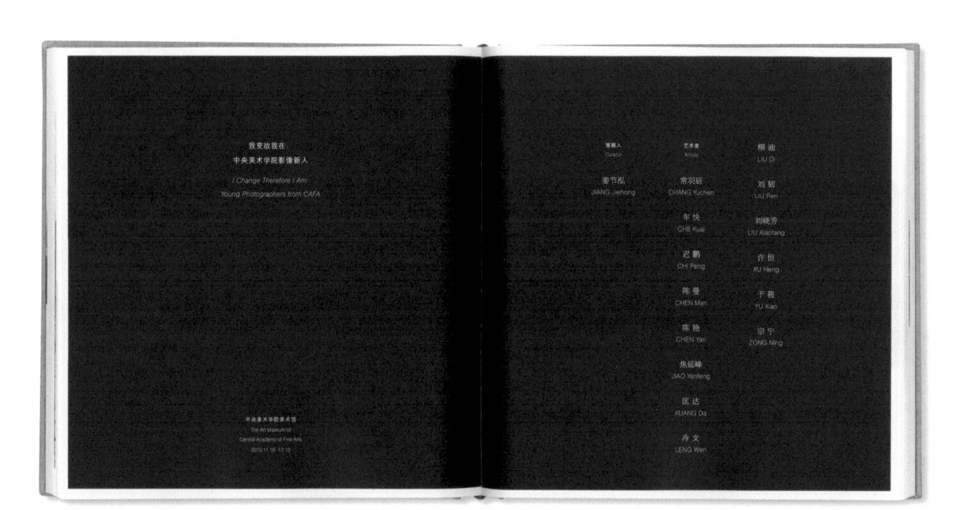

我变故我在
中央美术学院影像新人
I Change Therefore I Am:
Young Photographers from CAFA
中央美术学院美术馆
The Art Museum of
Central Academy of Fine Arts
2010.11.19–12.10
策展人
Curator
姜节泓
JIANG Jiehong
艺术家
Artists
常羽辰
CHANG Yuchen
车快
CHE Kuai
迟鹏
CHI Peng
陈曼
CHEN Man
陈艳
CHEN Yan
焦延峰
JIAO Yanfeng
匡达
KUANG Da
冷文
LENG Wen
柳迪
LIU Di
刘韧
LIU Ren
刘晓芳
LIU Xiaofang
许恒
XU Heng
于筱
YU Xiao
宗宁
ZONG Ning

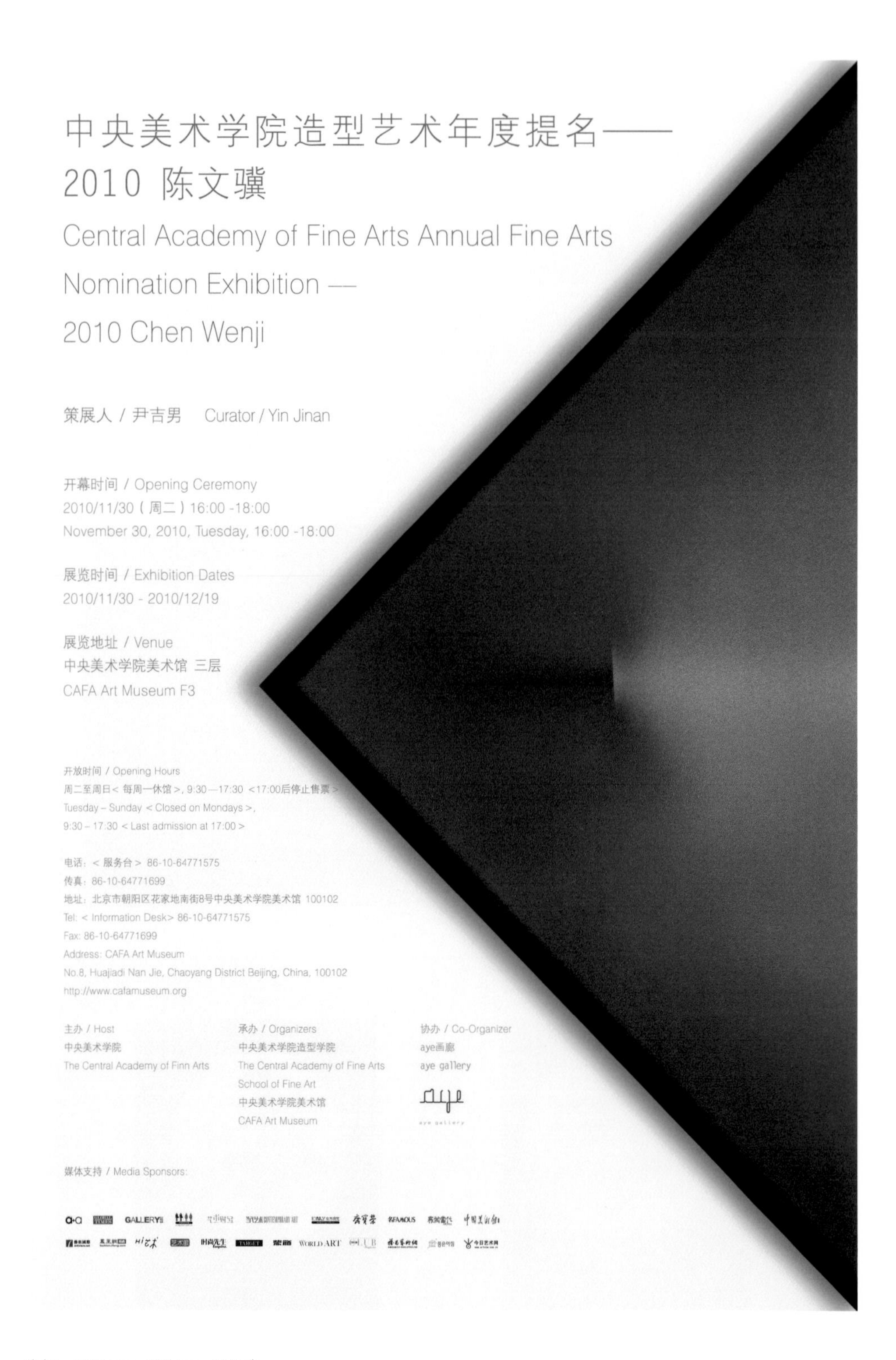

中央美术学院造型艺术年度提名——
2010 陈文骥
Central Academy of Fine Arts Annual Fine Arts
Nomination Exhibition —
2010 Chen Wenji
策展人 / 尹吉男 Curator / Yin Jinan
开幕时间 / Opening Ceremony
2010/11/30（周二）16:00 -18:00
November 30, 2010, Tuesday, 16:00 -18:00
展览时间 / Exhibition Dates
2010/11/30 - 2010/12/19
展览地址 / Venue
中央美术学院美术馆 三层
CAFA Art Museum F3
开放时间 / Opening Hours
周二至周日 < 每周一休馆 >, 9:30—17:30 <17:00后停止售票>
Tuesday – Sunday < Closed on Mondays >,
9:30 – 17:30 < Last admission at 17:00 >
电话：< 服务台 > 86-10-64771575
传真：86-10-64771699
地址：北京市朝阳区花家地南街8号中央美术学院美术馆 100102
Tel: < Information Desk> 86-10-64771575
Fax: 86-10-64771699
Address: CAFA Art Museum
No.8, Huajiadi Nan Jie, Chaoyang District Beijing, China, 100102
http://www.cafamuseum.org
主办 / Host
中央美术学院
The Central Academy of Finn Arts
承办 / Organizers
中央美术学院造型学院
The Central Academy of Fine Arts
School of Fine Art
中央美术学院美术馆
CAFA Art Museum
协办 / Co-Organizer
aye画廊
aye gallery
媒体支持 / Media Sponsors:
GALLERY
WORLD ART

海报，1000mm×700mm，200 张

陈文骥 **中央美术学院造型艺术年度提名——2010 陈文骥 /**

陈文骥油画廿四年 1986—2010

上海人民美术出版社，上海，2010

307mm×250mm×40mm，314 页

布面精装，裹纸质护封

2000 册

序言：苏新平

前言：潘公凯

文章：尹吉男、安静、陈婧莎、斯然畅畅

评论节选：栗宪庭、吕胜中、殷双喜、冯博一、尹吉男、宋晓霞

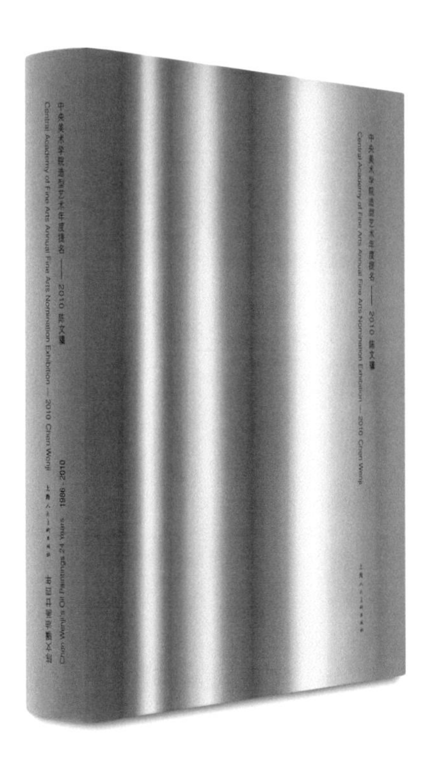

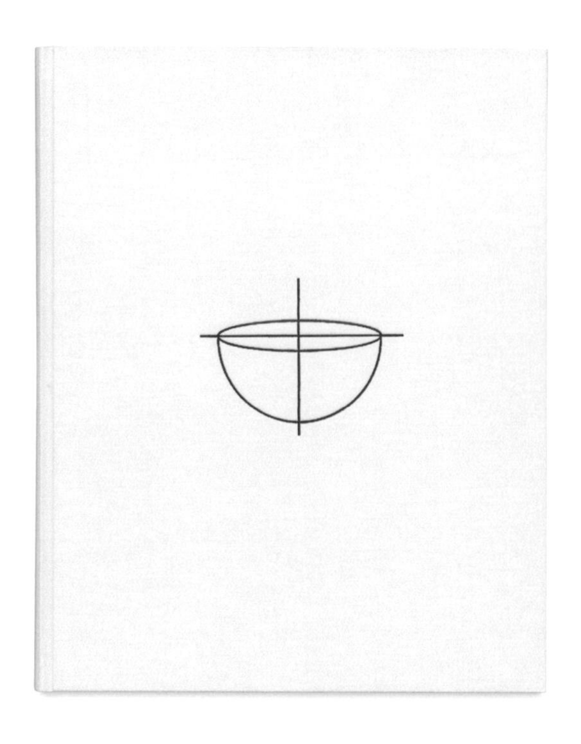

中央美术学院造型艺术年度提名 — 2010 陈文骥
Central Academy of Fine Arts Annual Fine Arts Nomination Exhibition — 2010 Chen Wenji

10

造型艺术年度提名委员会（按姓氏笔划排序）

王璜生 王少军 王华祥 刘小东 吕胜中 吕品昌 苏新平 展望 徐冰
高天雄 隋力 隋建国 谢东明 谭平

策展人 尹吉男

The Committee of Central Academy of Fine Arts Annual Fine Arts Nomination

Wang Huangsheng Wang Shaojun Wang Huaxiang Liu Xiaodong Lu Shengzhong [illegible]

项目负责：齐鹏
展览执行：马亮 吴鹏 徐世存 李英成
行政执行：孟子燕
教育推广：董慧萍
视觉设计：伊文
书籍设计：何浩

Project Executive: Qi Peng
Exhibition Implementation: Ma Liang, Wu Peng, Xu Shicun and Li Yingcheng
Administrative Implementation: Meng Ziyan
Education and Marketing: Dong Huiping
Visual Design: Yi Yi
Book Design: He Hao

11

展览时间
2010/11/30 — 2010/12/19
展览地点
中央美术学院美术馆三层
主办
中央美术学院
承办
中央美术学院造型学院
中央美术学院美术馆
协办
a.m画廊

Exhibition Dates
2010/11/30 — 2010/12/19
Venue
CAFA Art Museum F3
Host
[illegible]

中央美术学院造型艺术年度提名 — 2010 陈文骥

Central Academy of Fine Arts Annual Fine Arts Nomination Exhibition — 2010 Chen Wenji

CAFA
Art
Museum

14

色彩总是那么单纯而又鲜亮。早期的色彩是为具象写实服务的，而近期的色彩是和抽象理念相互匹配的，并沉迷于那种变化的微妙。从物体到物象的变化，在陈文骥的艺术中体现的极为明显。那些微弱的抒情因素已经被平静的叙述所冲淡。这是2009年11月我在东站画廊看陈文骥的新作时所得到的初步印象。

陈文骥一直独立于任何艺术运动与潮流之外。他和大多数出生于20世纪50年代的中国艺术家不同，他没有参加20世纪80年代中期的新潮美术运动。但他一直在创作令人印象深刻的作品，被批评家栗宪庭称之为“一种静悄悄的革命”。

无论是毛泽东时代的画家，还是后来的新生代画家，都是以人物画为中心的。人们经历了精英化的政治生活和平民化的经济生活，从集体经验到个人经验的转变，明确地表现在艺术中。人物是评价和表达的对象。在毛泽东时代，以人物为中心即是以政治为中心，国家主义和集体主义是当时艺术的主要特征。而到了新生代艺术家那里，政治中心被解构了，但人物被置换了，生成了以自我和个人为中心的自传性叙述。

实际上，在毛泽东时代的艺术和新生代艺术之间有两个重要的衔接，那是在中国影响深远的两个重要潮流：前者是以吴冠中先生为代表的形式主义美术，后者是85新潮美术运动。吴冠中的贡献十分明显，他通过唯美的形式语言消解以往的政治主题，让人物在画面里缺席，尽管他的绘画是抒情性质的。与吴冠中同步的形式主义美术还有云南画派（丁绍光、蒋铁峰等），学院派的写实油画也从政治生活退回到平静的画室和家庭，衍生了唯美的人物画作品。而后来的20世纪80年代的新潮美术家则用西方化的文化观念来取代中国习有的政治观念。无论是唯美的人物画，还是形式主义的风景画和静物画，是流行于中国20世纪80年代抒情模式，这与流行于20世纪80年代后期的新潮美术的评论模式是不同的。

人物的缺席是陈文骥绘画中最显著的特征。与他同辈的写实画家如杨飞云、王沂东强调的是人物的在场（画室或家中），尽管已经远离了政治生活。新生代艺术家仍然强调人物的在场，个人生活多于集体生活。新生代的艺术是以自我为中心的人物画，可以认为是一种超越了评论模式（政治主题或观念主题）和抒情模式的叙述模式。

陈文骥的绘画具有高度的混合性，形式语言的写实性和观念的抽象性被统一在一个空间里。他呈现的混合性是多重的：人与物的混合性，历史与符码的混合性，物质与意味的混合性。这种混合性正是以Lee Ambrozy女士所指出的“二元性”为基础的。更重要的是，陈文骥的作品也同时具有人的缺席与在场的混合性。

必须看到，独立于任何潮流之外的陈文骥，为我们贡献了一种新的模式，即场景与物象的叙述性模式，更准确地说是一种观念意味的写实形式，打破了抽象与具象的形而上学界限。这是他与秉承抒情模式从事写实绘画的吴冠中最本质的区别。

Colors are always so pure and vibrant. Early color in Chen Wenji's works was at the service of his realistic depictions; however, recently color has become an accessory to his abstraction, and he is infatuated by its subtle changes. [illegible]

陈文骥及其艺术的“混合性”

尹吉男

Chen Wenji's Art and Synthesis

Yin Jinan

15

物体性与极少主义

1967年6月，艺术批评家迈克尔·弗雷德（Michael Fried）在《艺术论坛》（Artforum）上发表了其重要论文《艺术与物体性》（Art and Objecthood）。在这篇论文中他试图改变我们对于观看行为的看法。他在这篇文章中以批判性的方式探讨了贾德（Donald Judd）、斯特拉（Frank Stella）和莫里斯（Robert Morris）等艺术家的富有挑战性的新作。这些人也就是格林伯格后来称之为的"极少主义者"，而弗雷德却将他们蔑称为"大白话主义者"（literalists）。因为他们对物品进行"如实"的阐释，并且强调整体性和三维空间中的"特定物体"。他的理论与现代主义者一脉相承，即认为艺术的基本前提是其非物体性。而陈文骥的一些近作就与这个时期的极少主义艺术家的形式主义气质有类似的地方。在下文中我将继续探讨弗雷德的争论，特别是他对于艺术的"剧场性"的否定。

弗雷德看到了"大白话主义"和现代主义绘画在两方面的对峙：第一，几乎所有绘画的关系性特征（即这样的观念：同一件作品中的不同元素共同创造着意义），也就是格式塔（gestalt）。第二，绘画幻觉感的不可逃避性。大白话主义艺术家通过三维空间创造了"特定的物体"，他们"强调整体、整一和不可分割的价值，艺术作品要尽可能地靠近观者，成为'一个物体'、'一个特定的物体'"。弗雷德发现了以形状作为基本特征的物体和以形状作为绘画媒介之间的矛盾（在陈文骥的艺术中，画布的形状是一个十分重要的方面），并且他还发现，大白话主义者所拥护的物体性（作为艺术的物体）和现代主义者认为的绘画要"挫败或者克服自身的物体性"的概念之间存在着矛盾。同样，

stress my personal ideas, or force myself to complete some kind of self-perfecting principals. I don't wish for my works to be merely epistemological vehicles."[4]

Objecthood and Minimalist Art

In June of 1967, art critic Michael Fried published the essay "Art and Objecthood" in Artforum magazine, and it was significant for the way it changed the way that we perceive the act of viewing. His essay critically examined the challenging new works by artists such as Donald Judd, Frank Stella and Robert Morris, later termed "minimalists" by Clement Greenburg, but whom Fried termed "literalists" for their literal interpretations of objects and focus on "wholeness" and the creation of "specific objects" in a three-dimensional space. His criticism takes its root in the modernist notion of art, which is that the primary condition of art is its being defined by its non-objecthood. Chen Wenji's recent work share some formalist qualities with minimalist artists of this period, and through examining Fried's argument, specifically his discussion of the "theatricality" of art, I will expound on how this relates to Chen's later works.

Fried identifies the literalist case against modernist painting on two primary counts: first, the relational character of almost all painting (the notion that different elements within a single artwork work together to create meaning), also known as the gestalt, and secondly, the inescapability of pictorial illusion. Through their tendency to utilize three-dimensional space to create "specific objects," literalist artists "assert the values of wholeness, singleness, and indivisibility—of a work being, as nearly as possible, 'one thing,' a specific object."[3] Fried finds a contradiction in shape as a fundamental property of objects and shape as a medium of painting (the shaped canvas also a significant element for Chen Wenji's works), and he identifies the incongruity between the literalists' espousal of objecthood (objectsas art) and the mission of modernist painting to "defeat or suspend its own objecthood." A similar

4 Peng Jun, [illegible] — An Prelude to Chen Wenji, in Art Reach magazine, vol 1, 2008, pg 30.

3 迈克尔·弗雷德：《艺术与物体性》[illegible]

3 Michael Fried "Art and Objecthood" (1967) in Art and Objecthood, Chicago: University Press, 1998, pg [illegible]

陈文骥的艺术也涉及到物体性的概念，对他来说，画布的形状是作品的重要构成部分（例如他的作品《涵·九识》），尽管他利用三维空间的方式不同于极少主义者，但是他的作品却通过幻觉效果而具有了三维特征，这是他创作的一个重要方面。

弗雷德认为，物体性与现代主义的艺术观念相对立，因为它追求"一种新的剧场，而当前，剧场就是对于艺术的否定"。同样，剧场性的问题对于我们如何看待陈文骥的作品来说至关重要。他的作品被放置在展览环境这个体制之中，也必定包含了观者的体验，这种体验不像现代主义绘画那样传递出了一种作品内的整体系统感（无论是图画的还是非图画的）。和极少主义者或者大白话主义者类似，陈文骥的作品是空间的功能、光线和观者的视域综合作用的结果。这也就是弗雷德所反对的"剧场性"。然而，陈文骥的作品与贾德等例证性的极少主义者的作品不同，贾德利用了大型的雕像般的材质给他的作品增加了一种"舞台效果"，以不同的方式和空间关系角度与观众进行互动。而陈文骥则不然，他的作品在视觉上已经提前给观者设定了位置，减弱了极少主义作品所追求的互动特征。但这并不是说陈文骥的那些带有幻觉感的作品与展览空间无关，因为作品的特定布置和对空旷空间（以纯白色的墙为背景）的需求是一致的。例如，作品《又一切》之所以能够取得幻觉的效果就是因为被安放在了展览空间的角落里。

拓展绘画的界限，并且将绘画带入三维空间也是极少主义者的追求，贾德说："从本质上来讲，真实的空间比画面所营造的空间更加有力，也更加具体。"因此，他追求创造一种所谓的"特定物体"，即那种不同于雕塑的三维艺术作品。陈文骥的物体绘画所营造的虚空间与之有异曲同工之处。弗雷德指出，由于这些特定物品所暗示的"在场"，在大白

sense of objecthood is apparent in Chen Wenji's works that take the shape of the canvas as a fundamental property of the work, such as in the work *Nine Facets of Continess*, although not the same use of three-dimensional space as the minimalists, Chen's works imply three-dimensionality through affects of trompe l'oeil, which is a significant element in his creations.

Objecthood, to Fried, is antithetical to modernist notions of art because it is a plea for a "new genre of theater, and theater is now the negation of art."[5] This theatrical viewing aspect is likewise pivotal to how we appreciate Chen Wenji's canvases, an experience that includes a situational location within an exhibition environment and necessarily includes the beholder—this is distinct from modern painting, which offers an entire system (pictorial or non) within the work. Similar to the minimalists or literalists, Chen's work is the collective result of the various functions of space, light, and the viewer's field of vision. This is precisely the "theatricality" to which Fried is opposed. However, unlike illustrative minimalist works, such as Donald Judd's, whose use of large-scale, statuesque sculptural-like materials gives them a "stage presence" and interact with their beholder in different ways and from different relational angles in space. Chen's visual illusion pre-determines one specific location for viewing, reducing the interactive properties of space that minimalist work exploits. This is not to say that Chen's trompe l'oeil is immune to exhibition environs. The relative placement of his works, and their preference for spacious, pure whitewalls is obvious. In the case of *Neo-All*, the success of the illusion relies on this two paneled work's placement in a corner.

Expanding the finite limits of painting and bringing in the third dimension was also a goal of the minimalists. According to Judd, "Actual space is intrinsically more powerful and specific than paint on a flat surface," and he strove to create what were called "Specific Objects"—three dimensional artworks that were something other than sculpture. This is similar to the virtual space created in Chen Wenji's object paintings. Owing to the implied "presence" of these specific objects, Fried suggests

6 同上，第[illegible]页。

5 Ibid, pg [illegible]

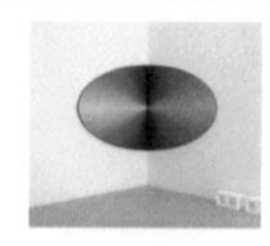

又一切（二联画）
Neo-All (Diptych)
[illegible]

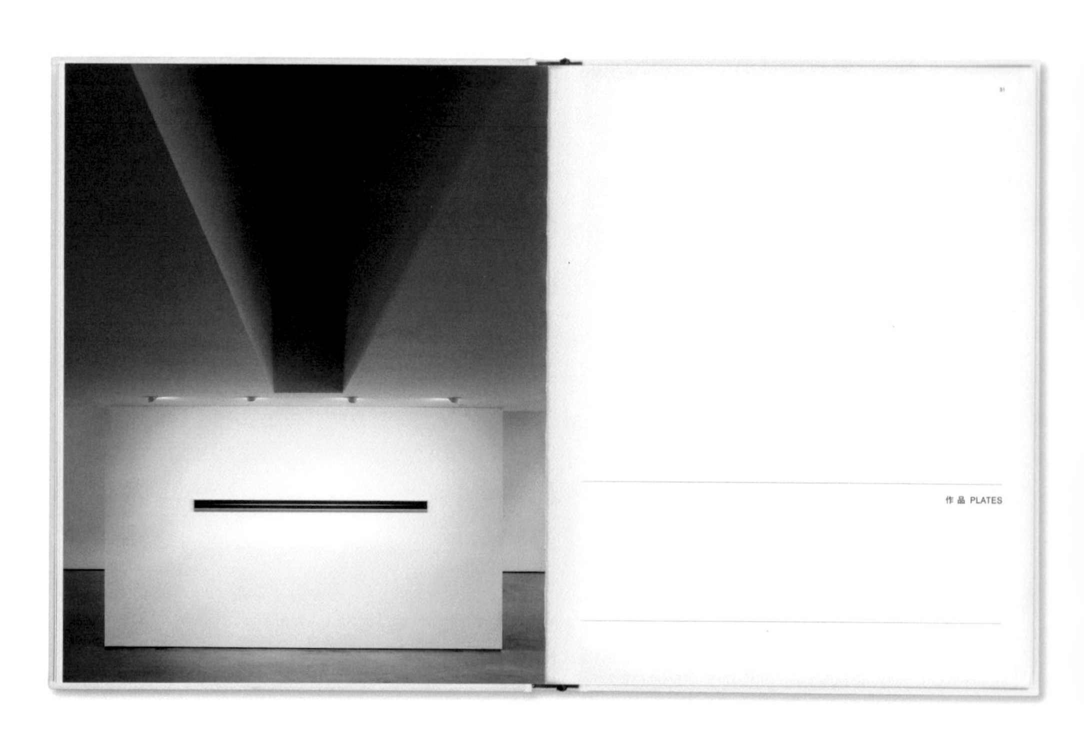

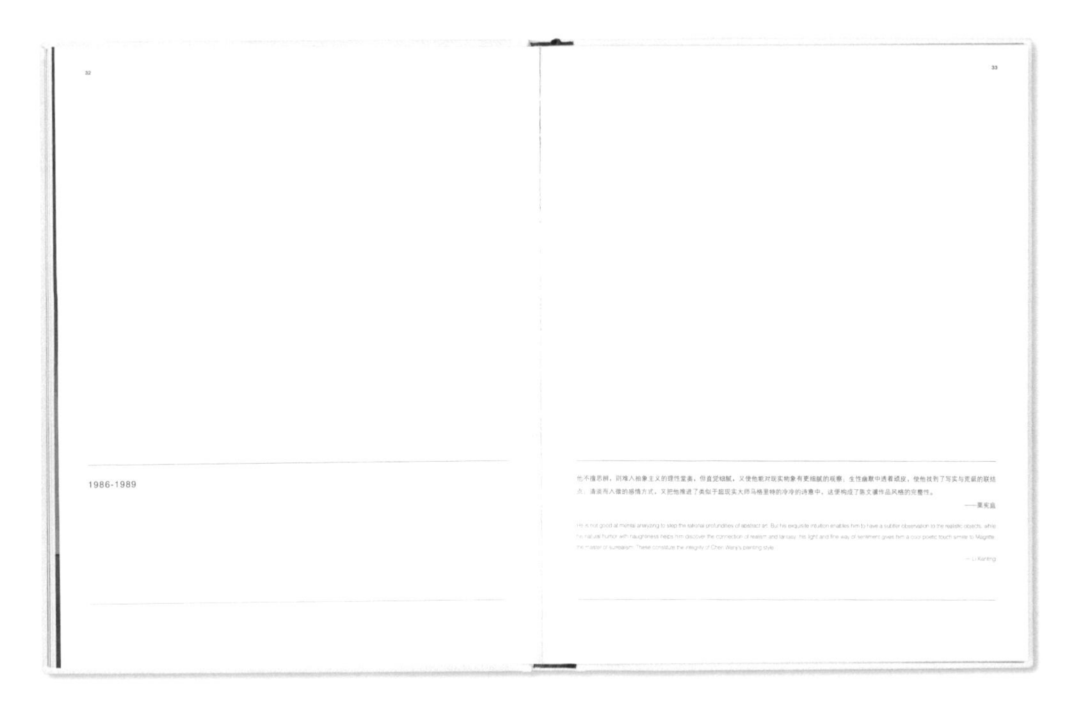
32

1986-1989

33

他不擅思辨，则难入抽象主义的理性堂奥，但直觉细腻，又使他能对现实物象有更细腻的观察；生性幽默中透着顽皮，使他找到了写实与荒诞的联结点；清淡而入微的感情方式，又把他推进了类似于超现实大师马格里特的冷冷的诗意中，这便构成了陈文骥作品风格的完整性。

——栗宪庭

He is not good at mental analyzing to step the rational profundities of abstract art. But his exquisite intuition enables him to have a subtler observation to the realistic objects, while his natural humor with naughtiness helps him discover the connection of realism and fantasy; his light and fine way of sentiment gives him a cool poetic touch similar to Magritte, the master of surrealism. These constitute the integrity of Chen Wenji's painting style.

— Li Xianting

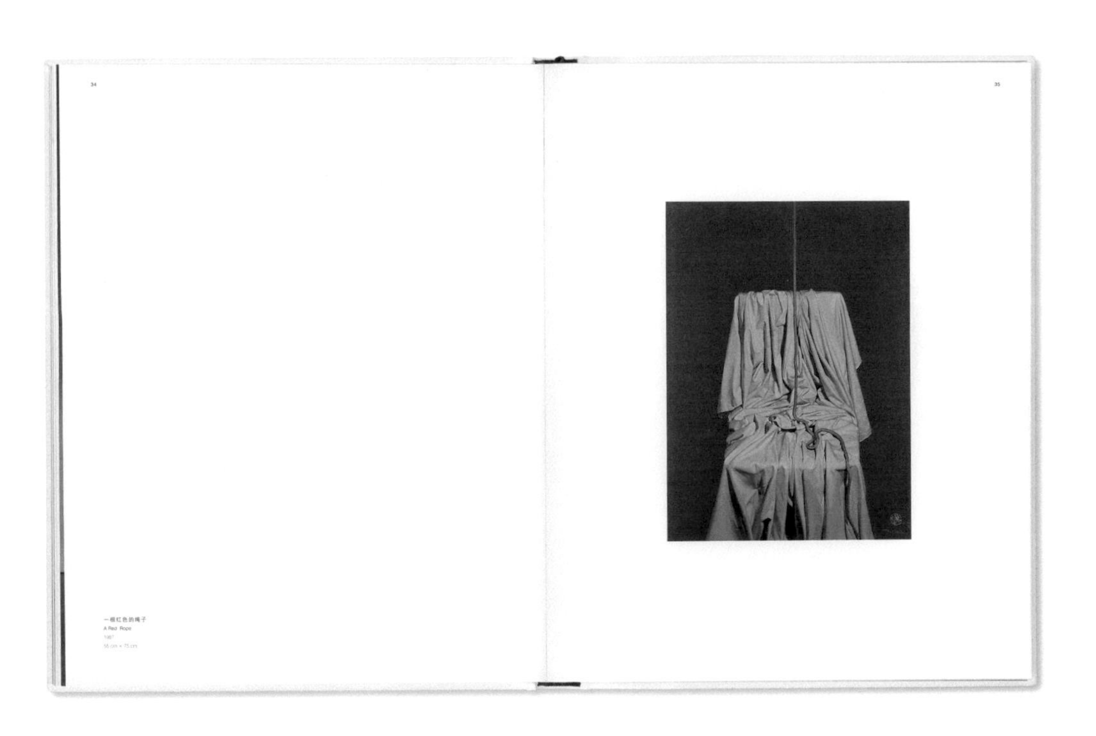
34

一根红色的绳子
A Red Rope
1987
55 cm × 75 cm

35

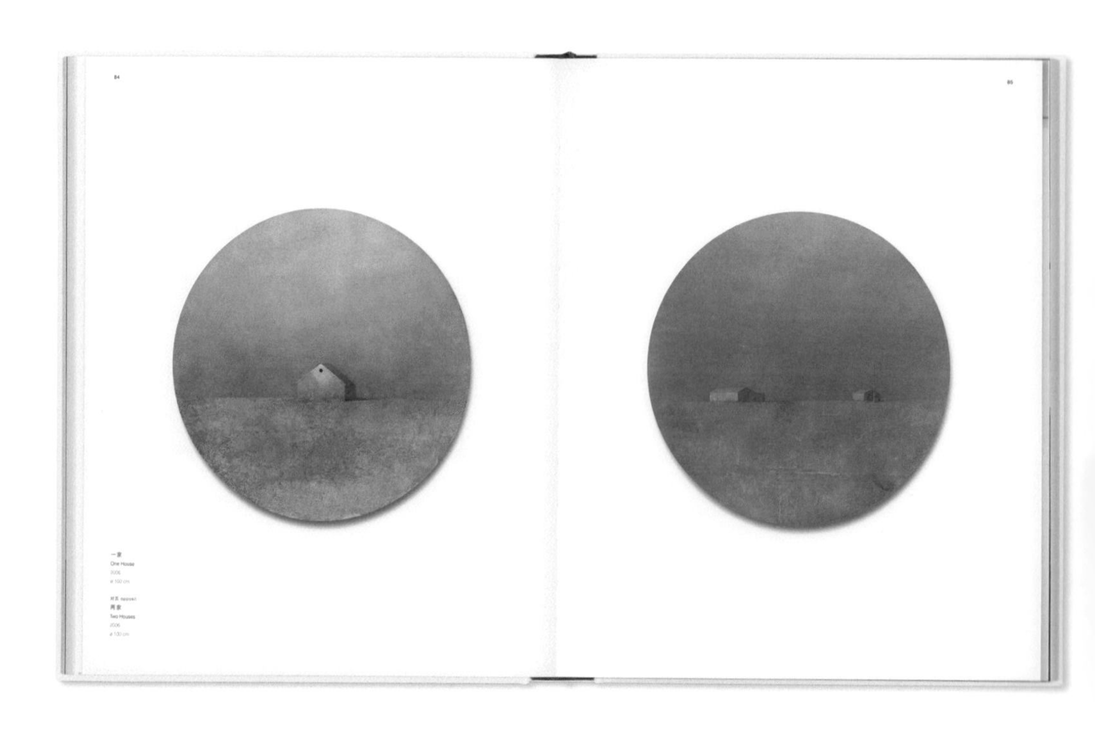

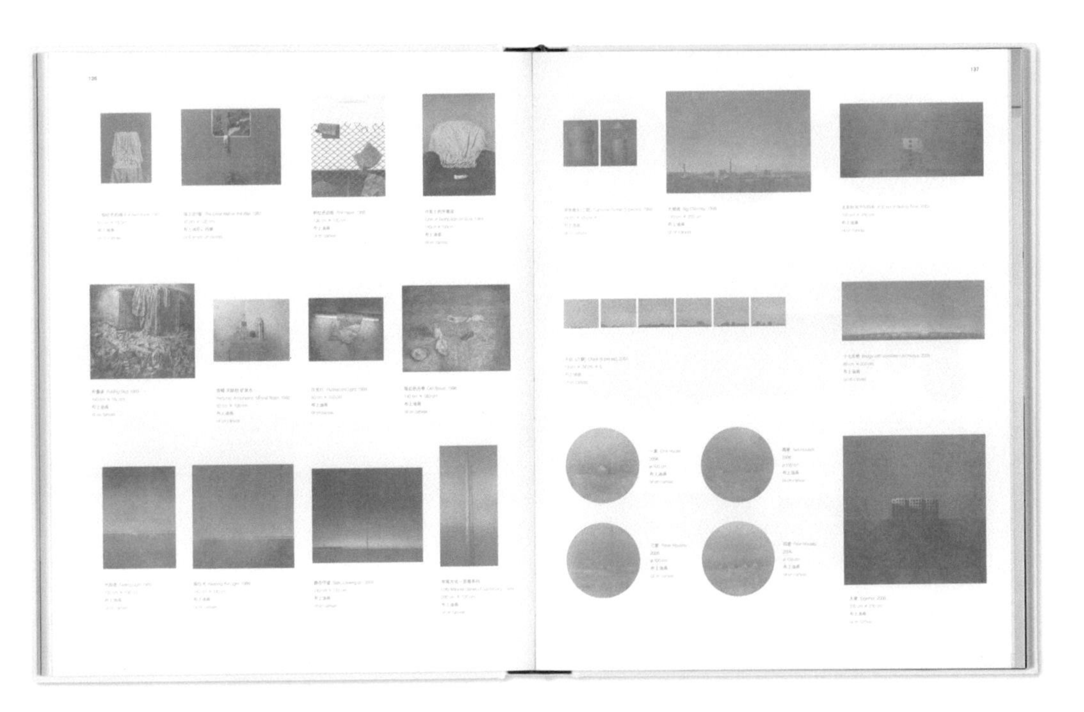

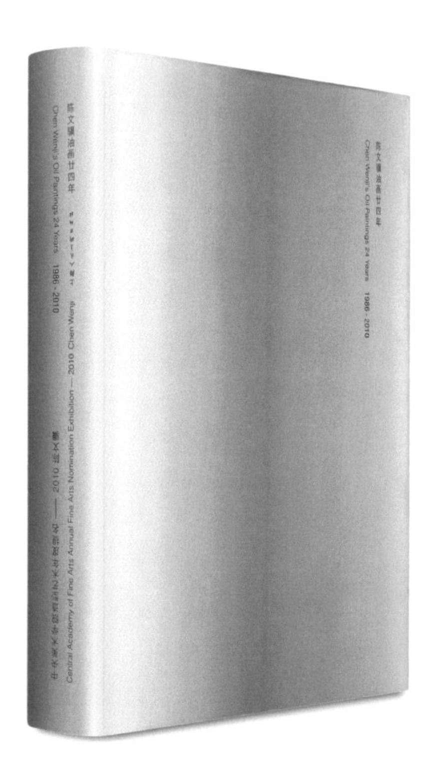
陈文骥油画廿四年
Chen Wenji's Oil Paintings 24 Years 1986 · 2010
Central Academy of Fine Arts Annual Fine Arts Nomination Exhibition — 2010 Chen Wenji

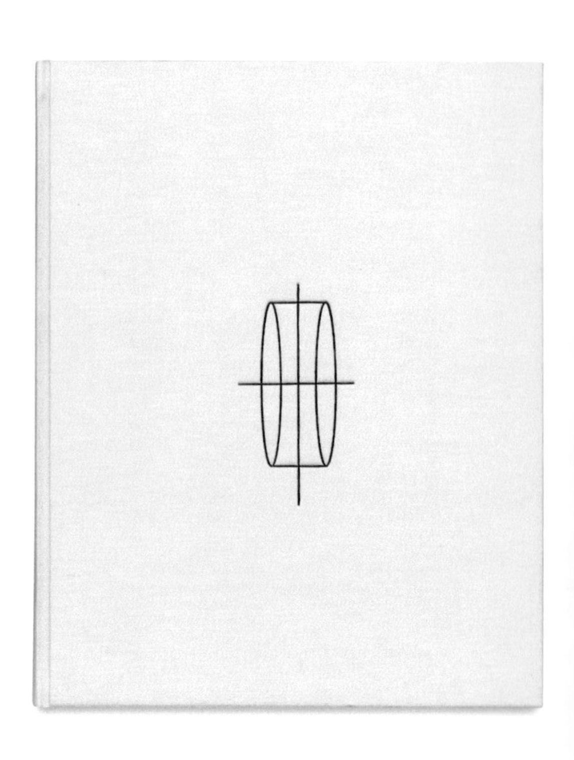

2
1986-2010
3
陈文骥油画廿四年
CHEN WENJI'S OIL PAINTINGS 24 YEARS
上海人民美术出版社

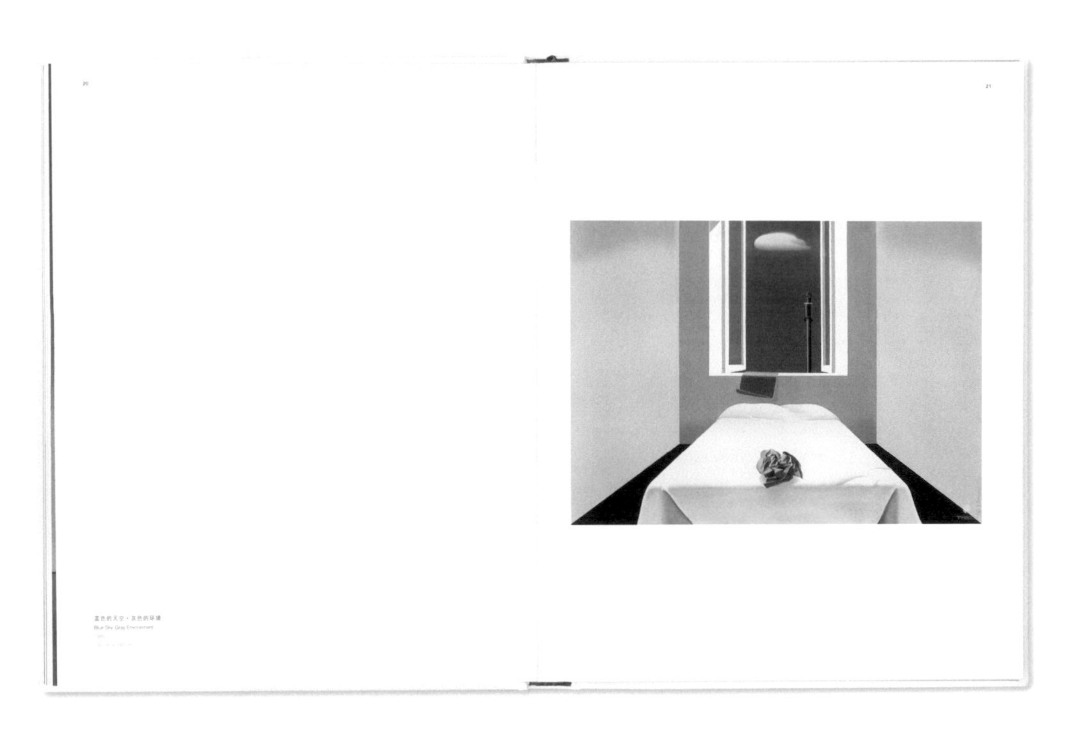
20

21

蓝色的天空・灰色的环境

Blue Sky Gray Environment

34

红色的领巾

Red Scarf

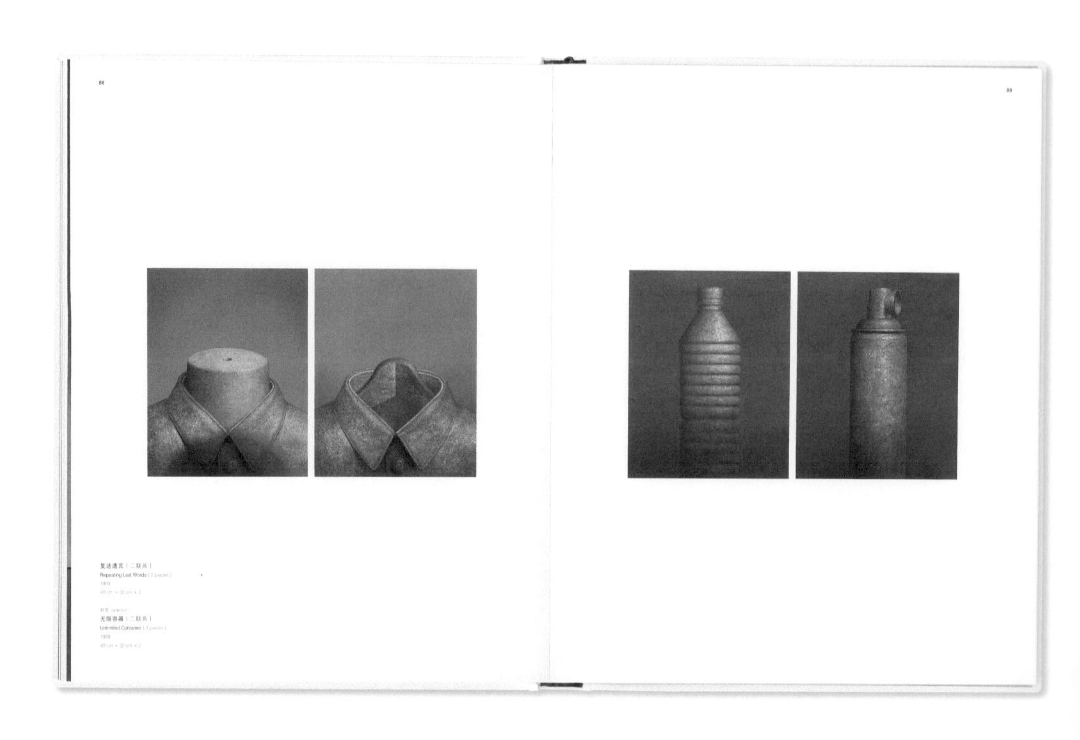

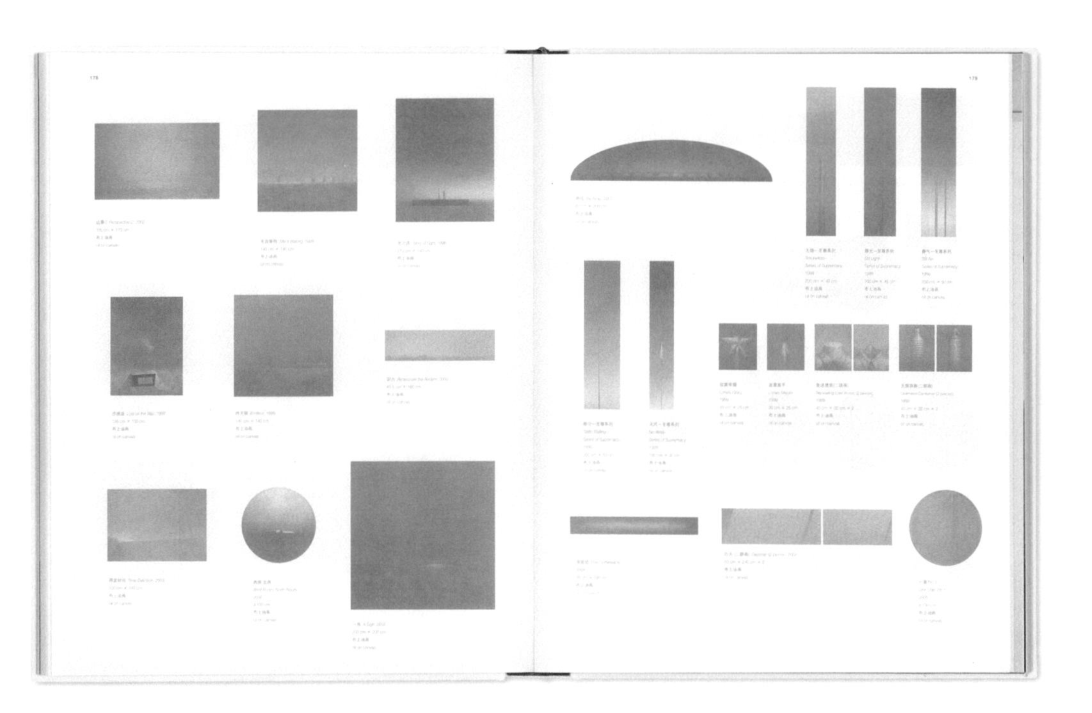

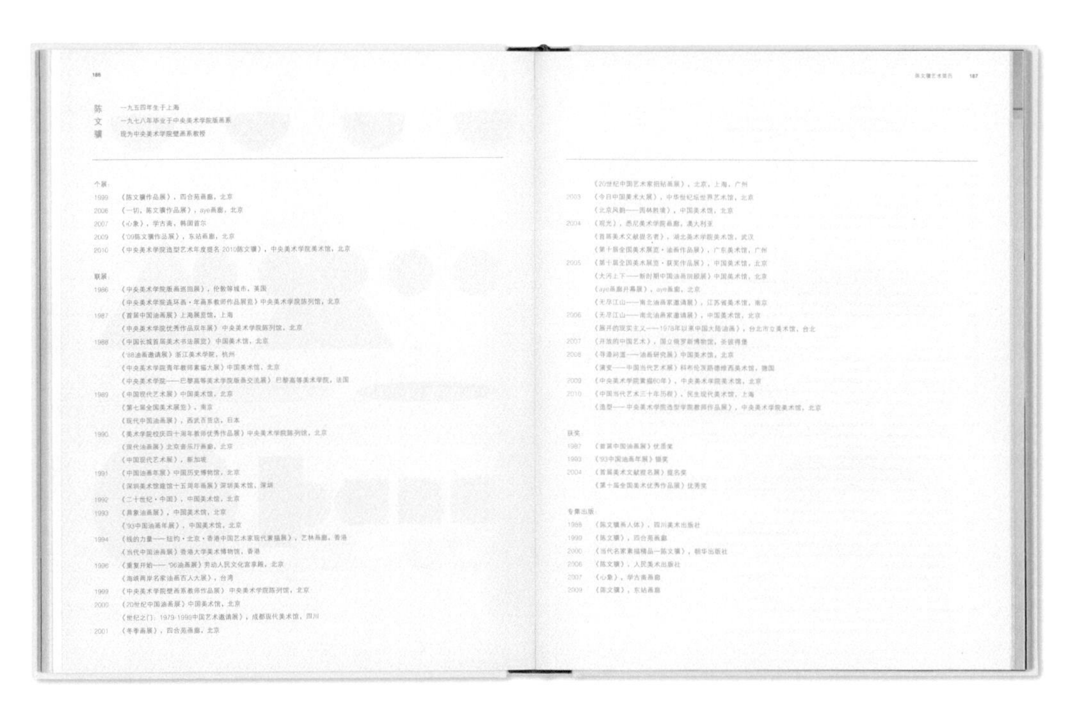

186

陈文骥

一九五四年生于上海

一九七八年毕业于中央美术学院版画系

现为中央美术学院壁画系教授

个展：

1999 《陈文骥作品展》，四合苑画廊，北京

2006 《一切，陈文骥作品展》，aye画廊，北京

2007 《心象》，学古斋，韩国首尔

2009 《'09陈文骥作品展》，东站画廊，北京

2010 《中央美术学院造型艺术年度提名 2010陈文骥》，中央美术学院美术馆，北京

联展：

1986 《中央美术学院版画巡回展》，伦敦等城市，英国

《中央美术学院连环画·年画系教师作品展览》中央美术学院陈列馆，北京

1987 《首届中国油画展》上海展览馆，上海

《中央美术学院优秀作品双年展》中央美术学院陈列馆，北京

1988 《中国长城首届美术书法展览》中国美术馆，北京

《'88油画邀请展》浙江美术学院，杭州

《中央美术学院青年教师素描大展》中国美术馆，北京

《中央美术学院——巴黎高等美术学院版画交流展》巴黎高等美术学院，法国

1989 《中国现代艺术展》中国美术馆，北京

《第七届全国美术展览》，南京

《现代中国油画展》，西武百货店，日本

1990 《美术学院校庆四十周年教师优秀作品展》中央美术学院陈列馆，北京

《现代油画展》北京音乐厅画廊，北京

《中国现代艺术展》，新加坡

1991 《中国油画年展》中国历史博物馆，北京

《深圳美术馆建馆十五周年画展》深圳美术馆，深圳

1992 《二十世纪·中国》，中国美术馆，北京

1993 《具象油画展》，中国美术馆，北京

《'93中国油画年展》，中国美术馆，北京

1994 《线的力量——纽约·北京·香港中国艺术家现代素描展》，艺林画廊，香港

《当代中国油画展》香港大学美术博物馆，香港

1996 《重复开始——'96油画展》劳动人民文化宫享殿，北京

《海峡两岸名家油画百人大展》，台湾

1999 《中央美术学院壁画系教师作品展》中央美术学院陈列馆，北京

2000 《20世纪中国油画展》中国美术馆，北京

《世纪之门：1979-1999中国艺术邀请展》，成都现代美术馆，四川

2001 《冬季画展》，四合苑画廊，北京

陈文骥艺术简历 187

《20世纪中国艺术家招贴画展》，北京，上海，广州

2003 《今日中国美术大展》，中华世纪坛世界艺术馆，北京

《北京风韵——园林胜境》，中国美术馆，北京

2004 《观光》，悉尼美术学院画廊，澳大利亚

《首届美术文献提名者》，湖北美术学院美术馆，武汉

《第十届全国美术展览·油画作品展》，广东美术馆，广州

2005 《第十届全国美术展览·获奖作品展》，中国美术馆，北京

《大河上下——新时期中国油画回顾展》中国美术馆，北京

《aye画廊开幕展》，aye画廊，北京

《无尽江山——南北油画家邀请展》，江苏省美术馆，南京

2006 《无尽江山——南北油画家邀请展》，中国美术馆，北京

《展开的现实主义——1978年以来中国大陆油画》，台北市立美术馆，台北

2007 《开放的中国艺术》，国立俄罗斯博物馆，圣彼得堡

2008 《寻源问道——油画研究展》中国美术馆，北京

《演变——中国当代艺术展》科布伦茨路德维希美术馆，德国

2009 《中央美术学院素描60年》，中央美术学院美术馆，北京

2010 《中国当代艺术三十年历程》，民生现代美术馆，上海

《造型——中央美术学院造型学院教师作品展》，中央美术学院美术馆，北京

获奖：

1987 《首届中国油画展》优秀奖

1993 《'93中国油画年展》银奖

2004 《首届美术文献提名展》提名奖

《第十届全国美术优秀作品展》优秀奖

专集出版：

1988 《陈文骥画人体》，四川美术出版社

1999 《陈文骥》，四合苑画廊

2000 《当代名家素描精品—陈文骥》，朝华出版社

2006 《陈文骥》，人民美术出版社

2007 《心象》，学古斋画廊

2009 《陈文骥》，东站画廊

张方白 **张方白**

湖南美术出版社，长沙，2010

335mm×250mm×30mm，232 页

布面精装，裹纸质护封

1000 册

文章：陈孝信、莱纳特 · 乌特斯特伦、彭锋、孟禄丁、杨卫、王萌

访谈：邹跃进 / 张方白

评论节选：范迪安、栗宪庭、易英、彭锋、郭雅希、邓平祥

诗歌：陈仁凯

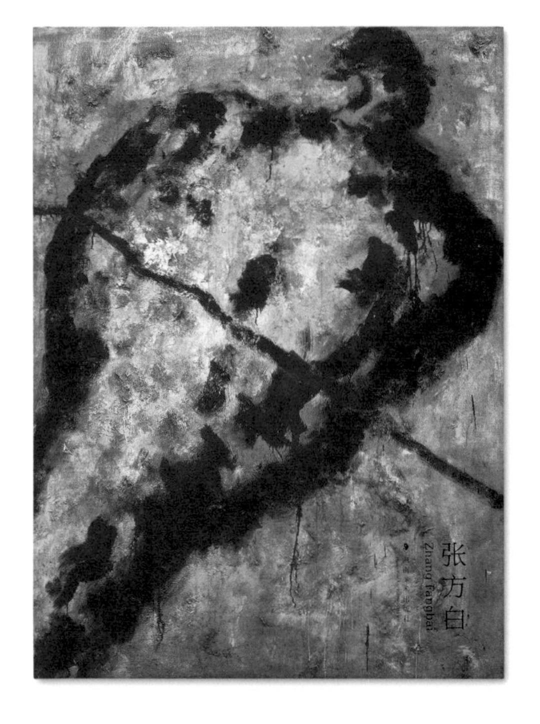

阳刚精神、悲剧性、文脉及其他
——论张方白

陈孝信

在风起云涌的中国当代艺术运动中，张方白并不是一登台就十分显眼的人物（指的是在20世纪90年代，如毛焰、曾梵志等）。可是，真正关注中国当代艺术的人们，渐渐地就会惊讶起来，这个矮个子、头发过早地开始谢顶的中年湖南佬竟然跑到了整个队伍的前列，而且，他那雄健有力的脚步声，越来越激动人心……

张方白，属虎，湖南衡阳人。我甚至觉得在他的生肖和出生地的“密码”里面，就已埋藏着他人生和艺术创作的许多秘密，这些秘密将伴随他的一生一世，也许永难解释清楚。

一、张扬与呼唤：阳刚精神

“男性雄风”之说来自贾方舟。贾方舟认为：张方白的画始终贯穿着一种自觉的男性意识和“我的思考”的特征，始终将“自我存在的确立”作为自己的艺术坐标（参见《美术文献》第20期，P20，2000年7月）。这个定位是可靠的。我甚至认为，在艺术上张方白是典型的“大男子主义”。从他所选定的形象和符号，乃至具体的创作方法，都可以充分地印证这一点。

张方白所选定的“新形象”主要是三类：鹰、男人体和塔。这三类“新形象”构成“男性雄风”或说阳刚精神的三个侧面：鹰在生前翱翔于广阔的天穹，从不成群结队，都是单独行动，堪称“百鸟之王”。在张方白的心目中，“鹰”展示了男性深深的孤独感和顽强的生命意志，而那硕大的身躯（像是一座铁塔），巴洛克式的宏伟造型更使男性们感到骄傲。所以，“鹰”成了张方白笔下最为成功、得意的男性隐喻（尤其是画家所着意勾勒的鹰头和鹰嘴，几乎就是活脱脱的男性阳具的翻版）。迄今为止，他已完成了上百只大大小小、姿态各异的“鹰”，而且正无有穷时。之所以反反复复画鹰，其奥妙就在于：只有“鹰”才是张方白“男性雄风”的第一主角。

其次才是人体。“男人体丰富的结构使我着迷”，“我要展现荒芜的人体”，“米开朗基罗的人体，则使我感到更为宏伟和复杂”（参见张方白：《画与存在》，同上书，P21）。由此可见，在人体方面，他自觉地以米开朗基罗为标杆，将自己的性别身份、特征当做创作的资源，在不断地开掘这个资源的同时，不断地印证自我的存在，自我的独立，自我的丰富、完满，从而进行一项“精神自塑”工程（贾方舟语），完成一幅幅男性世界的“精神肖像”。在这些作品里，肉体与精神的二律背反，竟奇妙地得到了暂时的“和解”：精神借助“荒芜的人体”得以自我呈现，而那硕大的肉体也由于精神性的“在场”变得光芒四射。用通俗的话说，男人体在张方白的笔下，就是一个大写了的“我”，一个充满了火热激情的“我”，一个足以傲视芸芸众生的“雄主”。自然，张方白偶尔也画女人体，可令人惊讶的是，那些女人体同样是雄风依旧，女性的妖娆和妩媚被一扫而空。女人体成了男人体的“变体”，成了男人体的另一个侧面，这倒暗合了上帝造人的神话，并且再次回到了米开朗基罗的“课题”上。（在他的《夜》与《昼》等作品中，不同样都是阳刚味十足的女人？）

据说，张方白选择画塔，是在他母亲去世以后（2004年），“想画一个塔来纪念她”。这些真正打动过画家的佛塔（埋葬那些圆寂了的高僧的），造型异常的朴素、简单，但却充满了一种精神力量。可是，画着，画着，

9

当，天赋从来就是必备的要素之一）。与此同时，他也得到了高人的指点。这位高人就是钟以勤。张方白后来有一个认识：认为钟以勤比罗尔纯深刻，比卫天霖凝重。此公生平坎坷，一生都不得志，但他最先肯定了张方白的造型能力。后来又有一位著名版画家王公懿也持同样的观点。这两位高人的肯定，极大地增强了张方白的自信心。此外，钟以勤的作画方式也对张方白产生了一定的影响，例如张方白如今喜用大粗笔，在蘸满颜料后又倾尽全力，把笔压进画面的方法即是得益于这位高人。

张方白在家乡的一所中专毕业后，工作了一段时间，还创作了几件主题性的作品。这些作品明显地受到了俄罗斯巡回画派的影响。后来，他又借机去“浙美”（如今的中国美院）游学了一段时间。这一次游学，对他以后的创作产生了深远的影响。原因是在“浙美”游学期间，他不仅是对中国画和书法产生了浓厚的兴趣，而且直接地受到了’85美术新潮运动的影响（当时的“浙美”正是’85美术新潮的重要策源地之一）。张方白立志在艺术上有所作为的念头，应该是在这个时候起就已牢牢地扎下了根。

1987年，张方白考入了中央美院——中国最高的美术学府（正所谓“皇家美术学院”）。他进的是油画系第四画室，这也是中央美院油画系里相对宽松、活跃的画室。在四年正规的学院学习中，对张方白而言，最重要的收获是拓展了他的艺术视野，使他后来所作出的“选择与拓展”有了一个较为坚实的学理基础和背景。其次一个重要收获是素描。张方白在素描上下了很大的工夫，而且十分重视自己造型能力方面的训练。欧洲艺术史上的“大师素描”，对他以后的创作，起到了极为深刻的影响。例如，他后来的创作之所以没有倒向充满了酒神精神的表现主义（虽然一度有过向往），而选择了以造型力量取胜的阿波罗——日神精神，就足以表明素描对他的影响之深。

身处北京这样的政治、文化中心，又有四画室相对宽松、活跃的学术氛围，使得血气方刚的张方白产生了折腾一番的冲动。于是，在一场“特殊风波”中，他做了两件行为艺术作品，并轰动一时。他成了中国最早的一批行为艺术家。如果再坚持做下去，他就不是今天的张方白了。可他很快也就明白：行为艺术只能是偶尔一乐罢了，不能成为他的终生选择。他的终生选择是早就确定了的视觉艺术，而且主要是架上创作。他必须坚守这一被国内外许多“后现代艺术家”所鄙夷的领域，并且认定了：只有在这一领域里，他才能真正地有所作为。

不久，在中央美院画廊他举办了平生的第一个个展（1991年）。展出的作品《实在》系列与《空鸣》系列都有一种早熟气息，与一副稚气未脱的“学生腔”几乎联系不上。尤其是《实在》系列，已经可以从中看到张方白以后的一些创作方向，例如简洁有力的造型、对边缘线的强调、用笔的方法、虚实关系的把握，以及相对单纯的色彩关系等等。《空鸣》系列的构思更是别出心裁，为了表达画家对一首乐曲的无限沉醉，竟然在画面上画满了一排排飞扬着的豆芽似的音符，而在画面深处，那一位弹奏者的身影则隐约可辨。整个画面充满了一种与“通感”有关的想象力，让我们如闻其声，如见其人。

为何在以后的创作道路上，张方白又放弃了已引起较大反响的《实

14

在》系列？并让《空鸣》系列成了绝响？如今来回答这个问题，相对当时要容易得多。原因就在于这两个系列受西方潮流（例如意大利的“三C”、抽象表现主义）的影响较为明显，属于一种相对被动的选择，而不是从自身文脉考虑的一种主动性选择。被动与主动之间虽只差了一个字，但出来的结果，留在文化史上的价值却是大相径庭，怎可同日而语？！张方白的创作尽管具有一种早熟的气息，但却没有迅速蹿红。相反，他从中央美院毕业后，自觉地处在了一个相对寂寞的沉潜期，这也正是他的明智之处。

张方白在圆明园作了短暂停留后，便进了天津美院，在那里他执起了教鞭。

天津美院成了张方白艺术生涯中的一块福地。别的不说，单从他的艺术创作来考察，正是在这个沉潜时期，他完成了由“被动选择”向“主动选择”的关键性转变，而这一转变的有力证据便是人体和《鹰》系列的创作，尤其是后者。

后来，他再次回到了北京（2001年），艺术创作也进入了真正的成熟期（同时也可以认为是高峰期）。

小巷
80cm×60cm
布面油画

15

24

the minds of spectators: Is it a sharp-edged sword, a resonant yell, or cracking lighting, or a big exclamation mark of tragedy? The background in the painting, much more chaotic and elusive, always created a sensation to us that there is a strange, threatening being or power, from which we cannot escape.

In brief, the eagle-themed works, after being altered, represent an advancement and a greater level of maturity. The "eagle as a hero in tragedy" has become a signature of Zhang's art. Throughout the history of images both east and west, this is unique (if there were are that were said to be similar, it would be the fish and bird with one eye wide open under Ba Da Shan Ren's brush. I used to believe that it has an appeal of a spiritual totem.

In fact, the "pagoda" series is equally overflowing with connotations of tragedy. The purpose of Buddhist pagodas was to honor the dead, thus, what the "pagoda" contains is the soul of the deceased. Zhang did his utmost to amplify the simplicity of the "pagoda", the sound modeling, towering sensation and pressing existence, all of which are hidden references to the enormous grief of death, the candor of death and the glory of death.

Why has Zhang focused on the "tragic hero" with such special passion? The answer lies in the depth of the times that we are currently living in. He has witnessed a tragedy and cannot erase it from his mind; meanwhile, he made a series of thoughts on tragedy at a metaphysical level and finally chose to make an outburst in silence. Seen from a broader view of sociology, Zhang truly is a conscientious fighter in art history even though today, such fighters are few in number and "simple in quality".

III. Choice and Expansion: Contemporary Context

To what extent can the history of art (either Chinese or Western) be seen as a history of pioneering? And to what extent can it be said to be a history of choice? This is a rather intangible question. In some cases, pioneering means choice, and a choice of initiative is often partially pioneering.

Both the modern and post-modern history of art in the West, to a great degree, is a history of pioneering while the history of modern or post-modern art is, to a certain degree, a history of choice in China (the pioneering history of Chinese art seems to end in pre-Qin period, Han and Tang Dynasties and the periods of Song and Yuan).

What we have is a history of choice, and a choice made with initiative is partially pioneering in nature. Considering this, "how shall we choose? And how can we expand?"---this is an urgent question that confronts every artist in contemporary times. Just as Li Bai, the great poet, wrote in his The Hard Road: "Journeying is hard, journeying is hard. There are many turnings and which am I to follow?" It is justified enough to conclude that choice and expansion is a challenge in any country and at any time. It can be said that the position that an artist finally occupies in the history of art depends on this.

Likewise, Zhang was confronted with a "hard road".

In his early days, Zhang showed a rare talent for painting (talent is something that is essential in art). Furthermore, Zhang was also tutored by a master, Zhong

女人体写生
100cm×140cm
1990
布面油画

25

Yiqing. Later Zhang realized that Zhong is more profound than Luo Erchun and more somber than Wei Tianlin. Zhong lived a frustrated life never having achieved his ambitions. But it was he who first affirmed Zhang's ability in modeling. His judgment was later reconsolidated by the viewpoint from Wang Gongyi, a well-know print painter. The recognition of these two masters greatly boosted Zhang's self-confidence. Moreover, Zhong's style in painting also had a certain influence on Zhang's work. For instance, Zhang likes to use a bold brush, which he dips in a large amount of paint and then presses into the frame with all his strength. He owes much of this technique to Zhong.

After graduating from a junior college in his hometown, Zhang worked for a period of time, during which he produced several thematic works. IT is clear that his were influenced by the Russian Tour School. Later, he capitalized on an opportunity to temporarily study at the Zhejiang Art Institute (presently the China Academy of Art). The experience exerted a significant influence on his artistic creation, for he not only developed a strong interest in traditional Chinese painting and calligraphy but was also directly affected by the 1985 art resurgence (At that time the Zhejiang Art Institute was one of the major origins of the Movement). It was most likely that his intention to do something in art took a deep root in his heart during this period.

In 1987, Zhang passed the entrance examination for the China Central Academy of Fine Arts---China's top academy in fine arts (a "royal academy of fine arts). He was admitted to the fourth studio in the department of oil painting, which

32

实在之一
160cm×120cm
1991
布面油画

33

张方白的世界

莱纳特·乌特斯特伦

前几天，当我到居于北京东北部的张方白的新工作室拜访他时，这里的变化让我震惊：以前这个地方是个白酒厂，而现在却颇有文人花园的韵味。当然，为了不被老酒厂建筑所遗留的一些斑斑点点的"悲哀"分心，或者是为了不去留意窗外随风起舞的、脏破的方便袋或邻居家慢慢驶过的梅赛德斯－奔驰轿车，我们可能需要睁一只眼闭一只眼，或者有时需双眼同闭。但是，在另一方面，大池塘和花园都一览无余地展现在我们的视野之中，还有花园内雅致的树丛和鹅卵石所铺就的小路。

恰如学者一样，张方白为人谦逊，讲话轻声细语。他不批评艺术家同行，而总是可以发现他人的优点。（注意：这并不表示批评态度的缺失或者是缺少个人的意见，这是一种谦虚而又谦卑的处世方法。）如果他提出一个问题，他表现出的是诚实、真诚的兴趣——他对闲聊毫无兴趣。至于"夸夸其谈"这个词，我认为张方白甚至不解其意。

张方白的世界和措辞正如他的艺术一样：恰到好处，从不画蛇添足，几乎没有什么颜色感可言（如果从他的绘画来讲），但又是主题和表情的聚合体。在这些方面，他与一些中国经典绘画大师相似，尤其是宋代名家范宽和梁楷，不必提宋代另一个大家——牧溪。可能你认为我在牧溪的《六个柿子》（收藏于日本京都大德寺）与张方白的《鹰》（第11号，私人收藏，法国巴黎）之间找到独到哲学关系的这种做法有点桀骜不驯、处心积虑，我指的是主题的复杂程度、个人的但却是对待主题精化的恭敬态度、处理黑白两种颜色的方法、稀疏的灰色调和对柿子与鹰的处理方法。而这不仅是隐含的维度：在"柿子"一画之中含有非常虔诚的宗教色调和意图，这使得此画成为禅的象征；而在"鹰"中包含有东西方都具备的政治色彩。在下面，我将深入介绍。

再引入另一位中国早期的画家，在情趣和技术方面看来他也与张方白相似，他就是活跃于14世纪的因陀罗——另一位使用大胆粗犷的线条和表面上空洞的空间来展示其作品的禅画家。我还没有同张方白谈起这位绘画前辈，但是，我深信他们之间有很多共同点。

张方白的的确确是不打折扣的绅士和学者。尽管多年受苦，为衣食住行犯愁，他的才智却未受影响，他将自己的个人价值观付诸艺术之中，并赋予其生命。他儿时和青年时走过的道路与他现在的人生之路没有区别：做一个真实的我！

张方白在老酒厂的画室的墙上挂满了画，摆满了关于中国古代绘画的书籍，莫兰迪、宁布朗、八大山人、高其佩的艺术复制品，地方艺术年鉴（现在十分风行），画册，他自己的写生簿、绘画作品。在他的面前总有一个茶壶、一个茶叶筒、不少中国绿茶茶袋和一瓶高品质的中国黄酒。在地板上还有来自他老家湖南衡阳的朋友新送的礼物。

在巨大的画布上，一幅黑色的作品尚未完工，棕树丝和黑色的颜料混合在一起。在另一幅画布上画有一只鹰、一座塔，还有一幅取景于安徽省某个小村庄的街景，几个水桶、一堆毛笔（主要是大号的，通常被用来涂墙的毛笔，但现在却成为这位画家的重要工具），以及黑色、白色、灰色颜

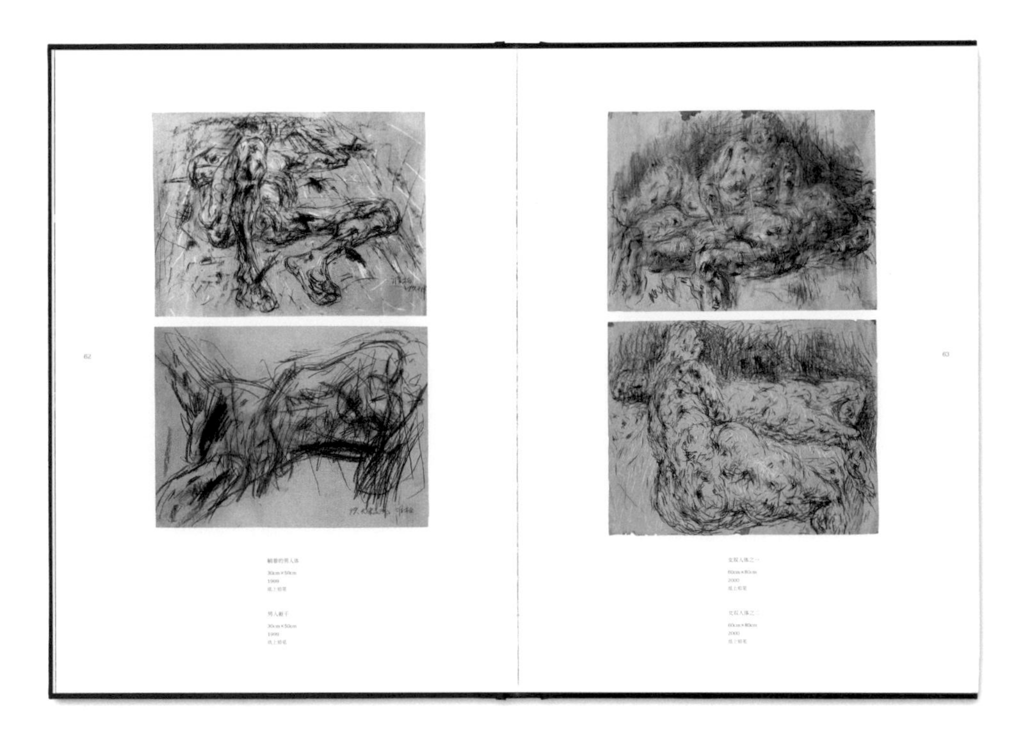

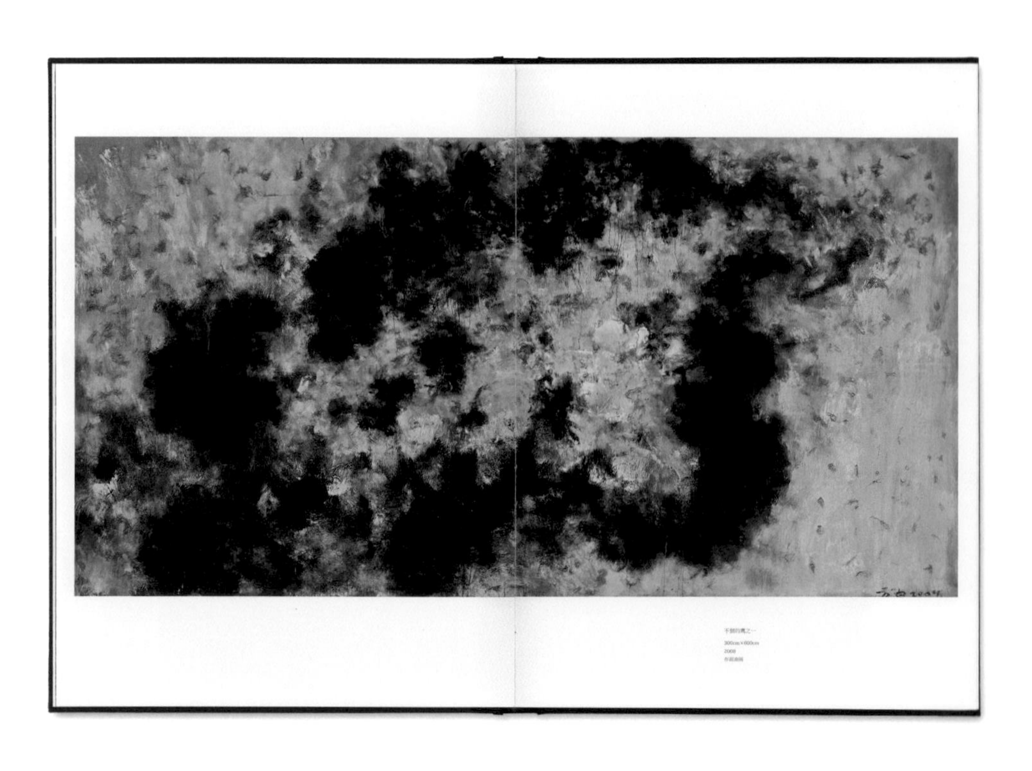

千朵的鹰之一

300cm×600cm

2008

布面油画

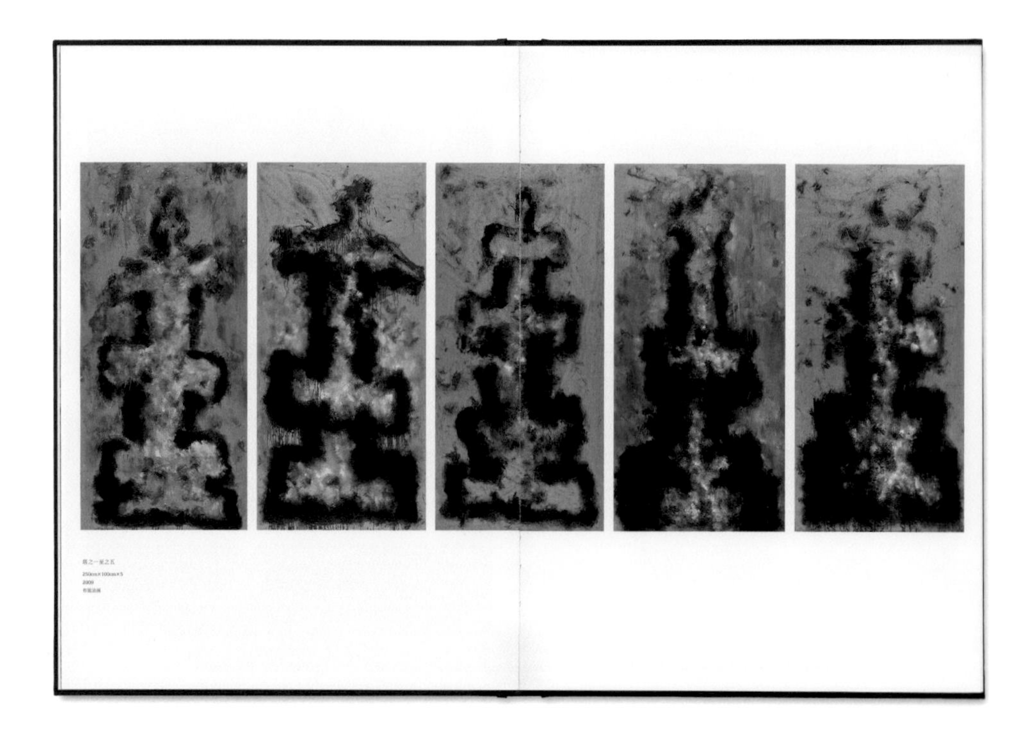

塔之一至之五

250cm×100cm×5

2009

布面油画

南方3

100cm×80cm

2010

布面油画

最近常见画家画以江南民居为题材的风景，从这些风景中，我们很容易看出一些共同的特征，比如，它们通常将中国水墨与西方油彩结合起来，追求一种轻松、明快、舒展、优美的感觉效果，画面充满诗情画意，引起我们对故乡、对故乡所象征的中国传统文化的浪漫遐想。这种风景画差不多成了中国当代绘画中的一种模式，成为中西画法融会贯通的一个范例。毫无疑问吴冠中先生是这种模式或范例的集大成者。不过，老实说，当这类流行的风景反复再三出现的时候，就难免给人一种甜熟甚至腻味的感觉。

与这类风景形成典型对照的是张方白近来画的江南民居。我可以毫不夸张地说，张方白的这些以民居为题材的风景几乎在所有方面都与那种流行的风景对立。与张方白的鹰一样，这些风景体现出厚重、滞涩、生辣、朴拙、内敛、粗犷等等感觉特征，引起我们对故乡、对故乡所象征的中国传统文化的反省和沉思。总的来说，与那种流行的风景常常给人一种水彩画的轻快感不同，张方白的这种风景更多地给人一种雕塑般的厚重感。尽管张方白的这种具有雕塑感的风景并不够漂亮，但却有力地抓住了我的目光，让我反复琢磨、玩味，以至于流连忘返。

这究竟是什么原因呢？

也许我可以用喜新厌旧或审美疲劳之类的美学常识来解释我对那类流行的风景的厌倦，对张方白的风景的眷恋。对于艺术（也许不仅是艺术）来说，流行就是这样一把双刃剑，它在让你获得观众的同时失去自身。逆潮流而上的张方白用一种陌生化手法解除了我们的审美疲劳，满足了我们追新求异的审美要求。

这种解释的确也说得过去，但用来解释我对张方白的风景的如此眷恋似乎过于轻描淡写了。凭我的直觉，张方白的画面中一定有某种更深层次的东西抓住了我。因为事实上与其说张方白的风景给我的感觉是“新”的不如说是“旧”的，正是那一系列与“旧”相关的感觉特征让我眷恋，这不仅因为它体现了故乡之故旧，而且因为它体现了中国传统美学的一种古老的审美追求：拙朴与老境。

与西方文化追求新巧不同，中国文化追求拙朴。大巧若拙是中国文化所蕴含的独特智慧，这种智慧体现在艺术中就是对熟巧的避免，对生拙的追求，所谓“画须熟后生”（董其昌），“画到生时是熟时”（郑板桥）。张方白通过长期锤炼形成的独特绘画语言，最能体现中国传统美学的那种独特的审美追求，它拉近了我们与历史的距离，或者说让历史重新回到了我们身边。这种厚重的历史感，正是那些追求轻快、漂亮的风景所缺乏的。

张方白的风景的厚重历史感，不仅体现在他的绘画语言上，而且体现在他的绘画题材上。我曾经对张方白画鹰的标本产生过疑问，如果张方白的绘画有意识地追求一种男性的力量，那么活鹰无论如何都要比死鹰更有阳刚之气。不过后来仔细想想，鹰的标本也许更适合张方白的绘画语言，更能体现那种生涩的苍茫感。通过张方白用自己独特的绘画语言的诠释，我们对鹰的理解被推向了一个新的境界，一种独特的精神力量的境界，它超出了动物力量和人格力量。如果说鹰所体现的动物力量和所象征的人格力量可以模仿的

沉郁的风景——读张方白《故乡》组画有感

彭锋

171

200

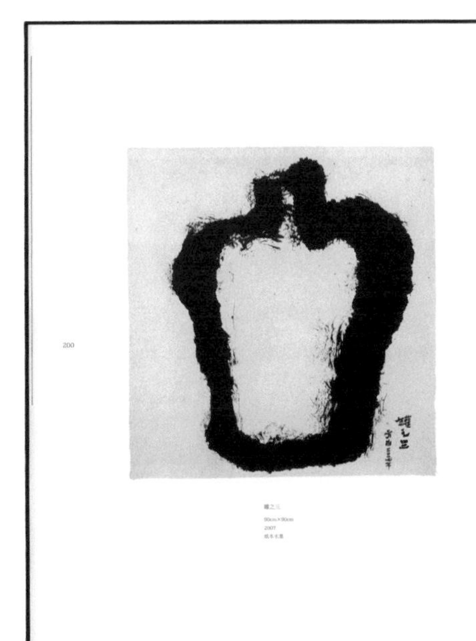

罐之三

90cm×90cm

2007

纸本水墨

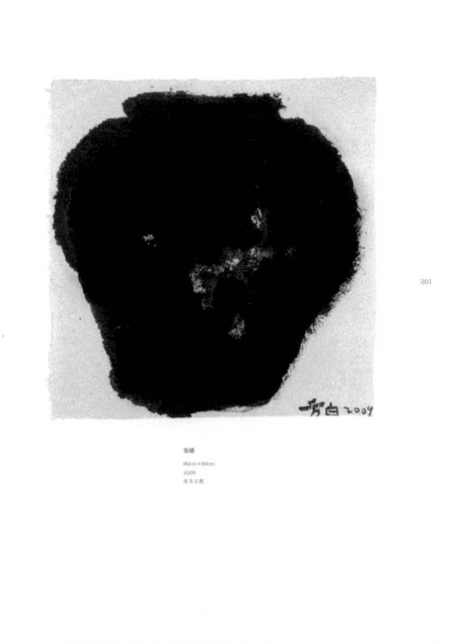

墨罐

90cm×90cm

2009

纸本水墨

201

朱昱 **朱昱**

长征空间，北京，2011

311mm×297mm×16mm，60 页

布面精装，裹纸质护封

1000 册

文章：秦思源

访谈：高士明、唐晓林、闵罕、刘潇、刘健伶、杨雨瑶、翁桢琪、张静、马楠、刘畑 / 朱昱

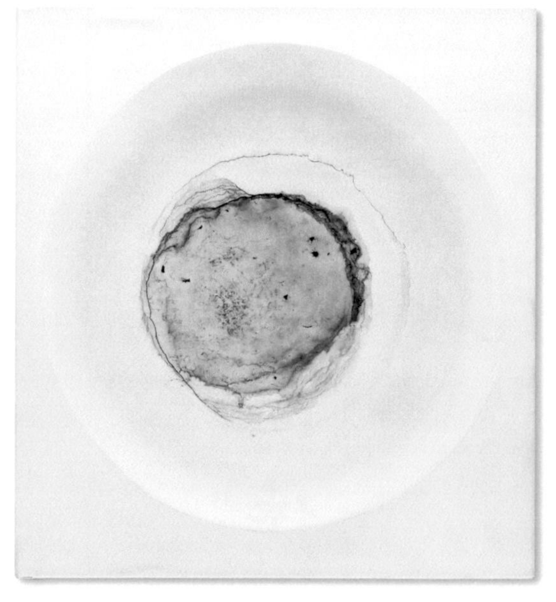

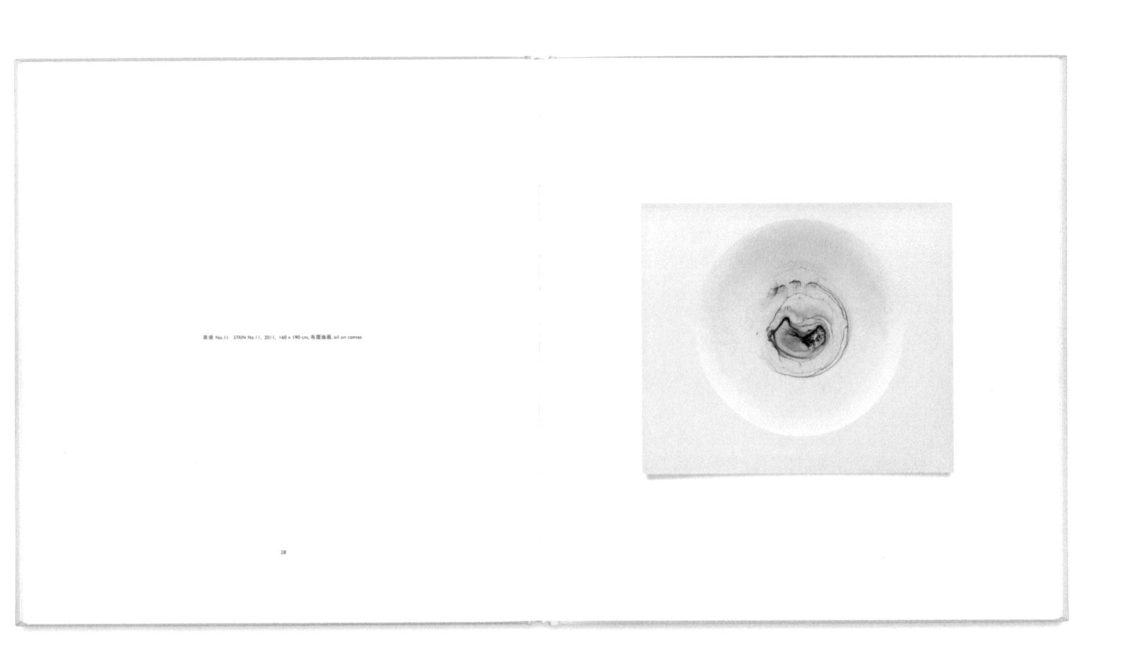
茶渍 No.11 STAIN No.11, 2011, 160 × 190 cm, 布面油画, oil on canvas
28

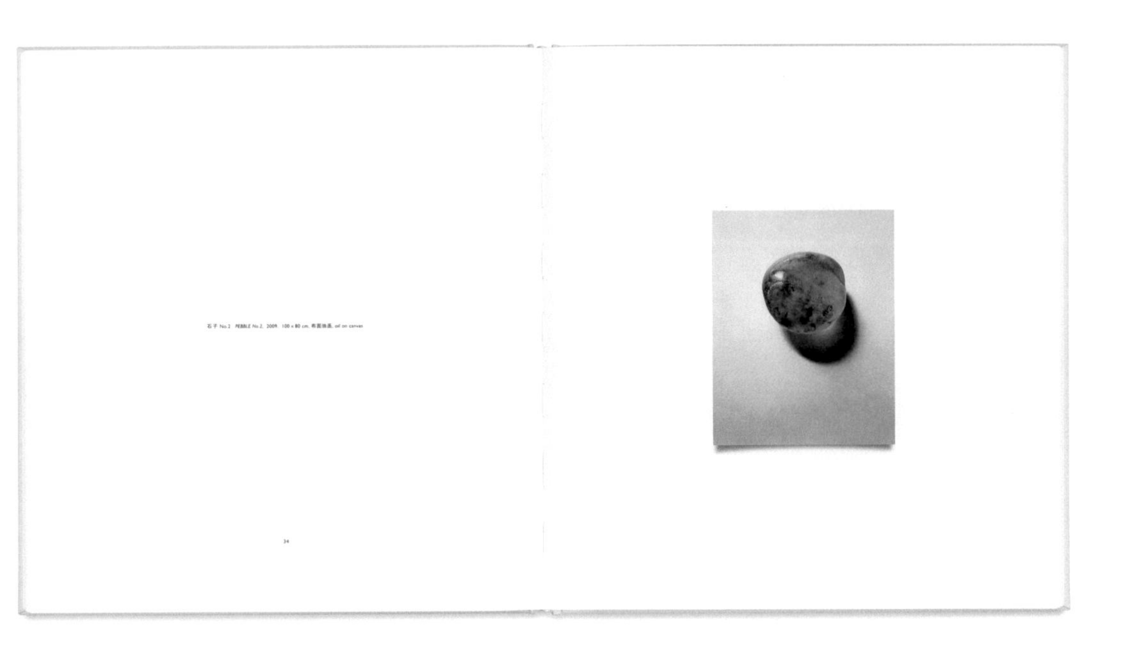
石子 No.2 PEBBLE No.2, 2009, 100 × 80 cm, 布面油画, oil on canvas
34

朱昱访谈

访谈者

[illegible]

[illegible]

[illegible]

of people's reception of them, the pieces do not leave a lot of space.

Gao Shiming: What do you mean by overdone? When a piece is overdone does it easily turn into spectacle?

Zhu Yu: Yes, that's what I am getting at.

Gao Shiming: In fact the series that includes *Dinner - Eating People* and *Sacrifice* is equivalent to a mass media myth. One feature of media myth is its fictiveness — it is pseudo-myth. But it is precisely through the mass media that a story attains truly mythic status. This topsy-turvy relation is quite interesting. The problem is that it is no longer possible to pursue authentic problems in art. For instance, take your current creative state when you make paintings. Judging from your description, this state is disengaged from art media and mass media. But actually it is not disengaged. It is still connected to the media by dotted lines.

Zhu Yu: It is offset by dotted lines. Authenticity is hard to come by. There is no possibility of getting back to the kind of authentic state you are talking about.

Gao Shiming: That is, you cannot pretend that the mass media is not present. At every moment it is present, and you are enfolded within it.

Proposals • Painting • Labor

Gao Shiming: For how many years have you been making this series of paintings?

Zhu Yu: I started the first painting in 2004. I finished *Sacrifice* and *Courtroom Record* in 2002, and began painting in 2002 or 2003. I had to paint until 2004 before I arrived at a method that I could work with and this has gradually changed over the years.

Tang Xiaolin: Are the "192 proposals" to be carried out? Perhaps what I should ask is how you leaped beyond the "proposal stage"? How did you enter into the painting phase?

Zhu Yu: My proposals and paintings are parallel phases. I used to paint in the daytime and write proposals in the evening. In one evening I would draw up two three of them. Before "United Nations" I had another piece called *Proposal Composed of 100 Proposals*. The precondition for this, as is often the case, was not initiative or purpose on my part. I got into that state because there were many exhibitions during that period, and each time an exhibition came around I would conceive a lot of proposals. However, one got the feeling that each proposal, in being carried out, would not really live up to expectations. One would still feel dissatisfied with it, or feel there was no great necessity into realizing it. So then I thought of bringing all these things out, to show a whole thought process. Since it was only a thought process there was never any intent to realize them. They could not be anything more than proposals, and that was the end point of my thinking about them. Actually from a proposal to a finished piece of artwork there is a still a long road to travel.

Gao Shiming: What do you mean by "long road"?

Zhu Yu: One thinks of many things, but why does one decide upon any particular idea to carry out? One is always vacillating between notions in the mind and its manifestation.

Gao Shiming: As you go from a proposal to an artwork, are you bringing it into reality? Are you putting it into practice? Are you objectifying it? Are you concretizing a concept?

Tang Xiaolin: Is there any particular proposal that you wish to realize? Why?

Zhu Yu: For each proposal in that pile of proposals, there was a wish to realize them, but once I was done wishing, it was over. The primary thing was that they fascinated me. While I was drawing them up I used to chuckle to myself. This was a kind of release. I was making paintings during the day and creating these proposals at night as a release.

Gao Shiming: Did you do your 192 proposals during the same period as your paintings? Did you paint in the daytime and write proposals at night?

Zhu Yu: Yes. Painting and creating proposals gave me two kinds of feeling. In the daytime I felt happiness — the blessedness of painting belongs to the real world, and in the evening a furtive kind of pleasure while creating the proposals.

Liu Xiao: "192 Proposals" reminds me of a distinction that Marshall McLuhan made between warm media and cold media. Proposals can be

seen as a cold medium, and painted works are a warm medium. Can the work be understood that way? A cold medium demands a high degree of involvement from the viewer, just like letter writing in the past, which demanded that words be carefully chosen. Today we read e-mail; we chat on qq or msn, so we do several things at once in a scattered way, lowering our degree of involvement. So when "192 Proposals" are placed in front of us as mere proposals, the degree of involvement is different than if they are actually carried out.

Zhu Yu: I understand what you are saying, but there is another level of relevance here. If the "192 proposals" are artwork in and of themselves, then such a proposal need not be carried out, it is a piece of artwork. Proposals may seem cold, but that does not necessarily change by making them into object form. Why does philosophy give us so many possibilities? Concepts that are visually demonstrated often end up being reduced in scope. Actually, the divergence between proposal and artwork is not equivalent to the difference between thinking and hands-on work.

Gao Shiming: As for your sense of joy while painting in the daytime, was that a false happiness? Or perhaps I should ask, should your daytime joy be described as more genuine, more reliable than your furtive pleasures in the evening?

Zhu Yu: This is not a problem of genuineness. Painting is like a farmer tilling the soil, whereas proposal drafting is like speculating in stocks. Perhaps the latter is also genuine, but they are wholly different states of being. This is not an issue of moral judgment.

Gao Shiming: The "192 proposals" you created in the evenings were also a piece of artwork – actually that was also a type of labor. Likewise your painting in the daytime was labor and it resulted in paintings. Would you say that such hands-on labor could actually be a kind of proposal? Many people would say that the paintings from the *Stain* series are different from the paintings of Liu Xiaodong or Yang Feiyun because Zhu Yu painted them. People still see these hands-created pieces, to which you devoted years of effort, as a kind of proposal — an even slyer proposal.

Zhu Yu: What a tragedy. When it comes to this, my only consolation is that I am making a bodily effort.

Tang Xiaolin: Lu Jie (Long March Space director) says that your paintings are very cunning. How do you view what he says? Lu Jie is equivalent to a promoter. Would you say that he takes a similar stance to the mass media in promoting your work?

Zhu Yu: I don't really know what he means by "cunning."

Gao Shiming: I think that when he says "cunning" he sees your painted works as proposals in another form — a long-term, spot-on proposal, but nevertheless a proposal. And you say this is tragic.

Zhu Yu: This is a new definition of proposal.

Gao Shiming: Whether or not this is tragic depends on whether or not it allows for hands on labor; or whether it flaunt its cleverness or indulge in furtive pleasure. As for painting, I would rather make distinctions based on the quality of the painting. We are still willing and able to look for gradations of quality, to say one painting is better than another. This is very important. Painting is constantly breaking through to new aesthetic frontiers; they bring old aesthetics and old perceptions into their circle of influence and re-energize them. This is called "the new within the new." This is not a difference of degree or creative adaptation; it is discontinuity that makes things new; it is a heterogenous space that opens within a system. If we think about it, the reading of Zhu Yu's paintings cannot escape the ghost of that cannibalistic Zhu Yu. The fact that these paintings were done by the cannibalistic Zhu Yu lends them a particular shade of meaning. We all think in this manner: one way or another, there is bound to be a connection there. Actually, *Dinner - Eating People* can empower your paintings with a subtle force; it can give them an aura.

Zhu Yu: There is danger in this also, so I have to be especially vigilant.

《朱昱》

2011年，长征空间总监卢杰邀请我为朱昱设计画册，从卢老板那儿我才知道，朱昱早已不做行为，回归架上了。我跟朱昱第一次见面约在长征空间展厅，下午要拍画，正好可以边看作品边谈。我先到，摄影师已开始布光调试。我是第一次看到朱昱的绘画，有点吃惊——极精彩——澄明，纯净，但有一种非常内敛的“狠”劲，直指人心。一会儿，朱昱来了，健康、阳光、明朗的笑容，戴无框眼镜穿帆布球鞋——这是把人脑搅拌做成罐头，从自己肚子上割下一块肉缝到猪身上，做《食人》《献祭》的朱昱吗？

后来熟了，朱昱有时也会到我的工作室来，说说书的事也聊些别的。若是天气好，我们还会到我工作室对面的朝阳公园里散散步，边走边聊，十分轻松惬意。有时赶上饭点，我们也会一起吃个饭，我发现他总是左手握筷，他说因为长时间非常细致地作画，肩膀会痛，右半边得省着点用。他本来喜饮酒，但觉得酒精对视力会有细微影响，也戒了。我知道，朱昱作品中那种非常“狠”的极致感是靠精神肉体的双重燃烧而来，因此感人，因此不易。

当具体谈到这本书怎么做时，朱昱说希望它看上去有点“不太正常”，我知道他说的“不太正常”绝非设计形式上的奇技淫巧，而是能跟他的作品有内在契合的稍稍出位。最终，我们决定用“贴片”的形式来呈现他的画作。这是一种非常古典的方式，把画作单独印刷在薄纸上，然后手工粘贴在书里。但与古典画册不同，我把装订成册的背基纸大大加厚，这一转换，使得书籍翻动时的手感变得殊为不同。这种貌似古典实为现代的书籍形态，如同朱昱的绘画，既同古典绘画的格调相联结，又充满了内在的锋利和挑衅。从书籍上讲，越郑重其事，越荒诞不经；落差越大，张力越大。

书的文本部分，同样遵循了古典样式，大留白小版心，中世纪圣经一般。朱昱早年五个著名的行为作品，转为黑白小图，置入其中，没有任何态度地平铺直叙。印刷时这些图片输出成了粗网线，质感类似老报纸上的新闻照，进一步模糊细节褪去血腥，只留纪实、客观，不渲染。尘归尘土归土，当年的喧嚣今天的艺术文献，书籍是最好的归宿。

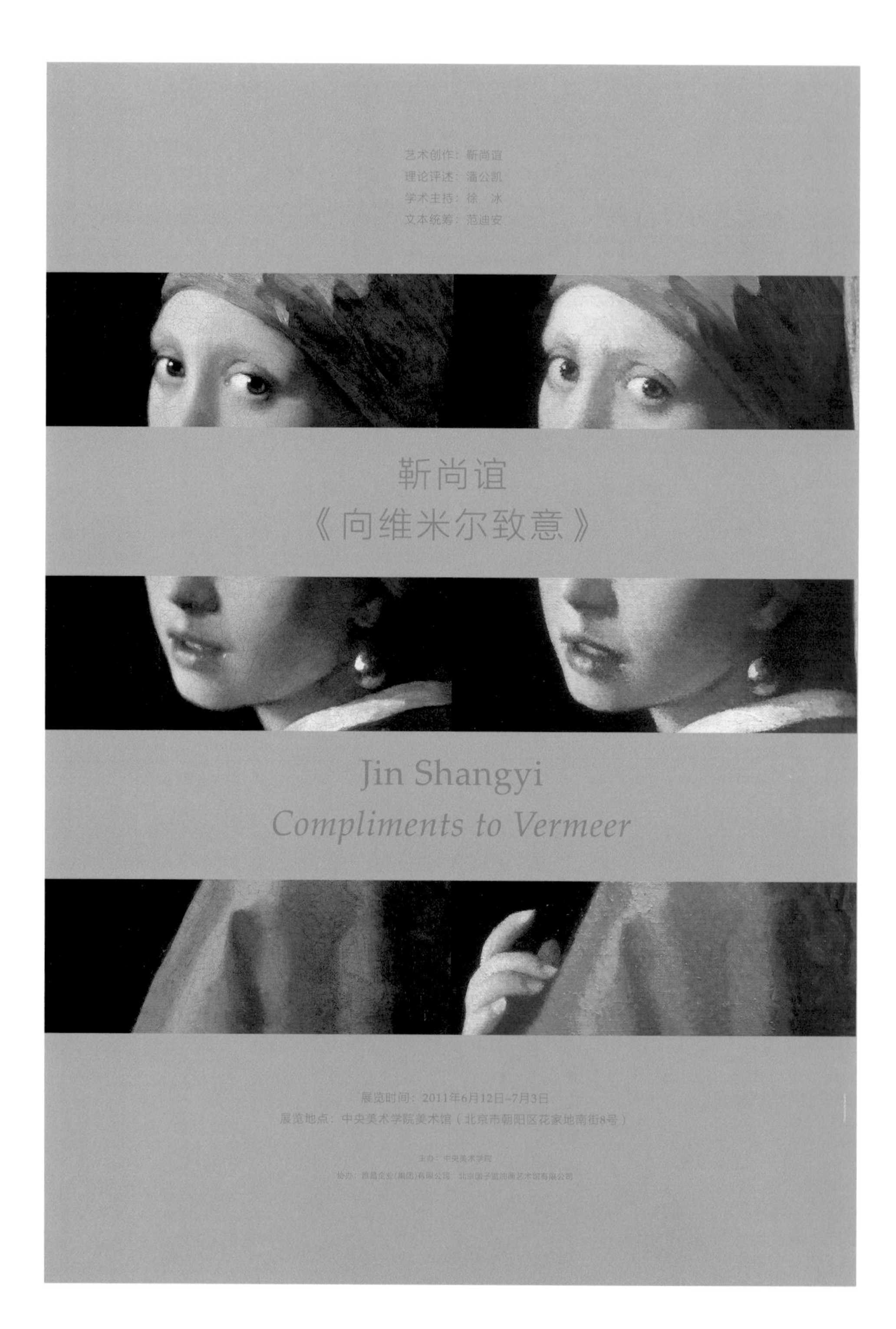

海报，1000mm×700mm，500 张

靳尚谊 **向维米尔致意**

江苏美术出版社，南京，2011

290mm×208mm×15mm，106 页

布面内衬海绵精装

3000 册

主编：范迪安

前言：潘公凯

访谈：徐冰 / 靳尚谊

文章：简森 · 爱德华 · 考夫曼

文本写作：范迪安、徐冰

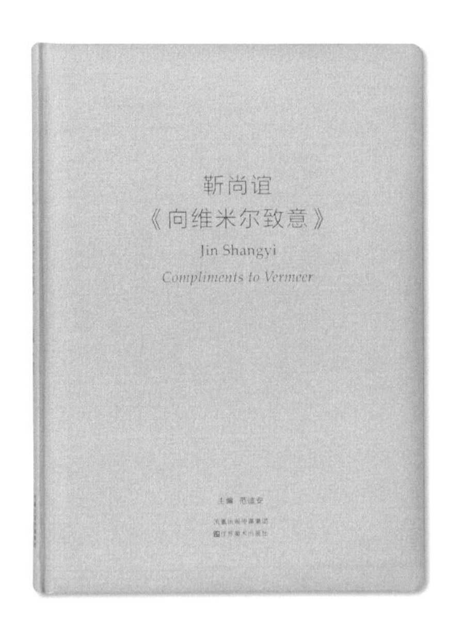

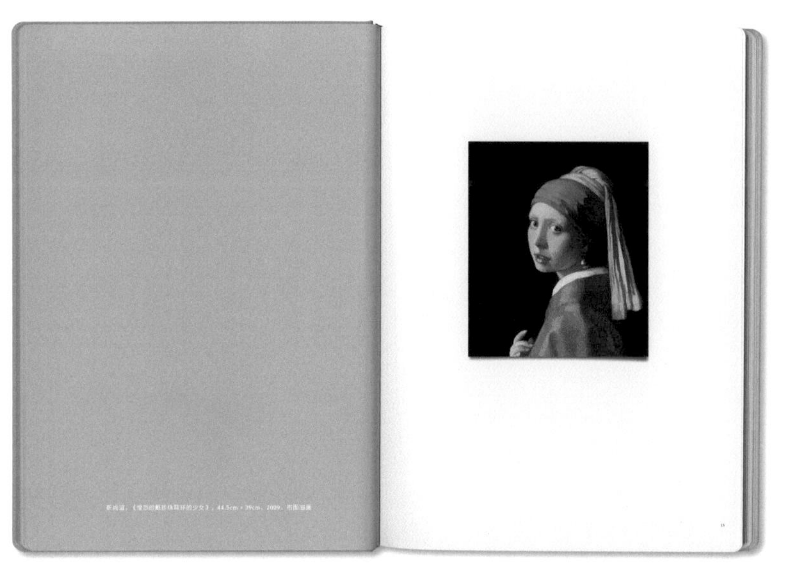
靳尚谊，《惶恐的戴珍珠耳环的少女》，44.5cm × 39cm，2009，布面油画

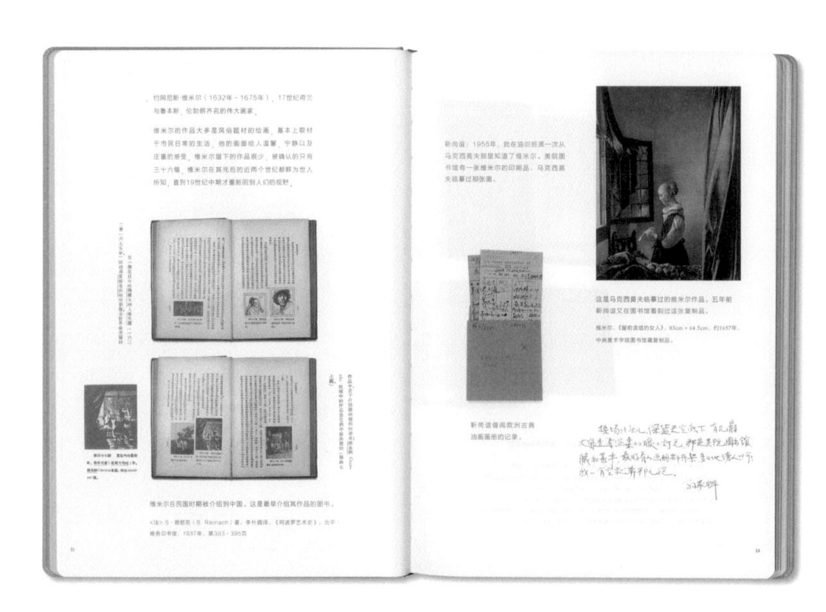
约阿尼斯·维米尔（1632年－1675年），17世纪荷兰与鲁本斯、伦勃朗齐名的伟大画家。
维米尔的作品大多是风俗题材的绘画，基本上取材于市民日常的生活，他的画面给人温馨、宁静以及庄重的感受。维米尔留下的作品很少，被确认的只有三十六幅。维米尔在其死后的近两个世纪都鲜为世人所知，直到19世纪中期才重新回到人们的视野。
维米尔在民国时期被介绍到中国。这是最早介绍其作品的图书。
〈法〉S · 赖那克（S. Reinach）著，李朴园译，《阿波罗艺术史》，北平：商务印书馆，1937年，第393－395页
靳尚谊：1955年，我在油训班第一次从马克西莫夫那里知道了维米尔。美院图书馆有一张维米尔的印刷品，马克西莫夫临摹过那张画。
这是马克西莫夫临摹过的维米尔作品。五年前靳尚谊又在图书馆看到过这张复制品。
维米尔，《窗前读信的女人》，83cm × 64.5cm，约1657年，中央美术学院图书馆藏复制品。
靳尚谊借阅欧洲古典油画画册的记录。

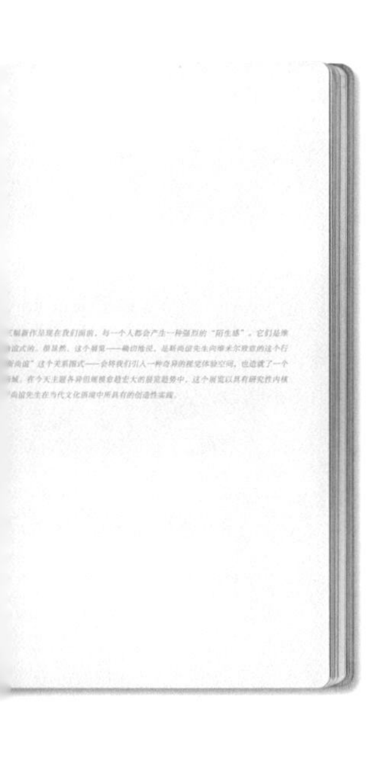
幅新作呈现在我们面前，每一个人都会产生一种强烈的“陌生感”。它们是维
谊式的。很显然，这个展览——确切地说，是靳尚谊先生向维米尔致意的这个行
靳尚谊”这个关系图式——会将我们引入一种奇异的视觉体验空间，也造就了一个
域。在今天主题各异但规模愈趋宏大的展览趋势中，这个展览以具有研究性内核
尚谊先生在当代文化语境中所具有的创造性实践。

在中国看维米尔

今天，在所有收藏了维米尔画作的博物馆，维米尔的作品总是挂在醒目的位置，以供人作朝圣般的观赏。在浩瀚的欧洲艺术经典中，维米尔的作品尺幅远不如那些“致广大”的鸿篇巨制，但以“尽精微”的神采占有不可替代的位置。它们吸引着观众走近前去，在近距离探索发现的瞬间，被它们所透射的光芒、静谧与和谐而打动，产生神来之笔的感叹和挥之不去的回味。
油画自西方传来，在中土植根已逾百年。中国画家对西方油画的接受与研习，一方面展现了吸收艺术新知的文化情怀，另一方面也显示了不同时代选择和借鉴的文化理想。很长时间以来，在一个以主题性、社会性创作为主导的艺术环境里，像维米尔这样的画家是被忽略的，只有文艺复兴盛期到法国浪漫主义那种雄强壮阔的绘画样式才得到足够的青睐。维米尔艺术在中国的命运——就像在西方也是逐步被发现和重视一样——经过了从寂寞到被社会认知的过程。摆在中央美术学院和其他美术院校图书馆里的维米尔画册，许多年总体上处在被冷落的位置，不能放上直接学习的前台，只有那些

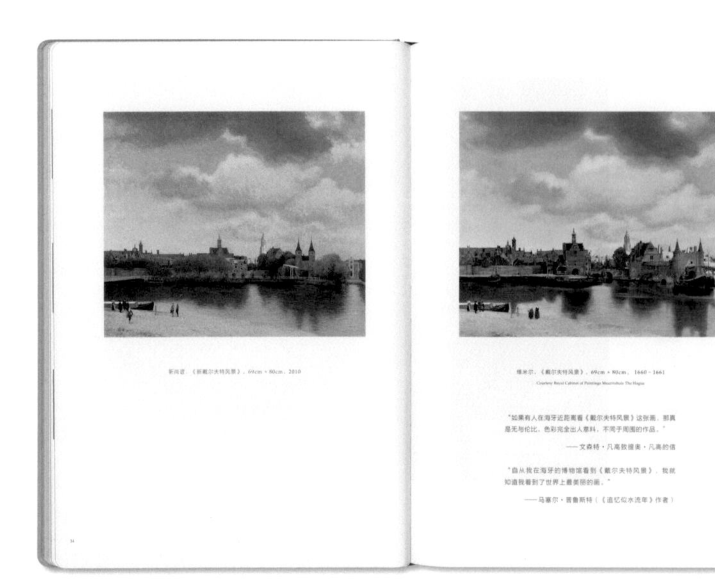
维米尔，《戴尔夫特风景》，69cm × 80cm，1660－1661
"如果有人在海牙近距离看《戴尔夫特风景》这张画，那真是无与伦比，色彩完全出人意料，不同于周围的作品。"
——文森特·凡高致提奥·凡高的信
"自从我在海牙的博物馆看到《戴尔夫特风景》，我就知道我看到了世界上最美丽的画。"
——马塞尔·普鲁斯特（《追忆似水流年》作者）

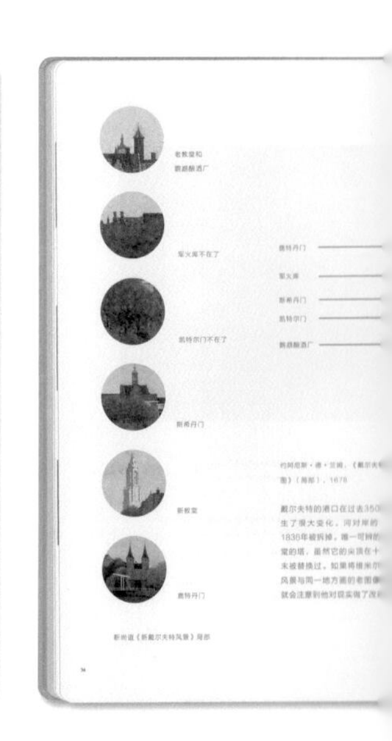

人类的文化史是对经典解读的历史。

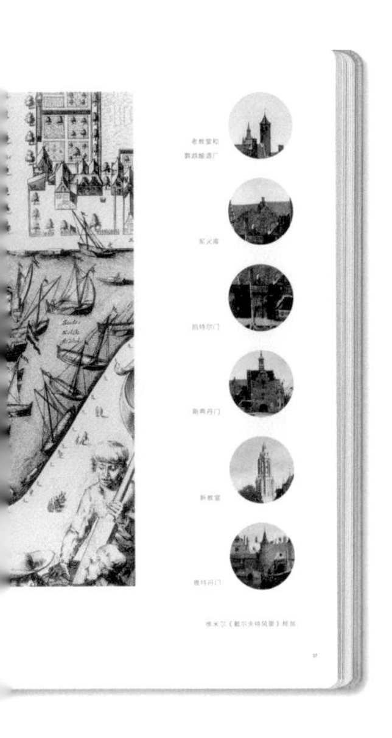

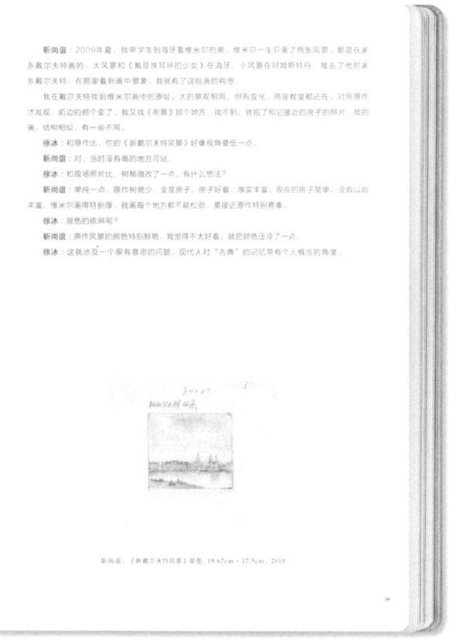

靳尚谊：2009年夏，我带学生到海牙看维米尔的画，维米尔一生只画了两张风景，都是在家乡戴尔夫特画的，大风景和《戴珍珠耳环的少女》在海牙，小风景在阿姆斯特丹。我去了他的家乡戴尔夫特，在那里看到画中景象，我就有了这组画的构思。

我在戴尔夫特找到维米尔画中的原址，大的景观相同，但有变化。两座教堂都还在，对照原作才发现，前边的那个变了。我又找《街景》那个地方，找不到，就拍了和它接近的房子的照片。我的画，结构相似，有一些不同。

徐冰：和原作比，你的《新戴尔夫特风景》好像视角要低一点。

靳尚谊：对，当时没有高的地方可站。

徐冰：和现场照片比，树稍微改了一点，有什么想法？

靳尚谊：单纯一点。原作树就少，全是房子，房子好看，厚实丰富。现在的房子简单，没有以前丰富。维米尔画得特别厚，我画每个地方都不能松劲，要接近原作特别费事。

徐冰：颜色的依照呢？

靳尚谊：原作风景的颜色特别鲜艳，我觉得不太好看，就把颜色压冷了一点。

徐冰：这就涉及一个很有意思的问题，现代人对“古典”的记忆带有个人概念的角度。

靳尚谊临摹作品欣赏

新尚谊：我临的最早一张印刷品是拉图尔的。50年代的中国还没有人研究维米尔。我看过拉图尔的印刷品，临过他的头像。

《临霍尔拜因 -1》，32cm × 24cm，1962，根据画册临摹

“后古典”的端倪

潘公凯

在我们研究中国这样典型的后发达国家的美术现代性问题的时候，由于历史境遇的特殊性，中国的近现代美术史显现出与西方近现代美术史大不相同的本土性特征。我将这种本土性特征看作是中国社会的巨变所表现出来的一种继发现代性特征。在西方发达国家的近现代历史演进过程中，艺术领域的前现代、现代、后现代这样一些历时性的阶段性特征是我们耳熟能详的。但从20世纪前半期直到70年代，由于中国特殊的社会政治背景，西方历时性的艺术演进阶段没有能够充分地在中国的土地上显现出区分——尤其是五六十年代至70年代，由于中国对外的封闭，可以说中国的艺术家们对于西方艺术领域当时所发生的变化几乎一无所知。随着80年代的改革开放，中国艺术家在国门迅速打开的突变中，面对蓦然展现在眼前的西方世界及其五花八门的艺术作品所感到的震惊与憧憬是可以理解的。似乎在一刹那之间，西方在其前现代、现代、后现代的历时性演变过程中的特别繁杂丰富的各种流派、风格、理念、论述和评判标准，都一窝蜂似的同时涌

高品位、高层次的学术价值。那么，还有什么因素决定了新风格新作品的学术价值呢？——是对于古典文化资源理解把握的深度，是艺术家深厚的学养和高难度的技巧训练，是艺术家将古典资源转化为现代营养时处理古今矛盾的艺术才华，是最终的作品所呈现出来的浑然一体的自身圆满性——而这正是“以古为新”的难度所在，同时也正是其魅力之所在。

在此次展览中，正是由维米尔作品转变而来的三件小尺幅的新尚谊先生的作，让我们看到了这种对古典精神的深刻领会，看到了高难度的技巧，看到了贴之后仍然浑然一体的自身圆满性，这真是我们的幸运！

当古典写实油画的原有功能已经大半被摄影所取代，当古典的写实技巧已为一种正在远去的文化资源之时，油画也不得不将重点从表现客观对象转向对身文化资源的再解读、再利用，将拼贴和组合作为一条新的改革创新之路——种在古典与当代，西方和东方的比照中形成的自觉的策略性思考，就是当今世尤其是后发地域的现代性的表征。新尚谊先生近期作品所做的探索，正是基于种策略性的自觉。因而，作为以古典写实技巧与拼贴组合为特征的新的油画风取向——“后古典”——将日益显现出它的潜在生命力。

2011年4月25

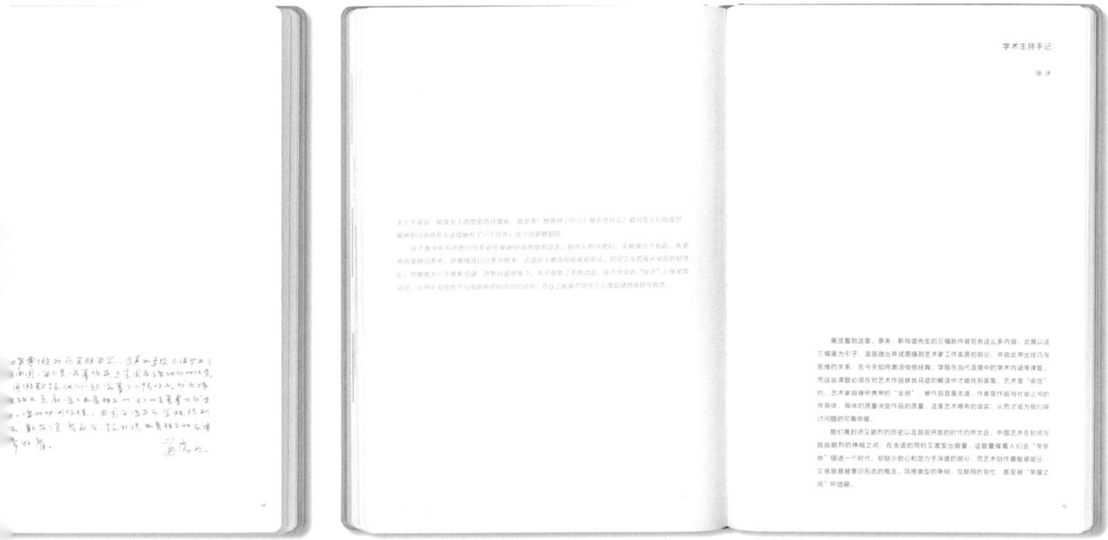

[illegible]

[illegible]

学术主持手记

徐冰

展览看到这里，原来，靳尚谊先生的三幅新作背后有这么多内容。此展以这三幅画为引子，层层拨出并试图摸到艺术家工作实质的部分，并由此带出技巧与思维的关系；在今天如何激活传统经典；学院在当代语境中的学术内涵等课题。而这些课题必须在对艺术作品蛛丝马迹的解读中才能找到答案。艺术是“命定”的，艺术家自身所携带的“全部”，被作品显露无遗。作者是作品与社会之间的传导体，导体的质量决定作品的质量，这是艺术唯有的诚实，从而才成为我们探讨问题的可靠依据。

我们离封闭又剧烈的历史以及自由开放的时代仍然太近，中国艺术在封闭与自由剧烈的伸缩之间，在失语的同时又激发出能量，这能量催着人们去“夸张地”描述一个时代，却缺少耐心和定力于深度的部分，而艺术创作最敏感部分，又很容易被意识形态的概念、风格类型的争辩、互联网的匆忙，甚至被“荣耀之词”所遮蔽。

从一位中国艺术家研究一位西方古典大师想到的

简森·爱德华·考夫曼

靳尚谊是一位典型的中国官方艺术家，他现任中国美术家协会名誉主席，全国政协常委，他是中央美术学院前任院长，经常应邀参与政府的各种与艺术相关的委员会，作品被国家博物馆和其他重要美术馆收藏。但是，他的声望不是出自于传统的中国画，而是来自于油画。几十年来他潜心研究欧洲古典学院派的油画艺术，最终成为中国油画人物画家的典范。

从中央美术学院美术馆正在为靳举办的展览上可以了解到他是如何研究欧洲的油画艺术的。与2005年在中国美术馆举办的“靳尚谊回顾展”完全不同，这个展览展出2010年他画的三张受17世纪荷兰大师约翰尼斯·维米尔启发的画。这些新画是对维米尔名作的临摹，对原画略加改动，靳尚谊说，改动是为了表达他对当代生活的感受。

新的展览《向维米尔致意》提出了若干问题。本文将从一个在西方工作的艺评人的角度来予以探讨。

维米尔作品

靳尚谊作品

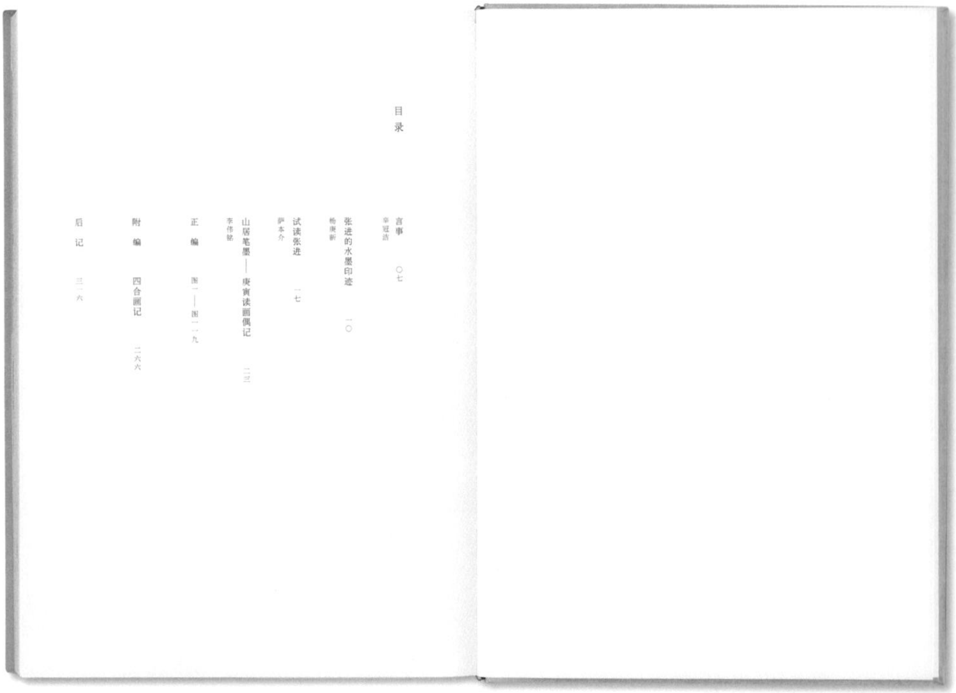

目录

试读张进

萨本介

画画儿分画形的，画感的，画理的，画画表面上比的是手在哪儿，其实比的是心在哪儿。

自由便是想画形便有形，想说理便有理，想传感便传感。一个人自个儿想什么，别人不知道，可画出来便露了『底』，尽管是露了底，也不见得所有的人都能掂量得出来。

绘画的虚实基本在于三块儿：天分、技术和境界，干了一段的人最好回头看看，三样里缺哪样或者哪样不足，不足可以后找补，如果缺，那就只有往下面的层里找了。

有人把事分成三份去做的，也有人把事分成五个方面去对付的，还有人可以把一件事分成几百、上千，甚至上万种可能去设想的。自己能实到哪儿，能虚到哪儿，自己心里最清楚，本来下边一层对上边一层的事儿就不容易理解，更甭说隔好几层了，紧上眼，慢张嘴，千万别不懂装懂。

写生是画画的一剂灵丹妙药，只画形，那只是皮儿，『写』和『生』说的其实都是传感。

写生不是画当初你脑子里的那个东西，也不是画你眼睛看到的那个景，而是画两个东西碰出来的『火花』。

看张进画的太阳，是不是会感到温度和耀眼？！

现抓会产生意外，意外会产生动感，生命感……求全不容易纯，夹杂的东西就多了。

进特别注意多方面的文化修炼，积蓄内美。

一九九四年，一个偶然的机会，张进一脚迈进了潘家园旧货市场，至今不能自拔。在我看来，偶然也是必然，他本来就是一个传统文化的坯子，与古玩有一种天然的亲和力，不掉进去才怪！记得一九九四年，我和张进一道参加包泡在怀柔交界河组织的雕塑公园奠基活动，那是我们第一次晤面，一路上几乎没有谈画，可一说到潘家园、古玩收藏，张进却异常兴奋，问这问那，一下子就拉近了我们的距离。后来每到周末，我们结伴去潘家园淘宝，交流心得体会，竟成了要好的朋友。现在潘家园几乎变成工艺品市场，难以再寻到有价值的古董了，可张进依然乐此不疲，周末照转，有漏照捡，至今十数年过去了，张进的收藏已十分可观。在他自己设计的架子上摆满淘来的物件，有石器、陶器、青铜器、玉器、瓷器、花板、绣片……令你目不暇接。他的藏品以残损者居多，别人看不上眼的残陶、断瓦，他都视为宝贝，以低价买进，回家清洗、拼粘、修补，竟能恢复得完好如初。经他修复的器物不下数十件，看着他修好的磁州窑大罐、马家窑彩陶壶真叫人垂涎！作为一位艺术家，他不以经济价值论高下，更看重藏品的艺术和文化价值。文物是历史文化的载体，是人类智慧的结晶，收藏远不是物的占有，而应该是学习社会、历史、文化、艺术的阶梯。张进特别注重感悟、体味这些藏品的内质，他从古陶俑到了深厚、古朴的品格，从民窑瓷片上学到了自由洒脱的画风，从民间刺绣中吸收了平面穿插的构成方法。张进还将陶俑作为写生对象，反复临摹，将厚朴、凝重、浑然的感觉，巧妙地移植到水墨画中。我认为，收藏不是目的，『藏以为学，学以致用』，把形式、美感转化到艺术创作中去，才是收藏的最高境界。张进做到了这一点，是值得称道和学习的。

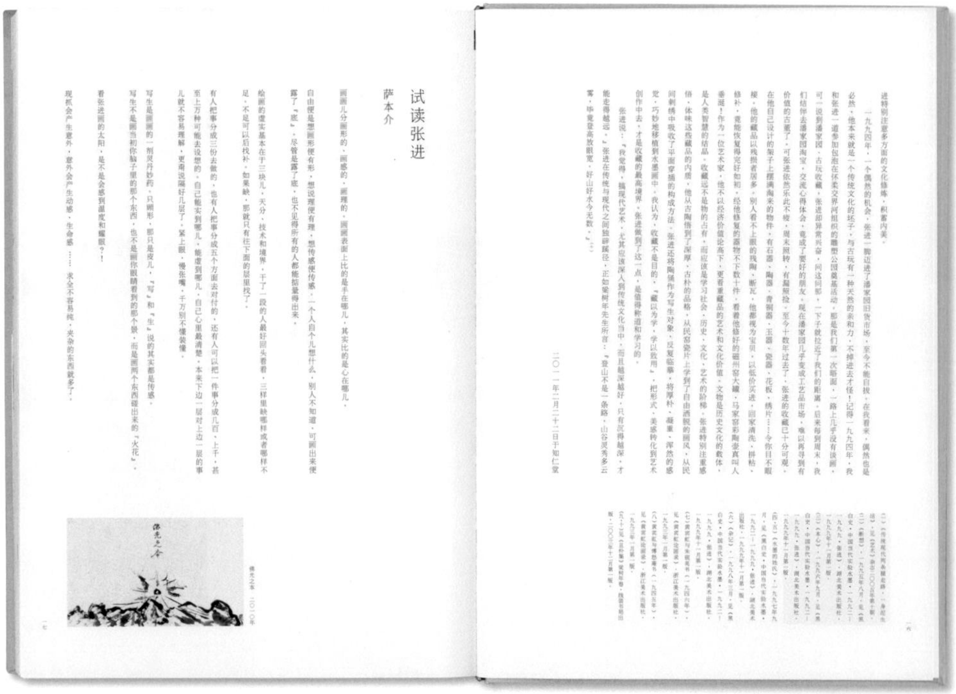

张进说：『我觉得，搞现代艺术，尤其应该深入到传统文化当中，而且越深越好，只有沉得越深，才能走得越远。』张进在传统与现代之间独辟蹊径，正如梁树年先生所言：『登山不是一条路，山谷灵秀多云雾，毕竟登高放眼宽，好山好水今无数。』(十)

二〇一一年二月二十二日于知仁堂

张进 **张进画集**

荣宝斋出版社，北京，2011

345mm×246mm×30mm，318 页

布面精装

1000 册

序言：辛冠洁

文章：杨庚新、萨本介、李伟铭

创作手记：张进

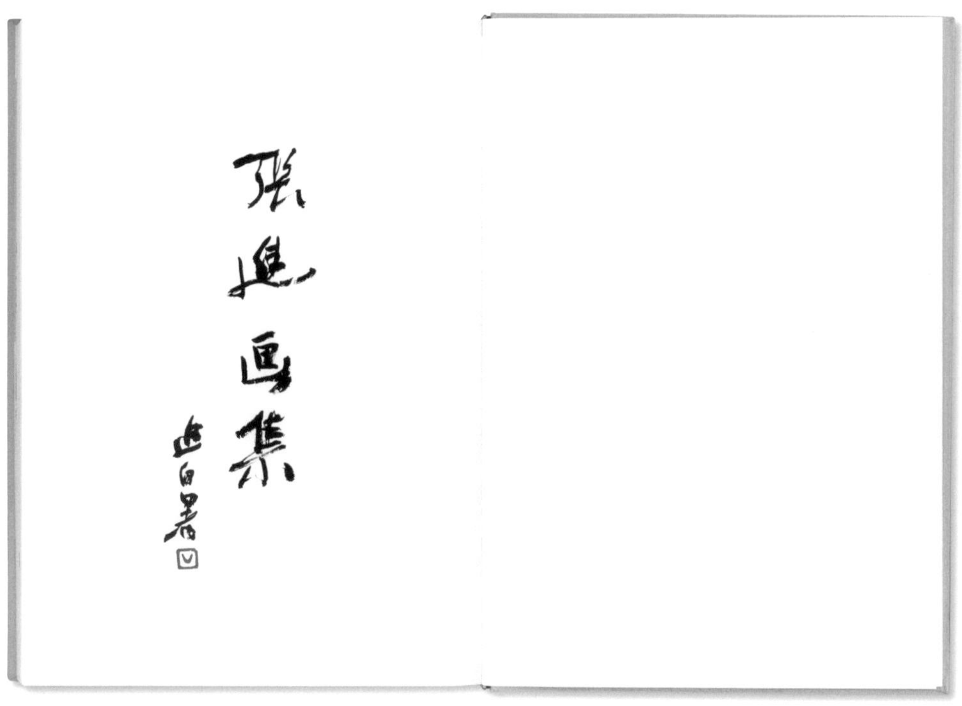

四合澗記

峰石岩写生　二〇一〇年

今年，是近五六年唯一没有参展任务的一年，故画时可以放得更松快些，艳卿曰：“平时真画得不轻快啊。”予答：“俗人总脱不了俗务。”所谓松快，实则还是想在原有的基础上画得更放松一些，只有心放下，才能手放松，笔才能真正提得起来。

（七月十七日）

（七月十八日）

头二日，未画。

（七月十九日）

三日，雨后云起，层峦叠嶂，取纸，写小幅二。

（七月廿日）

四日，早饭后，春林、华忠、艳卿对山写生，予赴春林荫凉处，取四尺六开（纸）写前门一幅，又写山嶂一幅，再写前门后，又对山路写生一幅，此四幅先以焦墨写，后以淡墨略在局部破染，逸丰毫大楷一（支），中锋行笔，画后置于墙上，线质有所提高，笔墨有所放松，如若不言形象，就线的结构关系而言，似乎比以前作品更为合理，自然，画面放松，唯在松中有紧，形散而神聚，下午太阳落山之际，山色晴淡，太阳金黄而不耀目，似佛光显现，其实佛光从何而来？还不是从自然而来，方有瑞象之说，人对自然的认识，亦决定人对宗教的，哲学的，历史的，艺术的态度，故眼前的光景也就崇高和神圣了。

七月廿一日

五日，取纸一小条，写崖檐垂下的二三条爬墙虎，构图由画面上边垂下，此法白石已有，但题材未有，后题打油诗：爬墙爬墙，青蔓几条，本来无心，却也逍遥，写生间，临艳卿带鲜于璜碑，此碑笔方，有魏风之貌气，又见振山带一柏英图册，上有翁同龢魏书四条，见此书多有神交已久之感，临数遍后，颇多启发，唐人写经亦每日必课，除此，又临汪士鋐《送秀野南归诗序》一遍及文徵明《新燕篇》数遍。

仰韶文化　半山彩陶罐

二八六　二八七

九五
观海
款识：太行观海，进写。高三二厘米，宽七〇厘米。

九六
地角
款识：进草记。宽三五厘米，高四四厘米。

《张进画集》

我跟张进1998年就认识了，但直到2011年他终于获得一个重要出版机会的时候，我们才有了设计上的合作。在之前的十多年中，我们“君子之交淡如水”，并不经常见面，但我必须得承认，在众师友当中，他是持续影响着我的一个人。

新世纪之初，艺术市场开始有点往上拱，艺术家中稍有钱的首选多是盖大房子。当时张进夫妻俩住在东四十一条的一个大杂院中，十平米一间屋，两张单人床呈L形贴墙角摆放，一人一张，其他空间则被张进收藏的高古器物填满。我跟张进开玩笑，问他是否也打算去当个“张大户”，他却正色道，人的物质空间大了，精神空间就会被挤压。有了大宅子，就不得不为这些物质的东西劳心费神，而他现在却可以不受任何羁绊地生活，想去哪儿拔腿就走，感觉整个世界都是自己的。

说这话，一晃十多年，但我至今还会时常想起。虽然现在明白了，其实形式不重要，问题不在具体的大小多少，而在价值观，但从另一个角度讲，这番话连同张进的生活实践，又何尝不是一种有着高古之气的生活格调。

从2011年起，我们先后合作了五部书，分别是《张进画集》、《中学生的现实主义》（302—303页）、《四合汲古》（304—305页）、《四合画稿》（358—359页）和《光洒太行》（370—371页）。我印象中张进似乎从没专门为设计的事来过我的工作室，我们的设计沟通，都是在一起逛古董摊中完成的，而寻觅品评古物的过程，即是张进跟我一起构想书籍品格的过程。这种结合，海阔天空却又天衣无缝。我曾问张进如何辨别古董真伪，特别是面对高仿赝品时。张进给我打了个比方：两件看上去完全一样的瓷器摆在面前，一真一伪，对的那件，釉面的感觉就像在脸上蒙了一层纱布，是透气的；而不对的那件，则像在头上套了个塑料袋。所以辨真伪往往就是一眼的事，看的是气息，而不是拿着放大镜去找书上描述的特征。这让我想起设计《张进画集》时的一个小细节：这本书封面很简单，只有“张进画集”四个字，用的是一款现成印刷字体。汉字跟西文不同，由于其间架结构的复杂性，单个字型很难做到无懈可击，而电脑字库的生成方式，更使得文字的“书写性”丧失殆尽。因此如果仅用少数几个汉字来构成封面，那么这几个字就需要特别讲究，书籍的格调往往系于一线。对于这种尝试，我一般会很谨慎，但这个项目我没犹豫，因为我知道张进除了画家身份之外还是一个造诣深厚的书法家，我想他肯定可以帮我一起对这些现成字块进行大刀阔斧的整肃，使其焕然一新、判若两“人”。当时我们为这四个字起码花了一上午的时间，实际上所做的调整却极其细微，如果不两相对照，甚至难以说清具体的差别。但我们又确实因此使“塑料袋”变成了“纱布”，彻底改变了文字乃至整个封面的气息，这大概就是东方式审美的奇妙之处吧。

我曾经看到一则对于审美格调的描述，四句：“古拙为上，大度次之，清秀再次，趣味最差。”这十六个字说得很妙，基本可以作为一把尺子，谁在哪一层一量便知。我一直觉得古拙在真正的日常实践中可以感知、接近，但最终还要交给时间来打磨，而“既雕且琢，复归于朴”则是接近的路径。“拙”不是肆意、荒疏或矫揉造作的磕磕绊绊，而是用尽心力雕琢却不留痕迹的“大巧”。

后来，张进送我一小条字，上书当年陈师曾题赠齐白石的“能似不能”。说得多好，这正是我对于格调这个事一直想说又不知该怎么说的感觉。

图像的游历——中国当代新工笔画家图文集萃

文化艺术出版社，北京，2011

290mm×220mm×16mm，160 页

布面精装，外裹函套

2000 册

艺术家：张见、姜吉安、雷苗、陈林、高茜、崔进、徐华翎、杭春晖、秦艾、郑庆余

序

"新工笔"是近年来出现的一种艺术现象，它的形成伴随着社会文化历史发展的必然逻辑，中西方文化艺术的不断深入交流，为它的出现提供了肥沃土壤。"新工笔"绝不是简单地使用中国传统绘画技法表达西方或当下的观念及话语，而是站在中西方哲学、美学及艺术史前沿成果的基础上，建构新的艺术方法、思想方法，并基于新的思想方法重新架构了工笔画的形式语言和材料语言，使"新工笔"形成了完整系统。

水墨历来被认为是中国艺术的显著特征，但传统写意水墨的直觉性、感性和偶然性特征与当代人生存感受存在着抵牾，且传统工笔绘画又因写意文人画的兴盛、照相机的出现而逐步被边缘化。在如此绘画语境中，"新工笔"的出现无疑改变了这一局面，使独具中国文化特色的绘画语言在当下社会语境中重现勃勃生机。

收录本书中的新工笔画家们身处当代艺术语境，作品面貌上相对比较成熟，呈现出独特鲜明的个人风格。姜吉安对光影语言的探索的鸡蛋与几何体；张见的荒原、云阵及棕榈风景；陈林的幻境与鸟；秦艾的动物与舞台布景；雷苗的瓶与花；高茜的静物；徐华翎的私密女子；郑庆余的透明少女；杭春晖的灰色小鸟；崔进的隐喻人物……这些作品不论在图式风格还是形式语言上都区别于传统工笔绘画。它们怀疑既有或既定秩序及思想；它们反思陈旧历史及其语言机制；它们逃离一切既有标准，在艺术领域内寻求绝对自由；它们在不断创新，不断寻求突破，在求新求变中，始终没有放弃[illegible]一贯坚守着的传统工笔技法。工笔不似写意，可激情地挥毫而就，它是于平静中慢慢沉淀，沉淀了光阴，沉淀了文化韵味。

作品可以传递情感、观念及想法，但因为其特有的艺术语言影响了表达的有效性，而文字的使用恰好弥补了这一遗憾。文字记录历史，抵抗流逝的光阴；文字记录艺术状态，承载艺术家的心得体会。观者正借文字走进画家内心，走进艺术家精心营造的艺术世界。本书中相关作品的评论及对画家的采访，正为观者提供了一次与"新工笔"艺术零距离的接触。"新工笔"画家群体面貌也由这优秀画作和诚挚文字勾勒而出。

本书旨在通过对新工笔史料性资料的搜集与整理，结集出版，肯定并支持新工笔发展及已取得的成就，也表明传统工笔画传承与延续至当下的可行性。更为重要的是，"新工笔"创作以质疑与怀疑为切入点，可为时下中国当代艺术的发展提供有效表达经验。"新工笔"在语言、价值、观念层面的创新，也会因不断地深入探索而广受观者喜爱，也能经受时间与历史的洗练，在艺术领域走得更远！

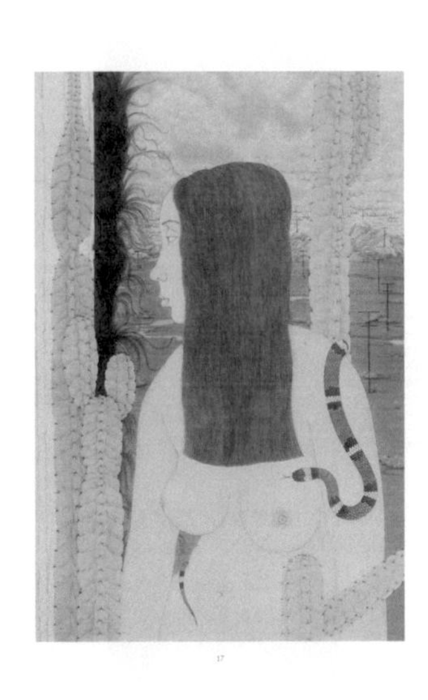

绘画与现场

姜吉安

《隔壁》系列延续了《影的铍法》系列的创作思路，这个思路源于传统扇面绘画的启发。从传统绘画方法论资源来看，传统绘画没有完全停留于绢和纸的平面，绘画与物和环境结合的例子很多，比如卷轴、扇面、瓷器绘画等。实际上，卷轴的画面内容不完全包含卷轴画的全部意义，其中蕴含的“观看方式”是卷轴画的一个重要组成部分；扇面的打开与关闭，包含人的行为介入，这种行为是扇面艺术不可缺少的组成部分。这种与画面之外的关系所蕴含的方法论意义在中国传统艺术家那里并没有体现为一种方法自觉，这种“关系”也没有作为方法独立出来，而主要做为一种依附与衬托关系存在。但这个线索无疑是存在的。

在绘画图式越来越陷入脸谱化、抒情、现实、超现实、抽象、构成等现代和前现代陈词滥调的中国艺术状况下，艺术家如何能够获得一种思想和艺术的解放感，使主体能够相对显现，又能使绘画保持存在的理由？无疑需要在各种传统资源和当代生活直觉中去发现和去蔽，并展开另外的可能。绘画与物和环境紧密结合的方法意义，运用于绘画实践则可将艺术家从图式经营的困境中解放出来，从而建构一种新的绘画意义。

《隔壁》系列，呈现一种“绘画与展览现场的关系”，这个关系有三个层次，如果拗口地用文字表述就是：展览馆中的……展览中的……展览，观众的观看，画面中的窗户，窗户中的另外一个展览构成关系的三个层次或者简化为：展览的展览的展览。在此，观众“看展览”这个行为情景本身成为《隔壁》系列作品的内在结构，《隔壁》组织了绘画与现场的“关系”，这个“关系”使展览馆、展览、观众都成为作品结构中的一部分。

（图28）
隔壁1
绢本设色
110cm×110cm
2010年

（图29）
隔壁2
绢本设色
162cm×52cm
2010年

（图30）
隔壁3
绢本设色
100cm×110cm
2010年

31

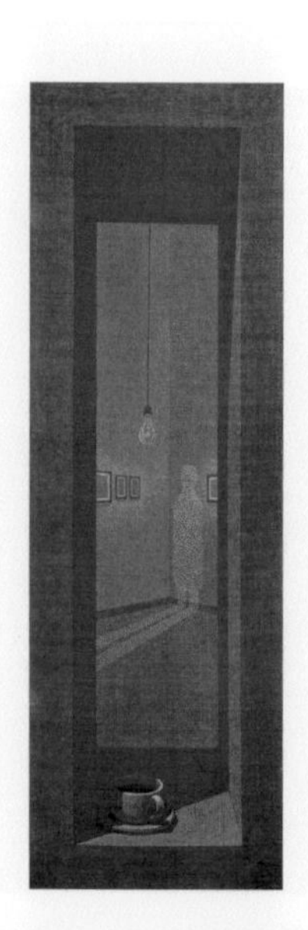

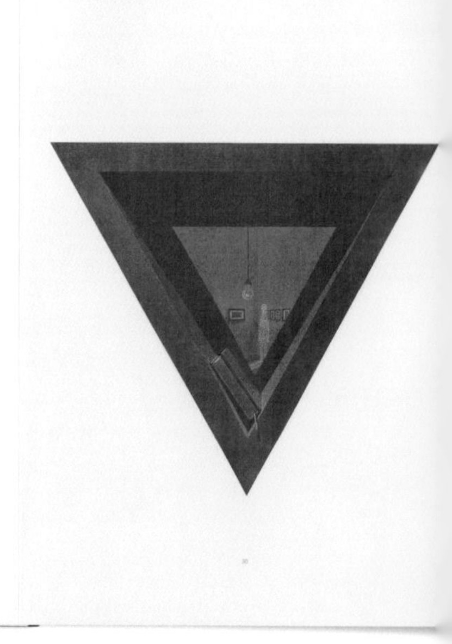

“静心”之缘起

徐华翎

《静心》系列创作于2009年初，当时席卷全球的金融危机正呈愈演愈烈之势，艺术界尤其是当代艺术市场受到了很大的冲击，整个艺术市场从2008年疯狂的状态一下沉静了下来。我记得在2008年的时候从一个小侧面可以反映当时艺术市场的红火，工作室的租金一涨再涨，供不应求，很多刚毕业的学生和外地来京的艺术家都投入这股浪潮之中。很快地，由于经济危机的爆发，很多工作室都空了出来，人去屋空。当时我的想法是，整个艺术圈重新洗了一下牌，有很多人在2009年以后可能就不会从事艺术创作了。

虽然金融危机给艺术界造成一些负面影响，但也还有积极的一面。它让很多人的头脑都冷静下来，不再那么狂热甚至机械地去做一些东西，而是静下心来，重新审视自己和自己的艺术。就我自身而言，沉得住的时候往往能创作出好的作品。所以从这个角度来说，金融危机给艺术家一个重新审视自己的机会。这个时候便有了《静心》的创作想法。

其实“静心”是一个比较抽象的概念，我还是选择了我喜欢的青春期女性作为载体，并借用了现在流行于年轻女性中的瑜伽运动来呈现我的这个想法。“瑜伽”也是这些年在中国渐渐热了起来，虽是舶来品，但与中国文化有异曲同工之妙。我自己闲暇时也会做做瑜伽，调整身心状态。练瑜伽最重要的其实也是“静心”，把烦恼暂时统统忘掉，使自己与自然融为一体，得到真正的休息和放松。这个出发点与我想要表达的东西是一致的。

女孩的形象选择也是我有意为之，我一直希望自己所表现的形象是带有时代特征的，也就是当代女性形象。化浓妆、刺青、穿性感内衣是这个时代年轻女性的表象特征，她们喜欢表现自己，虽有东方女性的含蓄但也有张扬的一面。《静心1》选择的就是这样一个形象。她有着时髦的发型，眼睛画着浓浓的妆，手臂上纹着一只小鸟。这样一个时髦的女孩本应跳脱恣意，现在正静静地屈膝坐在那里，看起来很安静。这种形象和行为的微妙反差也是我想带给观者的。

我的画面总给人以幽淡、朦胧的感觉，这也是我对这个年龄段女孩的一种感觉。她们很丰富，但是你无法完全捉摸透她，总是有雾里看花的感觉。这件作品我也一如既往地选择了高调的色彩来表现，白色的瑜伽服、若隐若现的蕾丝、淡粉色的皮肤和素色的绢本色。唯一跳脱出这种色调的是黑色的头发、重重的黑眼圈和手臂上的刺青。画法基本上采取了没骨的画法，线被隐到很弱，这也是为了整体的画面效果。虽然整件作品看起来颜色很淡，但每一处都经多遍渲染而成，这也是传统的“薄中见厚”，画面看起来会更有层次、厚度和质感。

《静心5》是《静心1》同一系列的作品，但体现的是另一种感觉。画面中一样是个时尚的女孩，她染着头发，戴着长长的睫毛，涂着红色的指甲，蓝色和粉色的运动衣裤看起来很自然，不去刻意雕琢。她的膝盖受伤了，她虽然安静地站在那里，但微微低下的头和攥起的拳头已经使画面变得不安静了。通过这些对比，我想传达的是一种再出发的意味。

这件作品里的线条更少了，我始终觉得线条在我的作品里并不适用，它使我要表达的东西简单化了。而在“虚”里去寻找变化，是我一直在技法上作的尝试。传统工笔画勾线填色的技法，容易使画面有“刻、板”的弊病，过多地运用线条也会使画面的距离感减弱。所以，线条的消解在我的画面里一直是个重要的因素。

（页102）
静心1
绢本设色
160cm×100cm
2003年

（页103）
静心5
绢本设色
160cm×100cm
2003年

100

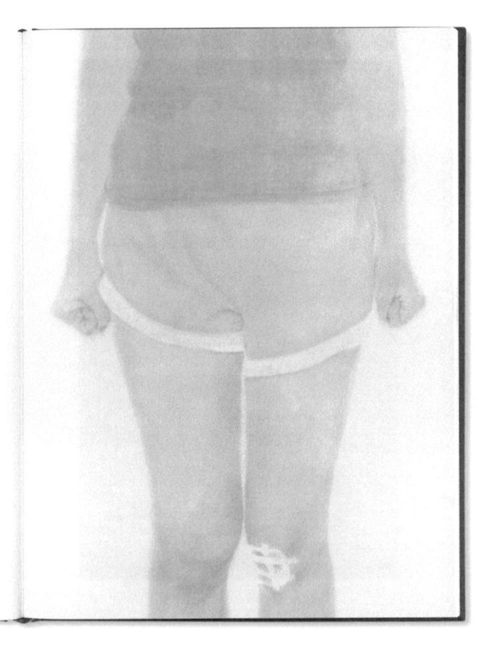

基于艺术语言转变后的自足性

（节选）

刘礼宾

迄今为止，杭春晖的艺术创作大体上经历了三个阶段。早期主要是以“玩具熊”为表现对象，在人为设定的“熟悉”空间中，在作者审视的目光下，惯常作为“可爱”代表的玩具熊显得暴戾、自闭，在自娱自乐中，“玩具熊”近乎与外界隔绝，只与所处空间相呼应，无任何与画外相沟通的倾向。杭春晖描述的是一个成年人眼中的童话世界，在纯洁、可爱的表层下面，暴力、杀戮在沉默地进行着。本应带来希望的“光柱”凝固为边界，彰显着被划定的活动范围。这一时期，杭春晖娴熟地运用传统工笔画的晕染、构线技法，营造出一个颇具超现实主义色彩的个人世界。我想这样一个世界，正是杭春晖转折之后所营造的创作空间的真实反映。在这个阶段，杭春晖还画了几张带纹身的男人半身像，“玩具熊”变成了“男人”，血凝固成了纹身——前者是伤口的排泄物，后者是伤口的凝固。身处复杂的社会状态中，作为个体的生存，杭春晖常常会有无力感、孤独感，这种感觉需要宣泄，否则他可能变得歇斯底里。于是在杭春晖的前期创作中，“伤害”变成了一个重要的主题。对于杭春晖来讲，这样的描画中交织着一种很矛盾的心态，里面潜藏着一种不可名状的情绪。人们对于伤害本来就保有一种恐惧，同时又会有一种自得其乐的内部欣赏，有时候因为人把环境给屏蔽掉了，然后在内部获得一种完满的自足，这种自足里面有疗伤，也有对伤害的一种迷恋和欣赏，一种抚摸伤口的快感。从杭春晖的这段自述中，我们不难发现他早期绘画创作时的状态。

杭春晖创作的第二个阶段以“假山石”为主题。假山石是他第一期作品中用于渲染现场的道具，现在独立出来，成为独立支撑画面的“主角”。杭春晖做此转变可能源自偶然的机缘，但持续下来，则是一个有意的选择了。从艺术家的精神状态来讲，杭春晖做此转变，明显超出了以往个体化的内视，而进入对物象的外向化分析（这与他前期对“玩具熊”以及“男人”的观察方式有相似之处）。以往他描画对象，无论是“玩具熊”还是“男人”，都十分精细。现在的“假山石”在双向晕染技法的处理下，变得含蓄而灵动。新近杭春晖又将这一技法用于创作系列作品“男头像”。在这些作品中，艺术技法和表现对象之间实现了精妙的融合。淡墨素雅，层次丰富，细节叠出，人物精神状态趋于存在的本真，潜在呼应着时代的律动。在不断纯化艺术语言的过程中，杭春晖实现了艺术语言转变后创作状态的自足性，同时实现了对表现对象的深入挖掘，而这一挖掘又对提升时下的精神诉求具有潜在的作用，即以个体探索的方式进行着上文所说的“中国崛起情况下民族精神建构对工笔绘画的期冀”。

（对页）
蓝色
纸本设色
80cm×57cm
2011年

（页114-116）
阴谋
纸本设色
100cm×72cm
2008年

113

喻红 **黄金界**

CHARTA，米兰；长征空间，北京，2011

278mm×250mm×18mm，132 页

布面精装，裹纸质护封

2000 册

编辑：梁中蓝

前言：卢杰

文章：张晴、Alexandra Munroe、杜小贞

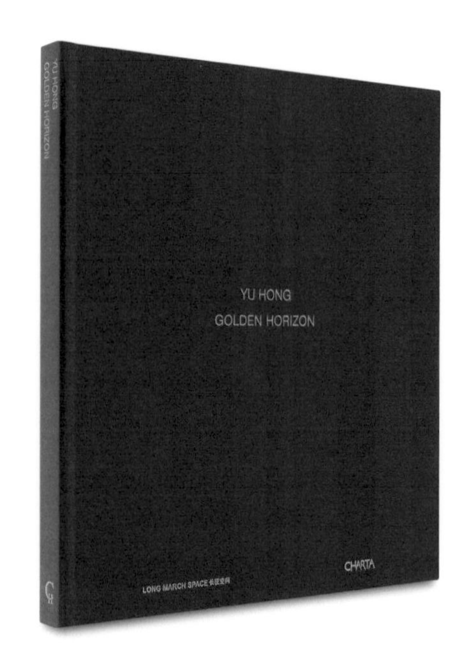

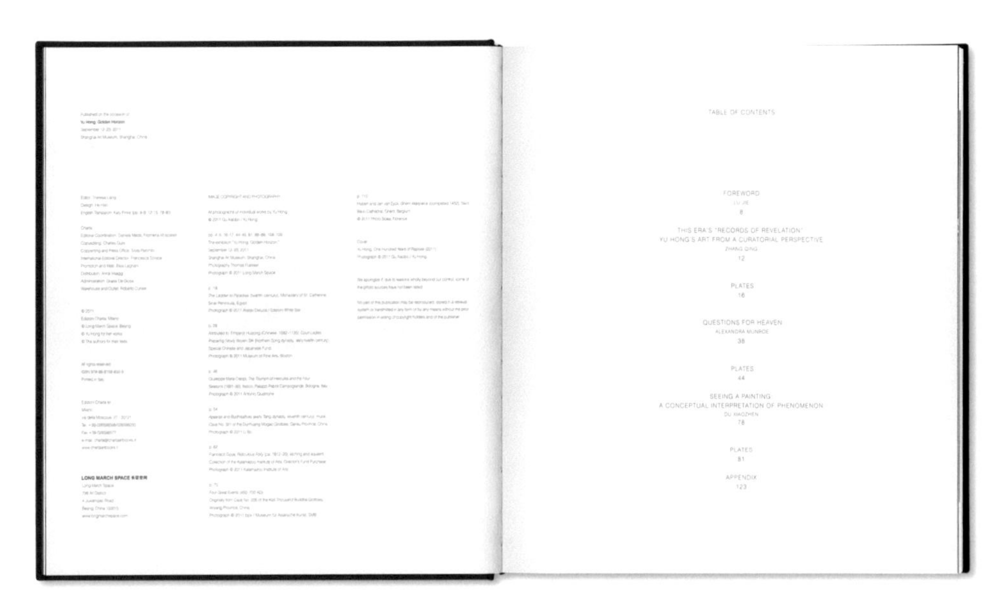

FOREWORD

With the work she produced in the eighties and nineties, Yu Hong—a principal representative of China's New Generation movement—shed glaring light on the political consciousness and social imagery of an era caught in the midst of a divide. From where we stand today, watchwords such as "avant-garde" or "feminism" are not nearly sufficient to convey the era's complex context and the significance of the individual within it—nor the way in which these elements were ultimately woven into the work of the artist inhabiting these changes and their impacts. Following this initial momentum, Yu Hong's "Witness to Growth" series (1999-ongoing) is an articulation of the connection between the individual and the "history" of the era, while the series "She" (2003-07) describes exquisitely—down to even the slightest tremor—the experience of the deeply complex relationship between a woman's inner life and her external world. In 2007, Yu Hong began work on the "Golden" series. After "Witness to Growth" and "She," this is without a doubt her most emblematic work to date.

Yu Hong completed *Us Two–Zhao Bo and Yu Hong 1* and *Us Two–Zhao Bo and Yu Hong 2* in 2007, followed by two more giant pieces in 2008: *Romance of Spring* and *Ladder to the Sky*. All four were exhibited in her 2009 solo exhibition "In and Out of Time" at the Guangdong Museum of Art. Next was "Golden Sky" at the Ullens Center for Contemporary Art, a solo exhibition marked by its distinctive "ceiling frescoes": four great masterworks titled *Altum* (2009), *Questions for Heaven* (2009-2010), *Natural Selection* (2010) and *Sky Curtain* (2010). Most recently, as both a retrospective of the past four years of creation and as an introduction to new work, "Golden Horizon"—Yu Hong's September 2011 solo exhibition at the Shanghai Art Museum—displays the above signature pieces alongside some of her newest works—including the ambitious *One Hundred Years of Repose* (2011) and two new series: "Spontaneous Motion" (2010-11) and "Wrestling" (2011).

The artist uses gold leaf extensively in her picture planes, prompting a connection between her work and both Western classical painting and traditional Chinese religious and court painting. However, the significance of this behavior runs much deeper: from a spectrum of divergent painting traditions—particularly from the practice of mural and fresco painting—Yu Hong derives a new and utterly unique painterly language and compositional syntax. Nothing could be a more accurate choice than the color gold as a symbol to convey the intensity of desire and loss of faith that characterize life in this era. Gold is a visual embodiment of Yu Hong's profound inquiry into the aberrations of our society and the individual's experience of life within it.

This catalogue brings together all of the works comprising Yu Hong's Shanghai Art Museum solo exhibition "Golden Horizon" and summarizes the past four years of Yu Hong's work on the "Golden" series. An essay contributed by the exhibition curator Zhang Qing, Director, Curatorial and Research Department, National Art Museum of China, Beijing, expounds upon the exhibition's curatorial concept and his research into Yu Hong's series. Alexandra Munroe, Samsung Senior Curator of Asian Art, Guggenheim Museum, and Du Xiaozhen, Professor of Philosophy, Peking University, Beijing, share research and analysis of Yu Hong's work from two distinct perspectives.

We are extremely honored to have the chance to collaborate with the publishing house Charta in jointly editing and publishing *Yu Hong: Golden Horizon*—in commemoration of Yu Hong's "Golden Horizon" exhibition at the Shanghai Art Museum—as we witness the continued growth of an artist of unique and central importance to the history of Chinese contemporary art.

Lu Jie
Founder
Long March Space

天井
Atrium, 2009
Acrylic on canvas, 500 x 600 cm
Composed of four canvases

QUESTIONS FOR HEAVEN

ALEXANDRA MUNROE

The expression that there is nothing to express, nothing with which to express, nothing from which to express, no power to express, no desire to express, together with the obligation to express.
—Samuel Beckett, *Three Dialogues*, 1949[1]

Contemporary figures are looking up, squatting, twisting, bending and stretching in a vast gold space. Yu Hong's painting, installed as a ceiling mural above our heads, makes us twist too, yearning to see what its figures see through their binoculars aimed at far distances. "They are searching for something," the artist says about her opus *Questions for Heaven* (*Tian wen*, 2009–10, see pp. 55–61).[2] The title is taken from a Chinese classic written by the poet Qu Yuan, an exiled imperial minister of the Warring States era who drowned himself in protest against corrupt rule. "The poem lists over 170 questions about the nature of the universe, how and why things happen, human and animal species, spirits and ghosts, society and history," says Yu Hong.[3] The poem, we come to understand, can be seen as a method for Yu Hong's probing observations of China today and, more broadly, of the existential conundrums of our contemporary global society.

Like much of her recent painting, the composition of *Questions for Heaven* is inspired by a work of art from a much earlier time and culture. In this case, her inspiration is a cave painting in Dunhuang, one of the largest and most important centers of early Chinese Buddhism that flourished along the western Silk Route from the fourth through the fourteenth centuries. Comprising some 500 cave sanctuaries embellished with all-enveloping mural paintings and sculptural tableaux dedicated to propagating Buddhist doctrine and cultivating faith among pilgrims, traders and monks who gathered at those crossroads, the Mogao Grottoes are among the greatest feats of religious art in the world. Yu Hong's inspiration is a scene dating to the Tang period depicting the Buddha preaching the Dharma in the heavens, seated regally on scrolling clouds set above trees and against a pale blue sky, flanked by bodhisattvas and a multitude of apsaras whose dancing postures and flowing scarves compose two long arcs emanating from the central deity (see *Apsaras and Bodhisattvas*, p. 54). She appropriates the crisscrossed assembly of lean figures moving in space, but positions them "to seem we are looking at the subjects through an invisible glass ceiling."[4] That is, she paints them as if they were actually celestial beings.

Yu Hong's use of China's poetic and religious sources in *Questions for Heaven* is far more than mere formalism; it is a radical critique of the chronological and conceptual divide between "traditional" and "contemporary" art forms and the assumed incongruity of such monolithic binaries as modern/pre-modern and secular/sacred. Yu Hong's real subject is temporality, how the present contains the past and predicts the future. Time is not linear but has multiple and coincident zones that the human imagination alone can penetrate and encompass. By appropriating history she can deconstruct it, collapsing historical time itself. Yu Hong's other great subject is faith. How, she asks, did people of the past believe so fervently? How were miracles made manifest? Stripped of dogma and religious practice, caught in everyday, momentary gestures wearing jeans and T-shirts, lost in golden spaces without horizon or place, what are humans today searching to know? And how can painting serve contemporary spirituality as it once served the religious cultures of Asia and the West, describing seen and unseen worlds? Like Qu Yuan's *Questions for Heaven*, Yu Hong poses the unknowable.

1
When above and below were not yet formed,
Who was there to question?[5]

Yu Hong was born in Xi'an in 1966, the year Mao Zedong launched the Great Proletarian Cultural Revolution, ushering in a decade of massive political chaos and social turmoil. Aimed at enforcing socialism and installing Maoist orthodoxy within the Communist Party, the Cultural Revolution became an all-out persecution of modern Western "bourgeois" elements and Chinese

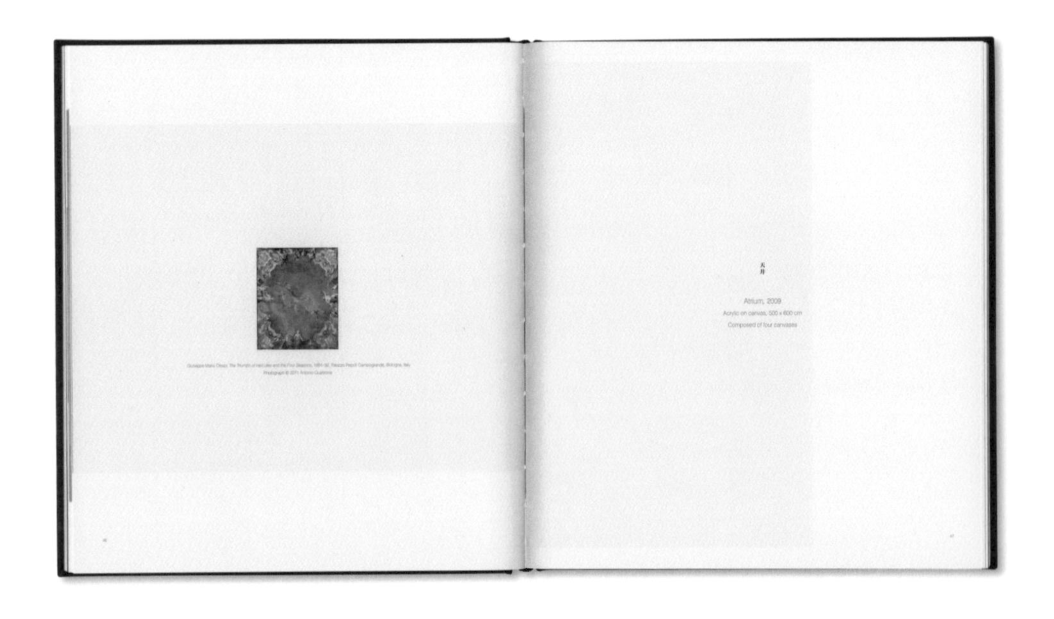
天井
Atrium, 2009
Acrylic on canvas, 500 x 600 cm
Composed of four canvases

Wrestling 3, 2011
Gold leaf, acrylic on canvas, 250 x 300 cm

Wrestling 4, 2011
Gold leaf, acrylic on canvas, 250 x 300 cm

《黄金界》

喻红的这个项目是从设计请柬开始的。2011年春，长征空间的老板卢杰给我打来电话，希望我能为喻红当年9月在上海美术馆大型个展《黄金界》的同名画册来做设计。出于对喻红作品的喜爱，我欣然接受了这一委托。但在画册之前，有个稍显棘手的问题需要先行处理——喻红这次个展开幕式计划特别邀请百余位重要宾客前来参加，其中有些还需从海外专程飞抵，为此主办方打算额外制作一枚有别于普通请柬的邀请函来彰显诚意。但卢老板说由于这次展览耗资巨大，恐怕在此项目上难有太多的预算，因此想听听我的意见，看有什么好的办法能够解决。

卢老板的疑虑不是没来由，就我目力所及，之前这类VIP请柬多是用夸张的形式和昂贵的材料所构建，牛皮、金丝楠木……极尽张扬之能事，似乎奢华繁复的程度与其所蕴含的价值之间有着不容置疑的正比关系。但说实话，即便不惜工本，这种"大金牙"式的设计也完全不适合喻红和她的展览，寻找另一种"价值感"对于这个项目而言根本就是必然——我遂请喻红根据展览海报上所选用的那件名为《失重》的作品专门绘制了一幅素描手稿，继而把这帧铅笔稿用单黑印在一款象牙色轻柔的薄纸上，仅150份，艺术家逐件亲笔签名、标号，之后夹在统一印制的普通请柬中寄出。一张成本几可忽略不计的印刷品（我估算了一下，150张单页用了约300元人民币）因为量身绘制、亲笔签名、少量且独一无二的编号而获得了艺术品属性，弥足珍藏；而另一方面，它又确实只是一张小小的印刷品，轻盈、疏朗而全然不会给人以任何急切和压迫的感觉，这种"不迎客来，不送客去""千里送鹅毛"式的优雅和超然正像我所看到的喻红。

再说画册。设计之初，喻红首先提出了她的疑虑：此次参展的几件主要作品并非首次发表，但在过往的出版物中，这些由若干单幅画作拼接而成的巨幅作品印在书里总感觉像一张轻飘飘的纸片而全无油画的质感和体量。她问我这种感觉是因何而起又该怎样解决。这个问题还真把我问住了。喻红说的感觉我能理解且亦有同感，但翻看之前的画册，似乎做法上也没什么不妥——依照惯例，印在书里的画作首先要沿作品边缘把翻拍的图片裁切规整，那么接下来排版的时候就势必要在单幅画作之间留有一定的缝隙，以示此作品为多拼而非整体绘制。但如此一来，本来彼此画框紧贴、有厚度有体量的油画势必就变成了几帧“照片”的排列组合，其轻飘飘、硬生生的感觉也就在所难免了。道理认清了，但具体怎么应对却让我一时无措，正巧世界设计师大会在台湾开幕在即，作为当年的参会者，我只能带着问题先行启程了。

设计师大会坐而论道乏善可陈，但台北美食令人流连。一天我和几个朋友坐在一爿街头小馆，在他们点餐的过程中我无意间注意到自己放在桌上的手——四指并拢，手指边缘直且不留缝隙，尽管如此，却仍然可以清晰地感知每根手指的弧度转折以及彼此的关系，这种体量感其实来自手指边缘的细微起伏和随之带来的光影变化。这一发现让我豁然开朗，紧贴在一起的由厚亚麻布包所裹的画框岂不与此同理！回到酒店，我打开没有裁切过的拍摄原图，沿画作边缘精准地勾勒出直边中的每一寸细微起伏，之后将两张画作紧对在一起，背后加一深色阴影——因画布厚度所带来的微妙变化和拼合时无法避免的细小参差在阴影的衬托下跃然纸上，虚拟的拍摄图像就在这样的抵力中又被反推成了“画”。

《黄金界》这本书中所收录的主要作品多为巨幅（《天梯》6 米×6 米，《天井》等 5 米×6 米），囿于书籍开本，不免要添加一些作品局部以作近观。但局部如何来选取是个问题。喻红作为学院艺术家的代表，其绘画技巧是十分精湛且独具个人语言特点的，但越是如此就越要更审慎地在设计布局上去处理局部呈现与整体画面的关系，以免“出戏”“跳戏”。在我看来，喻红在这批画中对每一个具体人物的精妙刻画，其用意都不在个体造型而是在群体的关系上——个体越生动则关系越传神，在这里，关键词只能是“关系”而非其他。为此，我在局部的选取上坚持不用“小”局部凸显画法细节，取而代之的是用“大”局部来强化对于“关系”的提示。如果把书籍比喻成电影，那么这部“电影”是不需要特写镜头的。

一年之后，我的老师谭平的个展在中国美术馆圆厅盛大开幕，美院同事去了很多，攒动的人群中又见喻红。喻红说她一直想找个机会当面感谢我，我给她设计的这本画册是她之前所有出版物中最使她满意的。“良言一句三冬暖”，这本是喻红的善意和教养，但对于一个设计者而言，又有什么能比听到这样的话更让人开心的呢？

滕菲 **寸·光阴**

偏锋新艺术空间，北京，2011

346mm×250mm×15mm，88 页

布面精装

500 册

文章：蒋岳红、吕胜中

对话：尹吉男 / 滕菲

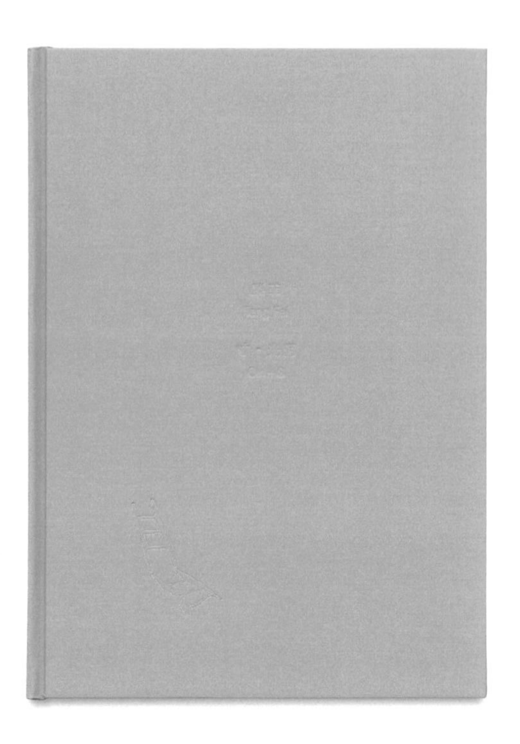

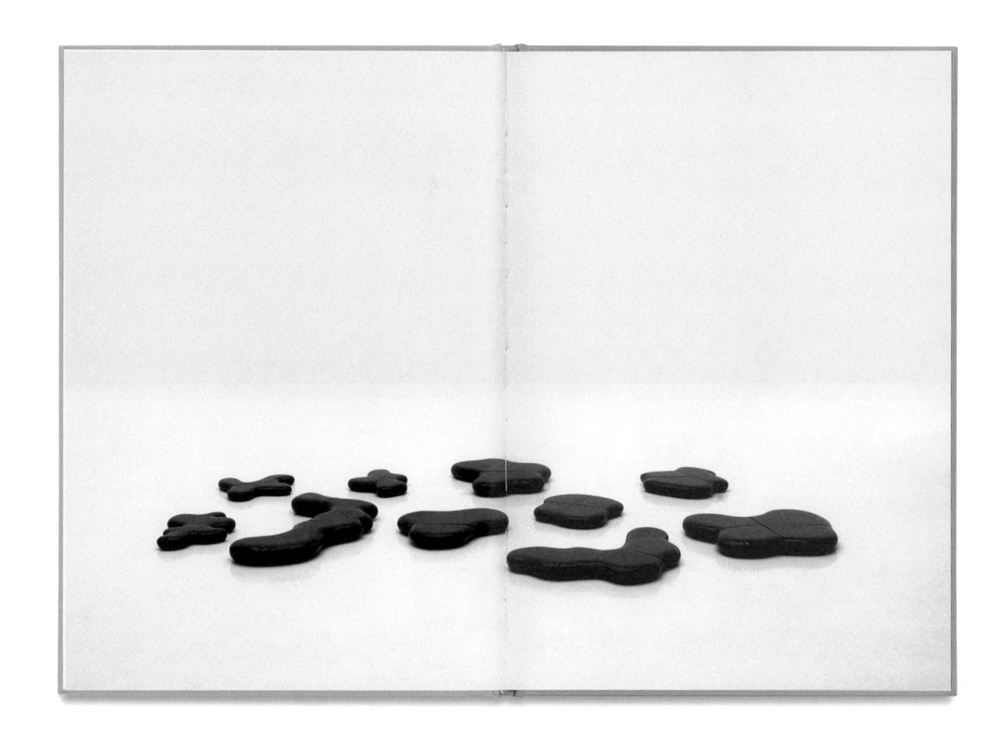

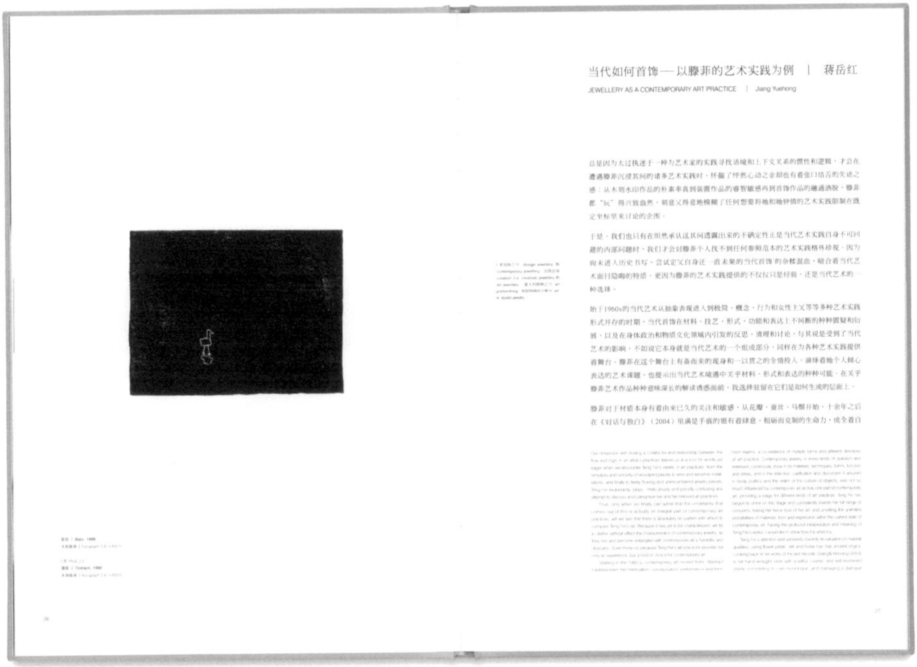

当代如何首饰——以滕菲的艺术实践为例 | 蒋岳红

JEWELLERY AS A CONTEMPORARY ART PRACTICE | Jiang Yuehong

总是因为太过执迷于一种为艺术家的实践寻找语境和上下文关系的惯性和逻辑，才会在遭遇滕菲沉浸其间的诸多艺术实践时，怀揣了怦然心动之余却也有着张口结舌的失语之感：从木刻水印作品的朴素率真到装置作品的睿智敏感再到首饰作品的融通洒脱，滕菲都“玩”得兴致盎然，刻意又得意地模糊了任何想要将她和她钟情的艺术实践限制在既定坐标里来讨论的企图。

于是，我们也只有在坦然承认这其间透露出来的不确定性正是当代艺术实践自身不可回避的内部问题时，我们才会对滕菲个人找不到任何参照范本的艺术实践格外珍视。因为尚未进入历史书写，尝试定义自身迁一直未果的“当代首饰”的杂糅混血，暗合着当代艺术面目隐晦的特质。更因为滕菲的艺术实践提供的不仅仅只是经验，还是当代艺术的一种选择。

始于1960s的当代艺术从抽象表现进入到极简、概念、行为和女性主义等等多种艺术实践形式并存的时期，当代首饰在材料、技艺、形式、功能和表达上不间断的种种置疑和拓展，以及在身体政治和物质文化领域内引发的反思、清理和讨论，与其说是受到了当代艺术的影响，不如说它本身就是当代艺术的一个组成部分，同样在为各种艺术实践提供着舞台。滕菲在这个舞台上有备而来的现身和一以贯之的全情投入，演绎着她个人倾心表达的艺术课题，也提示出当代艺术境遇中关乎材料、形式和表达的种种可能。在关乎滕菲艺术作品种种意味深长的解读诱惑面前，我选择驻留在它们是如何生成的层面上。

滕菲对于材质本身有着由来已久的关注和敏感，从花瓣、蚕丝、马鬃开始，十余年之后在《对话与独白》（2004）里满是手痕的银有着肆意、粗砺而克制的生命力，成全着白

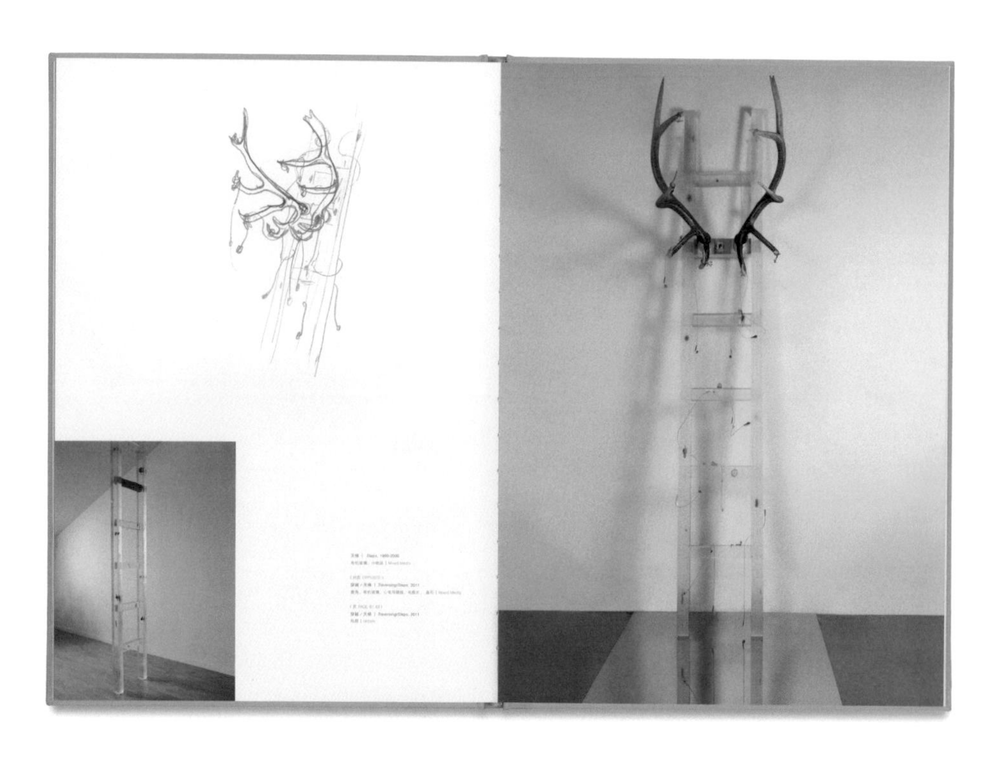

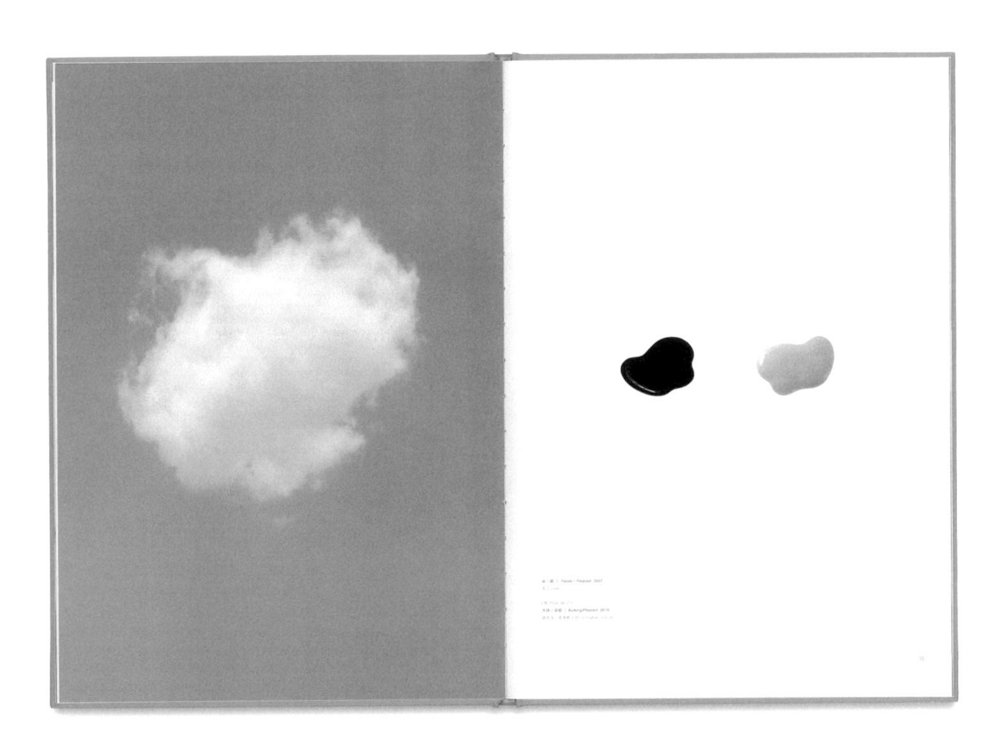

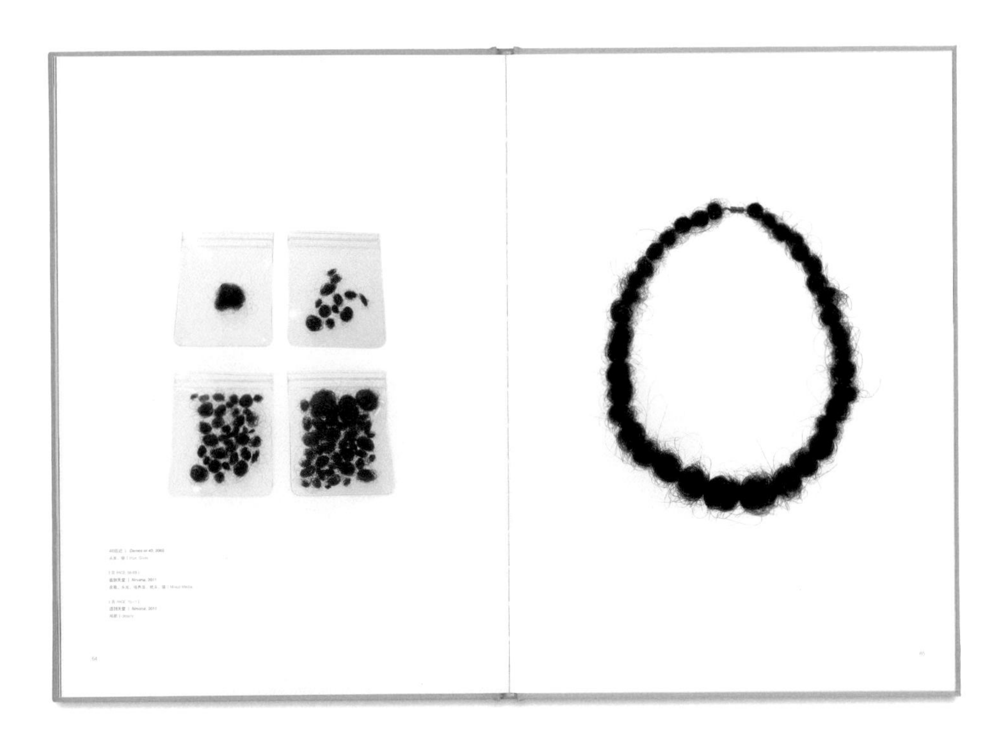

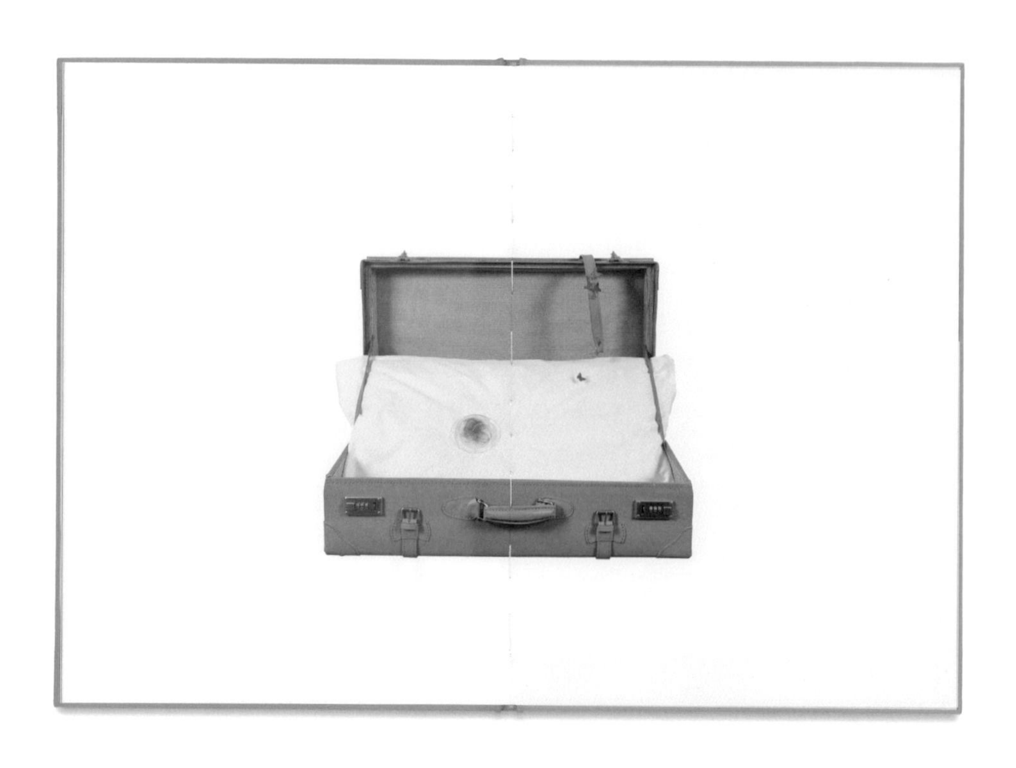

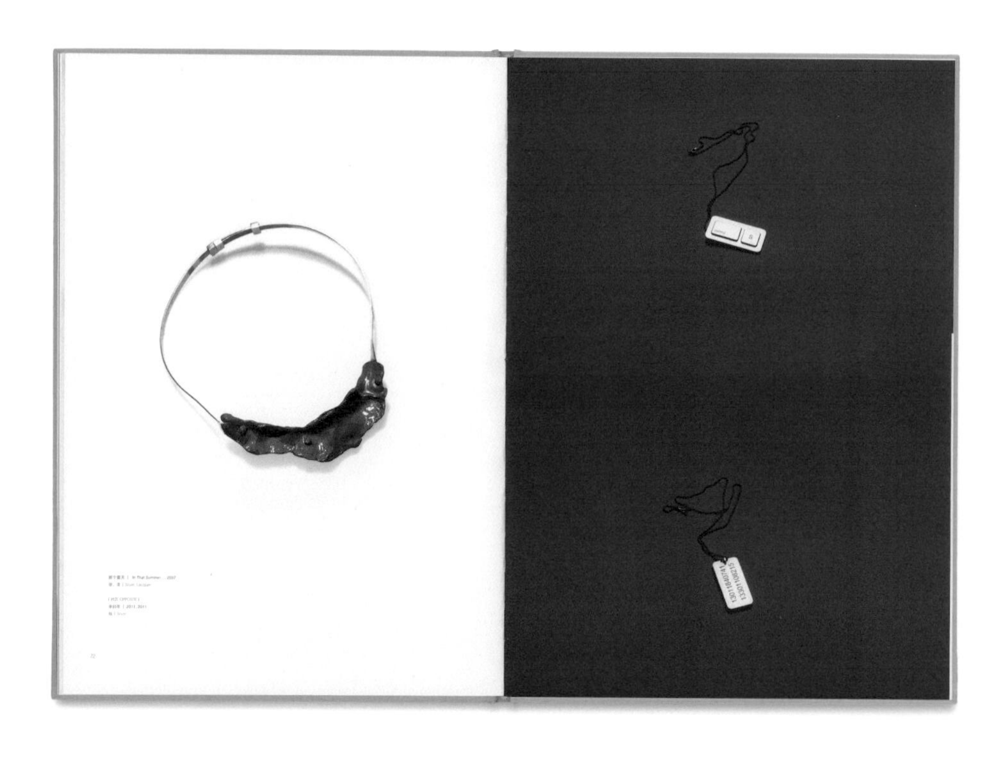

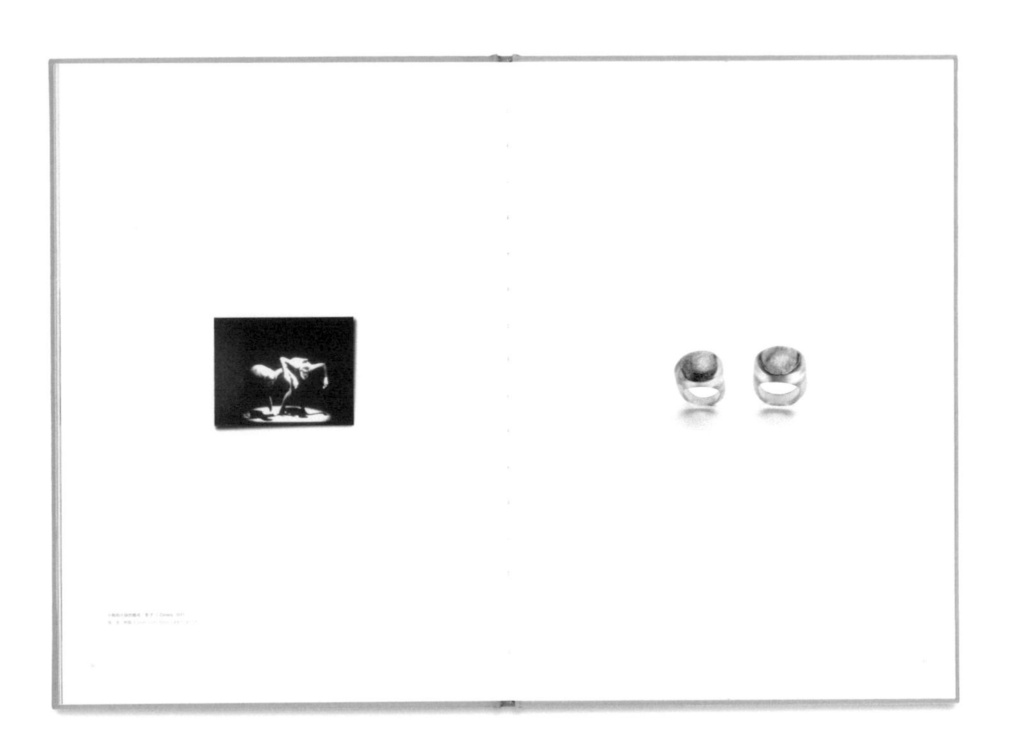

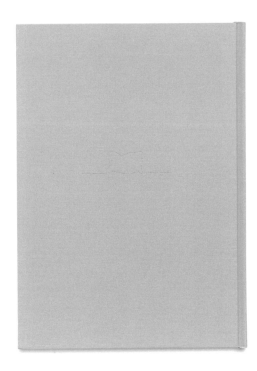

叶剑青 **观**

文化艺术出版社，北京，2011

353mm×253mm×31mm，254 页

丝面精装，裹纸质护封，外加瓦楞纸书匣（未展示）

2000 册

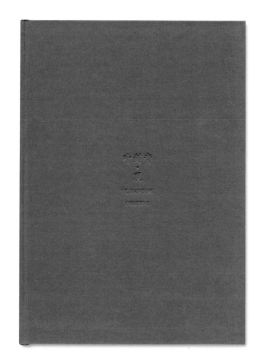

张进 **中学生的现实主义**

中国书籍出版社，北京，2011

13 本一套，210mm×140mm

平装，1000 套

文字：张进和他的学生们

内文版式设计：张进

中学生的现实主义
Realism of Middle School Students
我与面具
张进 编著
中国书籍出版社

中学生的现实主义
Realism of Middle School Students
我怕黑
张进 编著
中国书籍出版社

中学生的现实主义
Realism of Middle School Students
不能爱
张进 编著
中国书籍出版社

中学生的现实主义
Realism of Middle School Students
玩去
张进 编著
中国书籍出版社

中学生的现实主义
Realism of Middle School Students
解毒
张进 编著
中国书籍出版社

中学生的现实主义
Realism of Middle School Students
不怕黑
张进 编著
中国书籍出版社

中学生的现实主义
Realism of Middle School Students
防火墙
张进 编著
中国书籍出版社

中学生的现实主义
Realism of Middle School Students
瞎话儿
张进 编著
中国书籍出版社

中学生的现实主义
Realism of Middle School Students
打开缺点
张进 编著
中国书籍出版社

鬶
大汶口文化　白陶　山东出土　高十七厘米　最宽处十七·二厘米
五八
三孔高柄杯
大汶口文化　白陶　高二十一厘米　口径九·三厘米
五九

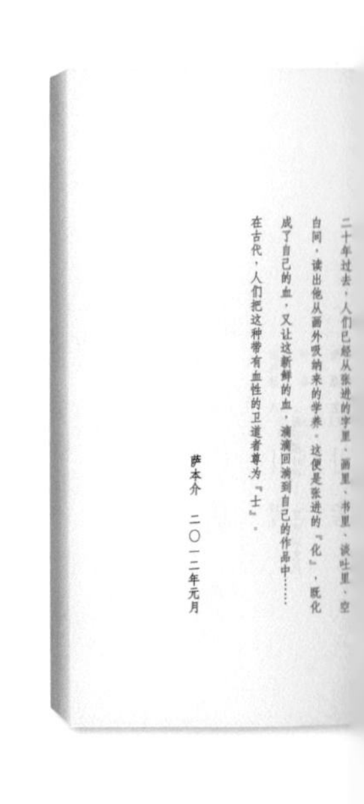
二十年过去，人们已经从张进的字里、画里、书里、谈吐里、空
白间，读出他从画外吸纳来的学养，这便是张进的『化』，既化
成了自己的血，又让这新鲜的血，滴滴回滴到自己的作品中……
在古代，人们把这种带有血性的卫道者尊为『士』。
萨本介　二〇一二年元月

武士俑
汉代　灰陶黄釉　陕西出土　残高十二厘米　底宽十·五厘米
二一八
农耕俑
东汉　红胎黄釉　河南南阳出土　高二十厘米　最宽处六·五厘米
二一九

虎形配件（另一面）
一四五

张进 **四合汲古**

高格出版公司，香港，2011

210mm×145mm

平装，两册，200 套

文章：萨本介、宗同昌

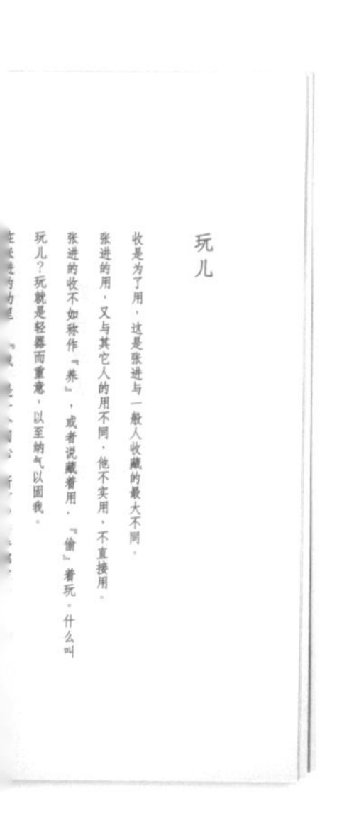

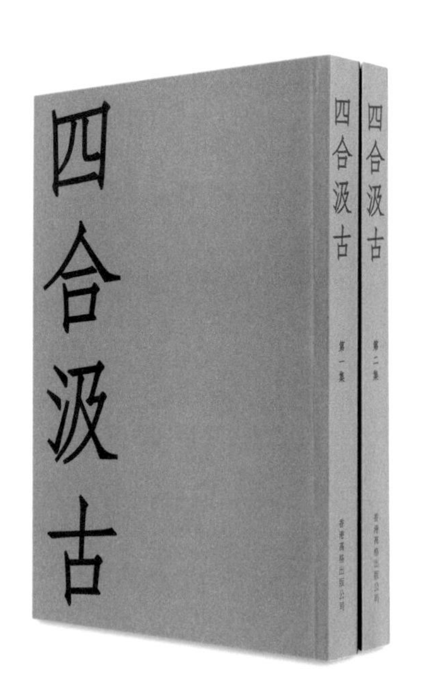

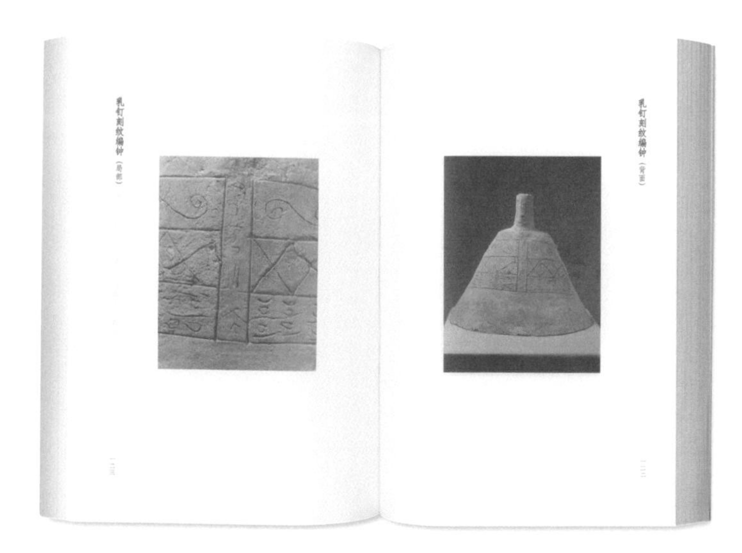

刘晓辉 **刘晓辉**

艺术家独立出版，北京，2012

161mm×113mm×30mm

共十册，外加纸质函套，一册经折装，九册活页裱头装

限量 600 函，每函艺术家签名、编号

编辑：刘晓辉、何浩

文字：刘晓辉

刘晓辉
LIU XIAOHUI

每天，我都在寻找可以画的题材，关注着围绕在我身边的那些景象。渐渐地，我越来越习惯于观察那些离我最近的、最常见的、那些仿佛随手就可以触摸到的场景。每天回来的路，父母的生活，甚至雾霾天气的色调，都成为了我关注的对象。我把红色的砖楼、地板的反光、金属的锅，还有灯光下的阴影都画在画布上。有时候我甚至有一种感觉，周围的生活是摆设出来的静物，在等待一个人把它们画下来。我不清楚它们到底是我在观察的过程中寻找，抑或是编造出来的，还是它们本来就在那里存在着，无需刻意挖掘与找寻。

也许，绘画就是主体被动地被呈现出来，无需思考。

对我而言，随着年龄和状态的改变，画出来的画也在变化。绘画对于我来说，像是一个“动词”，绘画真正吸引我的也就是这个变化——那不期而遇的生活所带来的变化。

每个时代都有它的烙印，昌盛与没落，混乱与活力。前不久，我看到一张明代帝王画像，且不论从技法上讲是否画得专业和精彩，画面确实传达出了那个时代的特质，与我们这个时代的气息完全不同。

我想，我的这些画，更像是一些信息，反映出这段时间我的生活。

刘晓辉

Everyday, I am in search of subjects to paint from what unfolds around me. Gradually, I become inclined to pay attention to the nearest, the most common scenes in my environment, as if they are as reachable as my fingertips. My daily drive to and from my studio, the lives my parents lead, even the color tones of this smoggy environment are all subjects of my interest. The red brick buildings, the reflections from the floor and the metallic wok have become subjects of my depiction on canvas. Sometimes I even have the feeling that what is around me are props for a still-life image waiting for someone to paint them on canvas. I am unsure whether they have been found and fabricated through our process of observation or they have always existed even without our intentional search.

Perhaps, painting is a representation of passive depiction of objects that does not require thinking.

My paintings transform with time and the state in which they have been produced. For me, painting is synonymous to the function of a "verb" and its transformation is what truly mesmerizes me – the unexpected changes that life presents.

Each era is marked by its prosperity and decline, chaos and vitality. Recently, I accidently stumbled upon a portrait of a Ming Dynasty emperor. Without commenting on the expertise techniques applied to this artwork, the image successfully conveys the unique characteristics of that particular era, which differs entirely from this current time.

I think my paintings are sources of information that reflect my life at present.

Liu Xiaohui

女人体之一 Female Body no.1, 2011
40 x 30 cm
木板丙烯
Acrylics on wood board

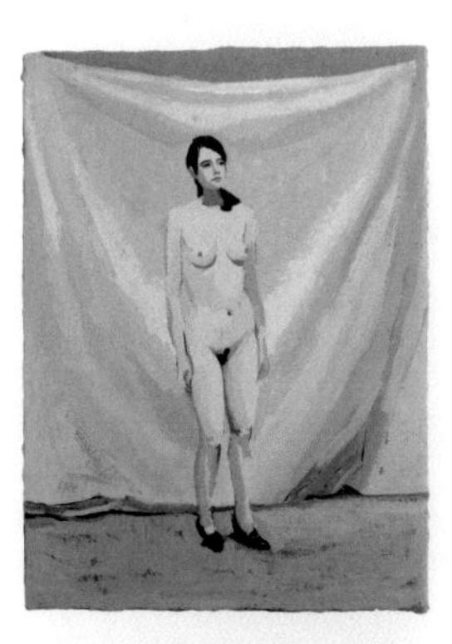

女人体之二 Female Body no.2, 2011
40 x 30 cm
木板丙烯
Acrylics on wood board

本书限量600册，每册由艺术家签名并编号。

刘晓辉

348/600

编辑：刘晓辉 何浩
概念 设计：何浩
摄影：杨超
印刷：北京雅昌彩色印刷有限公司
版次：2012年2月第一版

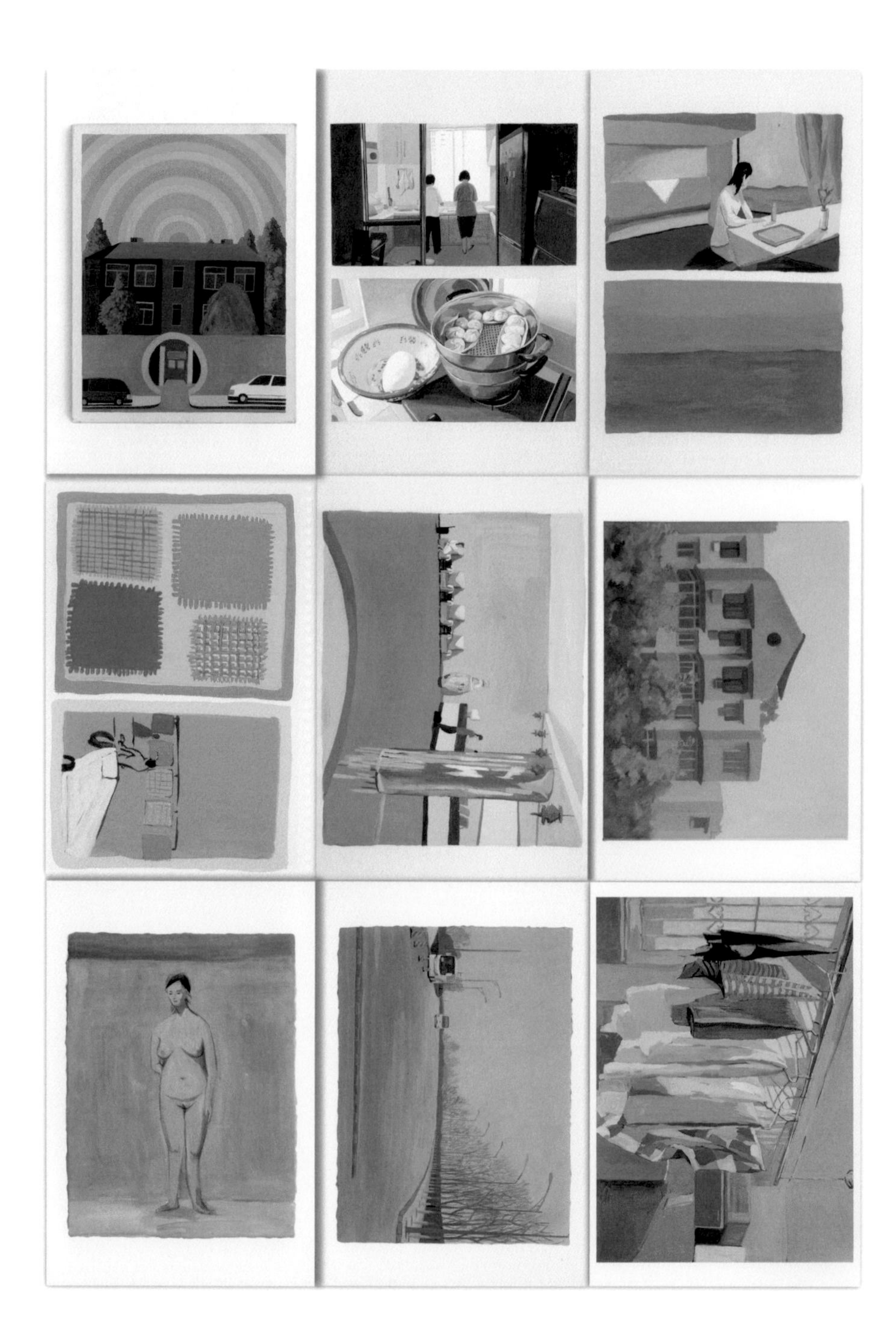

①

刘晓辉

One Day of a Model

17-panel

acrylic on canvas

31.5×41 cm

2011

模特的一天

17联画 布面油画

The Dinner

acrylic on canvas

晚餐

①

布面丙烯

刘晓辉

Dunhuang ①

Acrylic on paper

敦煌 ①

255×205 mm

纸上丙烯

刘晓辉

2011

Red ①

Acrylic

on

paper

红 ①

纸上丙烯

150 mm×220 mm

刘晓辉

2011

Youyi Hotel ①

Acrylic on paper

友谊饭店 ①

纸上丙烯

刘晓辉

2011

Chongqing Cable Car Ride

oil on canvas

重庆长江索道

1

51×40.5 cm 布面油画

刘晓辉

2011

①

One day of a Modle

egg tempera on board

20×26 cm

模特的一天

木板蛋彩

刘晓辉

2011

The Winter of the Circular Railway

egg tempera on board

环线的冬天 一 1

坦培拉木板

30×40 cm

刘晓辉

The Lives of My Parents

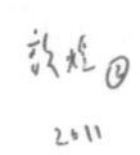

海报，1000mm×700mm，50 张

《刘晓辉》

过去十年间跟我合作过的艺术家中与我年龄相仿的不多，刘晓辉是其中一个。我跟晓辉同岁，算是过去的同学如今的同事——早年同在美院上学，后又都留在学校做老师，只是他一直在壁画系，我在设计学院。

虽然看似渊源不浅，但因为既不同届也不同系，在过去的日子里我们并无太多交集，直到2012年初因为他要出版一本小画集来找我设计时，我才真正有机会了解他的作品。

晓辉的画作跟惯见的“大”作品不太一样，尺幅小却数量多，作品不以“件”计而以“组”论。若干张小画结为一组，有点像连环画，有顺序，顺序中似隐含着欲说还休的松散情节，但拆开看单张画亦是自成一体的完整作品，抛开上下文也无妨。每组作品少则七八张，多则十几张，同组作品材质相同，但组别之间却差异颇大——纸上、画布、木板，丙烯、油彩、坦培拉，各种排列组合。

面对这些繁多而纷杂的小画，我一时有点无从下手，但有一点我跟晓辉有共同的直觉，就是这些小画放在一起最好不要弄得太“严肃”，要“有趣”一点。这种有趣，不是简单的形式趣味，而是像王小波说的——“有道理而且新奇”。我怎么能够通过设计把晓辉作品中那种“有顺序但也可以没有”、暗含着一点茫然忧伤的松弛感传递给读者呢？

随着晓辉把要收入书中的作品陆续整理完备，我的设计构想也渐渐明晰：九组作品，做成九本小册子，册与册之间无先后顺序，随意摞在一起纳入封套。单册中则依画序编排，采用信纸、明信片的黏结方式——连成一册又可轻易撕开，每组作品既可作为完整一册顺序翻阅，也可以随意分解，拆成散乱单页。所有画作均为单面呈现，背面则印此画的“背签”[1]，这样每一张拿在手里的单页，都是一幅正背完整的作品。

“背签”的介入对于此书至关重要：一方面，晓辉的画作数量繁多、材质

1.“背签”是指作者在画作背面的签名及手书作品信息，详尽一些的会包括名称、材质、尺寸、创作年代等。

各异且暗含编序，即便是一本“有趣”的书，清晰的信息标注亦不可少；另一方面，这本书在书籍形态和阅读方式上特有的率性、松弛又与印刷字符这类“异质”难以见容。复印背签的方式无疑轻松化解了矛盾——不借助任何外力，让作品自我呼吸、自我循环。更何况带着晓辉的“背签”的画作背面，本身就是包含着最丰富信息和最有效传达的“版面设计”。

最后，还需要一款展览海报，我并没有为此专门设计，而只是做了一张文字镂空的印版——用黑色往已经印好但未裁切的内文印张上轻轻一压，镂空文字中随机渗出的斑驳色彩便如万花筒一般在有序和无序间流转弥漫开来……

张慧 **张慧**

长征空间，北京，2012

296mm×268mm×20mm，112 页

布面精装，裹纸质护封

1000 册

文章：秦思源

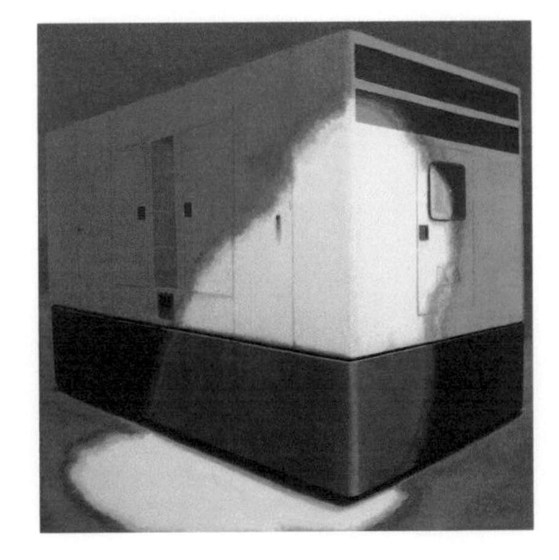

来自虚空

文国璋 **文国璋油画作品选集**

广东人民出版社，广州，2012

351mm×252mm×27mm，192 页

布面精装，裹纸质护封

1000 册

编辑：华艺廊

文章：孙景波

后记：文国璋

"塔吉克人的叼羊系列"人物素材——色麦克一
Kulaxi-I, figure portrait used for the Tajik Sheep Tussling Series
2008
72 cm × 53 cm
布面油画 Oil on canvas
"塔吉克人的叼羊系列"人物素材——色麦克二
Kulaxi-II, figure portrait used for the Tajik Sheep Tussling Series
2008
72 cm × 53 cm
布面油画 Oil on canvas
86
"塔吉克人的叼羊系列"人物素材
Figure portrait used for the Tajik Sheep Tussling Series
2008
布面油画 Oil on canvas
2008. Wen Guozhang
87

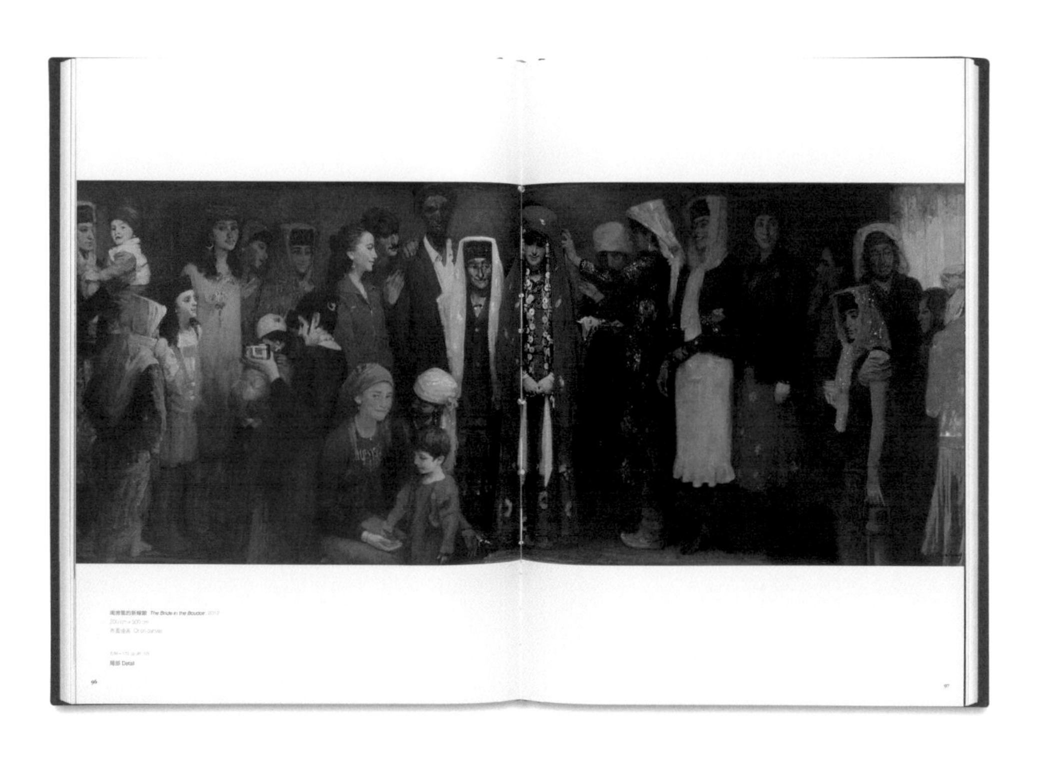
闺房里的新嫁娘 The Bride in the Boudoir 2012
200 cm × 500 cm
布面油画 Oil on canvas
局部 Detail
96
97

徐唯辛 **众生**

艺术家独立出版，北京，2012（第一版，2010）

275mm×215mm×15mm，176 页

平装，裹纸质护封

2000 册

文章：汪民安、王明贤、刘骁纯、李述鸿、廖廖

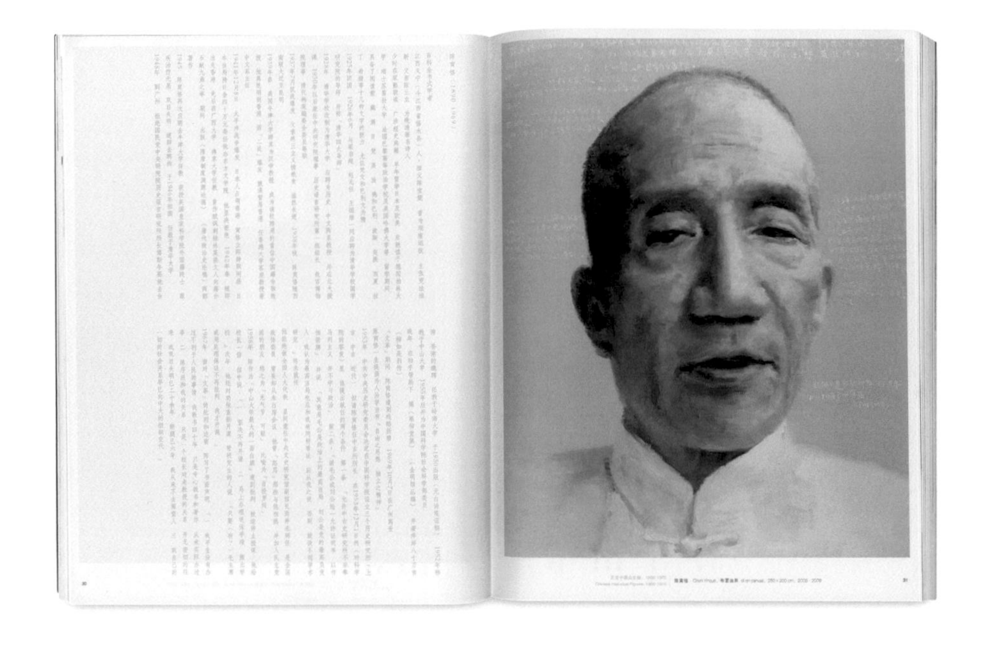

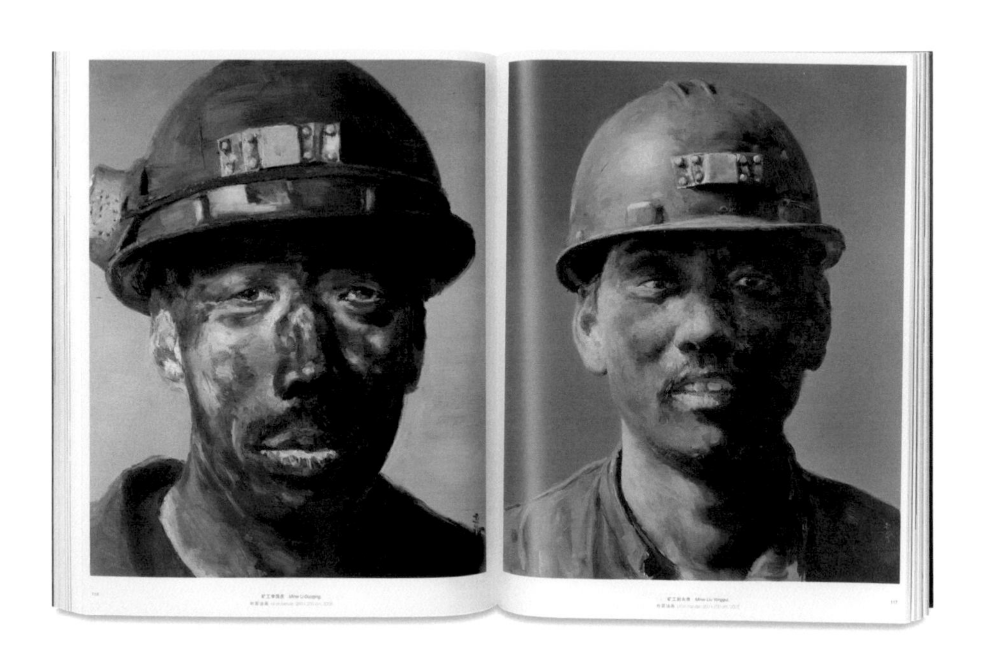

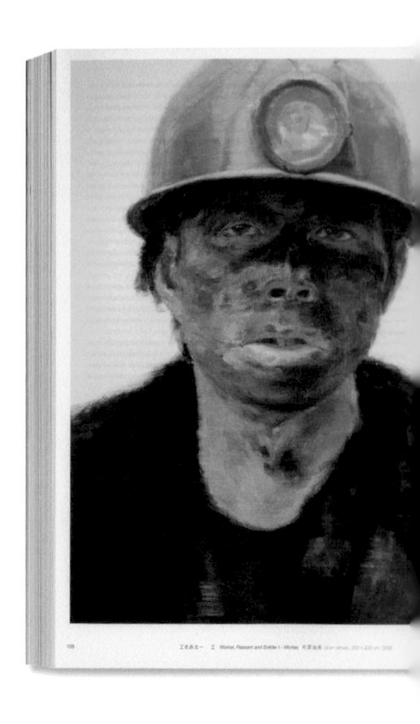

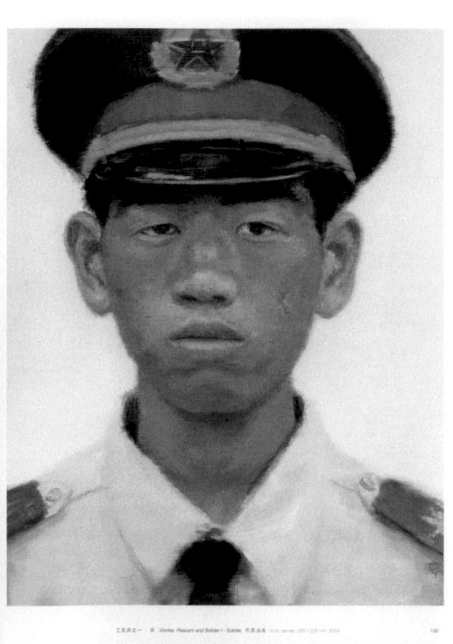

荒木经惟 **感伤之旅·堕乐园 1971—2012**

三影堂摄影艺术中心，北京，2012

311mm×226mm×23mm，230 页

布面精装，裹纸质腰封

限量 1500 册，每册编号、艺术家钤印

编辑：本尾久子、毛卫东

前言：荣荣和映里

文章：本尾久子、毛卫东、王莲

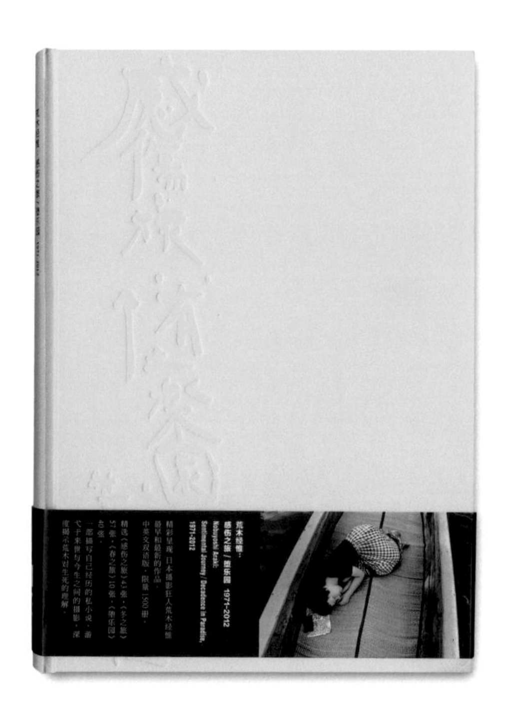

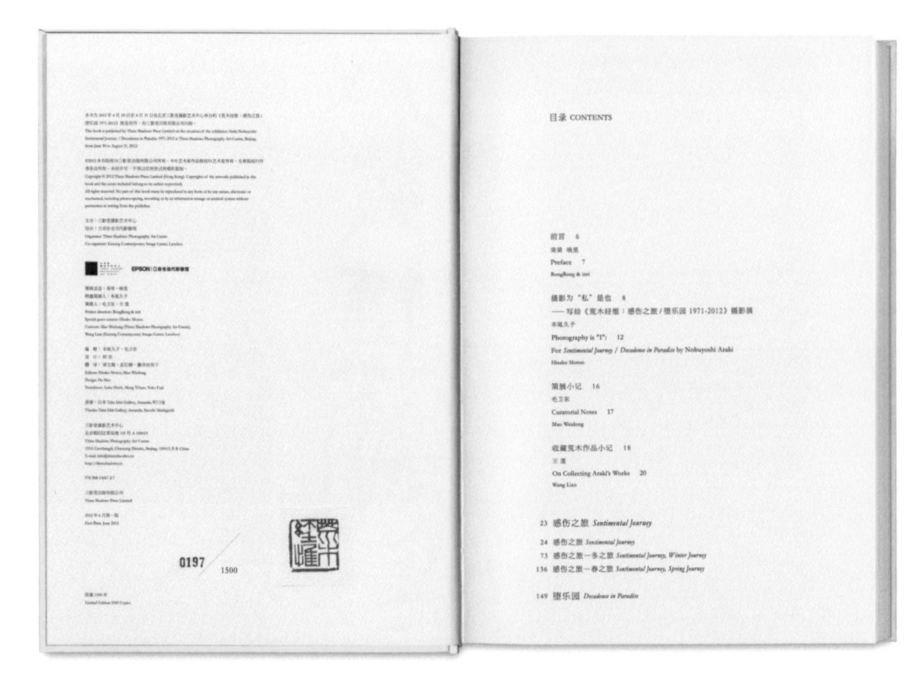

摄影为“私”是也

—— 写给《荒木经惟：感伤之旅 / 墮乐园 1971-2012》摄影展

说起荒木经惟摄影作品的起始点，正如我以前提到过的，是 1971 年自行编辑出版，限量 1000 本的《感伤之旅》，本书刊登的宣言原文，最早也是出自那本作品集的序文，这实际上也可以称之为荒木经惟对于“私写真”的宣言。《感伤之旅》用赤裸裸的黑白影像记录下了 1971 年荒木与阳子的新婚旅行，并且引起了不小的轰动。从那以后一直到现在，荒木经惟作为摄影艺术家的地位就被确立了下来，并且在他的作品里自始至终贯穿着“私＝写真”的概念。

结婚 20 年后，阳子被子宫瘤夺去了生命。在婚后生活的日子里，荒木经惟拍摄的绝大部分影像，都出自他们二人在寓所（东京世田谷豪德寺公寓）开始的婚后生活的点点滴滴，比如无比幸福灿烂的微笑，和爱猫奇洛嬉戏的身影，夫妇二人的旅行，被无数被摄体侵蚀的宽敞的阳台等等，直到阳子的病情把这一切幸福的光景都打破了，最后，阳子静静地躺在棺木里，面庞淹没在无数花海中（《冬之旅》）。在葬礼上，荒木经惟怀抱着的阳子那华贵雍容的遗容，也是荒木经惟在她生前留下的。

阳子出类拔萃的美丽是毋容置疑的，然而从她的内在所映衬出的魅力，通过不间断的快门声和一系列的化学反应得到了增强，并且清晰地反映在荒木的作品中。荒木经惟常常说：“照片是被摄体和时间组成的。……我们拍摄的不是空间，是时间。我们在框取的其实是时间啊。”

荒木经惟，1940 年作为长子出生于东京台东区三轮的一个普通市井家庭。父亲长太郎是位制作木屐的匠人。新吉原附近江户时期的净观寺成了他幼年玩耍的地方，据说那里葬着许多生前无亲无故的艺妓。他的父亲把摄影当成兴趣，这也深深影响了荒木经惟。他从都里上野高校升入千叶大学工学部摄影学科，在大学期间受到意大利新现实主义影响，在大学的毕业制作中创作了电影《阿幸》。1963 年毕业当年，便进入电通广告公司成为了一名摄影师。在 20 世纪 60 年代日本战后社会复兴这一热火朝天的大背景下，他拍摄的以市井儿童为主题的摄影作品《阿幸》一举获得了首届太阳奖。

荒木在电通公司担任广告摄影师时，也着实发挥了他的能力。一方面他有自由使用公司摄影棚的权利，做过秘书之类的文职，拍过很多女模特的摄影作品，其中就有后来 1972 年成为他的妻子的青木阳子，漫步于市井街道、抓拍大量人物等等，荒木经惟把精力投入到所有这些个人的作品中。

丧失的母亲在 1974 年也离开了人世，荒木主持了葬礼。“好像第一次看到母亲如此安详的面容，我凝视了许久。我看到了超越现实的东西，那正是母亲的‘死景’。”

荒木说，生（Eros）与死（Thanatos）都被包含在摄影中。（1993 年的摄影集《EROTOS》的标题就是荒木经惟把这两个词连结在一起组合而成。）他还说：“与爱的人死别使得摄影作品更加深化。”对于荒木来说，摄影是游弋于来世与今生之间的船。几年前查出身患前列腺癌的荒木，经过手术治疗得到康复后，对于生与死的理解更加大彻大悟了。

认识荒木的人都知道，他为人率直，待人亲切热情，语言幽默诙谐，腔调丰富，这样的印象深深印刻在每个人的心里。这也是为什么在和荒木的镜头对峙的一刻，大家都努力地使出全力去表现的原因之一。不只是拍人，在街头他也一样会放开手脚去拍摄。东京的街头对于他来说就像磁场，当他阔步其中时，不管是一丁点儿不和谐的感觉，还是擦肩而过的一瞥，都逃不过他的快门，有如雷光石火一般被他捕捉下来。这样的作品充满了魅惑与欲望，连小样都成了无需剪裁的有着完美构图的作品荟萃。可是，为什么在看他的作品时会有一种压迫胸口的感觉呢？

自从 1989 年夏天阳子获病入院以来，荒木拍摄天空的作品便多了起来。1990 年 1 月，在和阳子说了最后一声“谢谢”后，阳子安然离世。为记录下阳子的死，荒木拍下了她最后的身影。“在那以前我曾经说过，等我到了 50 岁我开始拍人像。其实教会我如何拍摄‘人像摄影’并一直给我拍摄机会的，就是阳子，直到最后

感伤之旅

Sentimental Journey

24

前略

もう我慢できません。私が慢性ゲリパラ中耳炎だからではありません。たまたまファッション写真が氾濫していたのにすぎないのですが、こうでてくる顔、でてくる裸、でてくる私生活、でてくる風景が嘘っぱちじゃ、我慢できません。これはそこいらの嘘写真とはちがいます。この「センチメンタルな旅」は私の愛であり写真家決心なのです。自分の新婚旅行を撮影したから真実写真だぞ！といってるのではありません。写真家としての出発を愛にし、たまたま私小説からはじまったにすぎないのです。もっとも私の場合ずーっと私小説になると思います。私小説こそもっとも写真に近いと思っているからです。新婚旅行のコースをそのまま並べただけですが、ともかくページをめくってみて下さい。古くさい灰白色のトーンはオフセット印刷で出しました。よりセンチメンタルな旅になりました。成功です。あなたも気に入ってくれたはずです。私は日常の単々とすぎさってゆく順序になにかを感じています。

敬具

荒木経惟

'90 1 27
'90 1 27

A splendid presentation of "the madman" of Japanese photography, Nobuyoshi Araki's earliest and most recent works
Chinese English bilingual edition, 1500 copies Limited Edition.
Selections from Sentimental Journey (43 images), Winter Journey (57 images), Spring Journey (10 images), Decadence in Paradise (40 images)
A narrative of personal experience, these photographs drift between this world and the next, intimately revealing Araki's understanding of life and death.

《感伤之旅·堕乐园 1971—2012》

2012 年是三影堂成立五周年，作为庆典，三影堂特别筹划邀请了日本艺术家荒木经惟的个展，展出他最早与最新的两组作品——《感伤之旅》和《堕乐园》。前者为荒木经惟的成名作，包括《感伤之旅》《冬之旅》《春之旅》三个部分，拍摄始于 1971 年，记录了他和妻子阳子的蜜月旅行以及二十年后阳子被子宫瘤夺去生命的一些平淡却能穿透人心的瞬间；后者为荒木 2012 年癌症手术后在与阳子共同生活过的自宅阳台上用花和玩偶的摆拍，阳子去世后，荒木曾无数次地站在这个阳台上对着天空按下快门……而几个月后，这栋房子将被拆除而不复存在。

之前曾有耳闻，荒木对书籍出版极为重视且要求颇高，为此荣荣和映里在去日本和他商讨展览事宜时，还特意带去我以前设计过的书，向他说明和推荐。事实上，荒木这些年书籍的出版大多由他最信赖的出版人本尾久子小姐负责，一年前本尾来北京时，我们曾一起在三影堂做过关于摄影图书跟读者的对话。她介绍了荒木以前出版过的书，我则展示了自己的几个摄影书籍的设计项目，虽然语言不通无法深入交谈，但我们彼此惺惺相惜，为一年后的合作埋下了伏笔。

展览时间定在 2012 年 6 月，我跟荣荣和映里在 2011 年的年底已开始不断地商量，我在三影堂非常仔细地观摩了原作，并依此寻求设计之道。《感伤之旅》系列幅面很小，仅为 12 英寸规格（比 A3 打印纸略大），明胶银盐工艺，曝光放大在纸基相纸上；《堕乐园》则为巨幅照片，光面相纸打印。两组作品不仅创作时间跨度大，尺幅和影像质感也殊为不同，在一本书中如何准确呈现似乎是个问题。

正当我反复思忖跃跃欲试，荒木的项目却因为沟通和一些技术性问题的阻碍没了下文。这本书再次被提起时，离开展已不到二十天。这个时间，不说设计，仅从印刷制作上看，恐怕也难以完成。制作一本高水准画册，所谓印制其实最耗时的不是印刷而是制版——一张图往往需要反复调试打样多次，才能达到或接近理想状态，这一过程被称为制版。最终当这些“准确信息”通过印版批量复制在一页页纸张上时，已是瓜熟蒂落，无须多虑。在过去的十多年中，彩色印刷的制版技术有了质的发展，实现了调图打样的全数字化管理，极大提高了这一环节的稳定性与工作效率。但黑白摄影为了保持影调的一致性同时不损失层次的丰润感，通常会采用一种已沿袭近百年的传统印刷方式——用黑色和一个专门调配的灰色两版套印而成。这种方式与彩色印刷的原理不尽相同，制版环节更是需要传统的全手工机械操作，无法数字化模拟，十分烦琐费时。如果荒木的项目要如期完成，此法行不通，必须另辟蹊径。

正如我一直坚信的，任何制约的提出同时也可能意味着新的契机，当我不得不根据所剩时间来重新思考、规划制作工艺的时候，最核心的设计概念其实就栖身其中并随之生长。当再次回到荒木的作品去体会其中意味时，我意识到传统工艺所确立的古典影像标准其实未必合适，荒木不同寻常的拍摄方式与照片哲学即意味着另一种影像格调的可能。我决定所有作品用单黑一色印刷，这种最简单的复制方式中所蕴含的直接、快速与不假思索的低技术正与荒木的作品相契合。但是，从技术上讲这又是一柄双刃剑，一方面，这样做可以大大缩短制版周期，同时也降低了印刷难度，但由于单黑印出的图片单薄干涩，如若使用，必须非常周密地调控其他因素来构建出新的关系——不加转换地直接使用低技术，其结果往往只能是粗劣荒疏。首先，我把图像印刷的网点密度大幅度提高，以解决层次粗糙质感干涩的缺陷；同时针对高密度印刷带来的反差减弱图像发闷的副作用，采用的办法是把一般黑油墨换为更黑的特黑墨。《感伤之旅》使用哑粉纸印刷，照片表面过哑光油，用来消除着墨部分与未着墨部分的质感差异，使照片与纸张融为一体且无一点反光；而《堕乐园》用光面铜版纸，图片全出血且满版过亮油——通过纸张和工艺把两组作品的质感完全拉开又相得益彰——《感伤之旅》没有一点光泽，深邃的黑有如时光隧道，铅般厚重深沉；《堕乐园》则在光线的反射下更加肆意绚烂，却暗含虚幻茫然……文本部分再换第三种纸，这是一种纤维很长的日本纸，手感厚实却重量极轻，略微有一点毛糙的触感但绵软温润，翻动时有一种缓步走在雪地上脚下沙沙轻响的感觉，轻柔洁净，冰冷温暖。

书的封面，我非常幸运地找到一款明黄色织物。说幸运，是因为这个颜色极微妙，可遇不可求。在东方人的感观世界中，恐怕没有什么能比这个极致的黄更适合传递“生”与“死”的终极意味。对于荒木经惟而言，他一生的全部作品又有哪件不是游弋于“生”“死”之间的船?

海报，1000mm×700mm，200 张

感傷旅・侑楽
荒木経惟

徐唯辛 **七个矿工**

美国芝加哥大学北京中心，北京，2012

291mm×207mm×4mm，24 页

平装，骑马钉

2000 册

编辑、文章：巫鸿

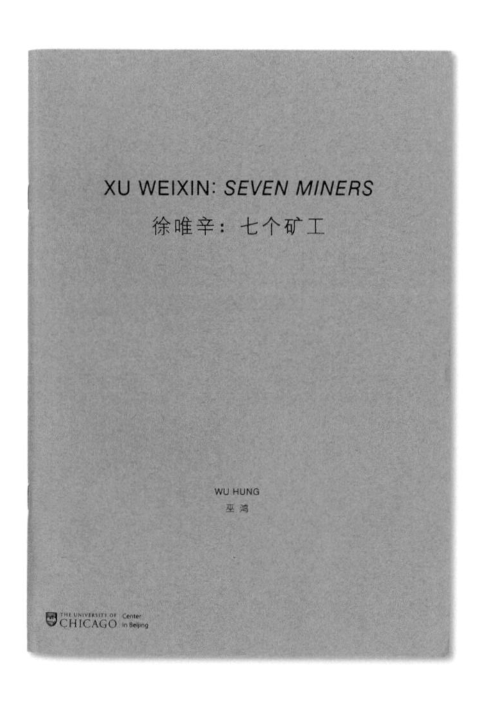

Xu Weixin: *Seven Miners*

Wu Hung

Realism is plural. Xu Weixin's realist painting differs from naturalistic and photorealistic images on the one hand, and from critical realist and revolutionary realist pictures on the other. Whereas Naturalism and Photorealism superimpose art onto life, Critical Realism and Revolutionary Realism promulgate political and moral doctrines through representation. Xu instead conducts in-depth observation of ordinary people in ordinary situations and depicts what he sees. Doubtlessly he hopes to reduce the gap between art and reality, but the purpose is to bring life into art, not to equate art with life. In this sense his vocation is close to that of Gustave Courbet (1819 - 1877), the French painter who first advocated Realist Art in reaction to Classicism and Romanticism. Courbet promoted "mimicking people and things in daily life that one can touch with one's own hand." Although he was a committed social reformer and even participated in the revolutionary movement of the Paris Commune, he did not push the agenda of "sublimating life" to his political cause.

The similarity between Xu Weixin and Courbet thus also implies a huge difference between them: separated by over a century, Xu's resurrection of a more direct and rudimentary type of "social realism" necessarily responds to more recent trends in realism after Courbet's time, especially Socialist Realism as the officially sponsored style in modern Chinese art. A general phenomenon in Chinese art since the 1970s, in fact, has been the rise of various rectifications of Socialist Realist Art, including Scar Art and Native Soil Art within the realist camp, aestheticism in academic art, non-representational abstract art, and contemporary Conceptual Art. The last three reject not only Socialist Realist Art specifically, but also the visual language and purpose of Realist Art in general. From here we can define Xu Weixin's historical position and understand the significance of his paintings: he is conducting experiments within Realist Art and reflecting upon this artistic tradition in the contexts of both modern painting and contemporary Chinese art. His recent "Coalminers" series best articulates such rethinking, as the pictures in the series all reject typification and idealization, and as his effort to "let the subject speak" liberates realism from the grasp of ideology. He portrays coalminers laboring in private-owned local coal mines, many of whom come from the countryside. They do not represent "workers" as an anonymous social class or "ordinary folk" in an abstract sense, but they are people of flesh and blood who have names and distinct personality. He demands that everyone see these people. He has therefore enlarged their images to the proportions that have been reserved for great leaders, so that no one can evade these men who are often kept invisible in art, even in Realist Art.

9. 18. 2012

I

Miner LIU ZHIXIANG

III

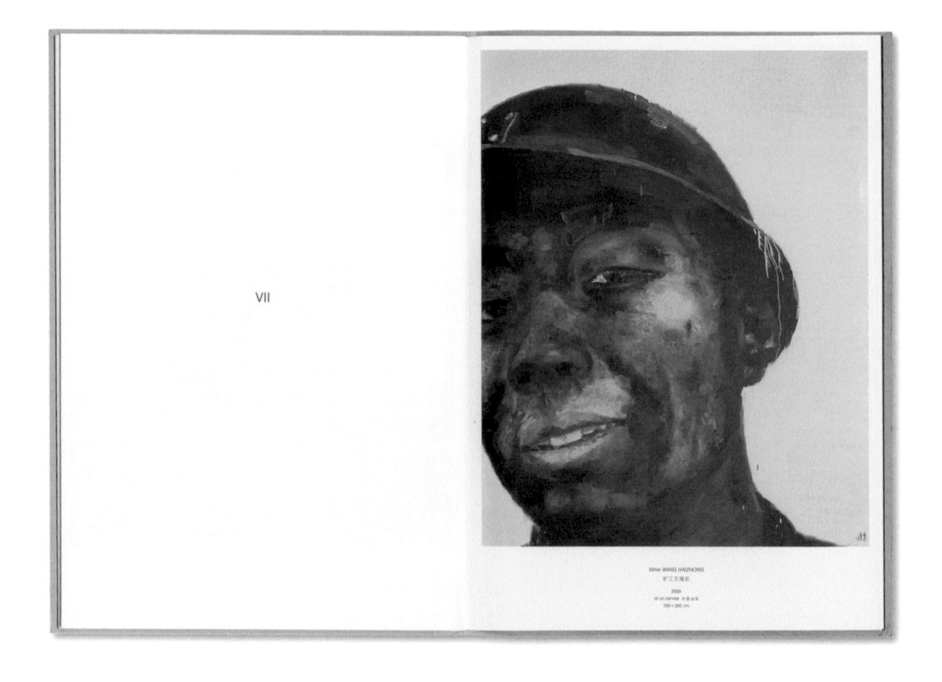
VII

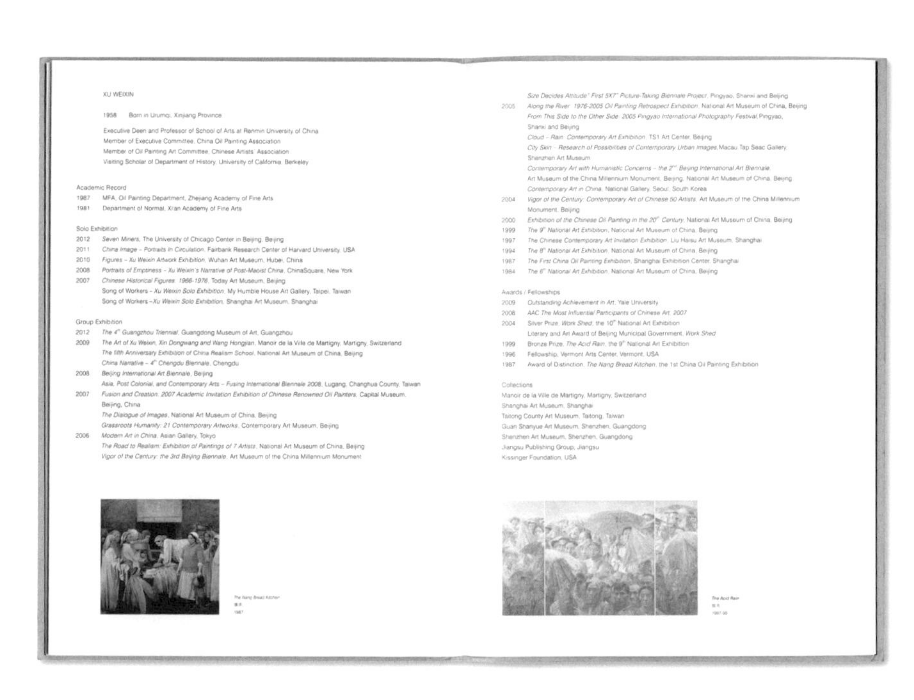

XU WEIXIN

1958 Born in Urumqi, Xinjiang Province

Executive Deen and Professor of School of Arts at Renmin University of China
Member of Executive Committee, China Oil Painting Association
Member of Oil Painting Art Committee, Chinese Artists' Association
Visiting Scholar of Department of History, University of California, Berkeley

Academic Record
1987 MFA, Oil Painting Department, Zhejiang Academy of Fine Arts
1981 Department of Normal, Xi'an Academy of Fine Arts

Solo Exhibition
2012 *Seven Miners*, The University of Chicago Center in Beijing, Beijing
2011 *China Image – Portraits In Circulation*, Fairbank Research Center of Harvard University, USA
2010 *Figures – Xu Weixin Artwork Exhibition*, Wuhan Art Museum, Hubei, China
2008 *Portraits of Emptiness – Xu Weixin's Narrative of Post-Maoist China*, ChinaSquare, New York
2007 *Chinese Historical Figures: 1966-1976*, Today Art Museum, Beijing
Song of Workers – *Xu Weixin Solo Exhibition*, My Humble House Art Gallery, Taipei, Taiwan
Song of Workers –*Xu Weixin Solo Exhibition*, Shanghai Art Museum, Shanghai

Group Exhibition
2012 *The 4th Guangzhou Triennial*, Guangdong Museum of Art, Guangzhou
2009 *The Art of Xu Weixin, Xin Dongwang and Wang Hongjian*, Manoir de la Ville de Martigny, Martigny, Switzerland
The fifth Anniversary Exhibition of China Realism School, National Art Museum of China, Beijing
China Narrative – 4th Chengdu Biennale, Chengdu
2008 *Beijing International Art Biennale*, Beijing
Asia, Post Colonial, and Contemporary Arts – Fusing International Biennale 2008, Lugang, Changhua County, Taiwan
2007 *Fusion and Creation: 2007 Academic Invitation Exhibition of Chinese Renowned Oil Painters*, Capital Museum, Beijing, China
The Dialogue of Images, National Art Museum of China, Beijing
Grassroots Humanity: 21 Contemporary Artworks, Contemporary Art Museum, Beijing
2006 *Modern Art in China*, Asian Gallery, Tokyo
The Road to Realism: Exhibition of Paintings of 7 Artists, National Art Museum of China, Beijing
Vigor of the Century: the 3rd Beijing Biennale, Art Museum of the China Millennium Monument
Size Decides Attitude" First 5X7" Picture-Taking Biennale Project, Pingyao, Shanxi and Beijing
2005 *Along the River: 1976-2005 Oil Painting Retrospect Exhibition*, National Art Museum of China, Beijing
From This Side to the Other Side: 2005 Pingyao International Photography Festival, Pingyao, Shanxi and Beijing
Cloud – Rain: Contemporary Art Exhibition, TS1 Art Center, Beijing
City Skin – Research of Possibilities of Contemporary Urban Images, Macau Tap Seac Gallery, Shenzhen Art Museum
Contemporary Art with Humanistic Concerns – the 2nd Beijing International Art Biennale, Art Museum of the China Millennium Monument, Beijing, National Art Museum of China, Beijing
Contemporary Art in China, National Gallery, Seoul, South Korea
2004 *Vigor of the Century: Contemporary Art of Chinese 50 Artists*, Art Museum of the China Millennium Monument, Beijing
2000 *Exhibition of the Chinese Oil Painting in the 20th Century*, National Art Museum of China, Beijing
1999 *The 9th National Art Exhibition*, National Art Museum of China, Beijing
1997 *The Chinese Contemporary Art Invitation Exhibition*, Liu Haisu Art Museum, Shanghai
1994 *The 8th National Art Exhibition*, National Art Museum of China, Beijing
1987 *The First China Oil Painting Exhibition*, Shanghai Exhibition Center, Shanghai
1984 *The 6th National Art Exhibition*, National Art Museum of China, Beijing

Awards / Fellowships
2009 *Outstanding Achievement in Art*, Yale University
2008 *AAC The Most Influential Participants of Chinese Art, 2007*
2004 Silver Prize, *Work Shed*, the 10th National Art Exhibition
Literary and Art Award of Beijing Municipal Government, *Work Shed*
1999 Bronze Prize, *The Acid Rain*, the 9th National Art Exhibition
1996 Fellowship, Vermont Arts Center, Vermont, USA
1987 Award of Distinction, *The Nang Bread Kitchen*, the 1st China Oil Painting Exhibition

Collections
Manoir de la Ville de Martigny, Martigny, Switzerland
Shanghai Art Museum, Shanghai
Taitong County Art Museum, Taitong, Taiwan
Guan Shanyue Art Museum, Shenzhen, Guangdong
Shenzhen Art Museum, Shenzhen, Guangdong
Jiangsu Publishing Group, Jiangsu
Kissinger Foundation, USA

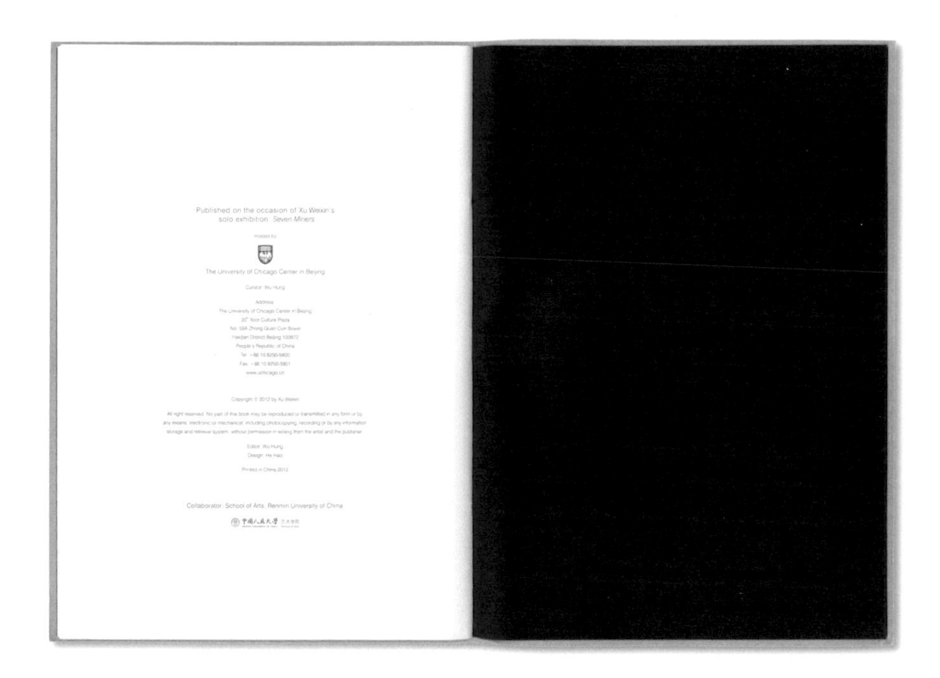

Published on the occasion of Xu Weixin's solo exhibition: *Seven Miners*

Hosted by

The University of Chicago Center in Beijing

Curator: Wu Hung

Address
The University of Chicago Center in Beijing
20th floor Culture Plaza
No. 59A Zhong Guan Cun Street
Haidian District Beijing 100872
People's Republic of China
Tel: +86 10 8250-5800
Fax: +86 10 8250-5801
www.uchicago.cn

Copyright © 2012 by Xu Weixin

All right reserved. No part of this book may be reproduced or transmitted in any form or by any means, electronic or mechanical, including photocopying, recording or by any information storage and retrieval system, without permission in writing from the artist and the publisher.

Editor: Wu Hung
Design: He Hao

Printed in China 2012

Collaborator: School of Arts, Renmin University of China

中国人民大学 艺术学院

十年·有声——中央美术学院与国际当代首饰

中国纺织出版社，北京，2012

242mm×233mm×29mm，286 页

布面精装

2000 册

序言：潘公凯

前言：王敏

访谈：蒋岳红 / 滕菲、姜节泓 / 参展艺术家

后记：滕菲

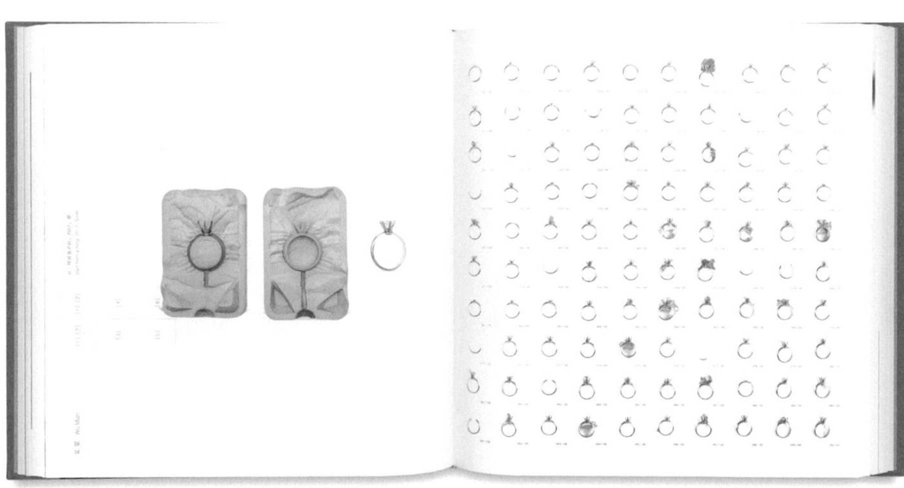

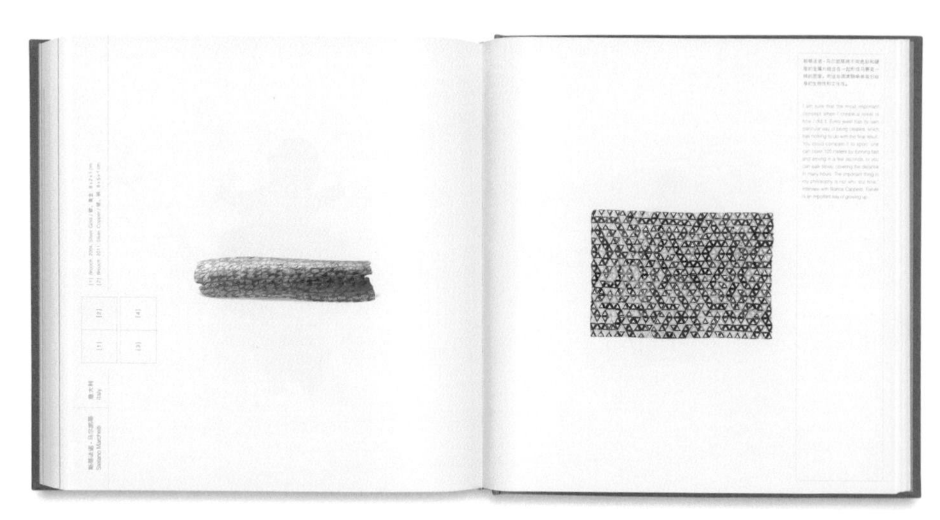

伊万·阿斯特弗克

[illegible]

Jivan ASTFALCK

[illegible]

'Design' and 'Art' are outdated and redundant concepts; I prefer to differentiate between 'designing for others' and studio practice which is [illegible] by an inherent enquiry and is informed by contemporary art strategies. While my work is informed by my understanding of already [illegible] artistic territories and strategies, it actively re-evaluates [illegible] and, in my view, succeeds in embedding ambiguity and cross-referencing in terms of the creative construction of visual metaphors and material transmission.

Furthermore and significantly, work such as mine must be understood within the context of a developing tradition of narrative jewellery. As a strand of jewellery making practice, narrative jewellery takes as one of its starting points the relationship between jewellery objects and meaning. A work of jewellery is not only a valuable object in terms of its craft, beauty, and materially quantifiable worth—all of which could be translated as 'proper' values—proper in the sense of belonging to the object itself, objectively as its own properties. Within narrative jewellery the object is acknowledged, and in fact embraced, as an object [illegible] values, values which find their source in emotional and psychological investments.

This led me to engage with the potential of cross-referencing in [illegible] and richly narrative found ready-mades, adapted and modified, [illegible] to found pieces and replicas that simulate 'founders' [illegible]

[illegible] a translator, for example, defining and interpreting the body to world, or really a storyteller introducing 'their world' through the language of the object to a body?

Jewellery practice brings to art and the discourse of cultural production a distinct sensibility of the relationship between making activity, application of skill and the transformation of materials, to conceptual considerations and contemporary criticality.

The work extends the traditional format of the wearable decorative object and refers dialogically to the body, either as its absence when exhibited, directly when actually worn or as a representation of a reflected enquiry about the body. One could argue that the body is always implied because of the physical nature of making. The hand-made crafts object (and those where a technological application has been used in a combination with the hand-made) carry the traces of the body's involvement, regardless of on which level of cultural production they are subsequently disseminated. However, I am interested in those body-related objects. They become a place, or idea of a place, where narratives are invested in body related objects with the aim to negotiate that gap, complexity, confusion, or conflict.

Thinking takes place in making, but my making signifies bodies all the same; despite thinking through the mediated layers of culture, history, and theory we are grounded in the physical—in our visceral sensations, like pain or desire, in our longings, fears, and our knowledge of the transience of bodily existence. Making narrative work that maps out the demarcation lines where body meets world happens at the node between complex referential networks, in between the physical world of making and the world of ideas, thought and language. My conceptualisation of the body exists in the many facets of the in-between, as a place where we create ourselves as individual subjects with distinct identities and where the activity of self-creation is attuned more to its ambiguities rather than its certainties, knowing quite well that the activity of self-creation is fraught with disappointment, porous and evasive.

Jack CUNNINGHAM

If I see your work within the context of narrative, in a sense, a visual representation of 'collections', they often appear naturally as a collection of various objects, or, seem to be made as an 'object assembly'. These collected, or made, [illegible]

刘商英 **未到达**

艺术家独立出版，北京，2012

239mm×227mm×8mm，48 页

布面精装

600 册

文字：刘商英

Sceneries on the road quickly pass by, they slowly become fragments of my recollections; they line up more or less to form another existence. I look for the real side of them to give me strength.
— Liu Shangying

艺术中间 2012

中间美术馆，北京，2012

232mm×232mm×20mm，168 页

布面精装

2000 册

前言：周翊

策展人阐述：梁汉昌、Tomas Vu、袁佐

后记：黄晓华

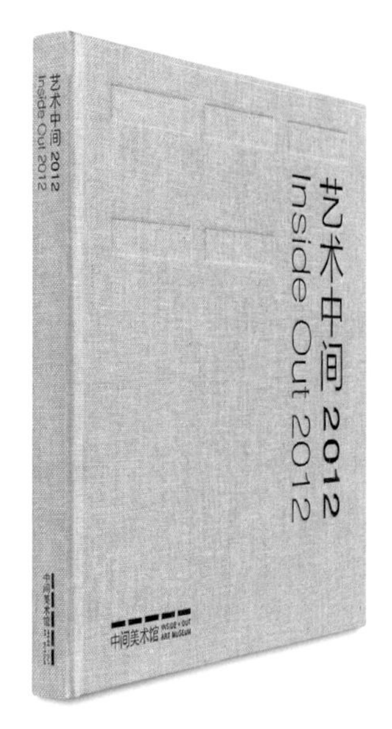

Brian Novatny

Shahzia Sikander

王蓬 **王蓬**

艺术家独立出版，北京，2013

288mm×225mm×21mm，134 页

布面精装，裹纸质护封

1000 册

编辑：王蓬、何浩

文章：章纳森·高得曼、鲍栋

访谈：林似竹 / 王蓬

王蓬：洞察，参与，与远见

章纳森·高得曼

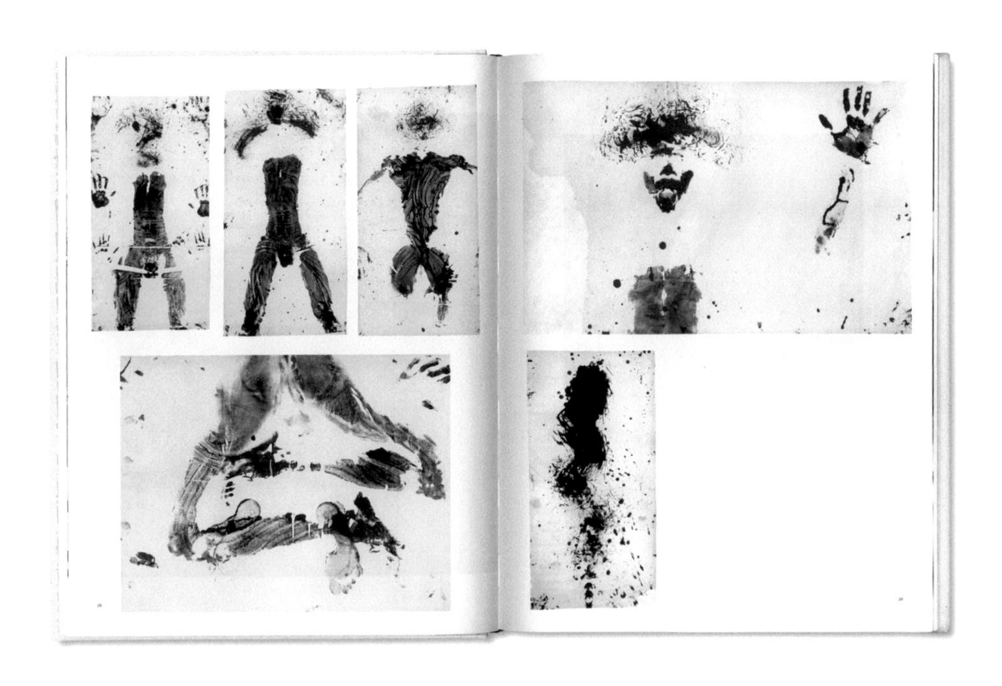

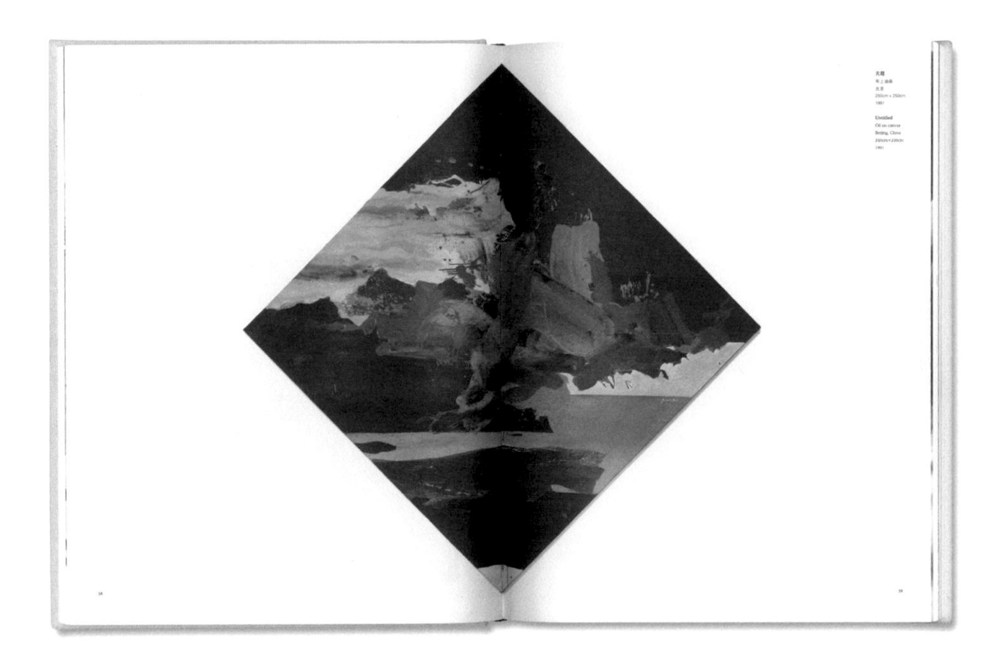

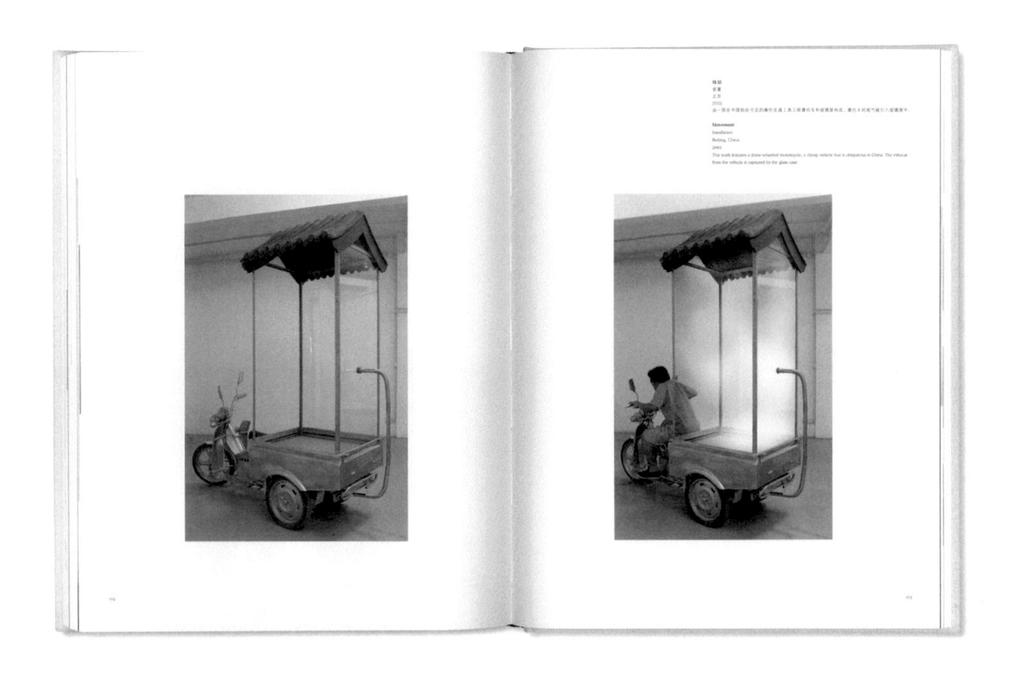
Movement
Installation
Beijing, China

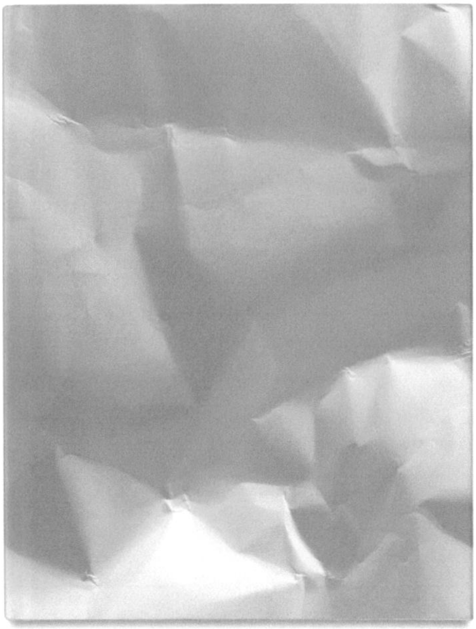

刘晓辉 **刘晓辉**

东站画廊，北京，2013

140mm×290mm×25mm，188 页

纸面精装，书脊包布

600 册

编辑：刘晓辉、何浩

文章：玛瑙

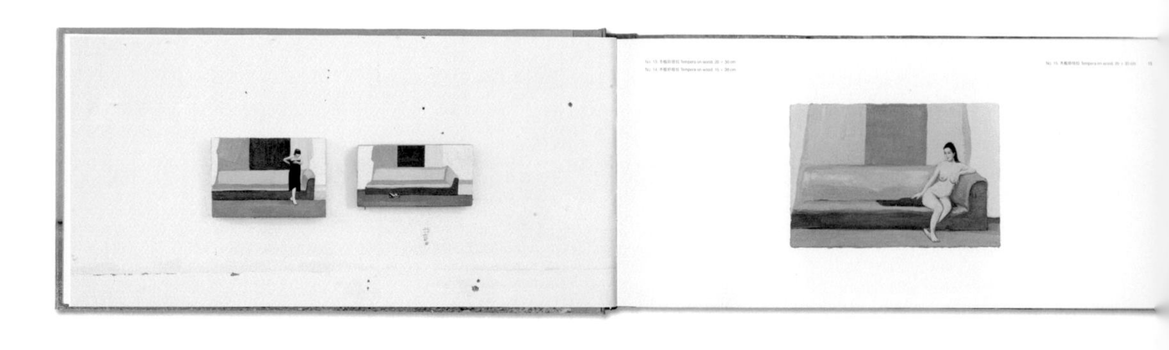

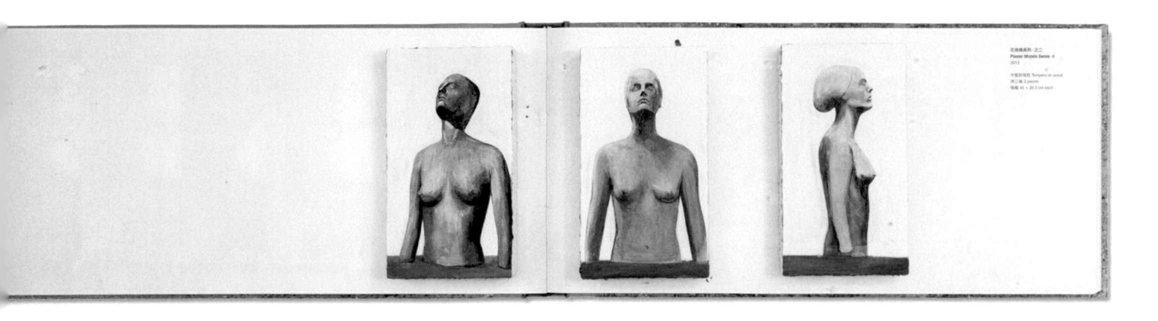

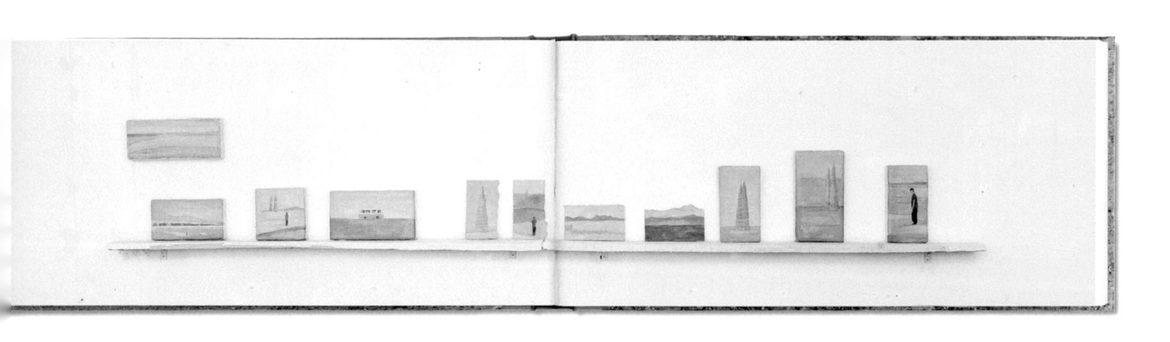

海报，1000mm×700mm，100 张

刘晓辉

Liu Xiaohui

开幕酒会：2013 年 6 月 29 日下午 4 点

展览日期：2013 年 6 月 29 日 - 8 月 4 日

周二至周日，10:00 -18:00

Opening: 4:00 pm, June 29, 2013, Saturday

Duration: June 29 - August 4, 2013

10:00 am to 6:00 pm (Closed on Mondays)

叶锦添 **梦·渡·间**

三影堂摄影艺术中心，北京，2013

263mm×223mm×19mm，120 页

布面精装，外裹护封

1500 册

前言：荣荣和映里

编辑、文章：马克·霍尔本

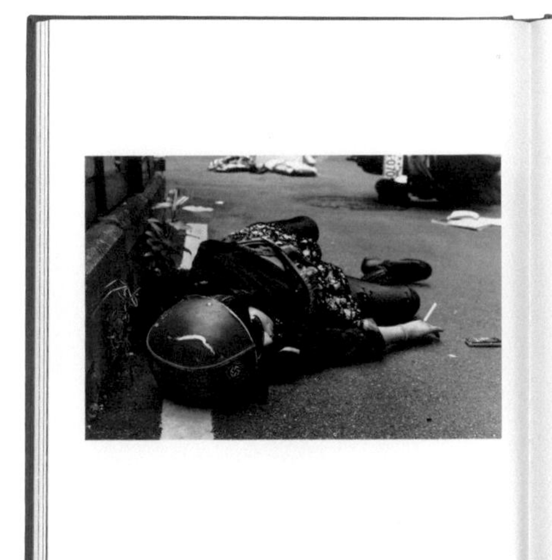

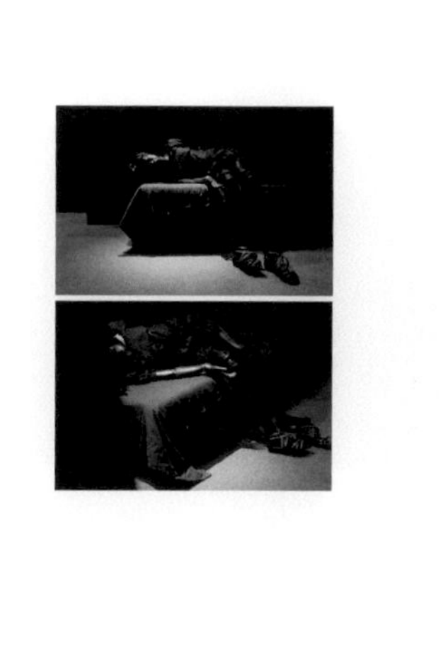

Tim Yip

Graduated from Hong Kong Polytechnic with degree in photography.
Photographer and writer for various Hong Kong publications.
Member of the Hong Kong Institute of Professional Photographers (HKIPP).
Illustrations were published and exhibited at Japan Comic Art Tokyo.
Won the Academy Award and Los Angeles Film Critics Award for Best Art Direction, and BAFTA for Best Costume Design for *Crouching Tiger, Hidden Dragon* in 2000.

Film

1986	Art director for *A Better Tomorrow*, directed by John Woo (HONG KONG)
1987	Art director for *Rouge*, directed by Stanley Kwan (HONG KONG)
1988	Art director for *Eena Bowl of Tea*, directed by Wayne Wang (HONG KONG, USA)
1989	Art director for Stanley Kwan's *A Port Moon* in New York (HONG KONG)
1990	Production and costume designer for *Ming Ghost*, directed by Chu Kang Chien (Taiwan)
	Nominated for Best Costume Designer at the Golden Horse Award
1991	Production and costume designer for *Autumn Moon*, directed by Clara Law Laecom
	Won Best Picture at Locarno International Film Festival in Switzerland
1992	Production and costume designer for *Temptation of a Monk*, directed by Clara Law (Hong Kong)
	Nominated at the Kaohsing Hong Film Award and Best Production Designer at the Golden Horse Awards
1995	Production designer and costume designer for *Red's Pavilion*, directed by Chen Kuo Fu in special arts (Taiwan)
1996	Costume designer for *Wolves Do Under the Moon*, directed by He Ping (Taiwan). Nominated at the 1997 Golden Horse Awards
1999	Production and costume designer for *Crouching Tiger, Hidden Dragon*, directed by Ang Lee (Taiwan, USA)
	Won Best Art Direction at the Academy Award and the Los Angeles Critics Awards in 2000
	Won Best Costume Designer at the BAFTA Awards and FCIM in 2000
2000	Production and costume designer for *What Time is it Over There*, directed by Tsai Ming-Liang (France)
	Won Grand Jury Award and at Best Director at the 2001 Chicago Film Festival
	Won Best Picture and Best Director at the 46th Asia Pacific Film Festival
	In competition at the 54th Cannes Film Festival
2001	Production and costume designer for *Double Vision*, directed by Chen Kuo Fu (Taiwan, USA)

《卧虎藏龙》*Crouching Tiger Hidden Dragon*, 1999

	Costume designer for *Springtime in a Small Town*, directed by Tian Zhuang Zhuang (China)
2002	Production and costume designer for *Baby in Love*, directed by Li Shao Hong (China)
2004	Production design/consultant for *The Restaurant Court*, directed by Tsai Ming-Liang (France, Taiwan)
	Production and costume designer for *The Promise*, directed by Chen Kaige (China)
2005	Production and costume designer for *Song Song & the Cat*, directed by John Woo in collaboration with the United Nations International Children's Emergency Fund and World Food Program
2006	Production and costume designer for *The Banquet*, directed by Feng Xiao Gang (China)
	Won Best Makeup and Costume Design at the 43rd Golden Horse Awards
	Won Best Art Direction the 13th Golden Bauhina Awards
	Nominated for Best Art Direction at the 26th Golden Rooster Awards
2008	Production and costume designer for *Red Cliff*, directed by John Woo (China)
2009	Key costume and image design for *The Message*, directed by Chen Kuo Fu & Gao Qun Shu (China)
2010	Production, costume and image designer for *Tai Chi Zero*, directed by Stephen Fung (China)
2012	Production, costume and image designer for *Tai Chi Hero*, directed by Stephen Fung (China)
2013	Costume and image designer for *Back to 1942*, directed by Feng Xiao Gang (China)
2014	Costume and image designer for *White Haired Witch*, directed by Jacob Cheung (China)

Performing Arts

1990	Costume and set designer for *Release*, Contemporary Legend Theatre (Taiwan)
1994	Costume designer for *The Life of Mandela*, Ba Gu Bata Dance Theatre (Taiwan), invited to Spoleto Arts Festival Aachen in Germany and the 1997 Spoleto Festival in South Carolina
1995	Costume designer for *Yuan Quan Homemade Gangs Tou "Hidden Taiwan"*
	Costume Designer for *Mountain is Loess*, Legend in Dance Theatre (Taiwan), invited to the 1998 Avignon Festival
	Costume designer for *Martin shadow Terrace*, choreographed by Susan Bunge Patricia, Commissioned by the Festival de Saint Yorre's le Hill
	Costume designer for *Perdition Volva Aven Las Feura d'La Lune*, Ta Gu Teeta Dance Theatre (Taiwan)
	Costume designer for *Renegate*, Teeta Dance Circle (Taiwan)
	Costume designer for *Shadow*, Contemporary Legend Theatre and directed by Richard Schechner (Taiwan)
	Costume designer for *Tumbleous Yesterday Song by the High Tang Yue Fu*, Music and Dance Ensemble, invited by the 1998 Avignon Festival in 1998, 2000 Adelaide Festival 2000–2001

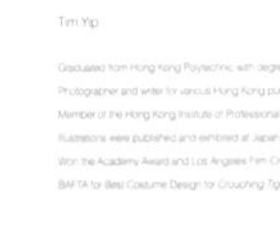

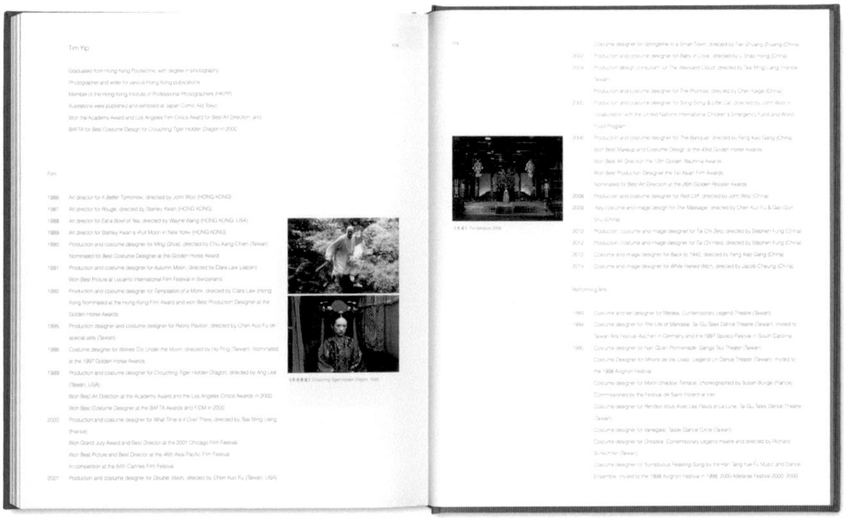

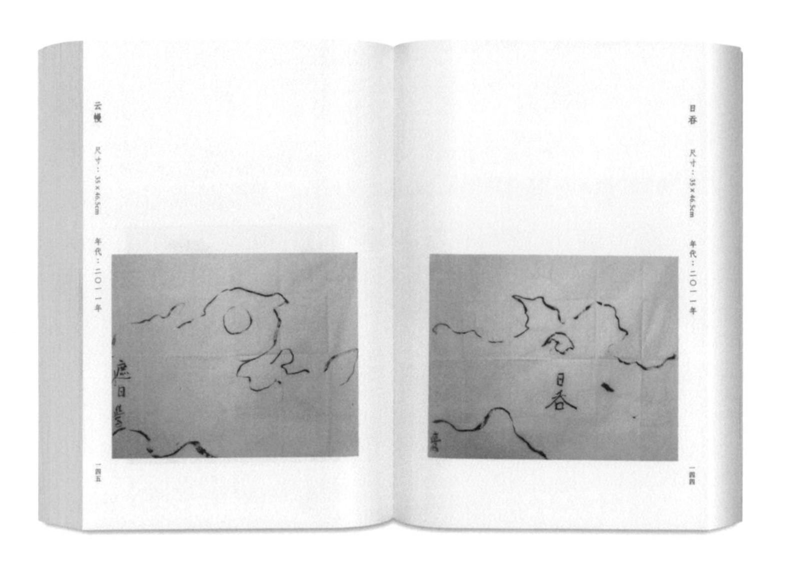

日吞　尺寸：35×46.5cm　年代：二〇一一年

一四四

云幔　尺寸：35×46.5cm　年代：二〇一一年

一四五

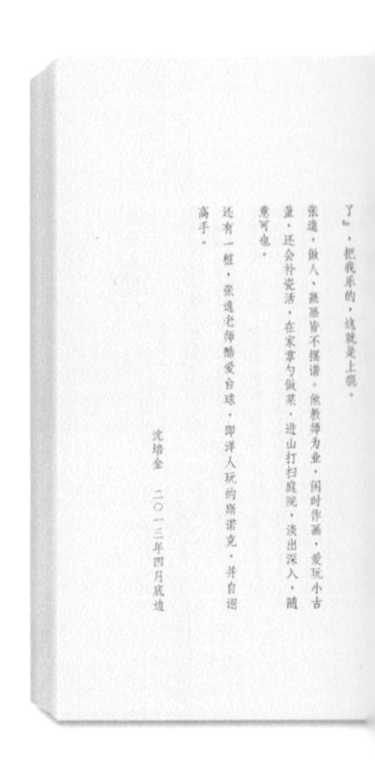

了』，把我乐的，这就是上乘。

张进，做人，濡墨皆不摆谱。他教师为业，闲时作画，爱玩小古董，还会补瓷活，在家掌勺做菜，进山打扫庭院，淡出深入，随意可也。

还有一桩，张进老师酷爱台球，即洋人玩的斯诺克，并自诩高手。

沈培金　二〇一三年四月底边

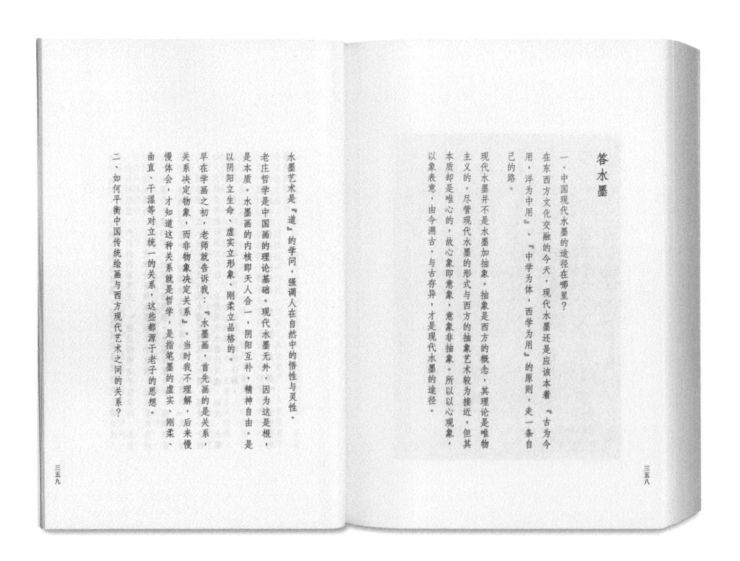

答水墨

一、中国现代水墨的途径在哪里？

在东西方文化交融的今天，现代水墨还是应该本着『古为今用，洋为中用』、『中学为体，西学为用』的原则，走一条自己的路。

现代水墨并不是水墨加抽象。抽象是西方的概念，其理论是唯物主义的。尽管现代水墨的形式与西方的抽象艺术较为接近，但其本质却是唯心的，故心象即意象，意象非抽象。所以以心观象，以象表意，由今溯古，与古存异，才是现代水墨的途径。

三五八

水墨艺术是『道』的学问，强调人在自然中的悟性与灵性。

老庄哲学是中国画的理论基础。现代水墨无外，因为这是根，是本质。水墨画的内核即天人合一，阴阳互补，精神自由。是以阴阳立生命、虚实立形象、刚柔立品格的。

早在学画之初，老师就告诉我：『水墨画，首先画的是关系，关系决定物象，而非物象决定关系』。当时我不理解，后来慢慢体会，才知道这种关系就是哲学，是指笔墨的虚实、刚柔、曲直、干湿等对立统一的关系，这些都源于老子的思想。

二、如何平衡中国传统绘画与西方现代艺术之间的关系？

三五九

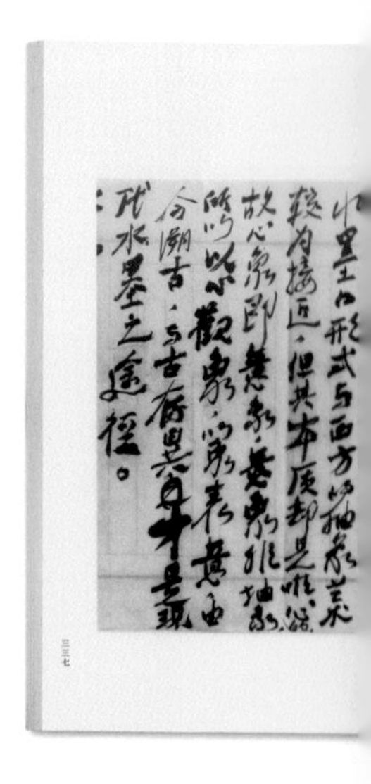

三三七

张进 四合画稿

高格出版公司，香港，2013

210mm×145mm，368 页

平装，190 册

文章：沈培金、萨本介

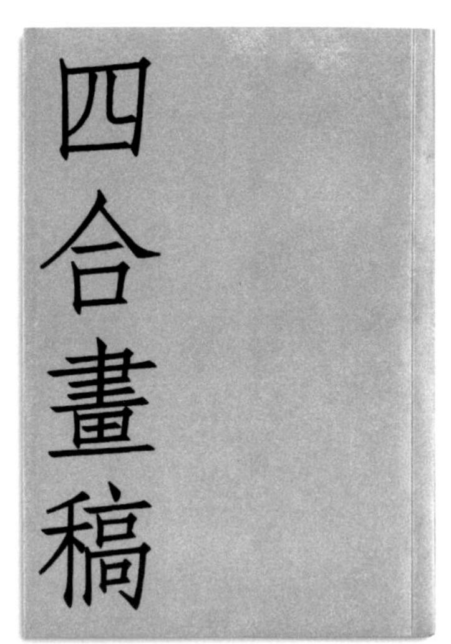

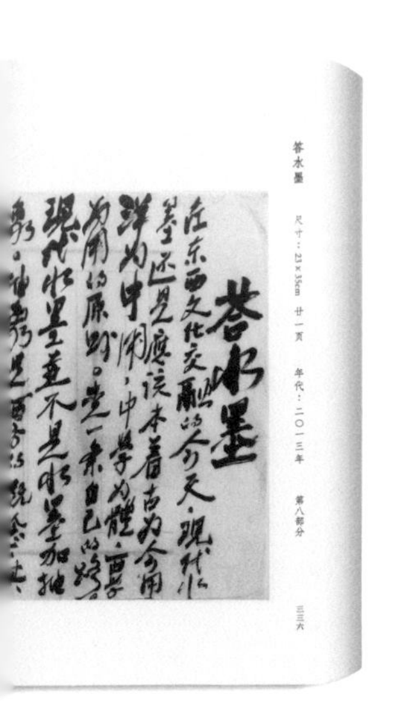

张杰　快乐的美术课：张杰的撕纸艺术

长春出版社，长春，2013

140mm×290mm×17mm，96 页

纸面精装，书脊包布

2000 册

编辑：中间美术馆

文章：周翊、曾庆豪、唐克扬、王纯信、张杰

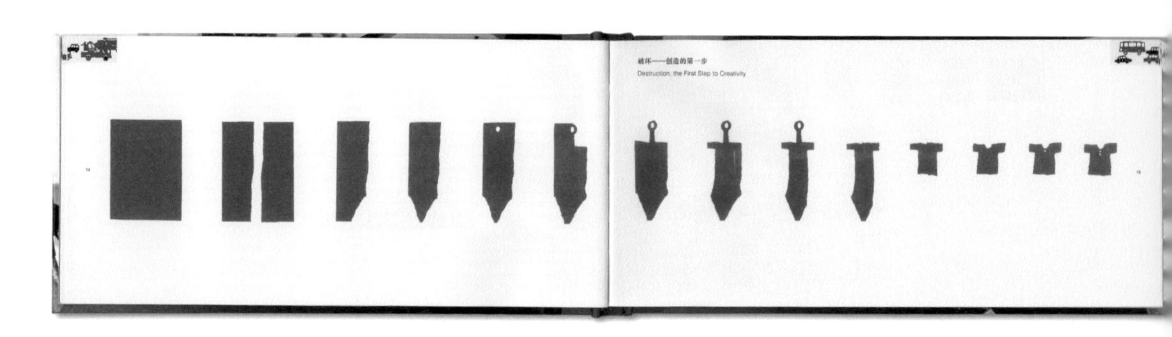

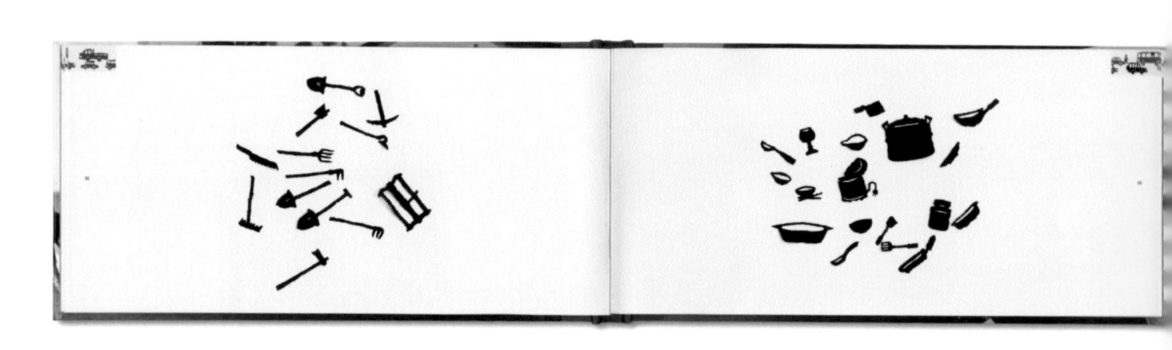

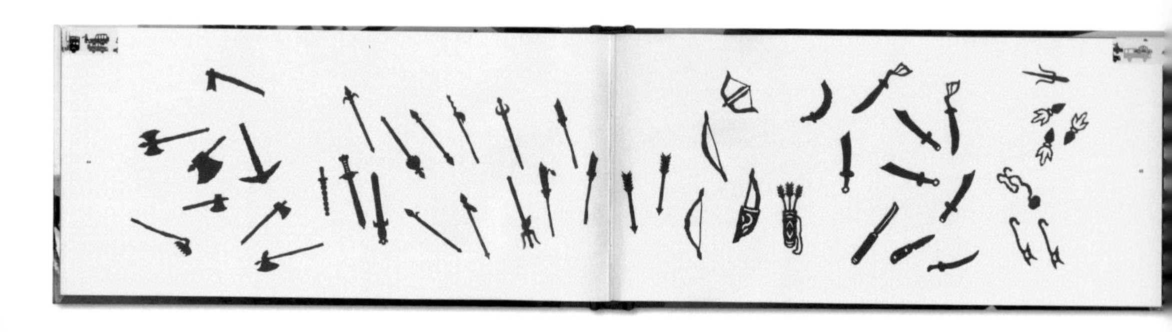

长白山风情：源于长白山区人民的传统生活。

Changbai Mountain Scenery
Taken from the lives of ordinary people who lived there since a long time ago

自主学堂

镇小人：快大茂镇中心小学在当地被人们称为“镇小”，镇小人是指快大茂镇中心小学的教师学生们。

Zhen Xiao Ren
Nick name refers to the Kuai Da Mao Town Elementary School, and the teachers and students who belong there

王纯信，男，生于1939年6月，吉林省通化县人。毕业于通化教育学院。曾任吉林省通化师范学院美术系主任、通化市美术家协会主席。其一生致力于长白山区民俗文化的发掘与传承工作，建树丰硕。王纯信先生于2013年4月2日辞世，享年75岁。

Wang Chunxin, born June 1939 in Tonghua County, Jilin Province. He was graduated from Tonghua College of Education. He was the art department chair of Tonghua Normal University, Chairman of Tonghua Artist Association. His life is dedicated to the research and preservation of folk culture in Changbai Mountain area. He passed away at 75 years of age on April 2, 2013.

勤奋的机遇——张杰“出山”纪实

王纯信

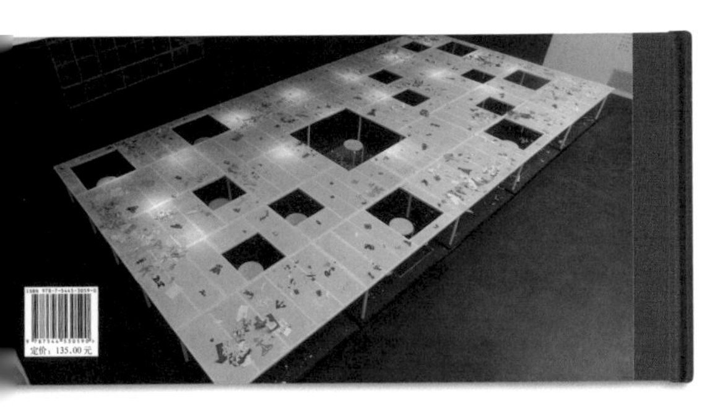

刘铮 **刘铮：摄影 1994—2002**

蝴蝶效应，北京，2014

262mm×220mm×25mm，232 页

布面精装，外裹护封

2000 册

前言：克里斯托弗·菲利普斯

文章：顾铮、杜曦云

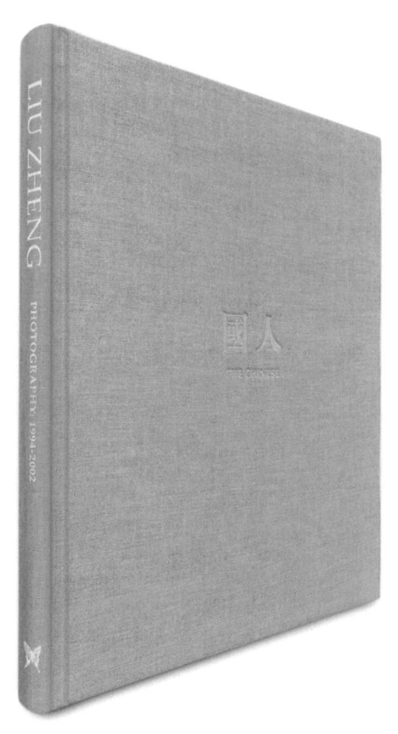

泥塑：地獄之門，四川豐都 Clay Sculpture: Gate of Hell, Fengdu, Sichuan Province, 1999

THE CHINESE

華山蒼龍嶺，陝西華山 Canglong Mountain Range, Hua Mountain, Shaanxi Province, 1999

6

前言

自上世紀90年代初以來，全世界的想象力都被中國的驚人轉變牢牢抓住，一個古老的國度如今正處在開足馬力的社會及經濟現代化陣痛之中，布滿摩天大厦的新興城市到處是光芒閃爍的高樓，精致的設計師别墅盤踞在長城脚下，火箭搭載中國航天員進入了外太空——全世界的大衆媒體衹要提起中國總不免會伴着這樣的景象。

劉錚的攝影作品是對這種歡欣鼓舞却一面倒的視覺印象的有力矯正。在大多數藝術家還在努力尋找屬于自己的聲音和視覺語言的年紀，他已經在安靜地想象并最終完成了一項了不起的攝影計劃　他稱之爲《國人》。劉錚在過去十年中爲這個項目拍攝了大量照片，這本書就呈現了他從中精選的120幅作品。他的作品以一種驚人地濃縮的視覺力量，清晰再現了中國過去那種曖昧矛盾的傳統如何在21世紀中生存下來。它們隱藏在日常的悲劇和荒誕的僞裝之下　如今突然呈現在你眼前。

劉錚在山西大同的煤礦小鎮長大，這是他生平中的一個關鍵事實。在貧瘠之地日復一日爲生存而掙扎，這在他作爲一名視覺藝術家的作品中留下了獨特而清晰的印記。1991到1997年間，劉錚就職于中國最大的報紙之一《工人日報》，擔任攝影記者。1994年他開始醞釀一個系列攝影作品，後來成爲《國人》。從某些方面看，這個計劃成長的土壤裏含有兩種後毛澤東時代的中國文學和視覺藝術潮流。一種是1970年代末及80年代初出現的“傷痕藝術”，顧名思義，這個藝術潮流着重于展現中國社會的陰暗面，特别是通過文革（1966–1976）時期的個人悲劇來加以揭示。同時，另一種是“本土”運動，將現實主義的目光轉向偏遠省份的普通百姓，追求一種迥异于美化工農兵的官方藝術的另類表達。這兩種藝術潮流一道，在上世紀80年代催生出第一批獨立紀實攝影作品，而紀實攝影很快就在中國實驗藝術領域裏占據了一席之地。

劉錚的攝影作品典型地捕捉到中國社會典型人物在現代化轉型時期的遭遇，且經常是在極端和意外的處境之下。截至目前，他拍攝的對象包括道士　街頭怪人、流浪兒童、易裝表演者、外省毒販、挖煤工人、和尚、犯人以及歷史博物館裏的古怪蠟像。劉錚最令人震撼的照片，是那些死者或瀕臨死亡者的肖像，其中包括年齡和社會地位不同的男女。

這些今日中國居民的肖像，讓人想起劉錚自己特别崇敬的兩位攝影藝術家的作品特點，即奥古斯特　桑德（August Sander）和黛安　阿勃斯（Diane Arbus）。和他們一樣，他運用精

7

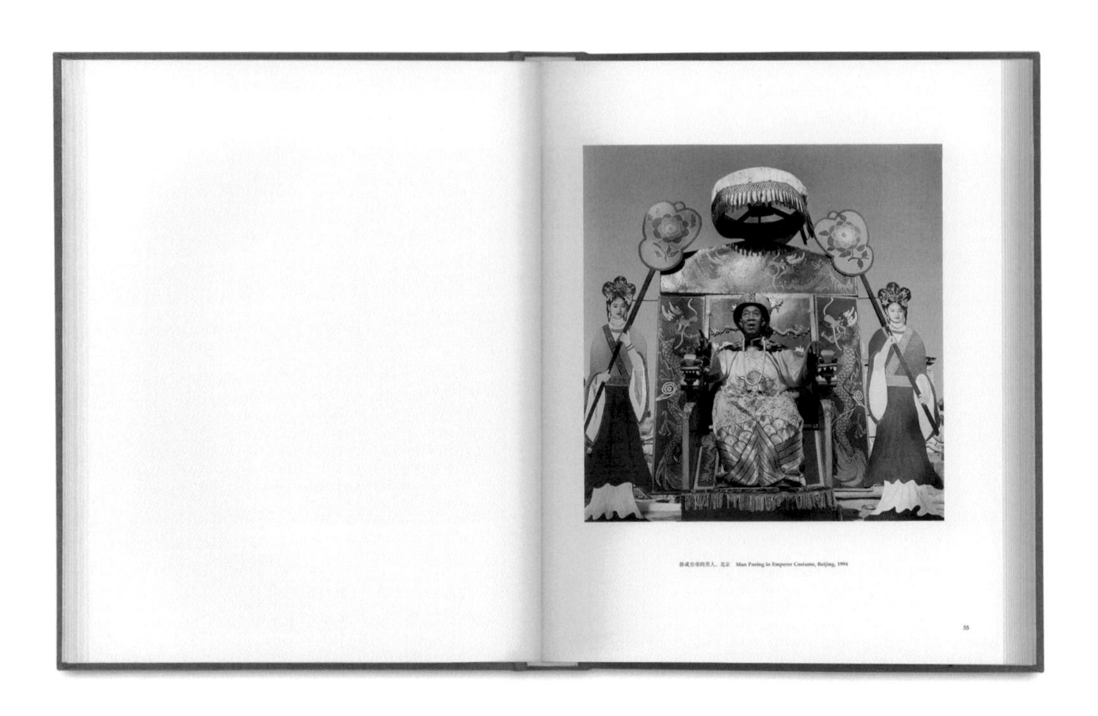
Man Posing in Emperor Costume, Beijing, 1994
55

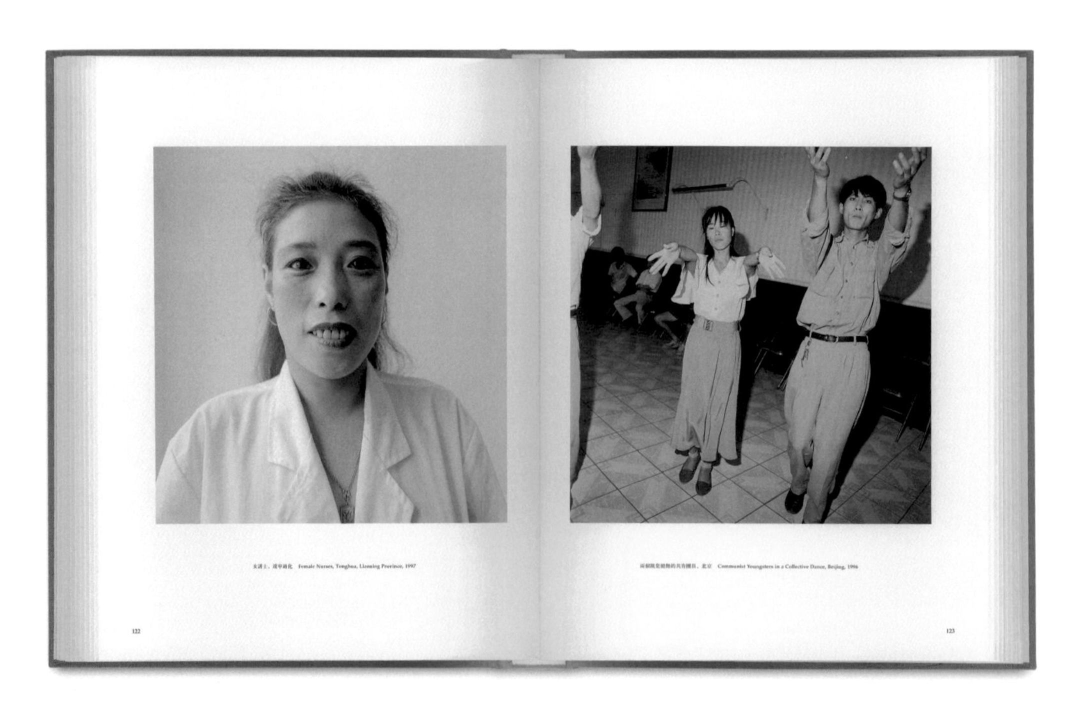
Female Nurses, Tonghua, Liaoning Province, 1997
Communist Youngsters in a Collective Dance, Beijing, 1996
122
123

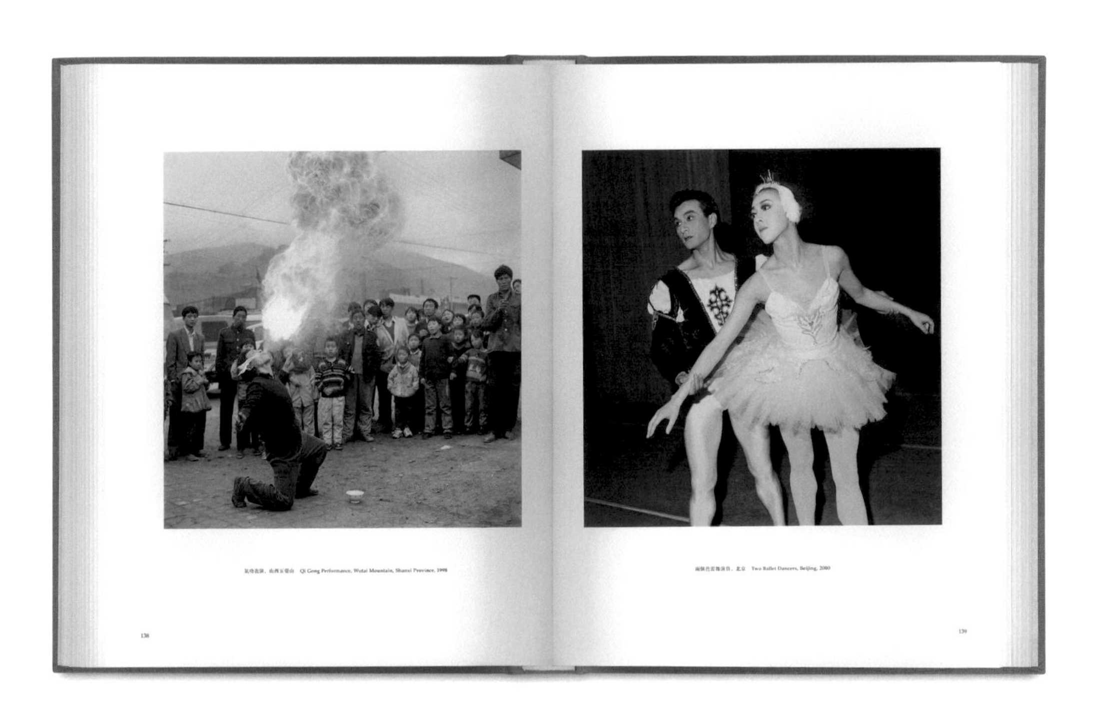
Qi Gong Performance, Wutai Mountain, Shanxi Province, 1998
Two Ballet Dancers, Beijing, 2000

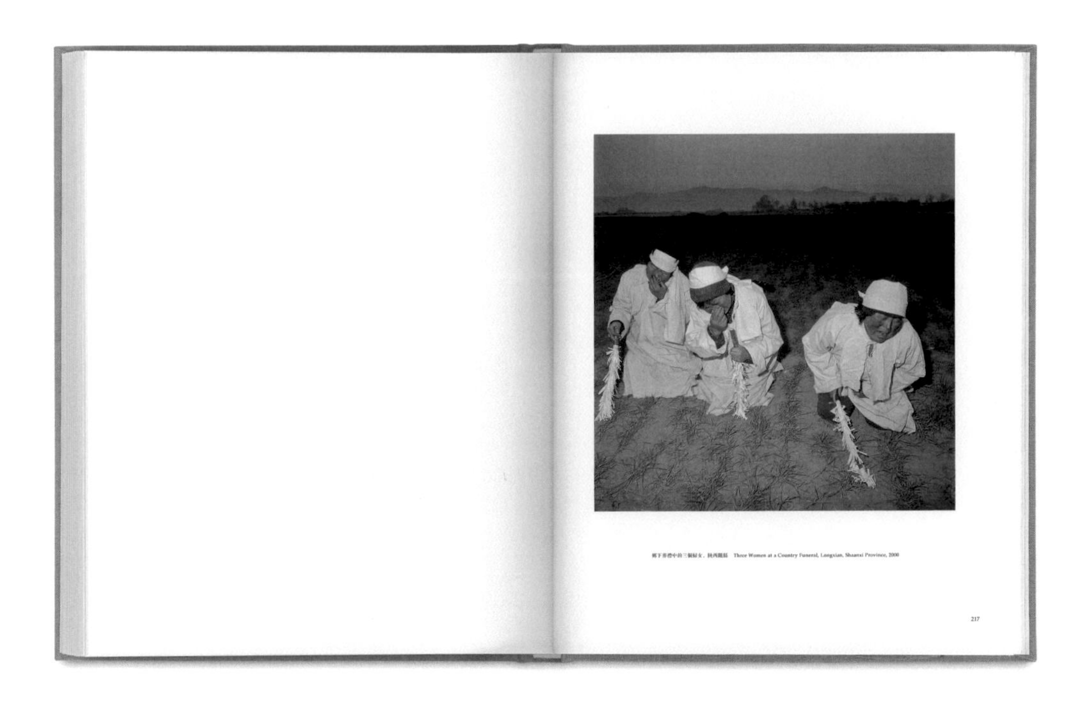
Three Women at a Country Funeral, Longxian, Shaanxi Province, 2000

陆亮 **陆亮**

诚品画廊，台北，2014

291mm×218mm×13mm，116 页

布面精装，外裹护封

500 册

文章：宋晓霞、陈文骥

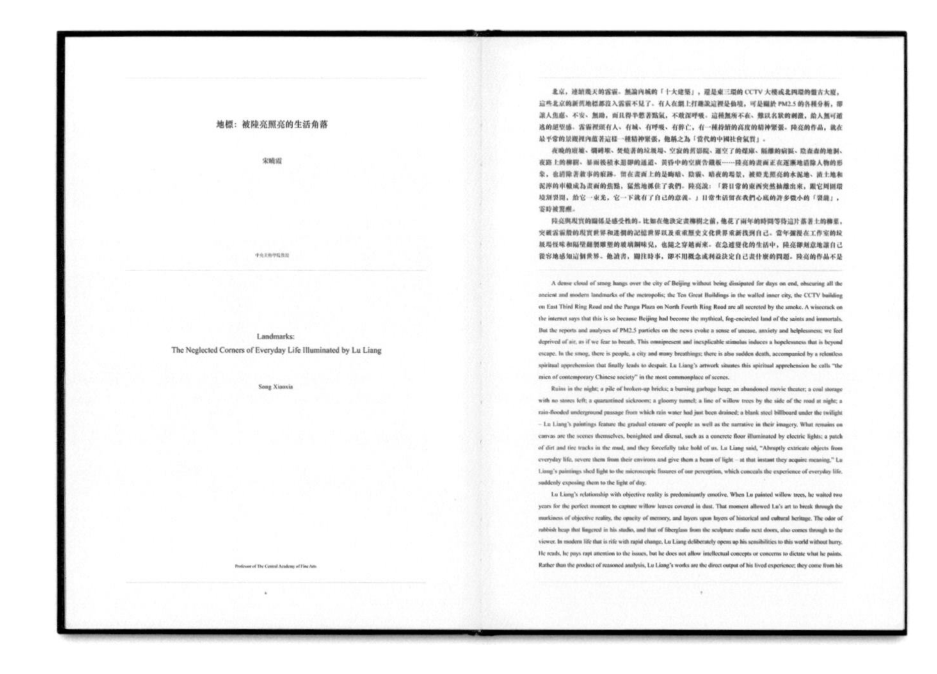

地標：被陸亮照亮的生活角落

宋曉霞

中央美術學院教授

Landmarks:
The Neglected Corners of Everyday Life Illuminated by Lu Liang

Song Xiaoxia

Professor of The Central Academy of Fine Arts

A dense cloud of smog hangs over the city of Beijing without being dissipated for days on end, obscuring all the ancient and modern landmarks of the metropolis; the Ten Great Buildings in the walled inner city, the CCTV building on East Third Ring Road and the Pangu Plaza on North Fourth Ring Road are all secreted by the smoke. A wisecrack on the internet says that this is so because Beijing had become the mythical, fog-encircled land of the saints and immortals. But the reports and analyses of PM2.5 particles on the news evoke a sense of unease, anxiety and helplessness; we feel deprived of air, as if we fear to breath. This omnipresent and inexplicable stimulus induces a hopelessness that is beyond escape. In the smog, there is people, a city and many breathings; there is also sudden death, accompanied by a relentless spiritual apprehension that finally leads to despair. Lu Liang's artwork situates this spiritual apprehension he calls "the mien of contemporary Chinese society" in the most commonplace of scenes.

Ruins in the night; a pile of broken-up bricks; a burning garbage heap; an abandoned movie theater; a coal storage with no stores left; a quarantined sickroom; a gloomy tunnel; a line of willow trees by the side of the road at night; a rain-flooded underground passage from which rain water had just been drained; a blank steel billboard under the twilight – Lu Liang's paintings feature the gradual erasure of people as well as the narrative in their imagery. What remains on canvas are the scenes themselves, benighted and dismal, such as a concrete floor illuminated by electric lights; a patch of dirt and tire tracks in the mud, and they forcefully take hold of us. Lu Liang said, "Abruptly extricate objects from everyday life, severe them from their environs and give them a beam of light – at that instant they acquire meaning." Lu Liang's paintings shed light to the microscopic fissures of our perception, which conceals the experience of everyday life, suddenly exposing them to the light of day.

Lu Liang's relationship with objective reality is predominantly emotive. When Lu painted willow trees, he waited two years for the perfect moment to capture willow leaves covered in dust. That moment allowed Lu's art to break through the murkiness of objective reality, the opacity of memory, and layers upon layers of historical and cultural heritage. The odor of rubbish heap that lingered in his studio, and that of fiberglass from the sculpture studio next doors, also comes through to the viewer. In modern life that is rife with rapid change, Lu Liang deliberately opens up his sensibilities to this world without hurry. He reads, he pays rapt attention to the issues, but he does not allow intellectual concepts or concerns to dictate what he paints. Rather than the product of reasoned analysis, Lu Liang's works are the direct output of his lived experience; they come from his

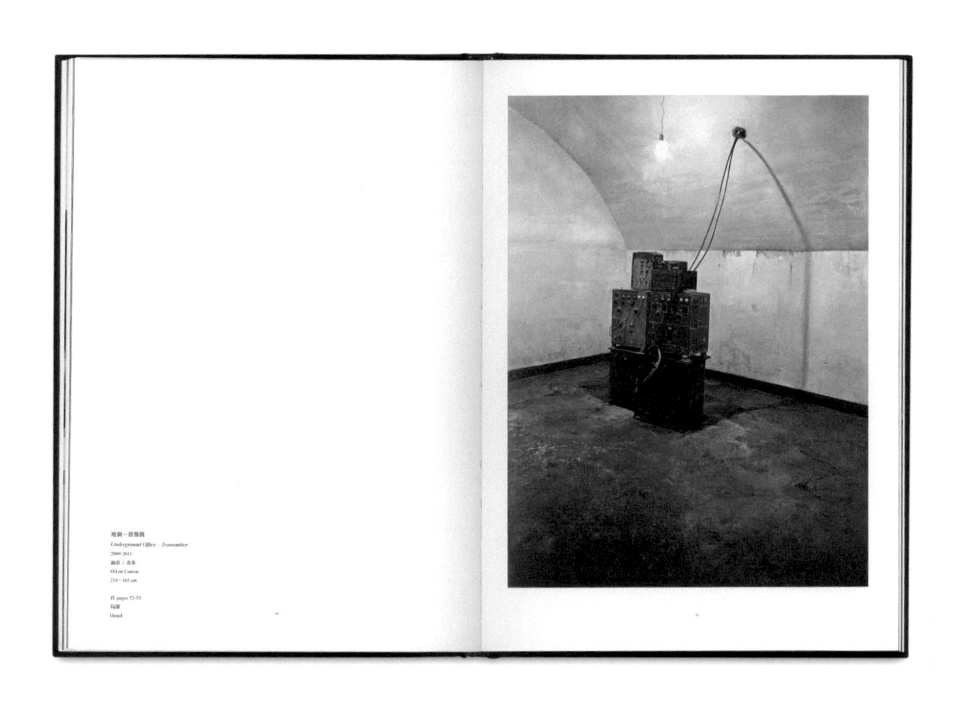
Underground Office – Transmitter
2009-2011
Oil on Canvas
210 × 165 cm
pages 52-53
Detail

彭斯　**景物斯和**

安徽美术出版社，合肥，2014

350mm×250mm×28mm，218 页

纸面精装，书脊包布

2000 册

文章：叶朗、彭锋、帕特里西奥 · 巴特西卡玛、四月（顾春芳）、何一鸣、李啸非

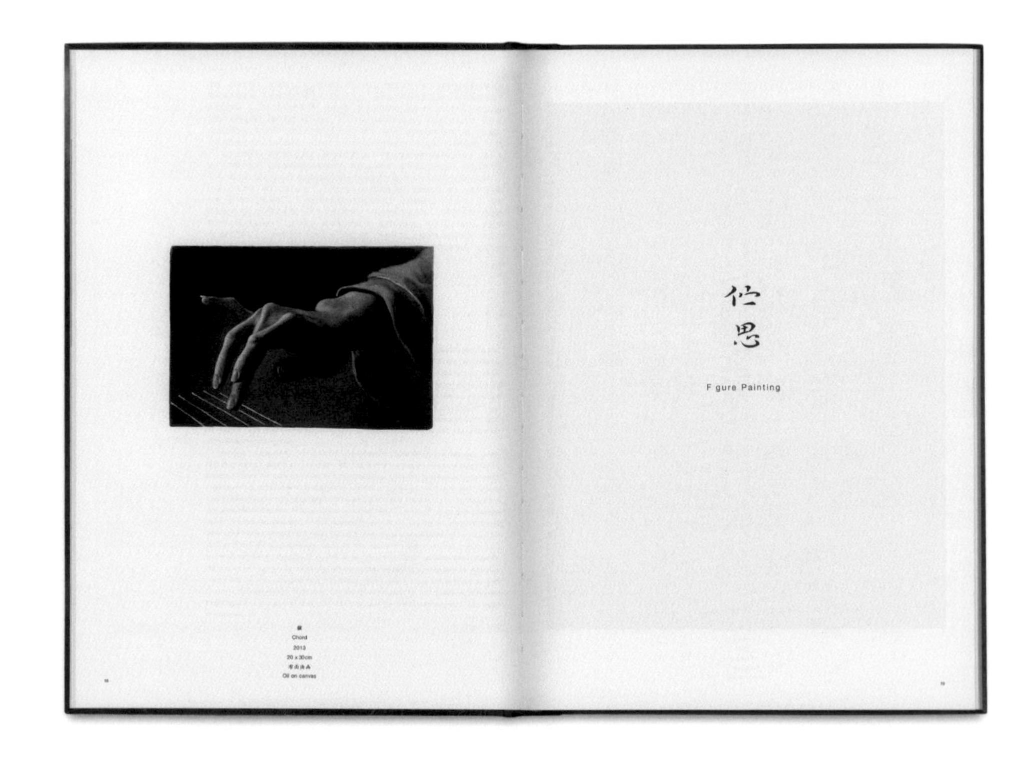

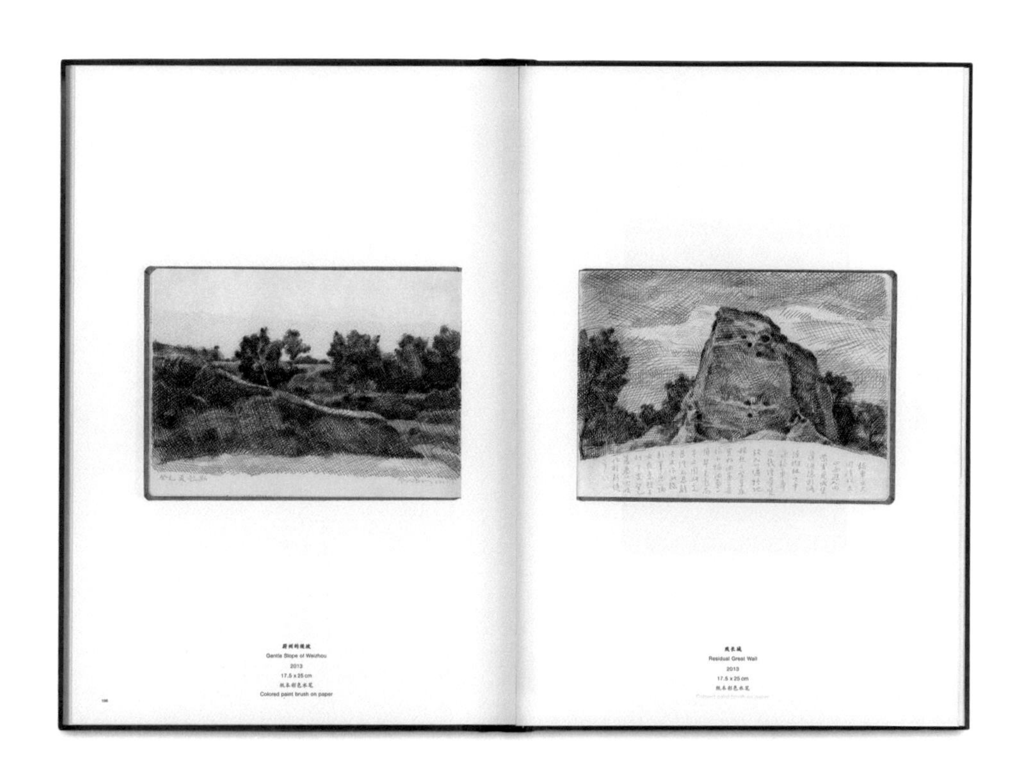
Gentle Slope of Weizhou
2013
17.5 x 25 cm
Colored paint brush on paper
Residual Great Wall
2013
17.5 x 25 cm

癸巳歲置一良鏨
名曰衡寒後植入
于齋室名典出於
船山先生句甚山

张进 **光洒太行**

北京荣宝拍卖有限公司，北京，2014

285mm×210mm×3mm，42 页

平装，骑马钉

500 册

文章：李抗

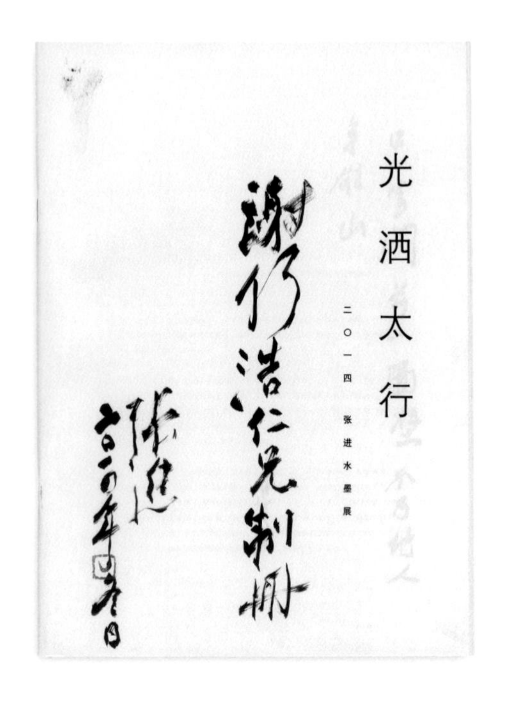

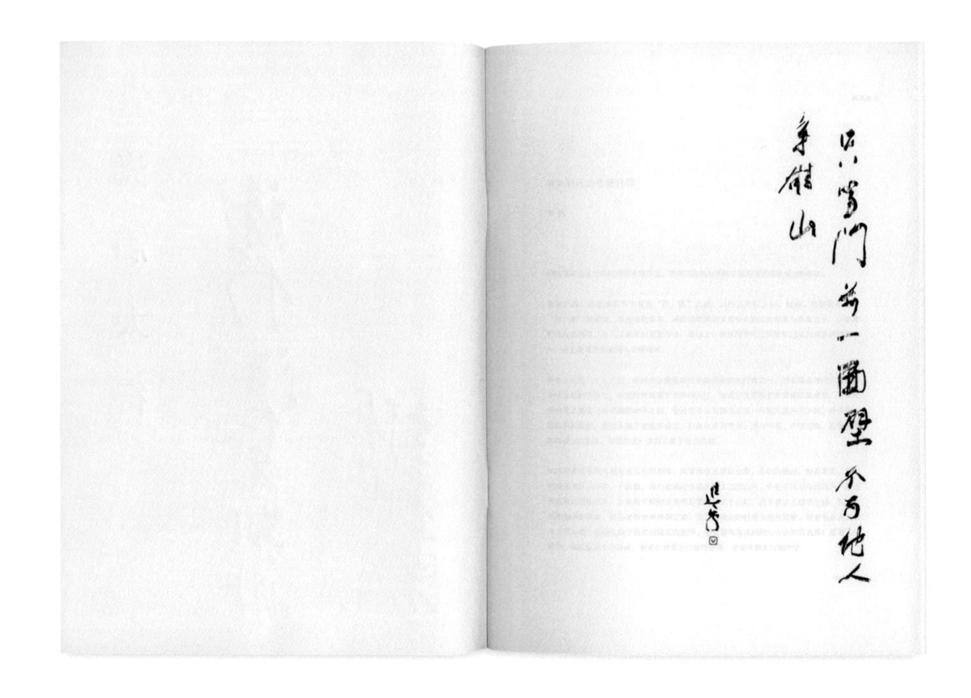

图版目录

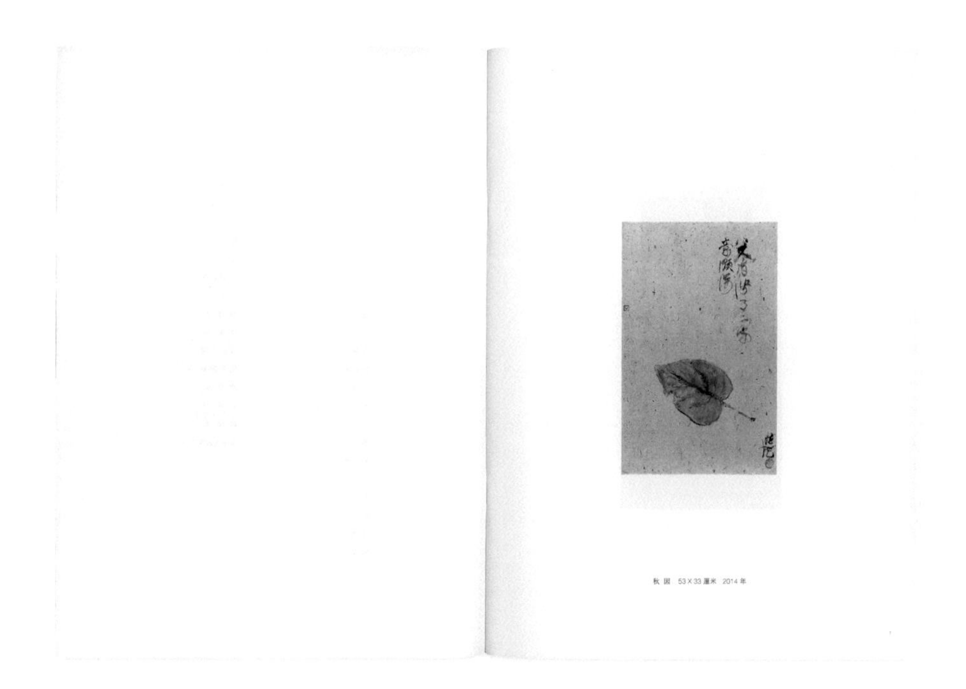

秋困 53×33厘米 2014年

唐小禾 程犁

唐小禾 程犁

人民美术出版社，北京，2014

壁画卷：346mm×251mm×34mm，306 页

油画卷：346mm×251mm×38mm，342 页

布面精装，外加瓦楞纸函套（未展示）

2000 套

主编：唐小禾

壁画卷序言：唐小禾

壁画卷文章：徐迟、王朝闻、唐小禾、周韶华、程犁、范春歌、皮道坚、孙美兰、刘骁纯、查世铭、墨吟、陈家琪

油画卷序言：孙景波

油画卷文章：彭德、王明贤、严善錞、陈履生、鲁萌

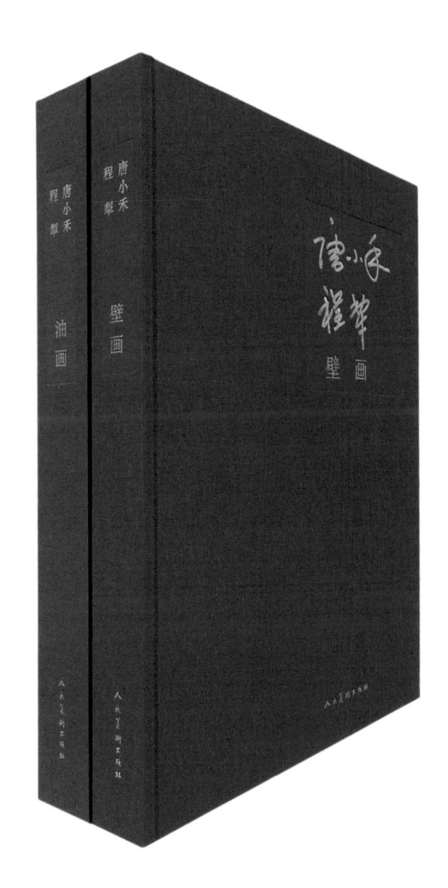

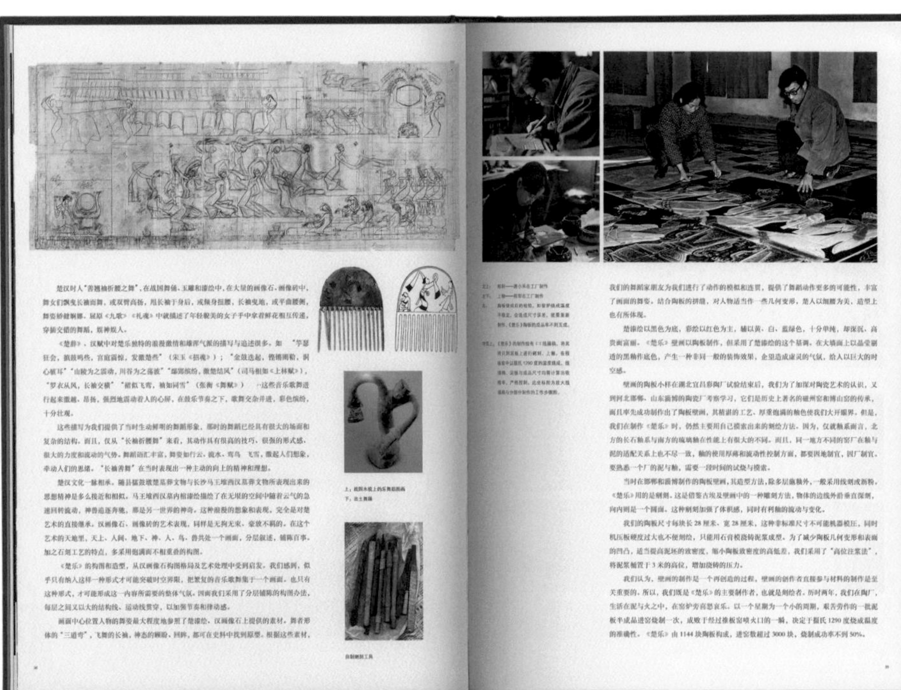

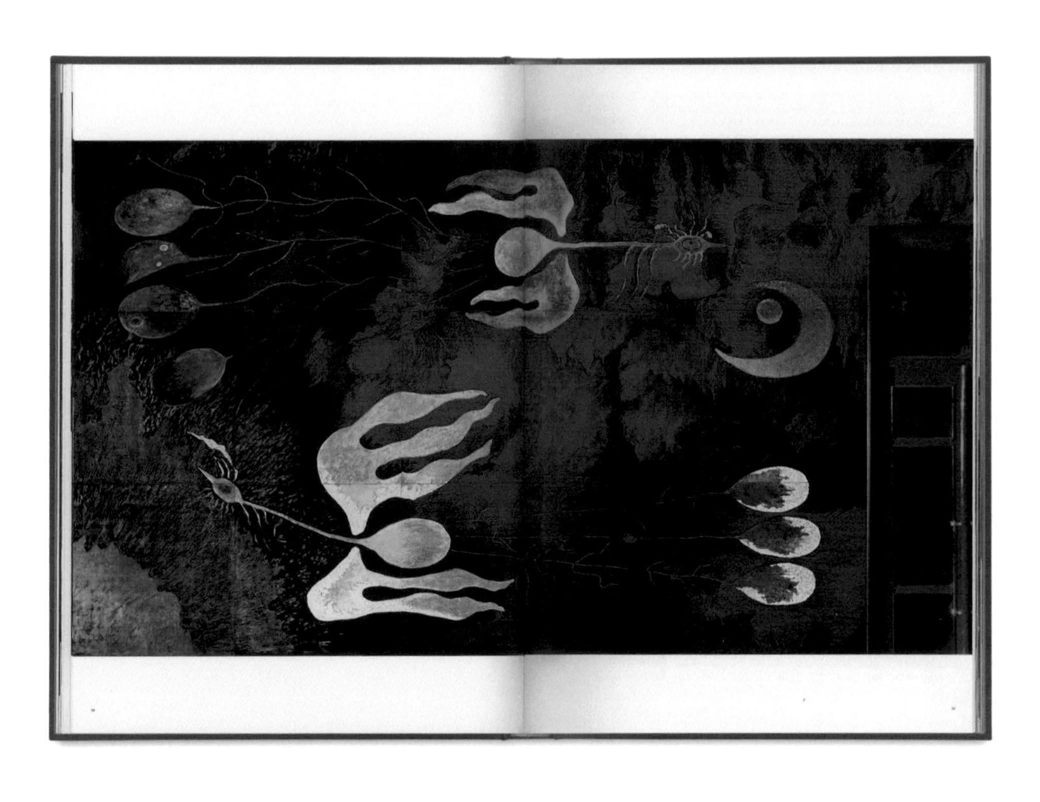

楚乐创作随想

程犁　唐小禾

地域，历史，人文的殊异形成了不同的艺术风格。

大溪文化的发现，说明早在新石器时代，长江流域就有一支与黄河流域仰韶文化并行的文化，这大约是我国南北文化各自的源头。正如北方民歌产生了《诗经》的现实主义精神一样，南方的神话孕育了《楚辞》的浪漫主义气质。

南国楚地湖泽棋布，云蒸霞蔚的平原，葱翠的高山，浩瀚的大江，养育了对神奇的自然善感而富于想象的人民，也形成了弥漫着神话色彩的巫官文化。由此而发展起来的楚文化是我们民族浪漫主义艺术的摇篮，也是我们民族艺术的宝贵传统。

1978年，在湖北随州擂鼓墩发掘出战国早期楚属国曾侯乙的墓葬，其丰富的陪葬物中竟然有一组庞大的编钟和大量金石丝竹管弦乐器，精美绝伦的65件编钟完整呈现，依音律排列三层，悬挂在由铜人托举的彩绘钟架上，其奇特的结构和造型、宏大的规模使人震惊，它已在地下屹立了2400年！与其陪葬的人与物无不与音乐歌舞相关，反映了墓主人的爱好。当我们站在先民们这个伟大的艺术创造面前时，耳际便轰响着充满原始艺术活力的乐音，"其词激宕淋漓，异于风雅"的"楚声"。

墓葬中的器物上有大量形态诡谲的漆绘，那黑底上红、黄、蓝、白色跳跃飞舞，寓意和表现着大自然中的生灵，日月星辰、风雨雷电，幻想与现实交相辉映，人们上天入地的窥探与流观，无一不透露出一个精力充沛、富于想象的民族的精神特征和那一时代的浓郁气息！这也是具有浪漫激情的楚艺术又一个丰富、生动的实例。

1979年，当我们接受了为武汉东湖宾馆宴会厅绘制壁画的任务时，很快地就想到了以曾侯乙编钟音乐和楚歌楚舞为母题进行创作，使这幅壁画具有强烈的浓厚的楚文化色彩，以展示我们地域的历史文化特点。

面对着壁画的墙面，我们设想以此为舞台，跨过2400多年的时空，这编钟、编磬、各种丝竹琴瑟伴着楚歌楚舞响起时，将会是何等动人的情景！

几乎是毫不犹豫地就决定了在画面上直接采用曾侯乙编钟庄重雄浑的造型，彩绘的钟架横梁是由腰束佩剑的铜人承托着，由下而上，由大而小的64件铜钟依次排列，有一种次序的美感，还有以怪兽为支架玉石编磬、流苏羽葆飞扬的建鼓、丝、木、匏、竹、八音俱全，这应当是一支具有浓烈雄楚风格的庞大的乐队，在它的伴奏之下，画面正中是一组少女散花舞蹈的场面，吉日良辰，金玉齐鸣，姣服芳菲，是对生命的祝福，也是春天的礼赞。

唐小禾　在大风大浪中前进，172.5 cm × 294.5 cm，布面油画，1971年　参加1972年"纪念毛泽东《在延安文艺座谈会上的讲话》发表30周年全国美术作品展览"

早期油画的代表作《在大风大浪中前进》

1966年6月28日，毛泽东从他家乡韶山"滴水洞"来到武汉。7月16日，武汉市正在举行一年一度横渡长江的群众性体育活动，参加者有五千之众，毛泽东乘一艘军用快艇来到游泳队伍中，江上一片沸腾！毛泽东下水游了起来，并对陪同者说："大风大浪也不可怕。人类社会就是从大风大浪中发展起来的。"要大家到大风大浪中去"经风雨，见世面"。毛泽东的这次公开露面和下水游泳，此后被赋予了特殊的政治含义，昭示了"文革"的开始。那段历史，就发生在我的身边，对领袖来到群众中，在大江里游泳，说了一些激励人心的话，我与大家一样感到兴奋不已。

事隔五年，1971年，得知要恢复全国美术作品展览，我在计划创作时，就选择这一个"本地的"题材。此前有人画过，但仅限于一个场面的记录，并没有赋予毛泽东畅游长江更深层的象征意义。而我这幅画的主题思想，正是从其意义与内涵上切入和构思的。

这年夏天，在素描稿确定后，我用9个日夜完成了这件近3米的油画（外地作品多是以火车托运的形式运到北京，由于画面尺寸受行李车厢车门尺寸的限制，不然这幅画我还会画得更大些）。时间虽然很紧，我还是坚持去武警部队，到那艘快艇采访和写生，到工厂、学校搜集人物形象素材。作品完成后，经各级审查，修修改改，花费了不少时日。展出时，画的题名定为《在大风大浪中前进》。

附　《新中国美术图史1966—1976》节选

王明贤　严善錞

"……唐小禾在1971年创作的油画《在大风大浪中前进》，在"文革"中也是广为流传，它的发行量可能仅次于《毛主席去安源》。《人民日报》、《解放军报》以半版的篇幅（《解放军报》甚至破例套色）给予刊登，《人民画报》和《解放军画报》则以中心画页登载，独幅画和其它形式的发表则不计其数。除了这幅作品在风格样式上颇受官方和大众的青睐外，其主题的重大也是一个不可忽略的因素。1966年7月16日，毛泽东以73岁的高龄在湖北武汉再次畅游长江，历时1小时零5分钟，游程近30华里，当时的传媒对此作了大量的报道。它一方面显示了他老人家当时的身体还十分健康，有足够的能力来领导中国的社会主义革命和中国的社会主义建设，另一方面，"大风大浪"一词，在革命的词典里，是喻意双关的，它常常被引伸为阶级斗争的风浪。为了纪念毛泽东的这一壮举，在随后的十多年时间里，每年的7月16日，武汉都要举行数万人的"横渡长江"和活动。

唐小禾的《在大风大浪中前进》，按他自己的说法，基本上属于一幅"急就章"，素描稿画了两天，画布上也只画了9天，人物的造型完全出自画家自己的想象，没有参照模特儿，从这点上，我们也可以看出画家扎实的造型功底。这件作品的创作缘起，也是为了参加《纪念延安文艺座谈会上的讲话》发表30周年的全国美展。这一主题虽然不是上级指定的，但唐小禾自己觉得确实有一种不可压抑的"革命的激情"想画这一题材。唐小禾是湖北艺术学院1965年的油画专业毕业，品学兼优。他的父亲唐一禾是中国近代的杰出的油画家，素描尤为精到，一派纯正的法国古典趣味，其功力、格调均不在名声远播的徐悲鸿之下。"文革"伊始，唐小禾便成了资产阶级"染缸"里培养出来的资产阶级接班人、黑帮子女，他自己也为之痛苦不已。但在各种舆论的渲染下，唐小禾和当时年轻人一样，从内心里还是激起了对毛泽东的无限崇敬之情。

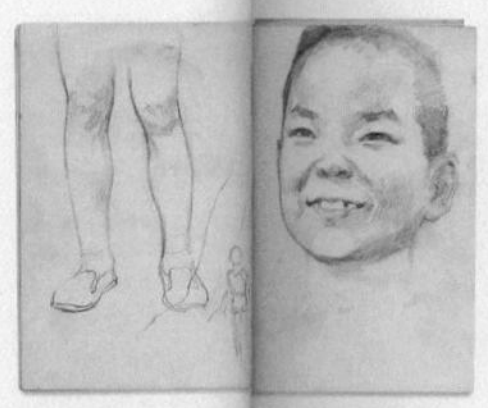

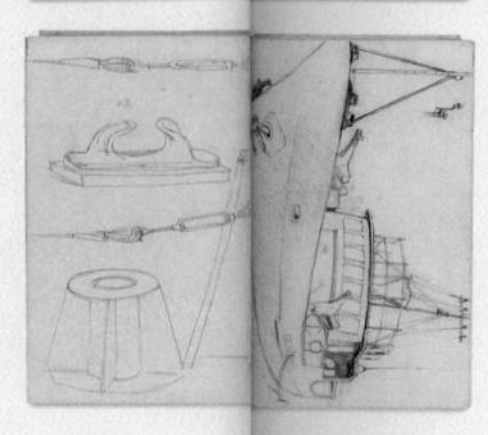

左：为创作收集的素材（速写）

右：《人民日报》《解放军报》1976年7月16日均以半版篇幅发表《在大风大浪中前进》与《在大风大浪中成长》

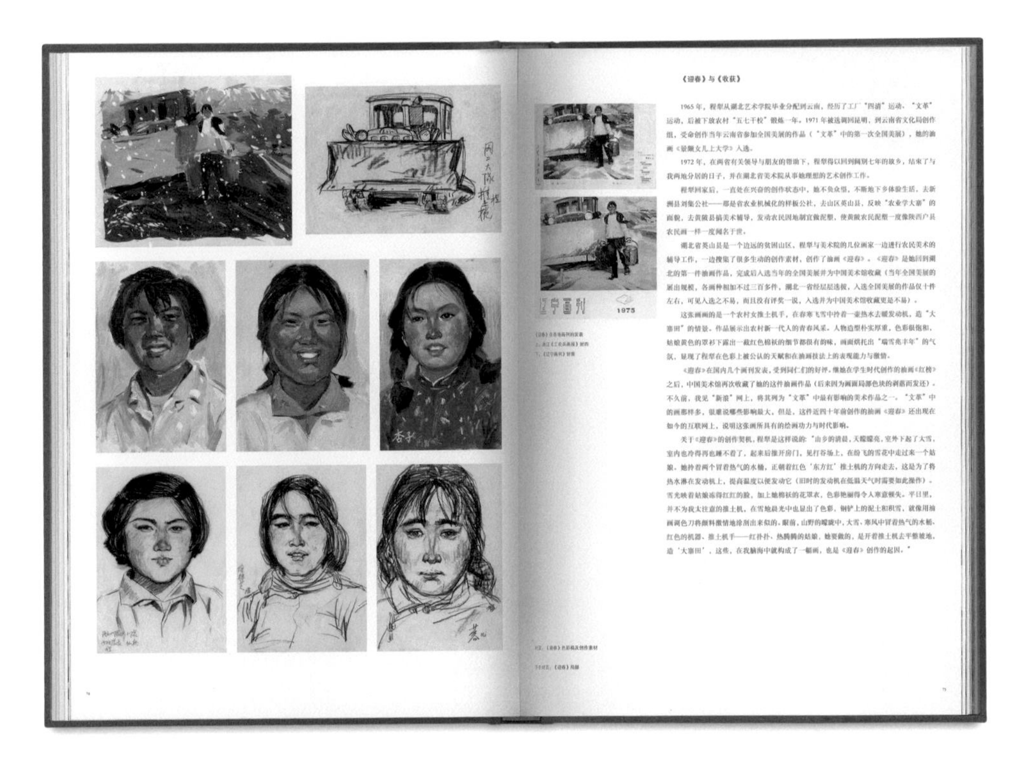

《迎春》与《收获》

1965年，程犁从湖北艺术学院毕业分配到云南，经历了工厂"四清"运动、"文革"运动，后被下放农村"五七干校"锻炼一年。1971年被选调回昆明，到云南省文化局创作组，受命创作当年云南省参加全国美展的作品（"文革"中的第一次全国美展），她的油画《景颇女儿上大学》入选。

1972年，在两省有关领导与朋友的帮助下，程犁得以回到阔别七年的故乡，结束了与我两地分居的日子，并在湖北省美术院从事她理想的艺术创作工作。

程犁回家后，一直处在兴奋的创作状态中，她不负众望，不断地下乡体验生活，去新洲县刘集公社——那是省农业机械化的样板公社，去山区英山县，反映"农业学大寨"的面貌，去黄陂县搞美术辅导，发动农民因地制宜做泥塑，使黄陂农民泥塑一度像陕西户县农民画一样一度闻名于世。

湖北省英山县是一个边远的贫困山区，程犁与美术院的几位画家一边进行农民美术的辅导工作，一边搜集了很多生动的创作素材，创作了油画《迎春》。《迎春》是她回到湖北的第一件油画作品，完成后入选当年的全国美展并为中国美术馆收藏（当年全国美展的展出规模，各画种相加不过三百多件，湖北一省经层层选拔，入选全国美展的作品仅十件左右，可见入选之不易，而且没有评奖一说，入选并为中国美术馆收藏更是不易）。

这张画画的是一个农村女推土机手，在春寒飞雪中拎着一壶热水去暖发动机，造"大寨田"的情景。作品展示出农村新一代人的青春风采。人物造型朴实厚重，色彩很饱和，姑娘黄色的罩衫下露出一截红色棉袄的细节都很有韵味，画面烘托出"瑞雪兆丰年"的气氛，显现了程犁在色彩上被公认的天赋和在油画技法上的表现能力与激情。

《迎春》在国内几个画刊发表，受到同仁们的好评。继她在学生时代创作的油画《红榜》之后，中国美术馆再次收藏了她的这件油画作品（后来因为画面局部色块的剥落而发还）。不久前，我见"新浪"网上，将其列为"文革"中最有影响的美术作品之一。"文革"中的画那样多，很难说哪些影响最大，但是，这件近四十年前创作的油画《迎春》还出现在如今的互联网上，说明这张画所具有的绘画功力与时代影响。

关于《迎春》的创作契机，程犁是这样说的："山乡的清晨，天朦朦亮，室外下起了大雪，室内也冷得再也睡不着了，起来后推开房门，见打谷场上，在纷飞的雪花中走过来一个姑娘。她拎着两个冒着热气的水桶，正朝着红色'东方红'推土机的方向走去，这是为了将热水淋在发动机上，提高温度以便发动它（旧时的发动机在低温天气时需要如此操作）。雪光映着姑娘冻得红红的脸，加上她棉袄的花罩衣，色彩艳丽得令人寒意顿失。平日里，并不为我太注意的推土机，在雪地晨光中也显出了色彩，铡铲上的泥土和积雪，就像用油画调色刀将颜料激情地涂刮出来似的。眼前，山野的朦胧中，大雪、寒风中冒着热气的水桶、红色的机器、推土机手——红扑扑、热腾腾的姑娘，她要做的，是开着推土机去平整坡地，造'大寨田'，这些，在我脑海中就构成了一幅画，也是《迎春》创作的起因。"

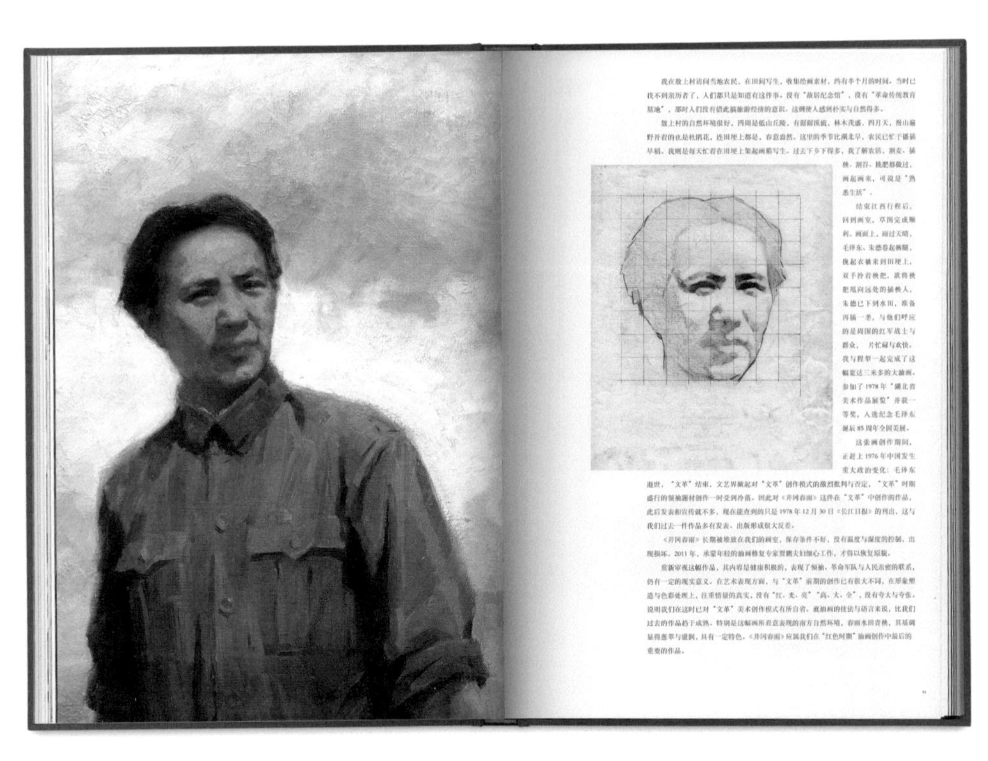

我在敖上村访问当地农民，在田间写生，收集绘画素材，约有半个月的时间。当时已找不到亲历者了，人们都只是知道有这件事。没有"故居纪念馆"，没有"革命传统教育基地"，那时人们没有借此搞旅游经济的意识，这倒使人感到朴实与自然得多。

敖上村的自然环境很好，四周是低山丘陵，有潺潺溪流，林木茂盛，四月天，漫山遍野开着的也是杜鹃花，连田埂上都是，春意盎然。这里的季节比湖北早，农民已忙于播插早稻。我则是每天忙着在田埂上架起画箱写生。过去下乡下得多，我了解农活，割麦、插秧、割谷、挑肥都做过，画起画来，可说是"熟悉生活"。

结束江西行程后，回到画室，草图完成顺利。画面上，雨过天晴，毛泽东、朱德卷起裤腿，挽起衣袖来到田埂上，双手拎着秧把，欲将秧把甩向远处的插秧人，朱德已下到水田，准备再插一垄，与他们呼应的是周围的红军战士与群众，一片忙碌与欢快。我与程犁一起完成了这幅宽达三米多的大油画，参加了1978年"湖北省美术作品展览"并获一等奖，入选纪念毛泽东诞辰85周年全国美展。

这张画创作期间，正赶上1976年中国发生重大政治变化：毛泽东逝世，"文革"结束，文艺界掀起对"文革"创作模式的激烈批判与否定，"文革"时期盛行的领袖题材创作一时受到冷落。因此对《井冈春雨》这件在"文革"中创作的作品，此后发表和宣传就不多，现在能查到的只是1978年12月30日《长江日报》的刊出，这与我们过去一件作品多有发表、出版形成很大反差。

《井冈春雨》长期被堆放在我们的画室，保存条件不好，没有温度与湿度的控制，出现损坏。2011年，承蒙年轻的油画修复专家贾鹏夫妇细心工作，才得以恢复原貌。

重新审视这幅作品，其内容是健康积极的，表现了领袖、革命军队与人民亲密的联系，仍有一定的现实意义。在艺术表现方面，与"文革"前期的创作已有很大不同，在形象塑造与色彩处理上，注重情景的真实，没有"红、光、亮""高、大、全"，没有夸大与夸张，说明我们在这时已对"文革"美术创作模式有所自省。就油画的技法与语言来说，比我们过去的作品趋于成熟。特别是这幅画所着意表现的南方自然环境，春雨水田青秧，其基调显得葱翠与滋润，具有一定特色。《井冈春雨》应属我们在"红色时期"油画创作中最后的重要的作品。

文金扬
绘画·教学·著作文献集
金扬
人民美术出版社

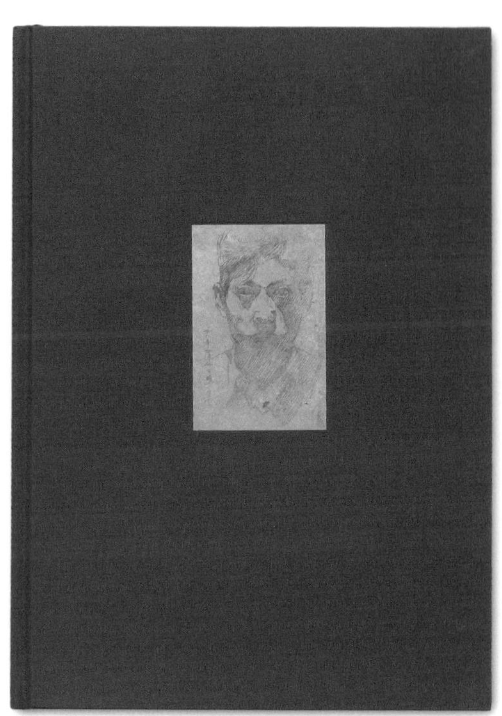

官厅水库
The Guanting Reservoir
1958
布面油画 Oil on canvas
89cm×127.5cm

著作手稿
Manuscripts

文金扬 **文金扬绘画·教学·著作文献集**

人民美术出版社，北京，2015

347mm×252mm×23mm，240 页

布面精装，外裹护封

1000 册

前言：范迪安

文章：冯法祀

刘商英 **空故纳万境**

广西美术出版社，南宁，2015

252mm×352mm×21mm，164 页

布面精装，外裹护封

1500 册

前言：范迪安

文章：罗贝尔·布贾德

艺术家陈述：刘商英

后记：一兰

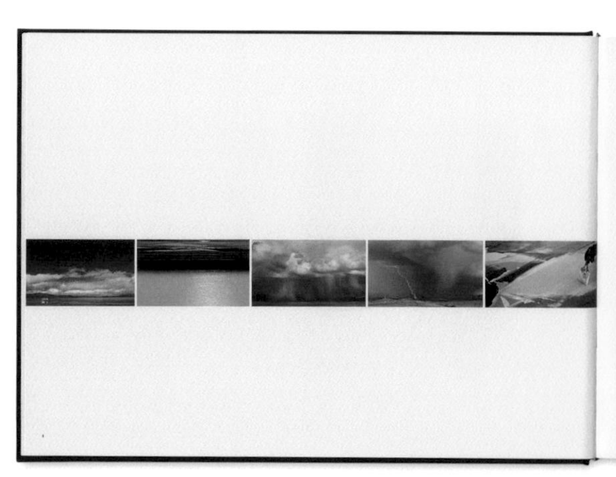

前言

范迪安

在全球图像时代到来的时刻，当代绘画所面临的挑战可以说越来越繁迫，也越来越鲜明，这不仅是一个中国性的问题，也是一个世界性的问题。因此，坚守绘画既需要画家树立坚定的理想与信念，也需要探求和探寻新的绘画思维与创作的方法论。从刘商英这几年的艺术探索中，可以看到他在解决这一课题上的用心与用力。

刘商英连续几年都利用学校的暑假时间前往西藏，他曾独自穿行藏北的无人区，也沿着冈底斯山脉在阿里地区进行长时间的踏访，这种“独步天涯”的体验为的是到大自然去找寻自我心灵的栖身之地，陶冶自己的情感世界，在某种程度上，也是为了寻找能够敞开自我心扉的创作现场。他这两年的脚步最终在玛旁雍错湖边停顿下来，把创作的营地驻扎在那片圣洁的湖水旁边。

玛旁雍错被称为世界的“圣湖”，她静静地平卧在雪域高原的群山之中，被称为“神山”的冈仁波齐山就在她的身边，将倒影投入她的怀抱。那是一片古老而寂静的世界，有着万古洪荒的时间积淀，显现了大自然的恒久生命与瑰丽神奇，那里又是佛教信仰者的朝圣之地，来自西藏地区和印度、尼泊尔的信徒每年都要前往神山圣湖做虔诚的礼拜，“转山”与“转湖”的身影为那片寂静的自然增加了一份信仰的神圣。那片大自然的“静”，既是“静寂”的“静”，与外物隔绝的安谧与寂寥让人只能听见自己身处海拔4600米之上的呼吸和心跳；它同时又是“静穆”的“静”，在呈现大自然的优美之外，也彰显了大自然崇高与肃穆的神性，让走近它的人从心底生发出一种精神的敬畏。在这样的风景面前，踏访者的身与心都会受到洗礼，都会感受到人与自然在精神上的一种契合，都会重新思考生命的价值与意义。但是，刘商英在这片自然面前不仅是一名观赏者、体验者，更是一名带着创作思考的跋涉者、探索者。他将数度踏访西藏的感受和体验集中起来，在玛旁雍错湖畔驻足，展开了巨幅的画布。他所面对的是自然风景，但他并无意去画一幅静态时间中的风景，在那片风景面前，“每天同一时间将画布固定在湖边，大部分时间我坐在那看，两个小时、三个小时…… 这样长久地注视同一个地方，会让我的观看更深刻，而所需一切随心才可能落地。那一刻，我好像忘记了自己的存在，忘记了艺术的存在”。“坐忘”是中国古代文艺创作论的一个经典术语，在大自然面前，在一片空白的画幅面前，既是身心投入的直观，也是去除遮蔽的沉思，“观”与“思”构成此时此在的关联，于神秘的契合中将自己带入一种无碍的状态，从而获得“返生”的感觉。

这种感觉是一个过程的结果，是一种将自己的心灵置于“空无”境地的状态，这种“忘”也是一个去除杂念，特别是搁置已有艺术经验的过程。胡塞尔曾经说过，要想达到事物的本质，就要将那些包裹在事物表面的遮蔽物“悬置”起来，使事物的本质在无碍的状态中澄明出来。当然，事物不会自动“澄明”，事物的澄明实是关于事物的思考者对各种遮蔽物的超越，在画家那里，就是真正地直面眼前的世界，达到“去蔽”的境地，使自己的思维和感受都处在“空”的原点，这个原点即是画意显象的新的起点。对于刘商英来说，他要追寻的不是那种可以用照相机捕捉下来的风景，而是要在画布

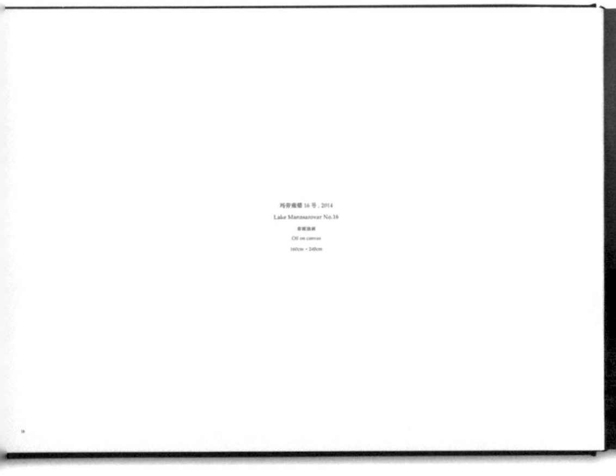

玛旁雍错 16 号，2014
Lake Manasarovar No.16
Oil on canvas
160cm × 240cm

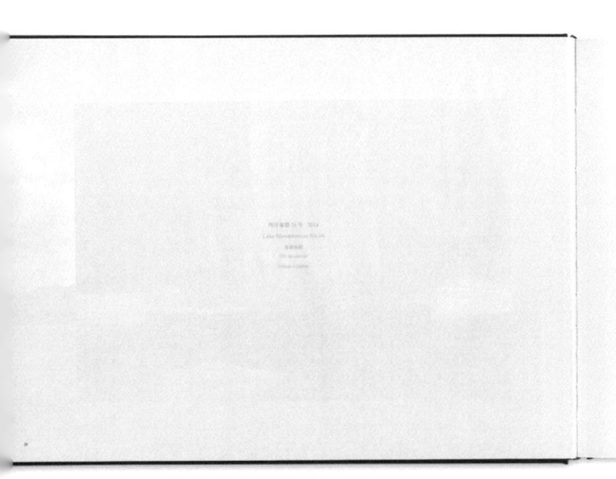

定价：380.00元

刘彦湖 **安敞庐印集**

上海书画出版社，上海，2015

351mm×252mm×41mm，362 页

皮面精装，外加瓦楞纸函套；筒子页装订

2000 册

主编：寒碧

文章：范迪安、卢辅圣、石开、曹意强

创作手记：刘彦湖

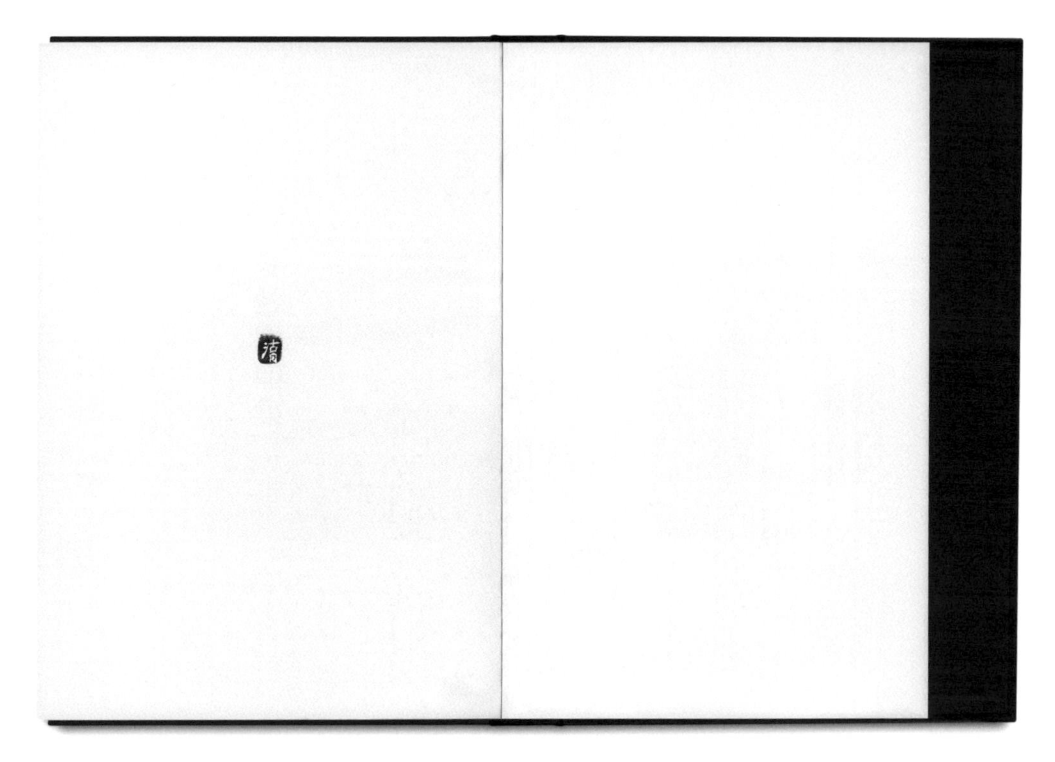

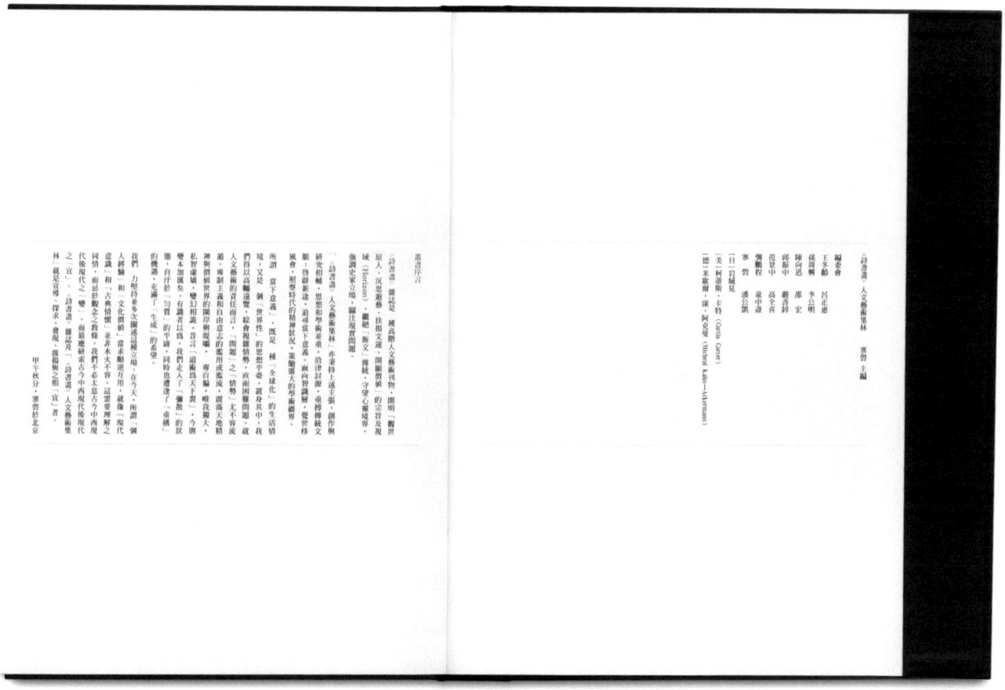

《詩書畫》人文藝術集林　寒碧 主編

編委會

王冬齡　呂正惠
孫周興　李公明
陳向迅　邵　宏
邱振中　嚴善錞
范景中　高全喜
龔鵬程　童中燾
寒　碧　潘公凱
〔日〕岩城見一
〔美〕柯蒂斯・卡特（Curtis Carter）
〔德〕米歇爾・康・阿克曼（Micheal Kahn-Ackermann）

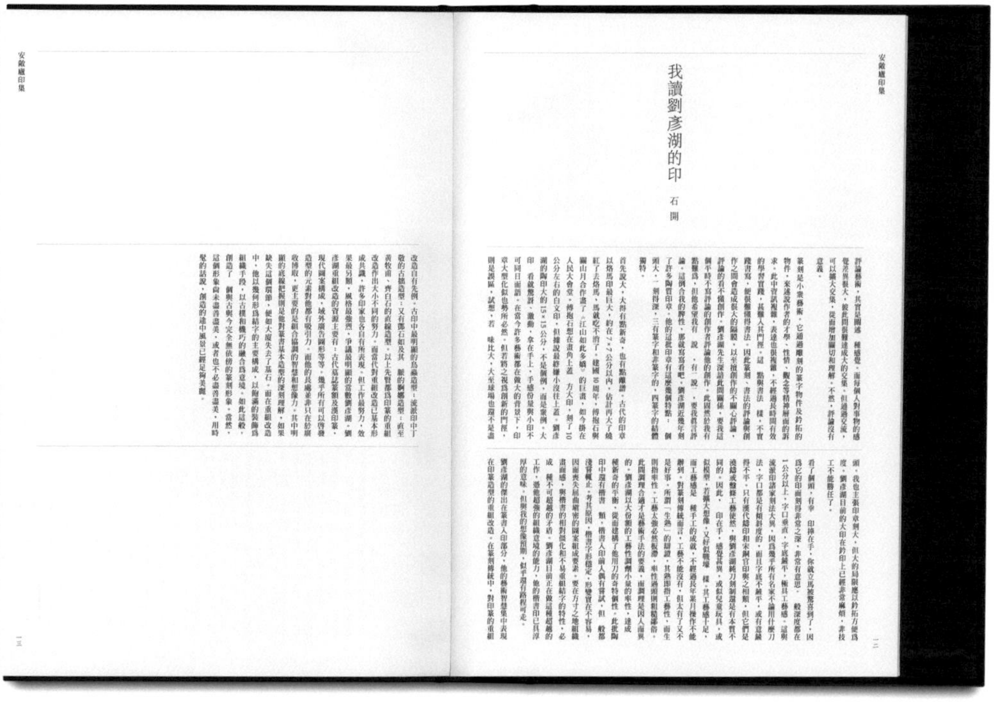

叢書序言

《詩書畫》雜誌是一種高階人文藝術刊物，闡明「觀世原人、沉思遊藝、扶揭文運、開顯價值」的宗旨及視域（Horizon），繼絕「斯文」傳統，守望心靈境界，強調史家立場，關注現實問題。

「《詩書畫》人文藝術集林」亦秉持上述主張，創作與研究相輔，思想和學術並重，沿津討源，重搏傳統文脈；啓闢新途，追尋當下意義，面向智識層，覺世移風會，照察時代的精神狀況，策馳廣大的學術疆界。

所謂「當下意義」，既是一種「全球化」的生活情境，又是一個「世界性」的思想平臺，置身其中，我們得以高瞻遠矚，綜會複雜情勢，直面困難問題。就人文藝術的責任而言，「問題」之「情勢」尤不容流遁，專制主義和自由意志的濫用或濫流，讚蕩天地精神與價值世界的關序與堤壩，專自偏，唯我獨大，私智盧矯，變幻相詭，昔言「道術爲天下裂」，今則變本加厲矣，有識者以爲，我們走入了「彌散」的狀態，自汙於「匀質」的平庸，同時也遭逢了「重構」的機遇，充滿了「生成」的希望。

我們力圖持並多次闡述這種立場，在今天，所謂「個人經驗」和「文化價值」當求順逆互用，就像「現代意識」和「古典情懷」並非水火不容，這需要理解之同情，而忌於觀念之教條，我們不必太息古今中西現代後現代之「變」，而最應研索古今中西現代後現代之「宜」，《詩書畫》雜誌及「《詩書畫》人文藝術集林」就是宣導、探求、發現、推揚與之相「宜」者。

甲午秋分，寒碧於北京

我讀劉彥湖的印

石開

改造自有先例，古印中最明顯的鳥蟲造型；浙派印中丁敬的古拙造型；又有鄧石如及其一脈的婀娜造型；直至黃牧甫、齊白石的直線造型，以上先賢都爲印篆的重組改造作出大小不同的努力，而當代對重組改造已基本形成共識，許多印家也各自有所表現，但工作最努力，效果最另類，風格最強烈，爭議最明顯的當數劉彥湖。劉彥湖重組改造的資源主要有：古代墓誌篆額及漢印篆，現代圖案構成，域外廣告圖形等等，幾乎所有可以啓發造型的元素對他都有吸引力，而他的長處並非只在於廣收博取，更主要的是組合協調的智慧和想像力，其中明顯的底線把握則是他對篆書基本造型的深刻理解，如果缺失這個環節，便如大廈失去了基石，而在重組改造中，他以幾何形爲結字的主要構成，以飽滿的裝飾爲組織手段，以古樸和機巧的融合爲意境，如此這般，創造了一個與古與今完全無依傍的篆刻形象，當然，這個形象尚未盡善盡美，或者也不必盡善盡美，用時髦的話說，創造的途中風景已經足夠美觀。

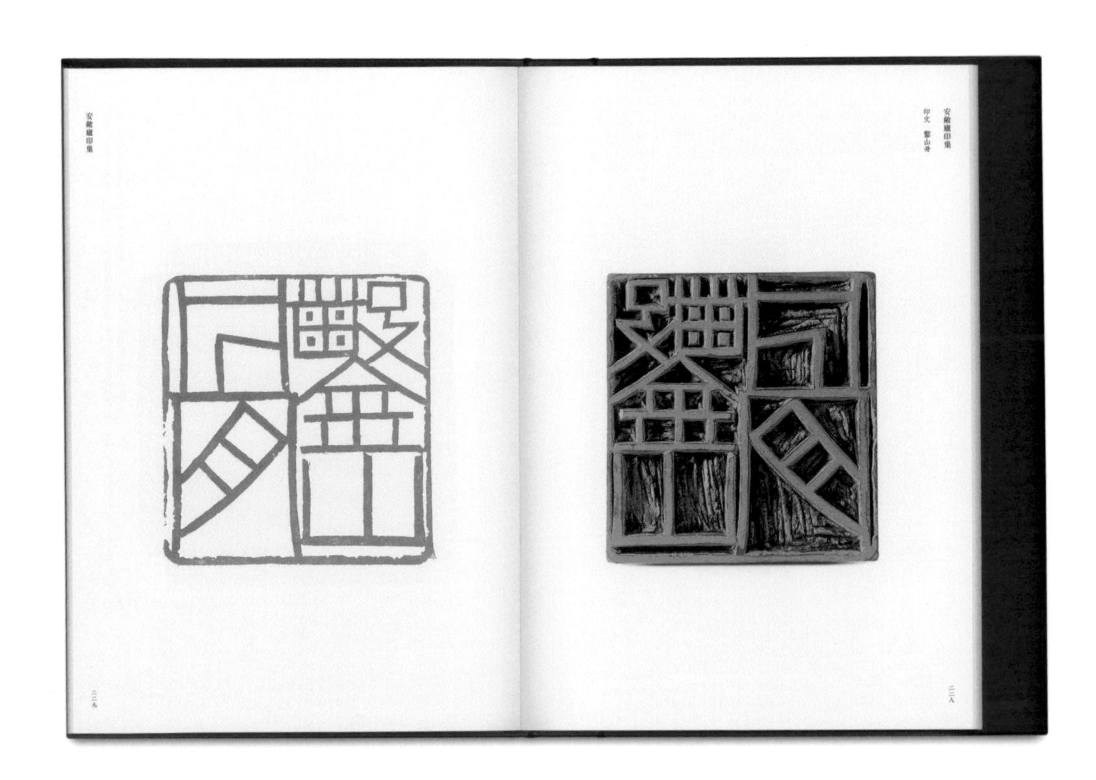

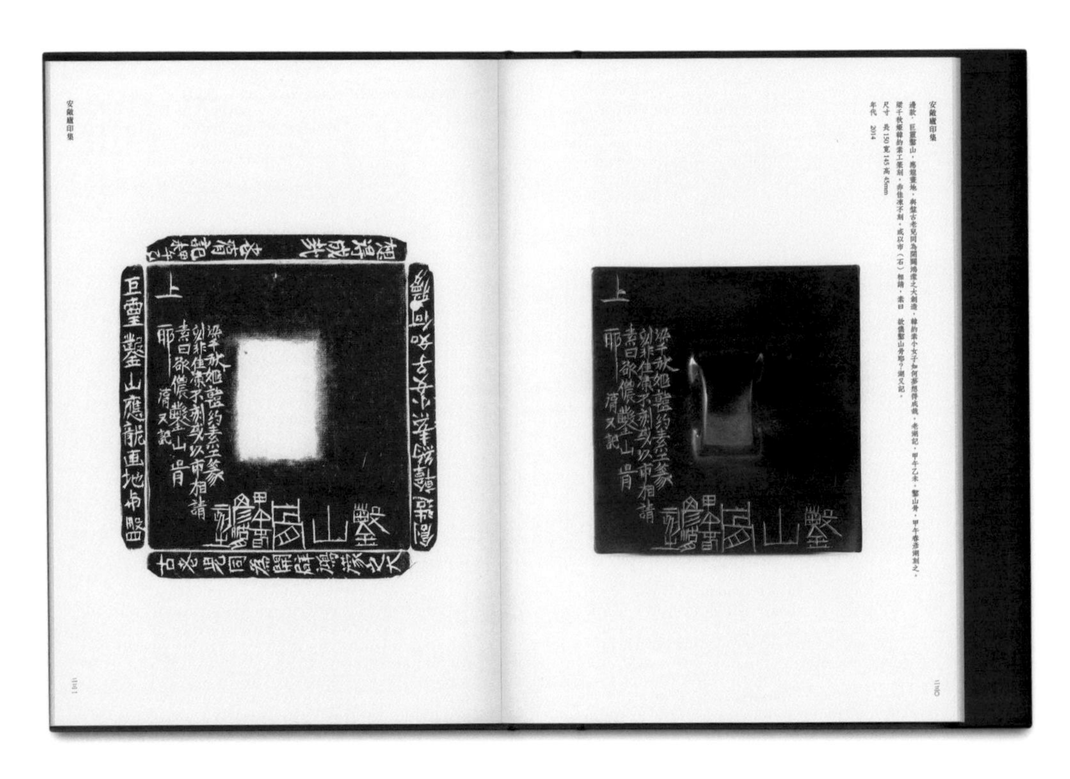

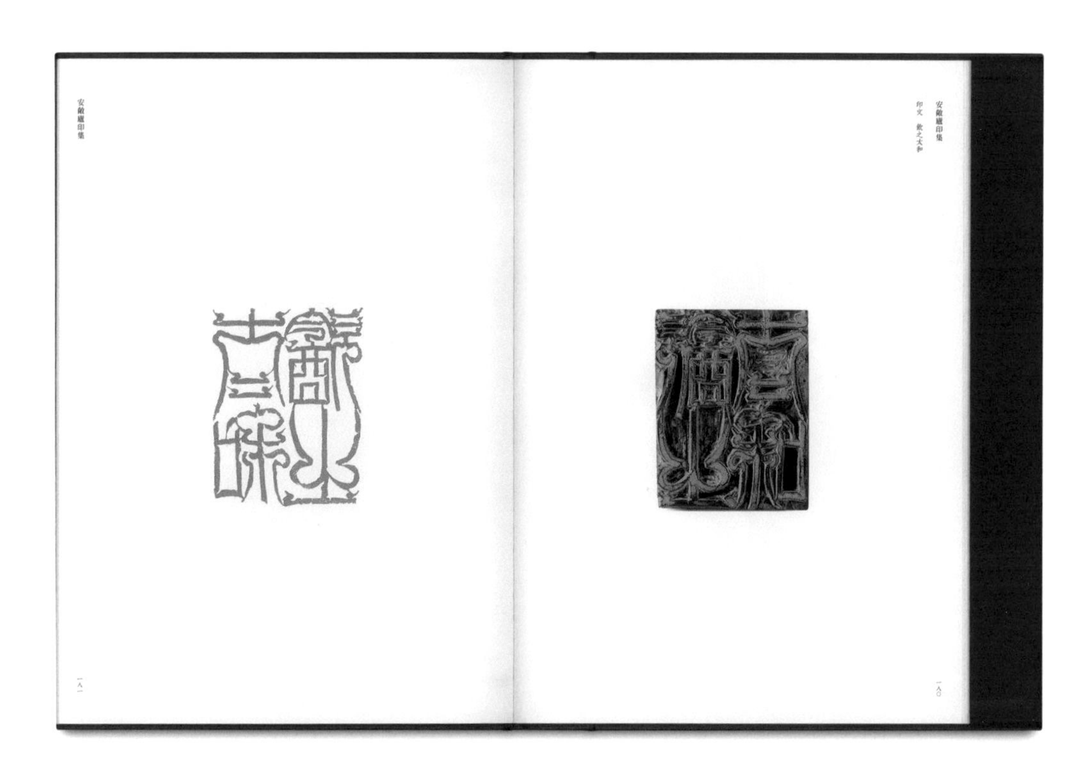

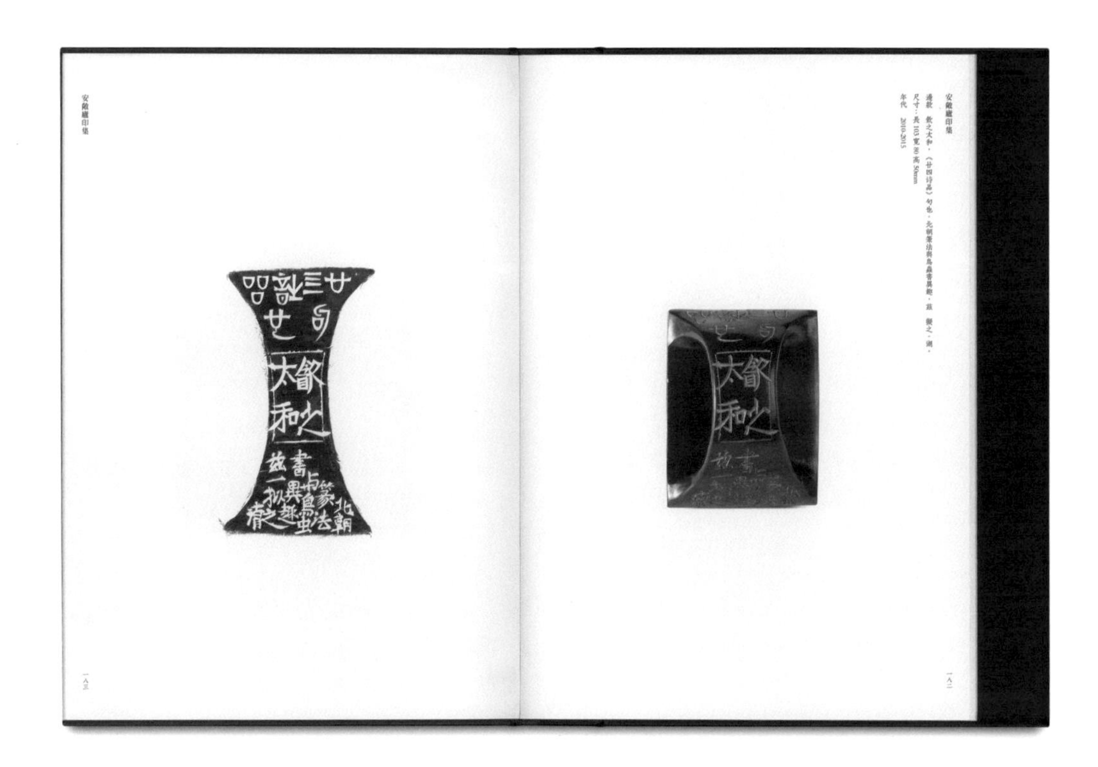

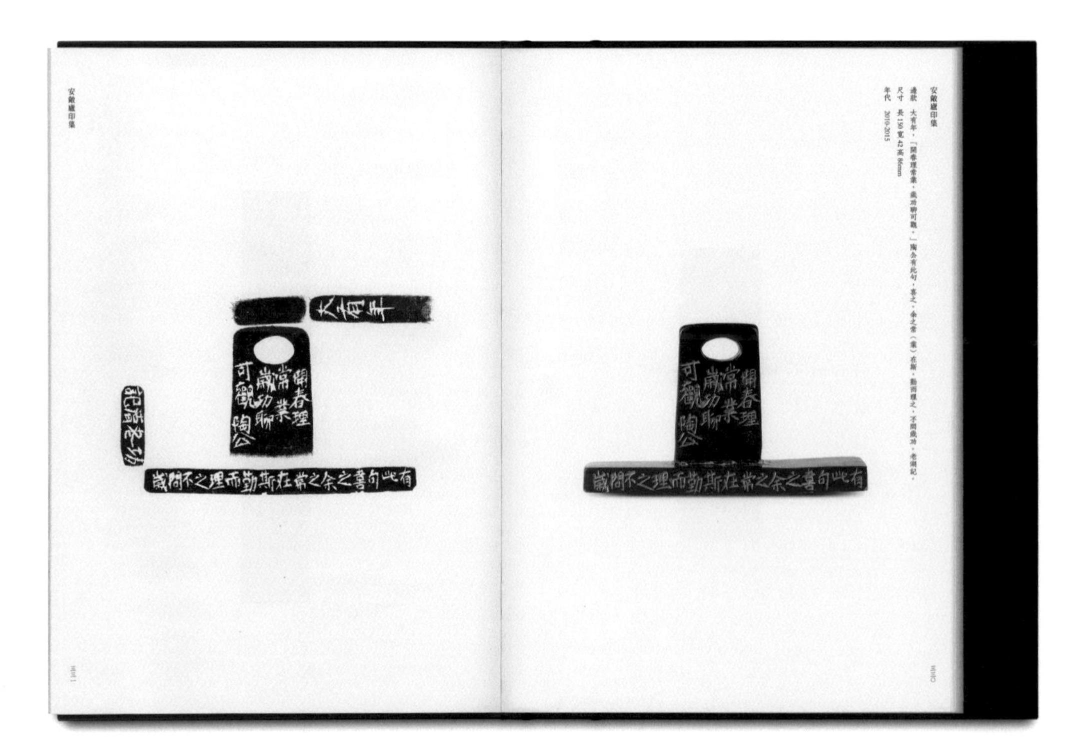

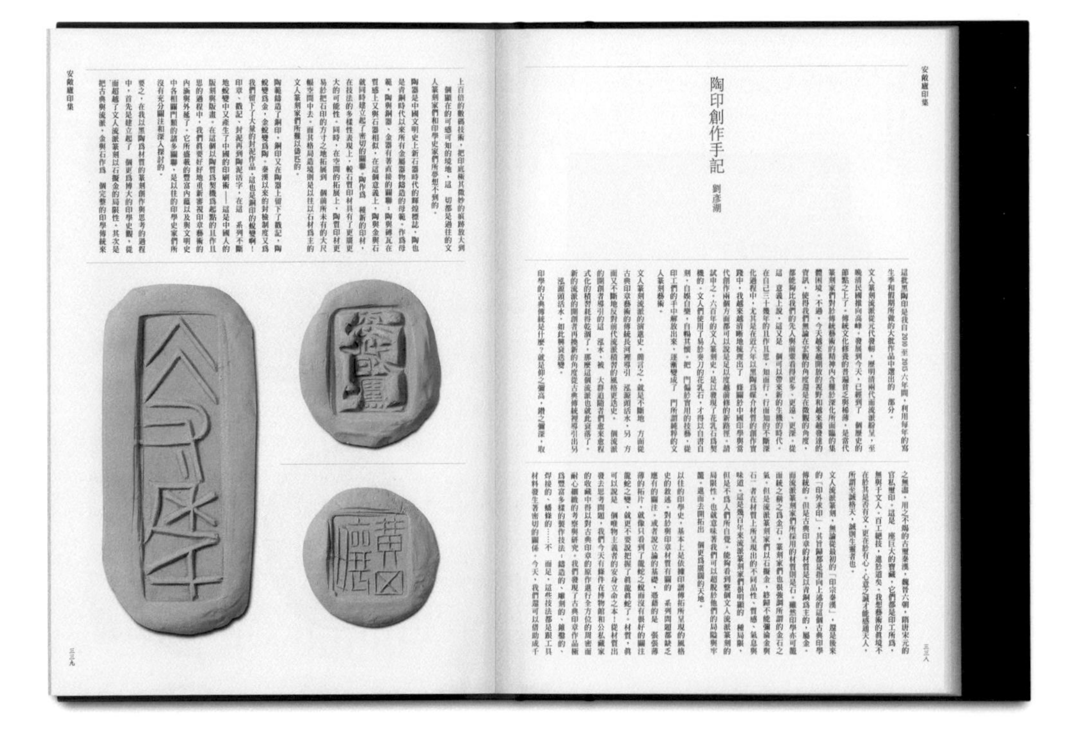
陶印創作手記
劉彥湖

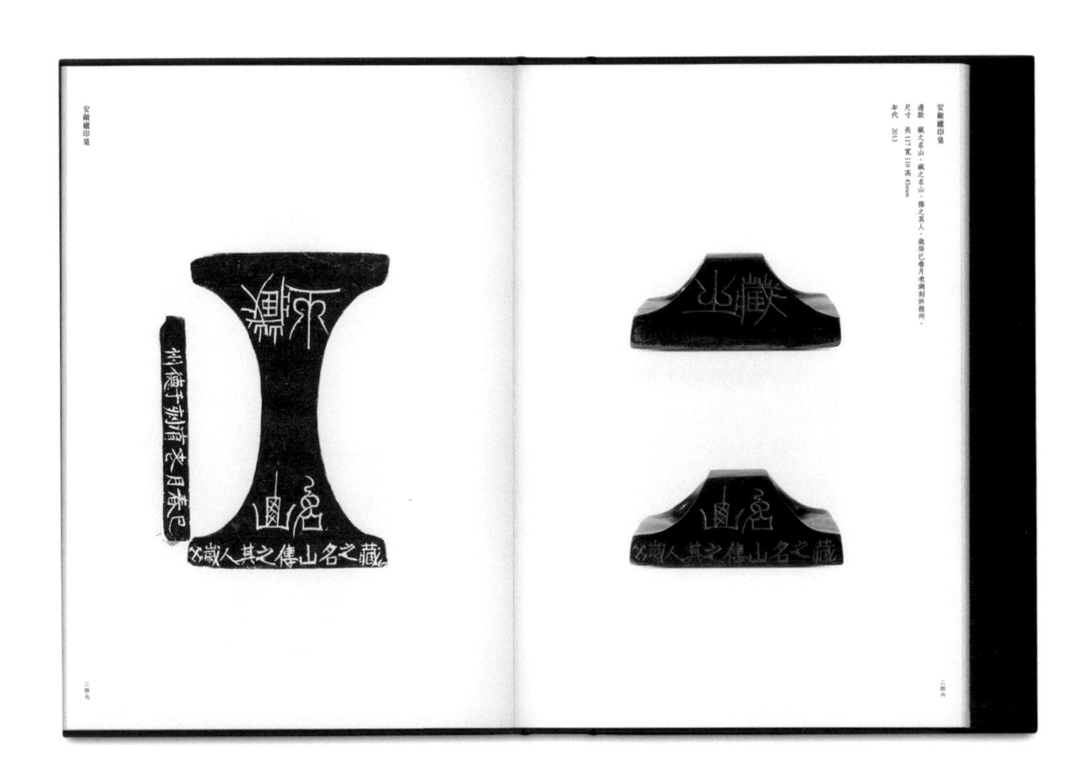

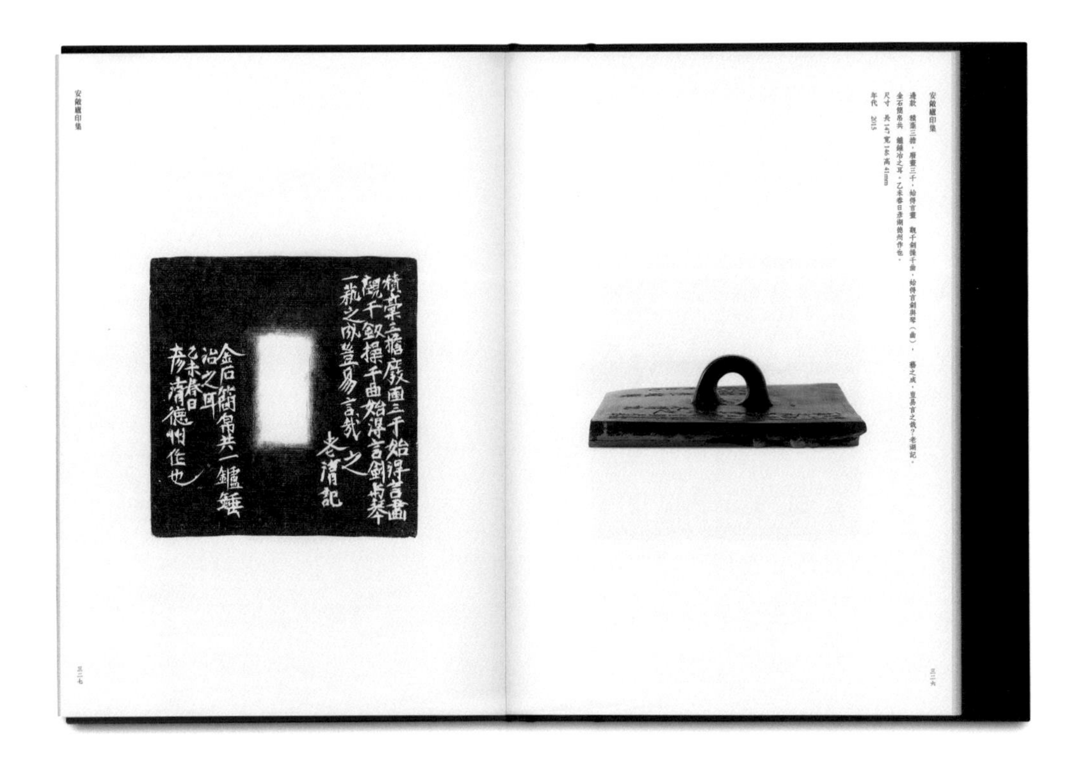

《安敞庐印集》

刘彦湖是我美院的同事，中国画学院书法专业的教授。虽然我们院系相隔，年龄又差着十几岁，但他却是我的一个十分谈得来的师友。我们在创作取向上有许多共通之处，他曾书一对自己集撰的三言联，曰："馆阁体，山林气。"此六字既是他的书法观，亦可视为我的设计观。我常想，要是当今书法篆刻界有谁真的可以做到入古出新的话，那刘彦湖一定是其不二之选。

2010—2015年的六年间，刘彦湖创作了一批尺幅巨大的陶印，《安敞庐印集》中所收录的121方即是其中精选。篆刻不同于书法绘画，尺寸的大幅延展绝非单纯的物理放大，其撼动的乃是文人篆刻的根本，为此刘彦湖给我讲过一则逸事：当年傅抱石、关山月合作为人民大会堂创作国画《江山如此多娇》，由于画幅宽广，本想为之钤盖一枚大印，但多次尝试制印不成，最终只得作罢。对篆刻而言，尺幅有赖于材料，而尺幅与材料的改变又会直接反映到作品的创作方法和形式语言上，有什么样的印底质感，就会有什么样的风格意趣。由此可见，材质与尺幅是认知这组作品的两个重要维度——其所带来的"大刀深刻"对于元明以降石质文人印章而言是颠覆性的，而这一认识则是开启设计的关键。

记得2015年暑假我接到设计委托进入这个项目时，作品的拍摄工作已经开始了一段时间。我看到的样片中，狠鸷深刻的巨型陶印在环境道具和光影、透视的映衬下成了雅集中的"文玩清赏"，看得出摄影师为此花了不少心思，照片本身也很唯美，但很遗憾，这与作品所要传达的气息南辕北辙。我坚决地否决了这个方向并提出了新的拍摄要求——彻底去除一切外部因素，只用平光不偏不倚完全正向地拍摄每方印章的六个面，要清晰还原印面字腔和刻制的刀法，以备我之后设计时从中选用。

我的设计构想直接明了，就是要用书籍使作品获得最大程度的客观呈现，以最真率的姿态和最简素的表情直面读者。拓本与原印需同页等大并置，表里互参，相互印证，而尺幅则是其中的核心要素——每方印章必须不打丝毫折扣地在书中等大还原。为此，我根据印章中最大的一方设定了书籍的开本，情理之中意料之外的是由此生成的书籍形态豁然有别于历史上的任何一册印谱，而完全呈现为一部当代艺术画册的样貌。之后在每页的排版中，我比对印花（压盖在纸上的朱色印文）和边款拓片，在电脑上精细测量规制了每一帧实物图片的尺寸，使之与原件大小得以完全吻合。

整本书的内文为筒子页形制，借其轻薄纤韧反衬巨印之如铸如凿、深心大力；书的外部则是黑色皮面硬装，挺括方直而无一丝“古籍”之气。其实所谓“馆阁体，山林气”，根本要解决的还是“体”与“气”的关系，书法、篆刻如是，设计亦然。

彭斯　**太素 | 流咏**

潄园，北京，2015

太素：215mm×285mm×9mm，88 页

流咏：215mm×285mm×9.5mm，96 页

平装

1000 套

编辑：彭斯、何浩

文章：彭锋、四月（顾春芳）、夏可君

Mountains-and-Waters and Landscapes: Thoughts on Peng Si's New Works

Peng Feng

Last winter, I visited Peng Si's studio. Although I had heard about his "change", I was totally impressed by the new look of his works in front of me. Looking at some "old" works on the wall again, you can not tell all these works are created by one same person. It occurred to me that two years ago Peng Si was tangled and he dealt with the entanglement on his own. Last April, his solo exhibition titled Harmonia, which was held in Beijing Fine Art Academy was the declaration of the end of the tangle. The work titled "Scenery by the Cao Zhi's Grave,"which is printed on the poster, depicts a warm and peaceful scene ,which is just to the theme of the exhibition. While right behind the peace, it is a tangle for three years. If this piece of work is examined with x-ray ,revision and starting over again and again will be seen.However,behind the calmness, it is a quandary for three years, which is not a stretch to say it is heart-stirring.I can not help but think of the ancient Greek sculpture of Winckelmann: "the general characteristics of Greek art is a kind of noble simplicity and a quiet grandeur, both in attitude and in the expression. like the deep sea stays in silence forever, no matter how rough the surface is.In the Greek art ,expressions display a great and quiet soul,which is in all the passions,though." Peng Si's works are as the rough sea with calm surface but turbulent inner side.

However, the reconciliation does not stop the pace of Peng Si's exploration. In fact, Peng Si is not the kind of artist who is to pursue different arts. The reason he is tangled and changes is that he aims to find a kind of art which is true to his heart. It is the self- driving, which is true to himself that motivates Peng Si to go on the road of exploring and self-denial.

As an artist born in 80's, Peng Si received a good education at Central Academy of Fine Arts .However, undeniably, our cultural traditions in today's college education have been impacted hideously, even beyond recognition and gone due to the influence of Western modernity. Peng Si, like most of his peers, learned to use "foreign language" in painting in school. But, unlike the senior artists, the growth of 80's artists meets the rapid growth of Chinese experience,hence they accordingly harvest the cultural self-confidence. Although Peng Si uses "foreign language" to paint, he loves Chinese traditional culture. In his spare time of painting, he plays the guqin, reads poems and creates calligraphy, which makes him look like a total traditional literati. After a careful study of the language and material of oil painting, Peng Si decides to return to his "mother tongue" Returning to the "mother tongue" means to swim back to the vast sea of Chinese culture. With obsession to the guqin, poetry and ancient calligraphy, Peng Si has his confidence not to be drown by the vast sea of Chinese culture. Even if submerged in the vast of sea of culture, he also can gain a carefree comfort.

The language corresponds with the world. From the "foreign language" point of view, the nature is landscape while it is mountain-and-water from the "mother tongue" point of view.For mountains-and-waters and landscapes, there are different painting techniques, different painting objects,and more importantly, there are two different ways of of living and life, two different world outlooks and values. Some sinologists who have in-depth study of Chinese and Western culture, recently begin to study the differences between Chinese and Western arts from the view of world outlooks. The research results by American Martin Powers and French Francois Jullien are now in fermentation within the scope of the world. Peng touches the same issue with his art creation, which forms the echo of theory and practice.

Today, people are already aware of the painting mode established in Renaissance, which is usually called the western painting mode and is actually based on a viewing mode. If this viewing mode can not be accepted, then the corresponding painting mode can not be accepted. In other words, the Western painting, which looks very natural to us, is actually the outcome that we accept Western viewing painting mode, which is a product of our acquisition and we are not born to view the world in that way. In other words, it is the

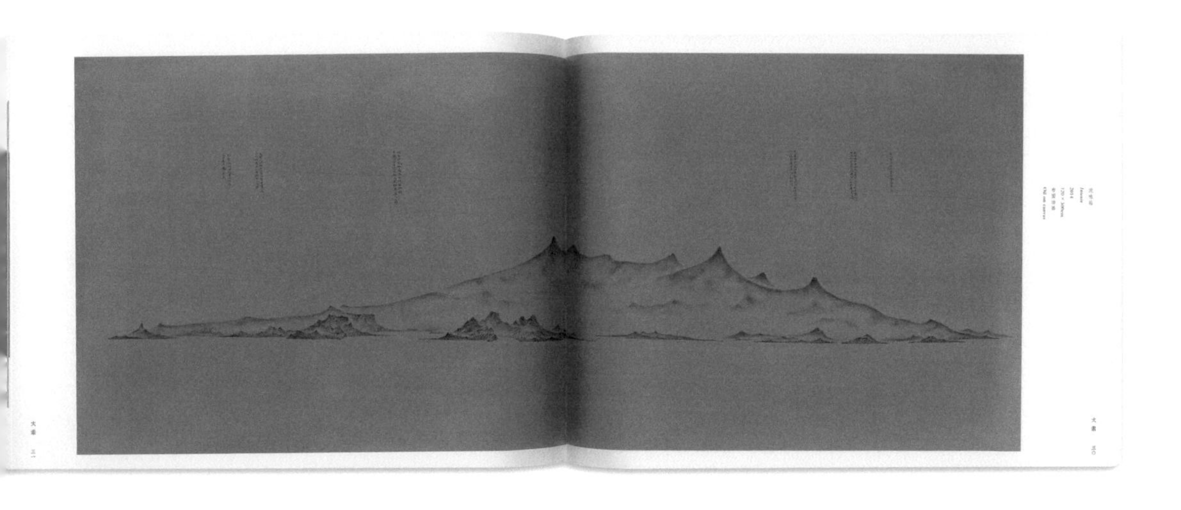

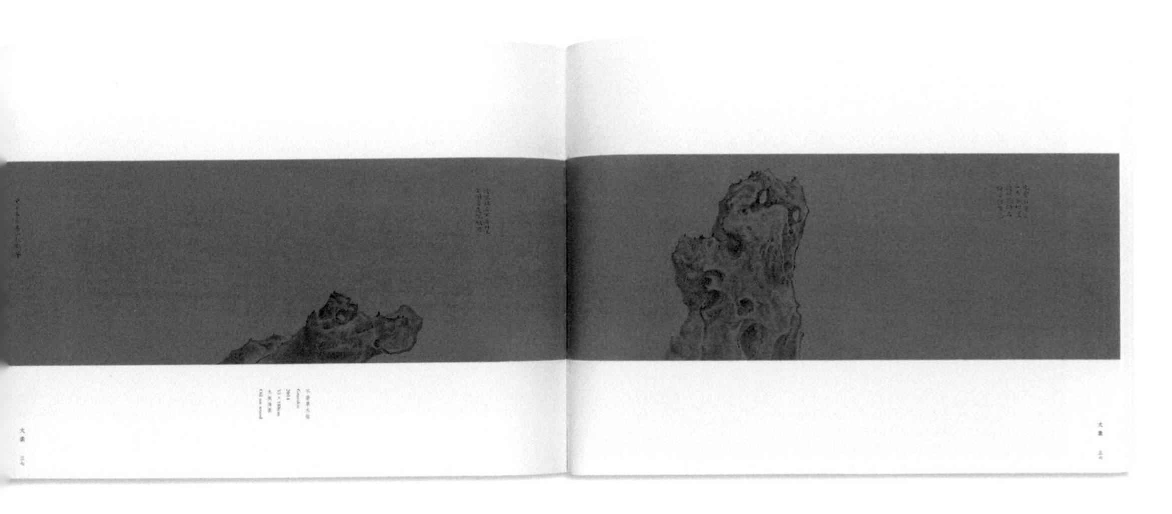

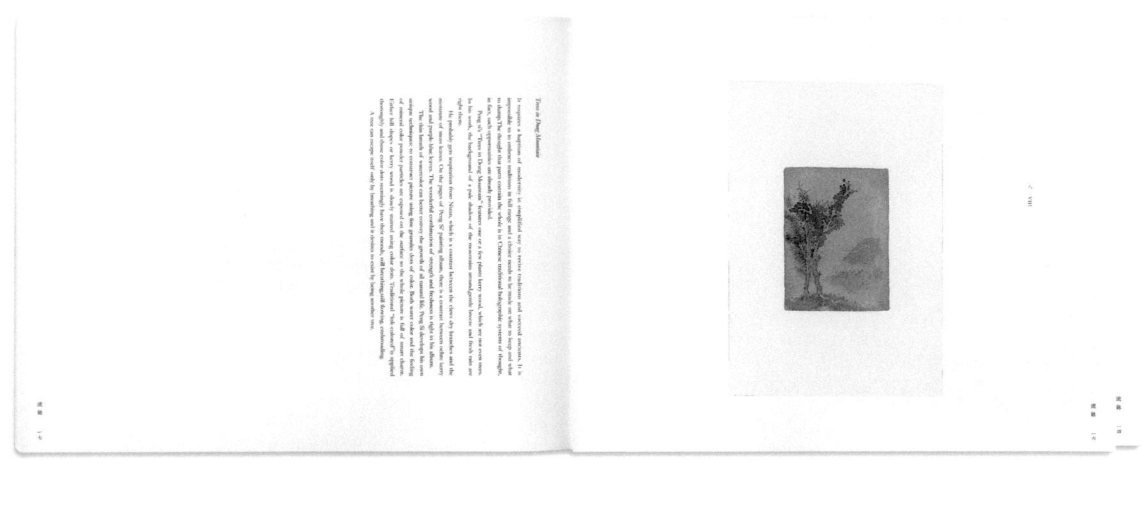

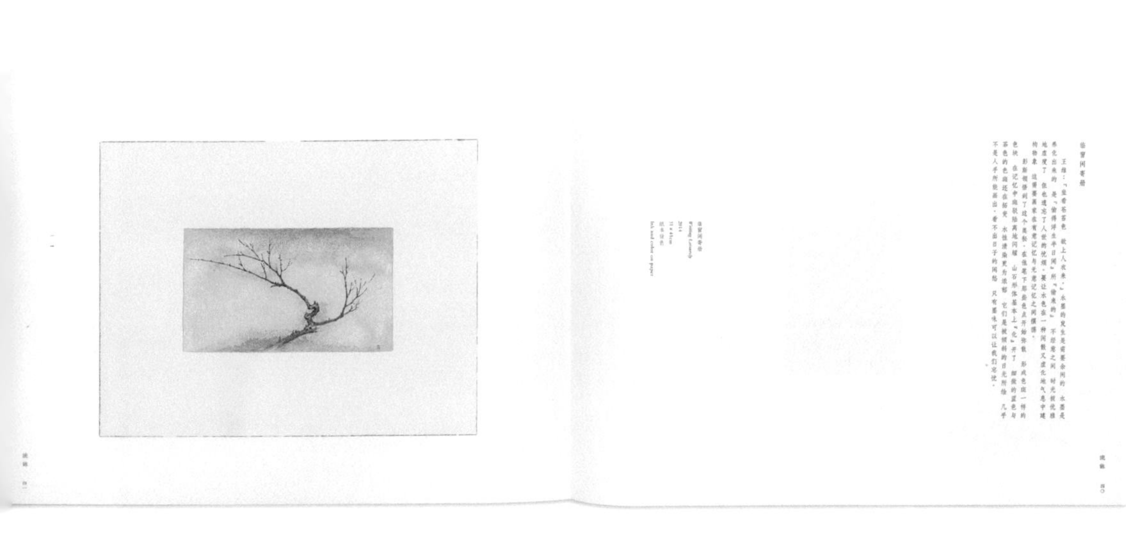

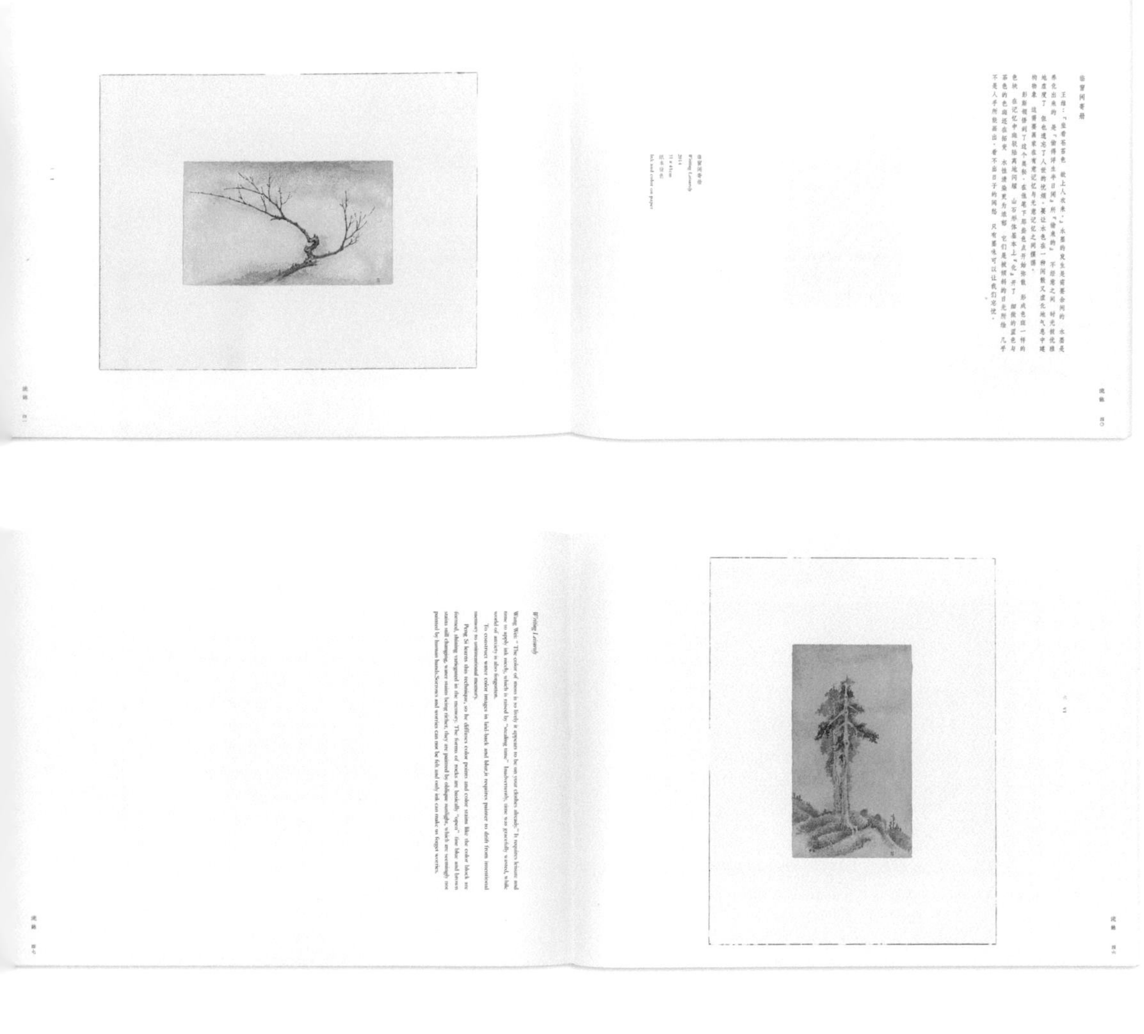

刘晓辉 **西西弗斯之谜**

天线空间，上海，2015

302mm×239mm×19mm，120 页

布面精装，外裹护封

1000 册

文章：贺婧

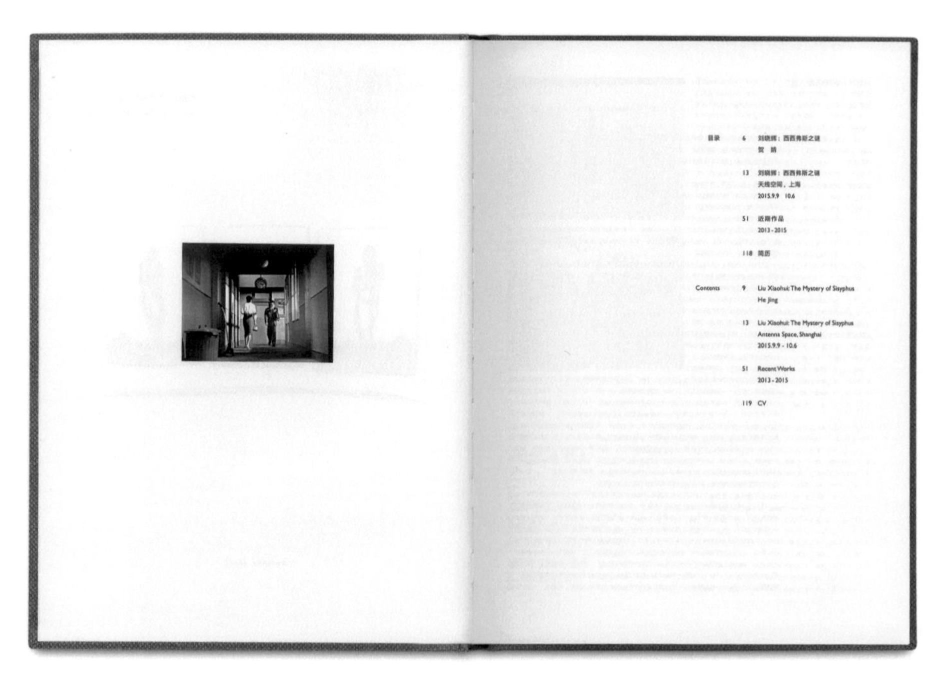

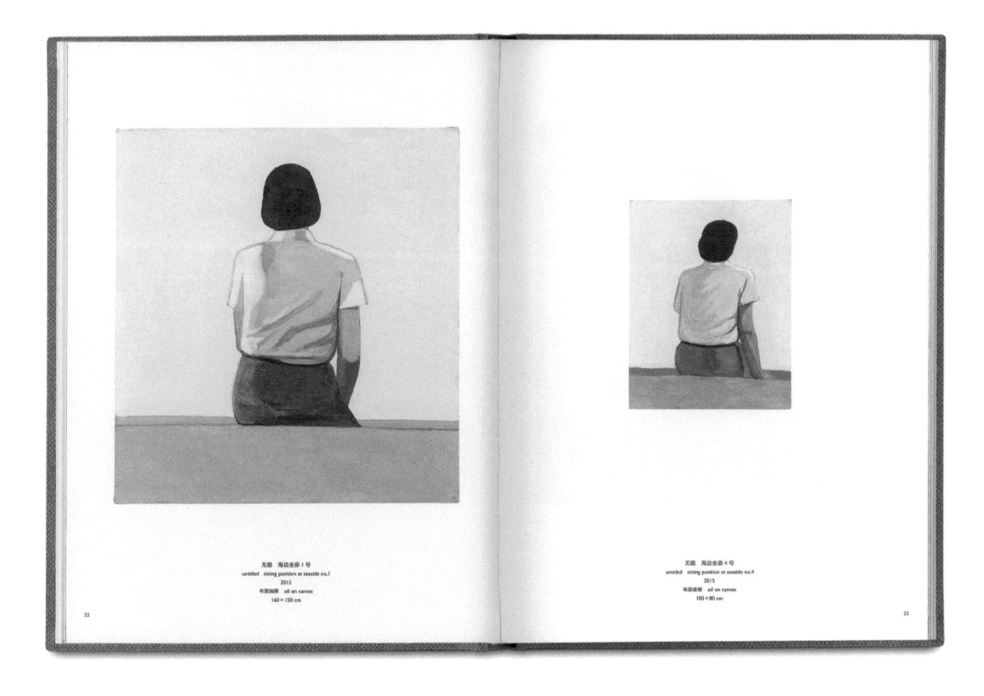
无题 海边坐姿 1 号
untitled sitting position at seaside no.1
2015
布面油画 oil on canvas
160 × 150 cm
22
无题 海边坐姿 4 号
untitled sitting position at seaside no.4
2015
布面油画 oil on canvas
100 × 80 cm
23

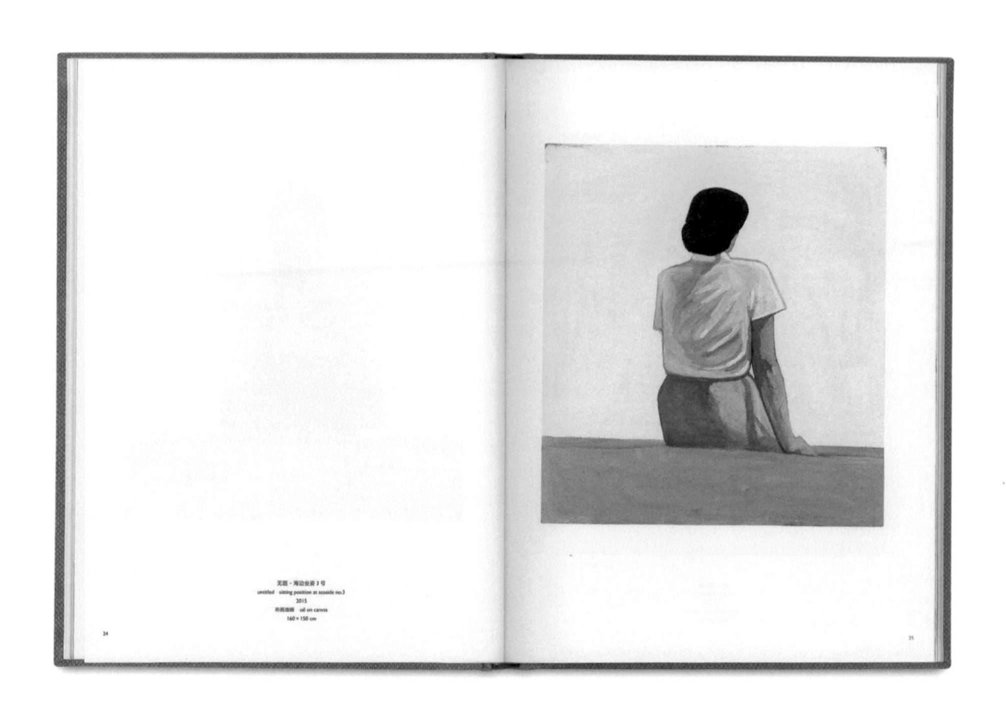
无题 - 海边坐姿 3 号
untitled sitting position at seaside no.3
2015
布面油画 oil on canvas
160 × 150 cm
24
25

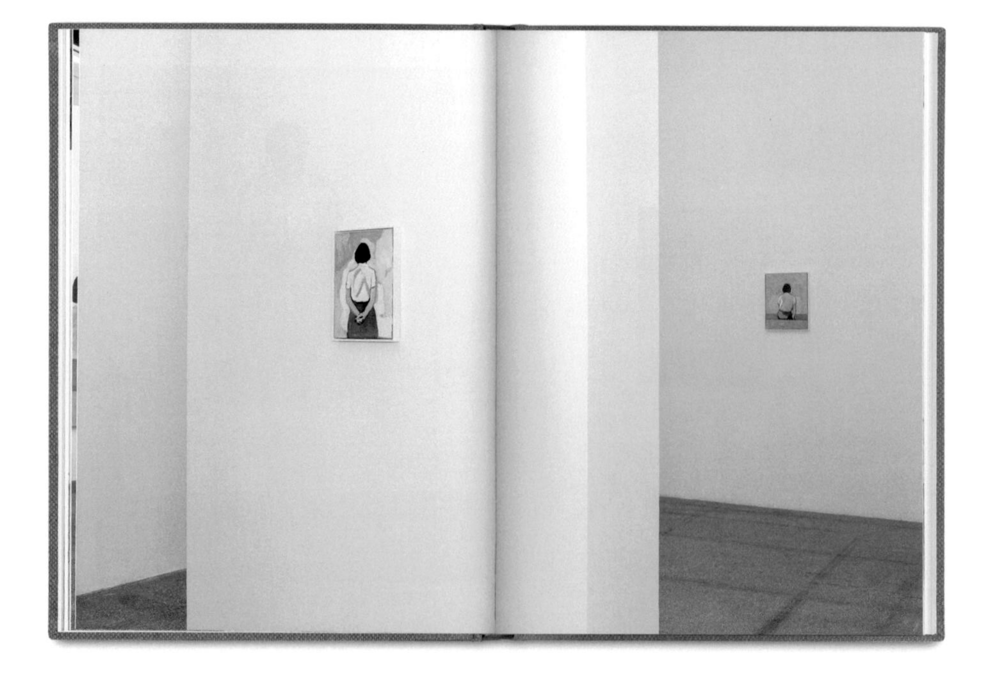

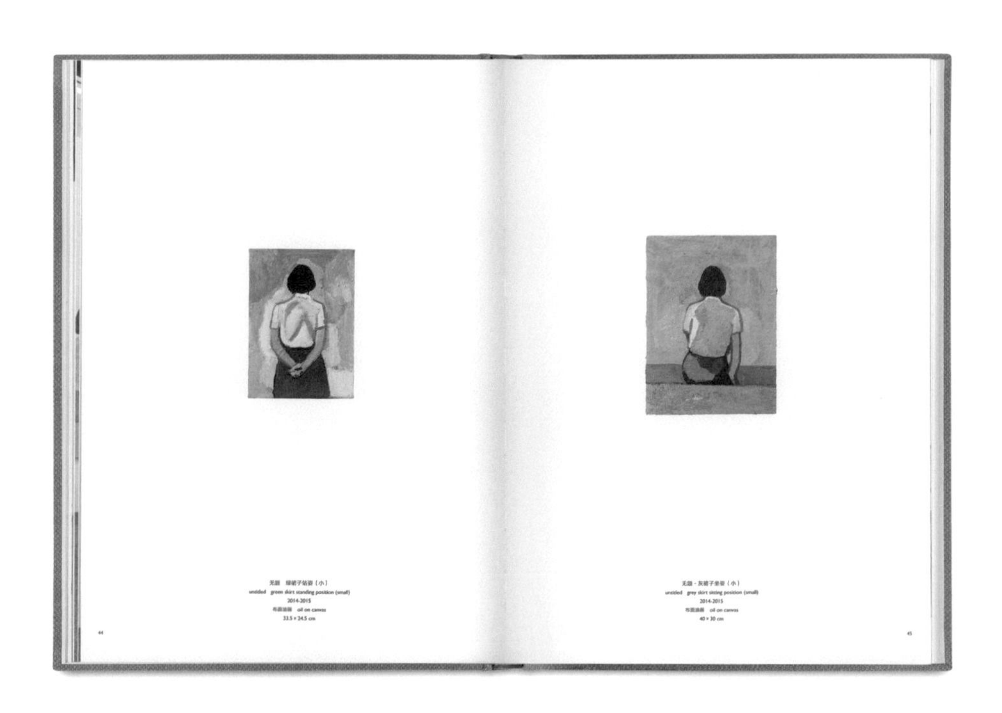
无题 绿裙子站姿（小）
untitled green skirt standing position (small)
2014-2015
布面油画 oil on canvas
33.5 × 24.5 cm
无题・灰裙子坐姿（小）
untitled grey skirt sitting position (small)
2014-2015
布面油画 oil on canvas
40 × 30 cm

姚璐 **遮蔽与重构——姚璐新山水**

中国民族摄影艺术出版社，
北京，2015
306mm×248mm×26mm，248 页
丝面精装，外裹腰封；筒子页装订
800 册
文章：顾铮、姚璐
对话：冯博一 / 姚璐

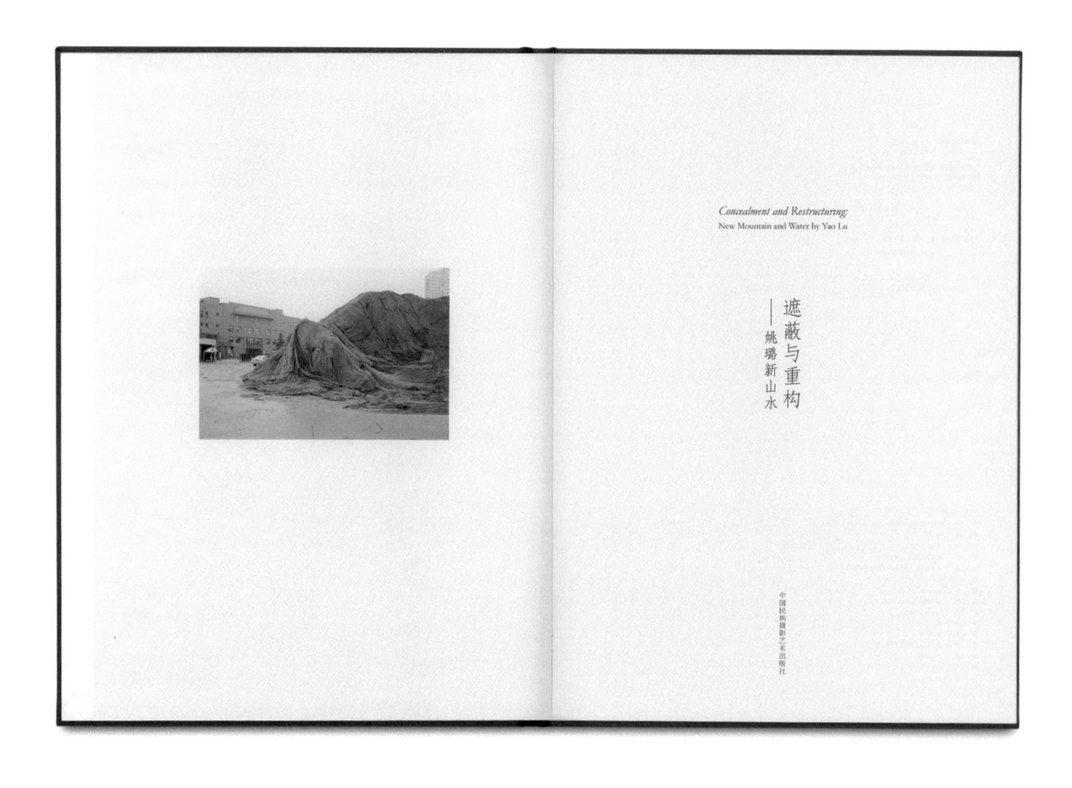

王强 **王强**

空白空间，北京，2015

304mm×248mm×19mm，130 页

布面精装，外裹护封

1000 册

文章：尤洋

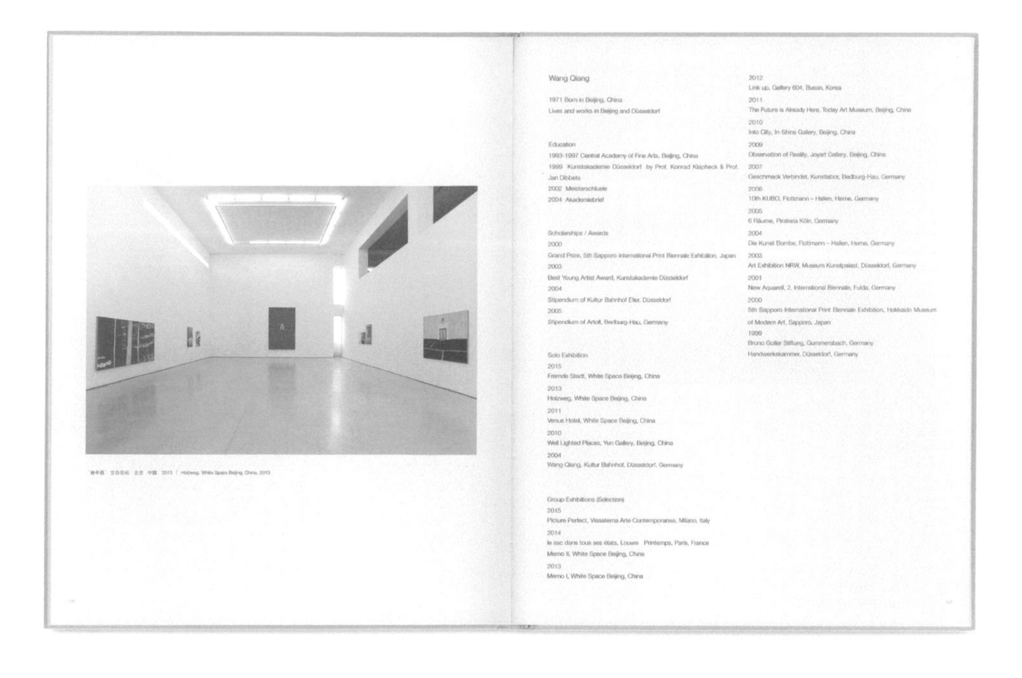

Wang Qiang

1971 Born in Beijing, China
Lives and works in Beijing and Düsseldorf

Education
1993-1997 Central Academy of Fine Arts, Beijing, China
1999 Kunstakademie Düsseldorf by Prof. Konrad Klapheck & Prof. Jan Dibbets
2002 Meisterschluele
2004 Akademiebrief

Scholarships / Awards
2000
Grand Prize, 5th Sapporo International Print Biennale Exhibition, Japan
2003
Best Young Artist Award, Kunstakademie Düsseldorf
2004
Stipendium of Kultur Bahnhof Eller, Düsseldorf
2005
Stipendium of Artoll, Bedburg-Hau, Germany

Solo Exhibition
2015
Fremde Stadt, White Space Beijing, China
2013
Holzweg, White Space Beijing, China
2011
Venus Hotel, White Space Beijing, China
2010
Well Lighted Places, Yun Gallery, Beijing, China
2004
Wang Qiang, Kultur Bahnhof, Düsseldorf, Germany

Group Exhibitions (Selection)
2015
Picture Perfect, Viasaterna Arte Contemporanea, Milano, Italy
2014
le sac dans tous ses états, Louvre Printemps, Paris, France
Memo II, White Space Beijing, China
2013
Memo I, White Space Beijing, China
2012
Link up, Gallery 604, Busan, Korea
2011
The Future is Already Here, Today Art Museum, Beijing, China
2010
Into City, In-Shine Gallery, Beijing, China
2009
Observation of Reality, Joyart Gallery, Beijing, China
2007
Geschmack Verbindet, Kunstlabor, Bedburg-Hau, Germany
2006
10th KUBO, Flottmann – Hallen, Herne, Germany
2005
6 Räume, Pirateria Köln, Germany
2004
Die Kunst Bombe, Flottmann – Hallen, Herne, Germany
2003
Art Exhibition NRW, Museum Kunstpalast, Düsseldorf, Germany
2001
New Aquarell, 2. International Biennale, Fulda, Germany
2000
5th Sapporo International Print Biennale Exhibition, Hokkaido Museum of Modern Art, Sapporo, Japan
1999
Bruno Goller Stiftung, Gummersbach, Germany
Handwerkskammer, Düsseldorf, Germany

王音 **礼物**

新星出版社，北京，2016

306mm×237mm×21mm，140 页

布面精装

2000 册

前言：田霏宇

文章：秦思源

对话：田霏宇 / 王音

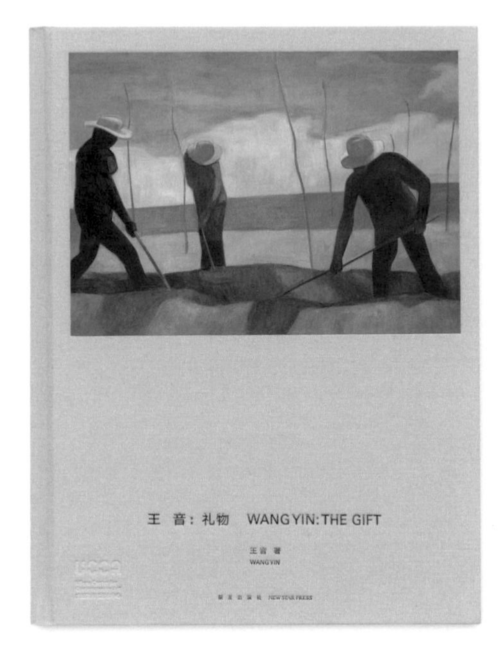

Preface

Towards a Chinese Oil Painting and Back Again

Philip Tinari

"Oil painting." Despite its European provenance, it is a designation and a category that exists in a specific relation to China, where it sits among a set of categories created in the fervor and furor of the early twentieth century. In China, Oil painting is the natural-born sibling of "National Painting," an attempt to tweak an august ink painting tradition for a belated modernity. As a concept, it stands indexically for the import of a whole range of ideas and practices, foremost among them "Science" and "Democracy," which were at the core of the May Fourth Movement's program of reform. And for Wang Yin, the stuff of oil painting—its materiality, its implicit tonality, its basic reliance on certain perspectival and compositional devices—is itself a document of China's emergent modernity.

In his last two decades of output, Wang Yin has engaged with styles that have slipped in and out of vogue throughout the century during which the medium has been widely practiced by Chinese artists. Like many painters of his generation, the first part of his career was spent acquiring, and then refuting, the realistic techniques of the Soviet canon as distilled from Italian and French precedents and later transmitted to the Chinese acolytes who formed the generation of his teachers. Later, after the epochal transitions of 1989, he became interested in what came before this Soviet orthodoxy, digging into the imagistic worlds of painters like Yan Wenliang (1893-1988), Ni Yide (1901-1970), Wu Dayu (1903-1988), Pan Yuliang (1895-1977), and others of the "1920s Generation" who trained abroad and subsequently brought the gospel of modern painting back to the upheavals of the motherland. He was interested not in the sophistication of these mostly well-heeled Republican-era figures who represent some of the most cosmopolitan presences in modern Chinese art history, but for what appeared to him, seven decades later, as the naiveté of their early experiments with an unfamiliar medium. A few years later, he began collaborating with untrained peasants in a performative attempt to realize the ideals of "art for the people" as postulated by Mao Zedong at Yan'an. In recent years, he has come to fuse these different registers into a single stylistic program, where the jouissance of the brushstroke and a distinctively muted palette are used to render scenes from everyday urban life, often carrying some dimension of metaphorical import, in a seemingly timeless, placeless register. Certain subjects—trains, minority nationalities, mangoes, solitary figures, suburban landscapes, allusions to earlier moments in modern Chinese (art) history—recur, but never in programmatic service to a particular serial or narrative agenda. Rather, they constellate, each one a document of and testament to his painterly vision.

Three things are important to remember about Wang Yin. First, he grew up in an "oil painting household," headed by a father who was locally famous for his skill and generosity in continuing the twentieth century project of transmitting painting technique to China. The idea of a Chinese oil painting, or for that matter the more basic question of the importance and urgency of art, was therefore not one that he arrived at as a soul-searching adolescent, but one that formed the backdrop of his childhood daily life. Thus implanted, these questions are then both transcended and internalized, becoming less subject matter than baseline intellectual premises that have grounded his subsequent investigations. That this household was physically situated in Qingdao—a city shaped by a colonial encounter with Germany that was as brief (1897-1914) as it was thorough, dotting the city with European architecture and infrastructure and creating a hybrid culture—goes even further toward explaining Wang Yin's heartfelt confrontation with his chosen medium.

前言

归程：朝向中国油画

田霏宇

"油画"，虽然起源于欧洲，却也是与中国存有特殊关系的一个名称和艺术种类。在中国20世纪早期的热潮与激变中，它与其他种种新事物一同被创造出来。在中国，油画是受国家美术系统承认的一种艺术种类，是将悠久权威的水墨画传统转折到迟来现代性的一种尝试。油画作为概念，索引式地证实了整体范围内的思想和实践的引进：其中最为重要的，是处于"五四运动"革新精神的核心——"科学"和"民主"。对于王音而言，油画的材料——它的物质性，它固有的音调，它对特定的透视方法和创作样式的依照遵循——本身就记录了中国涌现的现代性。

在二十年的创作生涯中，王音将一个世纪（在这一百年，油画这一媒介广泛地被中国艺术家所实践）中曾经风靡一时而后又默默无闻的风格在这种媒介中糅洽到一处。就像很多他这一代的画家，在最初的职业生涯中，王音先学习了苏联式现实主义绘画技巧——这些技巧萃取于意大利和法国先驱的作品，之后传播到中国追随者的手中，即王音老师那一辈人——之后又对此进行反叛。随后，在作为新纪元转捩点的1989年，王音开始对在苏联"正统"之前发生的事产生了兴趣，并挖掘曾于国外受过绘画训练、将现代绘画的"福音"带回激变中的祖国的艺术家的意象世界，比如颜文樑（1893-1988）、倪贻德（1901-1970）、吴大羽（1903-1988）、潘玉良（1895-1977）等"二十年代"艺术家。这些大多家境优越的民国时代人物（这些人物反映了在中国现代艺术历史中最为"世界主义"的呈现）的老到之处并非王音的兴趣所在，他所感兴趣的是，他们使用这种生疏的媒介早期经验的朴实，在七十余年后的自己面前呈现出的感受。几年之后，他开始与未曾受过美术训练的农民合作创作，以一种具有表演性质的尝试践行"为人民服务的艺术"——这个口号是毛泽东在延安时期提出的——的理想。近年来，他开始将不同的语调融合为单一风格化的实践，在此类创作中，"原乐"式的笔触和具有辨识度的柔和调色被用于描绘城市日常生活，常常以一种似乎丧失时空感的视觉语言，带入些隐喻的维度。被描绘的对象——火车、少数民族人物、芒果、孤独的身影、城郊风景，都暗指了在现代中国（艺术）早期历史中闪现的瞬间——王音将它们重现，但绝不让它们服务于特定的系列和叙事的目的。这些瞬间更像是被汇集到一处，分别记录了他画家式的视角，并对这种视角进行测试。

关于王音，有三件事情需要注意。首先，他成长于"油画世家"，他的父亲凭自身的绘画技巧，以及为了延续传入中国绘画技术这一二十世纪事业所尽的心力，在当地获得了名望。因此，中国油画的概念，或者就此而言更为基础的问题——艺术的重要性和紧急性，不是王音在寻找自我的青少年时期才面对的问题，而是形成了他童年的日常背景。自幼年耳濡目染，这些问题具有超越性和内在性的特征，成为了基本的知识前提，为王音之后的研究奠定了基础，而非仅仅是一些题材。这种自然的传递实际上发生于青岛——一个曾被德国殖民者短暂（1897-1914）塑形的城市，殖民者给城市遗留下了少量的建筑和基础设施，并创造了一种混合的文化——这也解释了王音面对他所选择媒介时的真诚。

第二件事情是他没有在中央美术学院的"专业"限制中完成学业，而是就读于学科界限更为模糊的、本质上属于跨学科的中央戏剧学院。在此，他依旧接受了同样的（苏联式的）技术训练，却吸纳了更为广泛的人文学科架构，总能着眼于如何习惯性地呼应上下文，并设置画布以外的情景。他写过一篇关于耶日·格洛托夫斯基（1933-1999）的论文，这位导演于

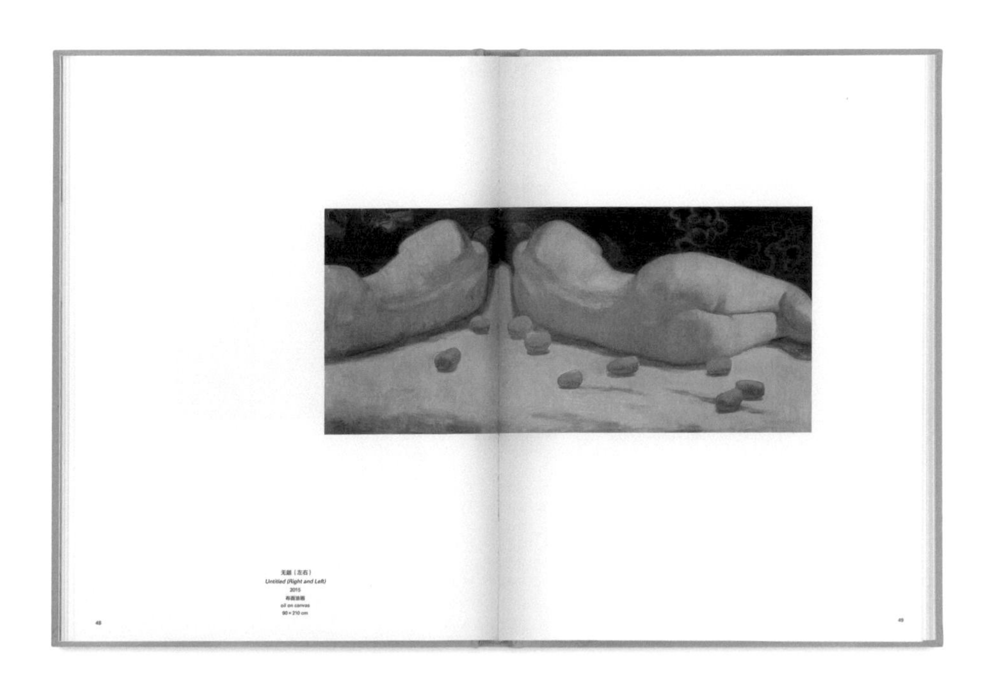

罗杰·拜伦 **荒诞剧场**

浙江摄影出版社，杭州，2016

291mm×237mm×22mm，130 页

布面精装，外裹护封

2000 册

序言：王璜生

文章：蔡萌、迪迪·博齐尼

后记：罗杰·拜伦

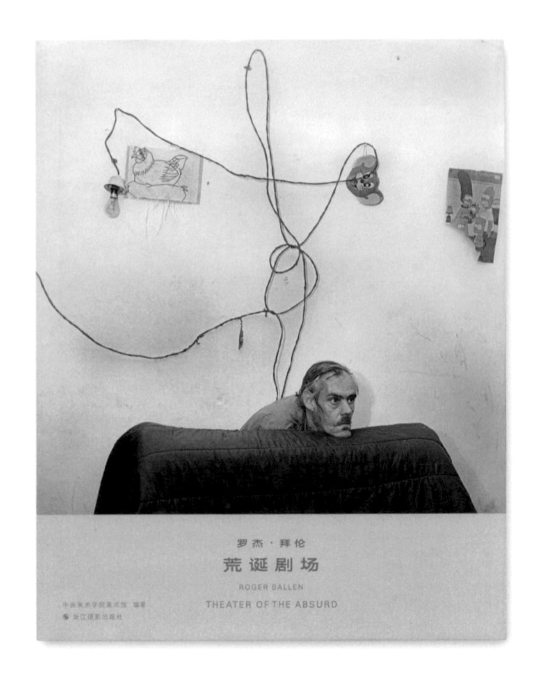

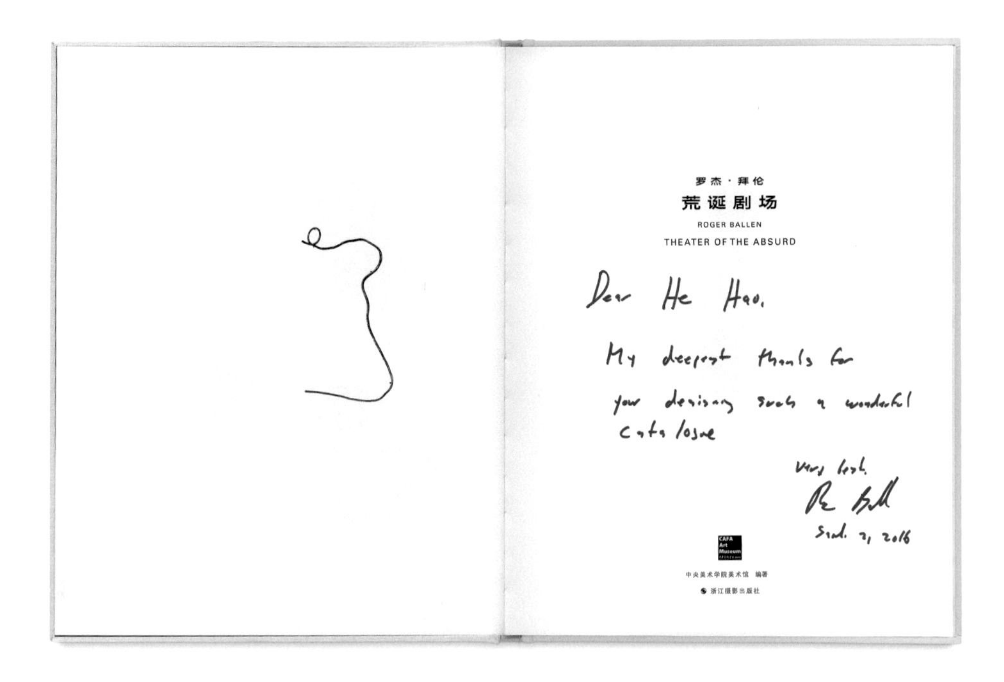

罗杰·拜伦

荒诞剧场

ROGER BALLEN

THEATER OF THE ABSURD

Dear He Hao,

My deepest thanks for your designing such a wonderful catalogue

Very best,

中央美术学院美术馆 编著

浙江摄影出版社

“罗杰·拜伦：荒诞剧场”序

王璜生

中央美术学院美术馆馆长

对于一位具有重大国际影响的艺术家/摄影家的认识和研究，是需要一个层层展开和深入的过程的。当我第一次从画册上看到罗杰 拜伦的作品时，就被一种奇特、荒诞、戏剧性的场面所深深震惊，这种震惊感不仅仅来自他作品中怪诞的人物形象和奇异的场景，更来自那种黑白质感而又说不清意义的影调，那画面结构和视觉关系的独特性，那琢磨不透的一种情绪和精神的奇幻感；而我就会进一步地不断追问，究竟是什么样的动机、生命经验、思想和现实，使罗杰 拜伦产生这样的表达，并以这样的表达方式一直持续而深入地展开？一位南非的艺术家，以一个“地质勘查员”的工作身份，在一个特定的历史年代，走遍了南非的几乎每一个村落。他就像一位“地质”工作者一样，对这片土地的地层做了一层层的挖掘、勘查，以他敏锐而奇特的镜头，将这一特殊时期南非底层民众的精神状态、生活情态和生存物态表现出来，深深抓住他们充满焦虑、迷茫、无奈、荒诞的心理及行为方式，结构出一个个罗杰 拜伦式的摄影画面。与其说是画面，不如说是“剧场”，一个巨大的上演着何止是南非，更指向于人类精神与行为、现实与情状的荒诞不经的“剧场”。于是，罗杰 拜伦又进一步将一组组静态而平面的镜头画面，结构成一场场能够让人们走进其中并不自觉地扮演“罗杰 拜伦式”人物角色的空间装置“剧场”。在这样的剧场中，我们也许不仅仅是参观者，不仅仅是演员，更可能是经历一次独特的精神体验和生命反思的艺术旅程。

感谢罗杰 拜伦为我们带来这样的一场视觉思考和精神体验的艺术盛宴！中央美术学院美术馆也有幸收藏了罗杰 拜伦一批珍贵的摄影作品，深深感谢拜伦！

感谢为本次展览提供赞助支持 出版、文献合作、空间与平面设计、展览组织协调、展示实施、公教推广宣传等工作的机构和个人！感谢本次展览策展人蔡萌！是大家共同而“认真”的努力，才得以将罗杰 拜伦的“荒诞剧场”搬到中国，在中央美术学院美术馆开演。谢谢大家！

2016年8月3日于北京中央美术学院

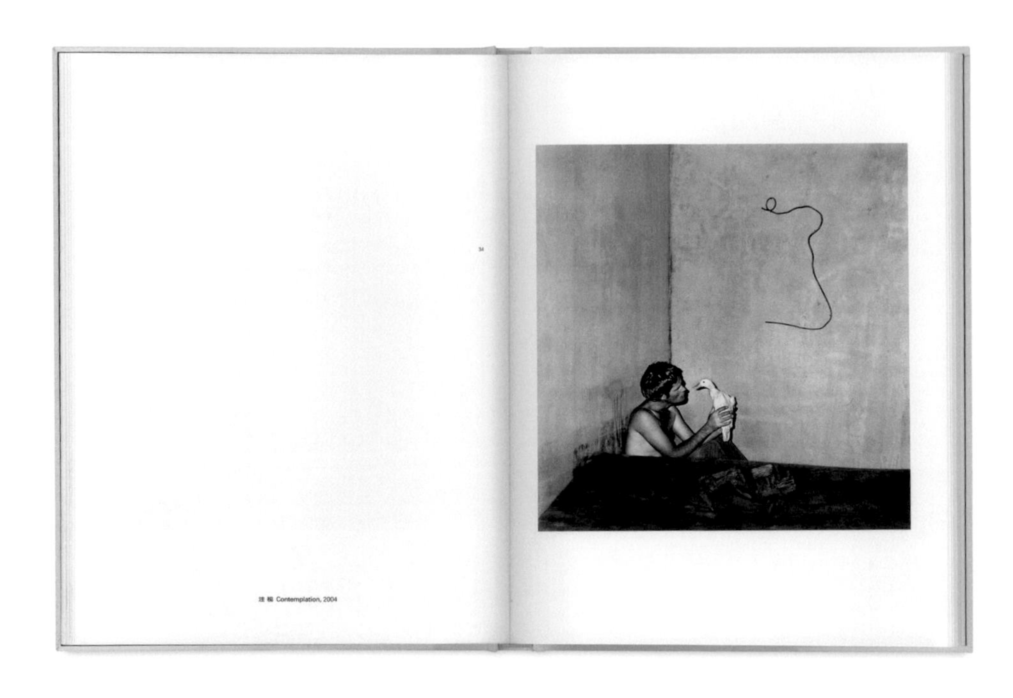

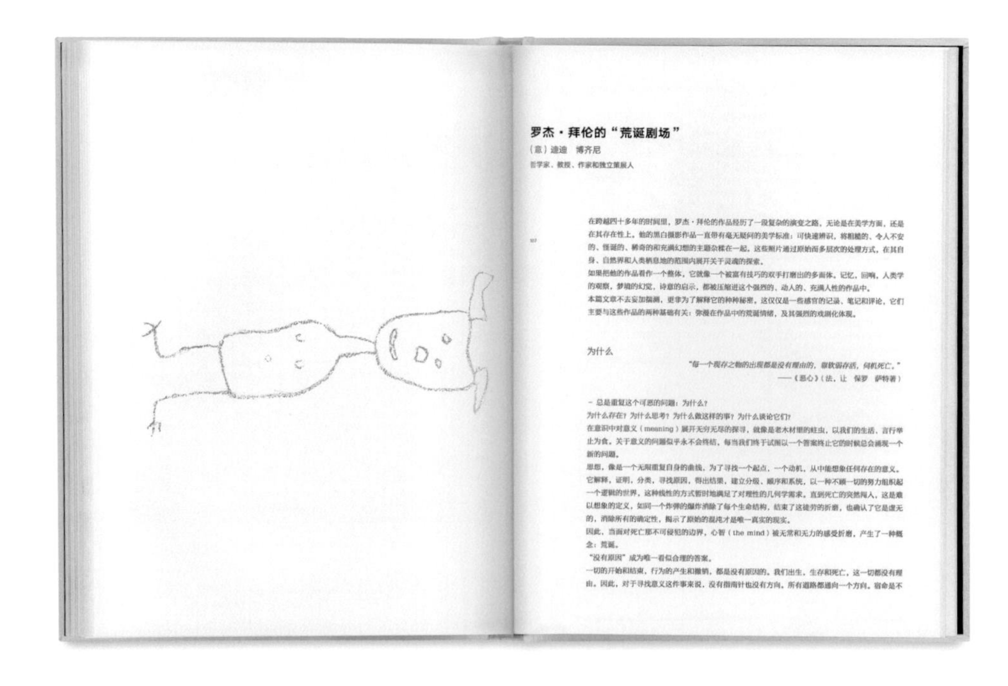

罗杰·拜伦的“荒诞剧场”

（意）迪迪　博齐尼

哲学家、教授、作家和独立策展人

107

在跨越四十多年的时间里，罗杰·拜伦的作品经历了一段复杂的演变之路，无论是在美学方面，还是在其存在性上。他的黑白摄影作品一直带有毫无疑问的美学标准：可快速辨识，将粗糙的、令人不安的、怪诞的、稀奇的和充满幻想的主题杂糅在一起。这些照片通过原始而多层次的处理方式，在其自身、自然界和人类栖息地的范围内展开关于灵魂的探索。

如果把他的作品看作一个整体，它就像一个被富有技巧的双手打磨出的多面体。记忆，回响，人类学的观察，梦境的幻觉，诗意的启示，都被压缩进这个强烈的、动人的、充满人性的作品中。

本篇文章不去妄加揣测，更非为了解释它的种种秘密。这仅仅是一些感官的记录、笔记和评论，它们主要与这些作品的两种基础有关：弥漫在作品中的荒诞情绪，及其强烈的戏剧化体现。

为什么

“每一个现存之物的出现都是没有理由的，靠软弱存活，伺机死亡。”

——《恶心》（法，让　保罗　萨特著）

- 总是重复这个可恶的问题：为什么？

为什么存在？为什么思考？为什么做这样的事？为什么谈论它们？

在意识中对意义（meaning）展开无穷无尽的探寻，就像是老木材里的蛀虫，以我们的生活、言行举止为食。关于意义的问题似乎永不会终结，每当我们终于试图以一个答案终止它的时候总会涌现一个新的问题。

思想，像是一个无限重复自身的曲线，为了寻找一个起点，一个动机，从中能想象任何存在的意义。它解释，证明，分类，寻找原因，得出结果，建立分级、顺序和系统，以一种不顾一切的努力组织起一个逻辑的世界，这种线性的方式暂时地满足了对理性的几何学需求，直到死亡的突然闯入，这是难以想象的定义，如同一个炸弹的爆炸消除了每个生命结构，结束了这徒劳的折磨，也确认了它是虚无的，消除所有的确定性，揭示了原始的混沌才是唯一真实的现实。

因此，当面对死亡那不可侵犯的边界，心智（the mind）被无常和无力的感受折磨，产生了一种概念：荒诞。

“没有原因”成为唯一看似合理的答案。

一切的开始和结束，行为的产生和撤销，都是没有原因的。我们出生，生存和死亡，这一切都没有理由。因此，对于寻找意义这件事来说，没有指南针也没有方向，所有道路都通向一个方向。宿命是不

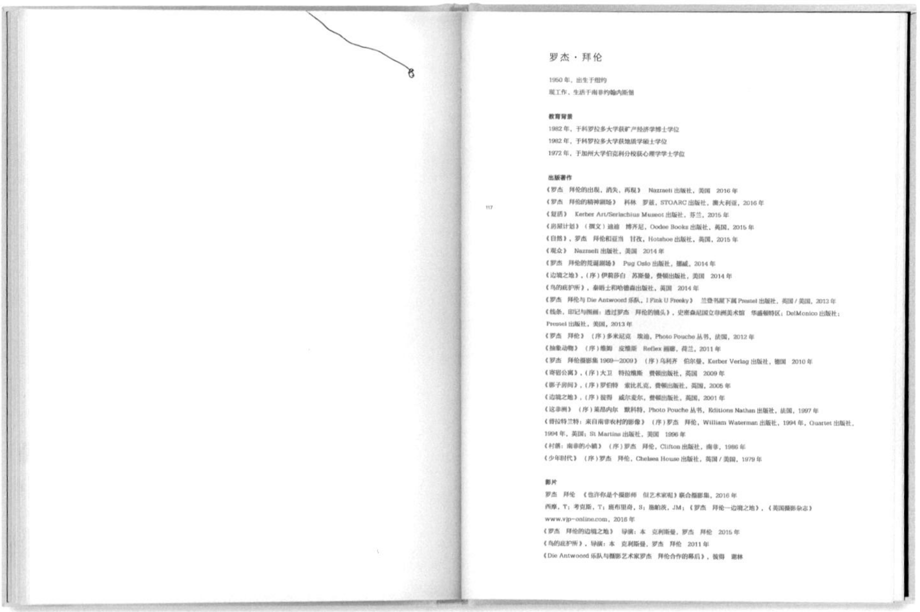

117

罗杰·拜伦

1950年，出生于纽约
现工作、生活于南非约翰内斯堡

教育背景

1982年，于科罗拉多大学获矿产经济学博士学位
1982年，于科罗拉多大学获地质学硕士学位
1972年，于加州大学伯克利分校获心理学学士学位

出版著作

《罗杰　拜伦的出现，消失，再现》　Nazraeli出版社，美国　2016年
《罗杰　拜伦的精神剧场》　科林　罗兹，STOARC出版社，澳大利亚，2016年
《复活》　Kerber Art/Serlachius Museot出版社，芬兰，2015年
《房屋计划》（撰文）迪迪　博齐尼，Oodee Books出版社，英国，2015年
《自然》，罗杰　拜伦和亚当　甘孜，Hotshoe出版社，英国，2015年
《观众》　Nazraeli出版社，美国　2014年
《罗杰　拜伦的荒诞剧场》　Pug Oslo出版社，挪威，2014年
《边境之地》，（序）伊莉莎白　苏斯曼，费顿出版社，美国　2014年
《鸟的庇护所》，泰晤士和哈德森出版社，英国　2014年
《罗杰　拜伦与Die Antwoord乐队，I Fink U Freeky》　兰登书屋下属Prestel出版社，英国/美国，2013年
《线条、印记与图画：透过罗杰　拜伦的镜头》，史密森尼国立非洲美术馆　华盛顿特区；DelMonico出版社；Prestel出版社，美国，2013年
《罗杰　拜伦》　（序）多米尼克　埃迪，Photo Pouche丛书，法国，2012年
《抽象动物》　（序）维姆　皮维斯　Reflex画廊，荷兰，2011年
《罗杰　拜伦摄影集1969—2009》　（序）乌利齐　伯尔曼，Kerber Verlag出版社，德国　2010年
《寄宿公寓》，（序）大卫　特拉维斯　费顿出版社，英国　2009年
《影子房间》，（序）罗伯特　索比扎克，费顿出版社，英国，2005年
《边境之地》，（序）彼得　威尔麦尔，费顿出版社，英国，2001年
《这非洲》　（序）莱昂内尔　默科特，Photo Pouche丛书，Editions Nathan出版社，法国，1997年
《普拉特兰特：来自南非农村的影像》　（序）罗杰　拜伦，William Waterman出版社，1994年，Quartet出版社，1994年，英国；St Martins出版社，美国　1996年
《村镇：南非的小镇》　（序）罗杰　拜伦，Clifton出版社，南非，1986年
《少年时代》　（序）罗杰　拜伦，Chelsea House出版社，英国/美国，1979年

影片

罗杰　拜伦　《也许你是个摄影师　但艺术家呢》联合摄影集，2016年
西摩，T；考克斯，T；班布里奇，S；施帕茨，JM；《罗杰　拜伦—边境之地》，《英国摄影杂志》www.vjp-online.com，2016年
《罗杰　拜伦的边境之地》　导演：本　克利斯曼，罗杰　拜伦　2015年
《鸟的庇护所》，导演：本　克利斯曼，罗杰　拜伦　2011年
《Die Antwoord乐队与摄影艺术家罗杰　拜伦合作的幕后》，彼得　谢林

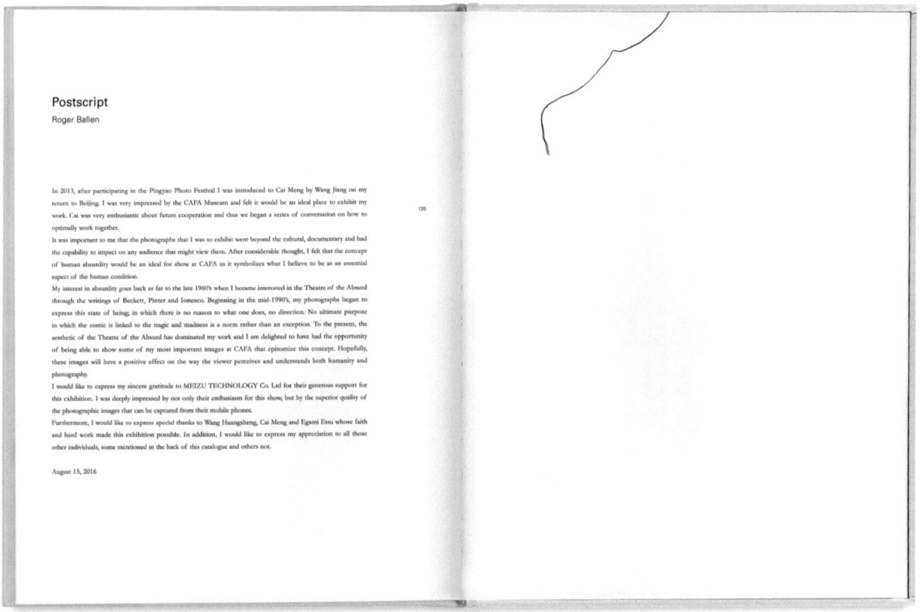

126

Postscript

Roger Ballen

In 2013, after participating in the Pingyao Photo Festival I was introduced to Cai Meng by Wang Jiang on my return to Beijing. I was very impressed by the CAFA Museum and felt it would be an ideal place to exhibit my work. Cai was very enthusiastic about future cooperation and thus we began a series of conversation on how to optimally work together.

It was important to me that the photographs that I was to exhibit went beyond the cultural, documentary and had the capability to impact on any audience that might view them. After considerable thought, I felt that the concept of human absurdity would be an ideal for show at CAFA as it symbolizes what I believe to be as an essential aspect of the human condition.

My interest in absurdity goes back as far to the late 1960's when I became interested in the Theatre of the Absurd through the writings of Beckett, Pinter and Ionesco. Beginning in the mid-1990's, my photographs began to express this state of being; in which there is no reason to what one does, no direction. No ultimate purpose in which the comic is linked to the tragic and madness is a norm rather than an exception. To the present, the aesthetic of the Theatre of the Absurd has dominated my work and I am delighted to have had the opportunity of being able to show some of my most important images at CAFA that epitomize this concept. Hopefully, these images will have a positive effect on the way the viewer perceives and understands both humanity and photography.

I would like to express my sincere gratitude to MEIZU TECHNOLOGY Co. Ltd for their generous support for this exhibition. I was deeply impressed by not only their enthusiasm for this show, but by the superior quality of the photographic images that can be captured from their mobile phones.

Furthermore, I would like to express special thanks to Wang Huangsheng, Cai Meng and Egami Etsu whose faith and hard work made this exhibition possible. In addition, I would like to express my appreciation to all those other individuals, some mentioned in the back of this catalogue and others not.

August 15, 2016

故事绘——中国美术馆藏连环画原作精品展

中国美术馆，北京，2016

334mm×264mm×28mm，250 页

布面精装，外裹腰封

1000 册

主编：吴为山

前言：吴为山

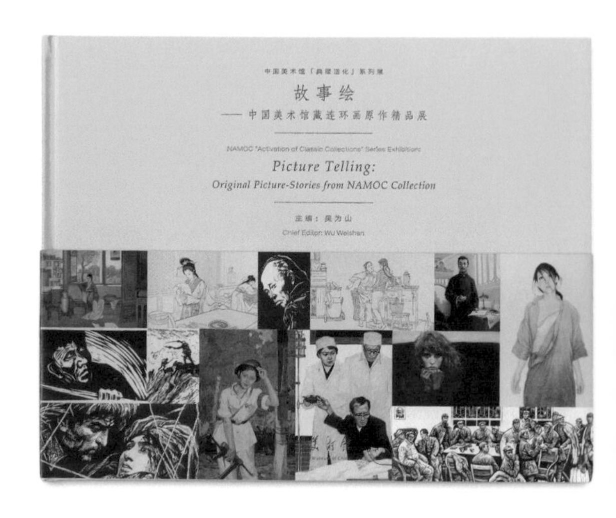

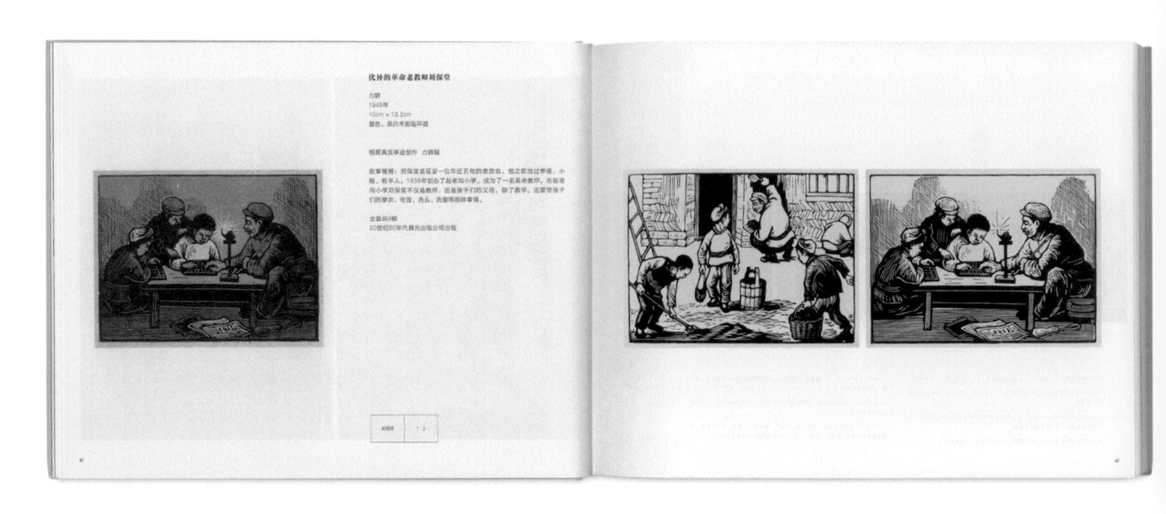

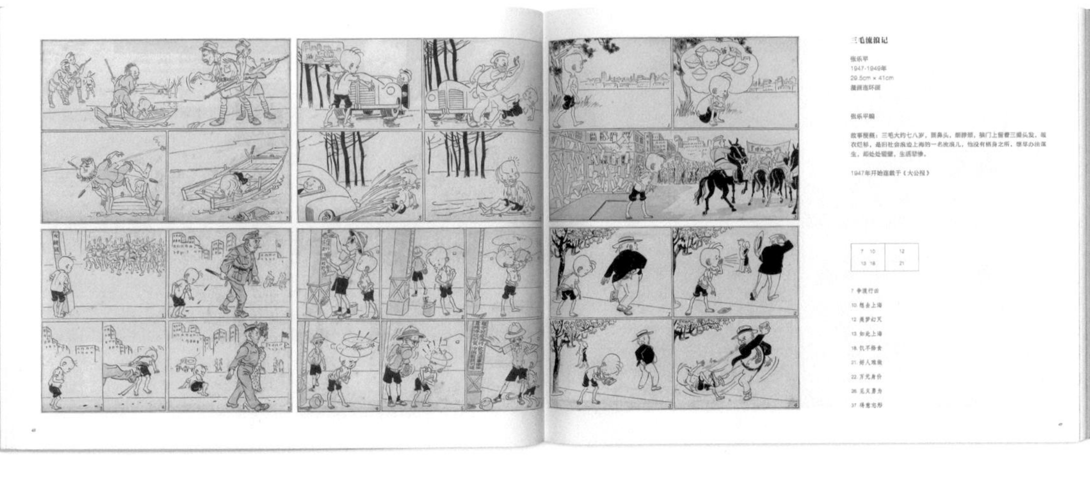
三毛流浪记
张乐平
1947-1949年
29.5cm × 41cm
漫画连环画
张乐平编
1947年开始连载于《大公报》

西厢记
王叔晖
1956年
22.3cm × 28cm
纸本白描连环画
根据[元]王实甫同名杂剧 洪曾玲改编
全套共128幅
1957年人民美术出版社出版

黄继光
董辰生
1969年
尺寸不一
水彩连环画
根据抗美援朝真实故事 李学鳌配诗
全套共21幅
1972年人民美术出版社出版

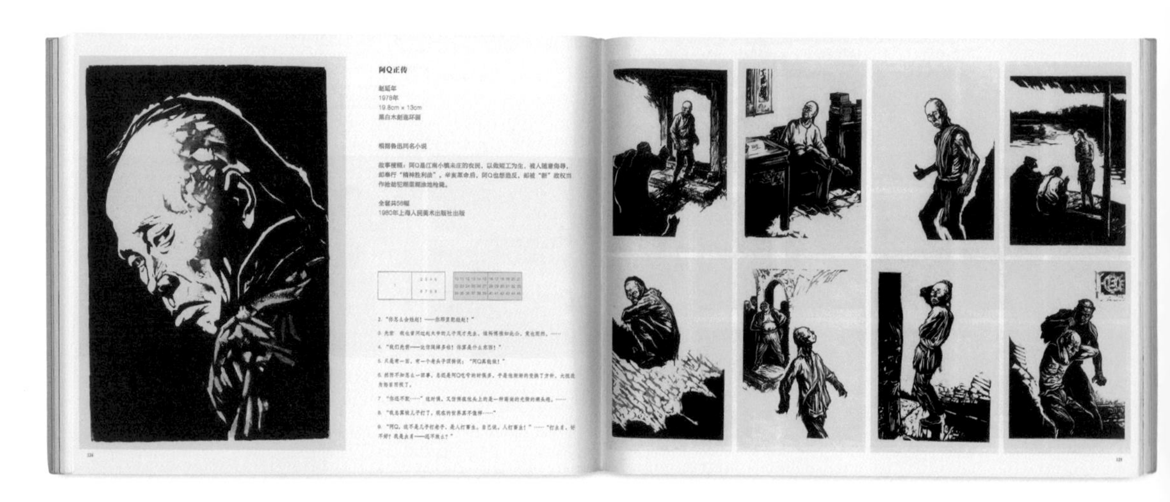

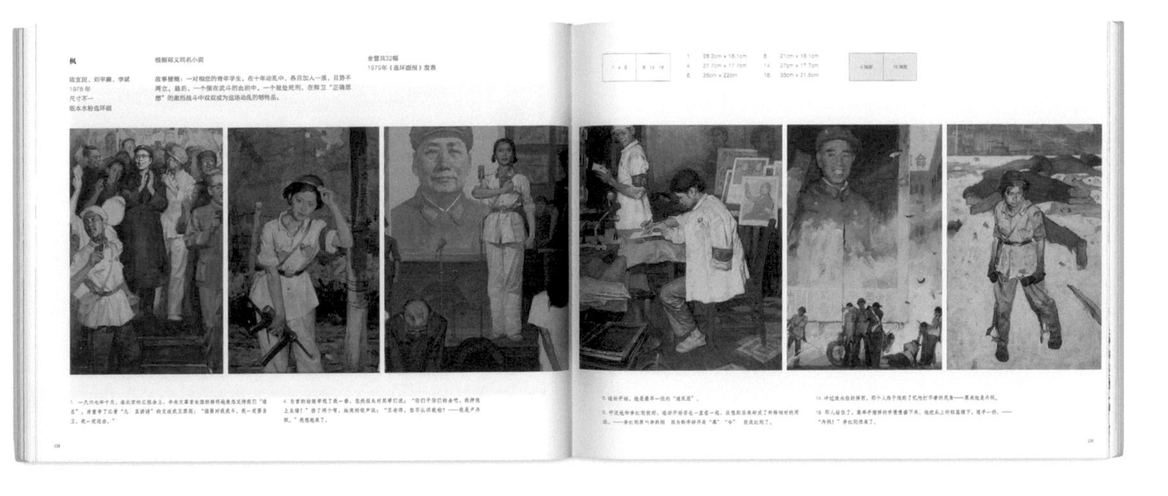

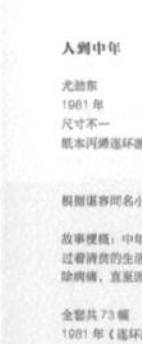

人到中年
尤劲东
1981 年
尺寸不一
纸本丙烯连环画
根据谌容同名小说 王素改编
故事梗概：中年眼科大夫陆文婷，虽是业务骨干，却收入微薄，过着清贫的生活，但她克服困难，排除干扰，全身心地为患者排除病痛，直至因超负荷的工作晕倒在手术台上，仍无悔无怨。
全套共 73 幅
1981 年《连环画报》发表

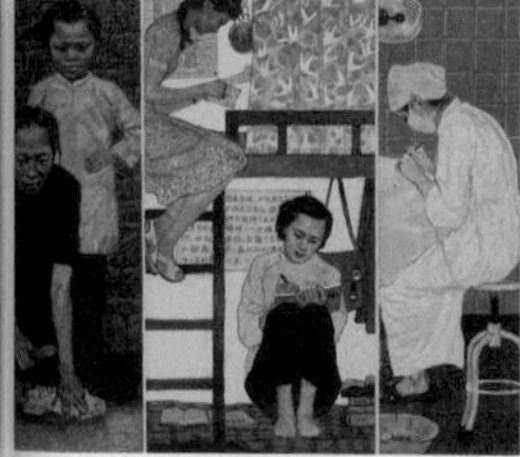

朝阳沟
贺友直
1979年
15.3cm × 23.8cm
纸本白描连环画
根据同名豫剧 廖德卿、胡廷楣改编
故事梗概：高中毕业生王银环到偏僻的山村朝阳沟落户，在农村艰苦生活的考验中，思想一度发生波折，后来经过朝阳沟支书和社员们的帮助教育，银环终于提高了觉悟，决心将自己的青春奉献给社会主义新农村。
全套117幅
1979年上海人民美术出版社出版

月牙儿
徐勇民、李全武
1982 年
54cm × 38cm（纸面尺寸）
纸本油彩连环画
根据老舍同名小说
故事梗概："我"从铁窗缝中望见熟悉的月牙儿，勾起往事，回忆不幸的身世——自幼丧父，母亲为生活历经挣扎，最终不得不出卖肉体。母亲老了，"我"生活无助，走上母亲的老路，沦为娼妓。
全套共 21 幅
1983 年《连环画报》发表

王功新 **王功新录像艺术 1995—2015**

OCAT 上海馆，上海，2016

306mm×248mm×26mm，248 页

布面精装，外裹护封

2000 册

主编：张培力

文章：黄专、皮力、Barbara Pollack

访谈：招颖思 / 王功新

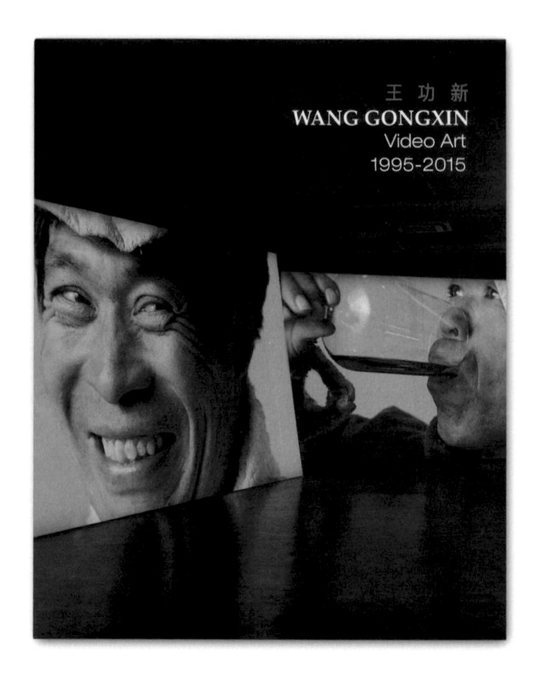

上拍的血色　2015

5 频道录像装置作品
5 屏幕　5 台 HD 投影机
5 台 HD 同步播放器
5 频道声音
8 分钟

Blood Stained Auction, 2015

5-channel video installation
5 screens, 5 HD projectors
5 HD synchronizers
5 sound tracks
08'00"

这件由五个巨幅屏幕搭建的有声影像装置作品呈现了一个“叠加”或“错乱”的“观赏”现场　演绎着一场“血色拍卖会”的场景。看似“老片儿”的资料文献碎片　不同的景　物　及错乱的时代人物　由五个屏幕按序逐渐推出　一幅著名的素描作品《血衣》中的人物和场景渐渐显现。《血衣》是五十年代中国画家王式廓的“革命的现实主义”绘画代表作　它真实再现了新中国成立后农民“斗地主”的大会场景。王功新将此作品中的基本人物和构图因素与当下生活中艺术品拍卖场景进行了叠加和重组。《血衣》中两种阶级矛盾的冲突　激愤的人群及悲痛控诉的妇女转化为拍卖场中激情抢拍的观众。通过将两种景观“混杂”的非线性叙事　再运用影像语言制造出近似文献“老片儿”的表现美感　作品缓缓地呈现出一个“历史”与“当下”之间既熟悉又陌生的叠加场景。五幅影像采用HD同步播放器控制　局部分割的景物被计时切换　当观众步入现场之中　艺术家试图营造出一个“心像”的“在场”。

The audible installation consisting of five large screens presents an "overlapping" or "disordered" "viewing" site that renders into a scene of *Blood Stained Auction*. Resembling the "old flicks" montaged with archival fragments, various scenes, objects and the personas of various periods, are presented on the five screens in order, the figures and scenes of the famous sketch *Blood Stained Shirt* are slowly revealed. *Blood Stained Shirt* is an epic "revolutionary realist" painting by the Chinese artist Wang Shikuo from the 1950s, it recreated the scene of "struggling against the landlord" at the farmers' gathering in the new China. Wang Gongxin mixes and reorganizes the prominent figures and compositional elements with the scene of an art auction of the present. The class struggle, the angered masses and the woman making an accusation in *Blood Stained Shirt* translate into the anxious audience at an art auction. With the "mix-up" of this non-linear narrative, and the representational aesthetics of the "old flick" rendered with archival footages from old films, the artwork slowly becomes an overlapping scene of familiar yet bewildering "history" and the "present". The five videos plays synchronously while the details of the segmented scenes and objects are switched in a timely fashion. As the viewer steps into the exhibition space, the artist attempts to create a "presence" of "a state of mind"

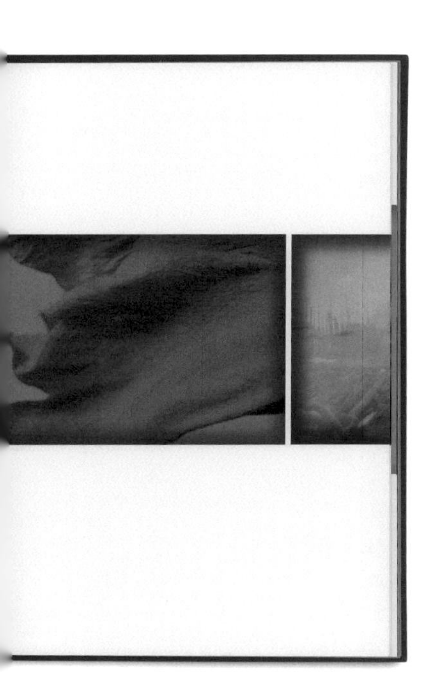

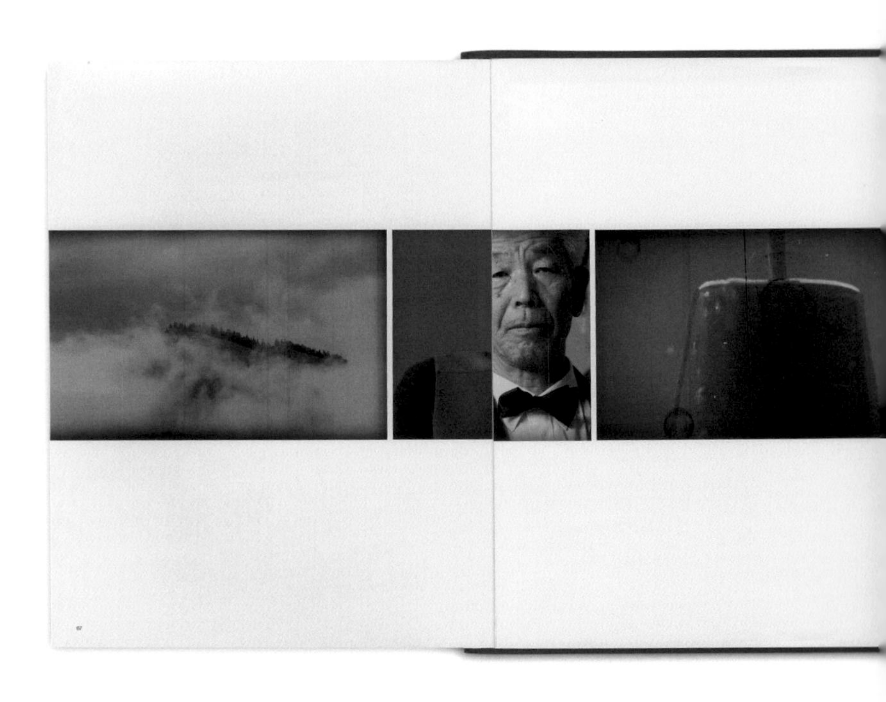

《王功新录像艺术 1995—2015》

我设计过不少展览画册，其中绝大部分都是在开展前完成的，一方面，开幕式上有画册似乎已经成了展览规格、仪式的一部分，之后再出总有点锦衣夜行的感觉；而另一方面，对于多数作品而言，画册与展览确实也是两种平行的阅读观看，在筹备制作上并无太多交集，同步进行倒也无妨。但王功新的这本作品集是个例外，画册在展览结束将近一年后才付梓出版，我则为这本画册的设计专程到上海去看了展览，这样的经历于我而言并不多见。

王功新的这部作品集虽名为《录像艺术 1995—2015》，但其主体部分实际上是他 2015 年在 OCAT 上海馆个展上推出的三件新作，而其中规模最大、所占篇幅最多，同时也是最难在书中呈现的，是那件名为《上拍的血衣》的多屏影像装置。影像装置不同于摄影、绘画、雕塑等静态作品，又不同于电影或单屏录像这样单纯的时间艺术，它是空间与时间交织在一起的复合性媒介，除非到现场亲身体验，否则很难真正把握住作品的特质，而任何对于内容认识的不足或偏差，最终都会在书中被放大，或空洞或荒疏，这样的例子不胜枚举。

《上拍的血衣》以王式廓油画名作《血衣》为缘起，作者把原画中农民斗地主的基本人物和构图因素与当下艺术品拍卖场景进行了叠加重组，《血衣》中阶级矛盾的冲突、激愤的人群及悲痛控诉的妇女转化为拍卖场中亢奋抢拍的观众，在两种景观混杂的非线性叙述中，作品缓缓地呈现出一个“历史”与“当下”之间既熟悉又陌生的蒙太奇场景。

记得那天下午我独自一人站在展厅中，面对五个错落成半环绕状排布的巨大屏幕思忖良久——五个屏幕同时播放但彼此又无线性的叙事关联，也就是说每一瞬间眼睛里都会同时看到五个“随机”组合的不同画面。但书籍是线性的，只要装订起来就有了天然的顺序；也是因为装订成册，通常一本书翻开只能同时看到左右两面，前后页面之间是无法随意组合、穿插、并置的。如何用书籍来呈现作品“五个屏幕跳跃性变换的动态组合关系”呢？

我的解决之道一如既往的简单——我为这件作品在书中设置了连续三组“开门折”拉页——书页打开之后可向左右两侧各再拉开一个折页，形似开门。这样一来，本来的左右两页拓展成为左右各两页，三组拉页正背共24个页码，分别以同样尺幅、位置排布五个屏幕上的录像截图，不同拉页的前后正背在翻动的开启闭合间产生了“无数”的排列组合关系，“纸上蒙太奇”油然而生，书籍语言本来是平面的，却又因此而“不平”。

此外值得一提的是像王功新、杨福东这样接受过长期现实主义油画训练的影像艺术家，其作品中画面独特的美学价值同样是应予被提示的，但这一特质在作品动态播放时往往会被其中的观念性所遮蔽。书籍对于作品静帧画面的选取与构建，正是对展览观看的有力补充。虽然设计的这部分工作总是隐性而不足为外人道的，但对于这样一本书而言却构成了其价值的另一个面向。

陈文骥 **陈文骥作品 2006—2016**

AYE 画廊，北京，2017

289mm×219mm×32mm，276 页

纸面精装，书脊包布，切口烫黑

1000 册

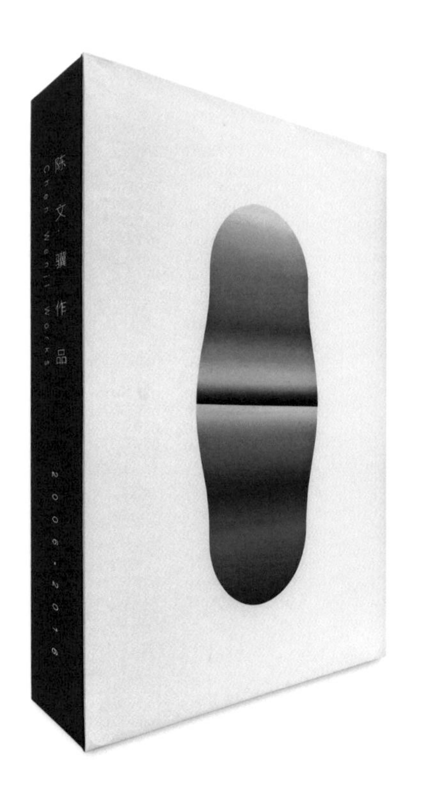

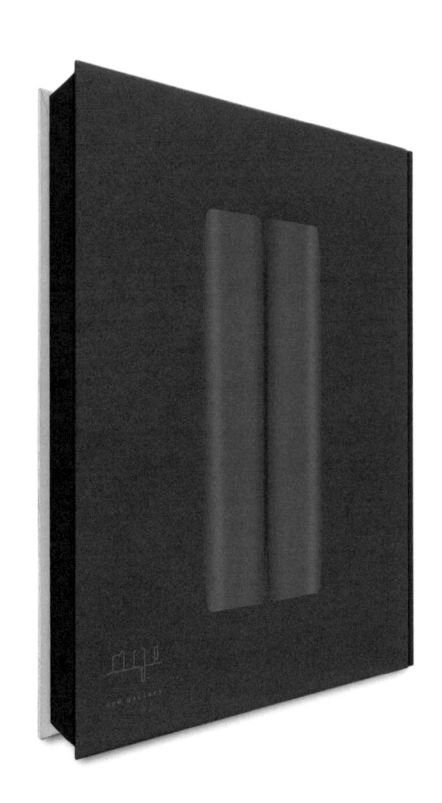

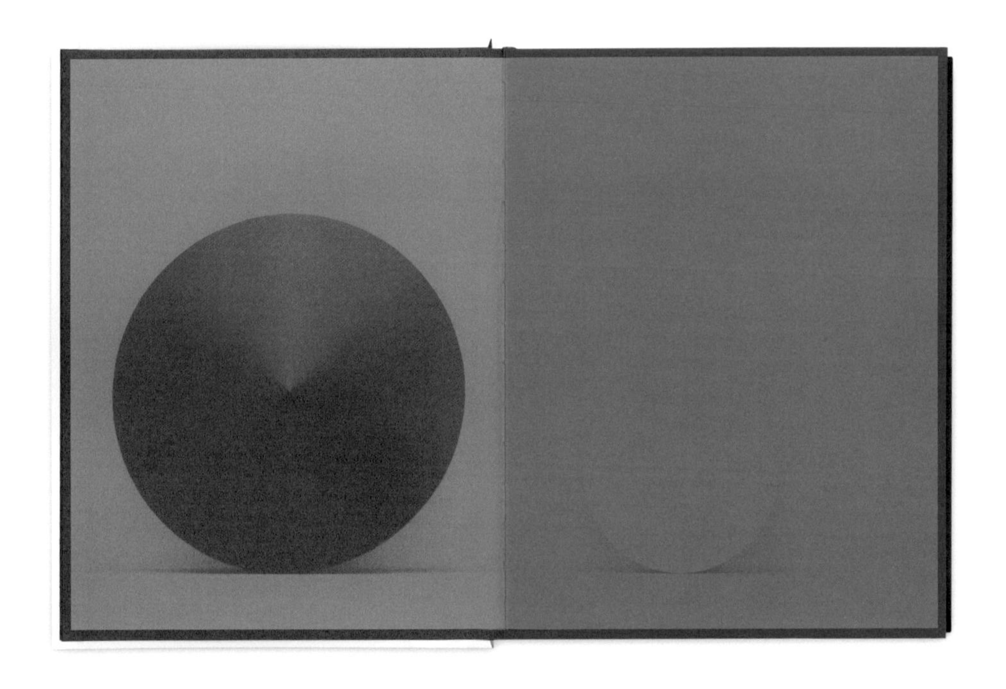

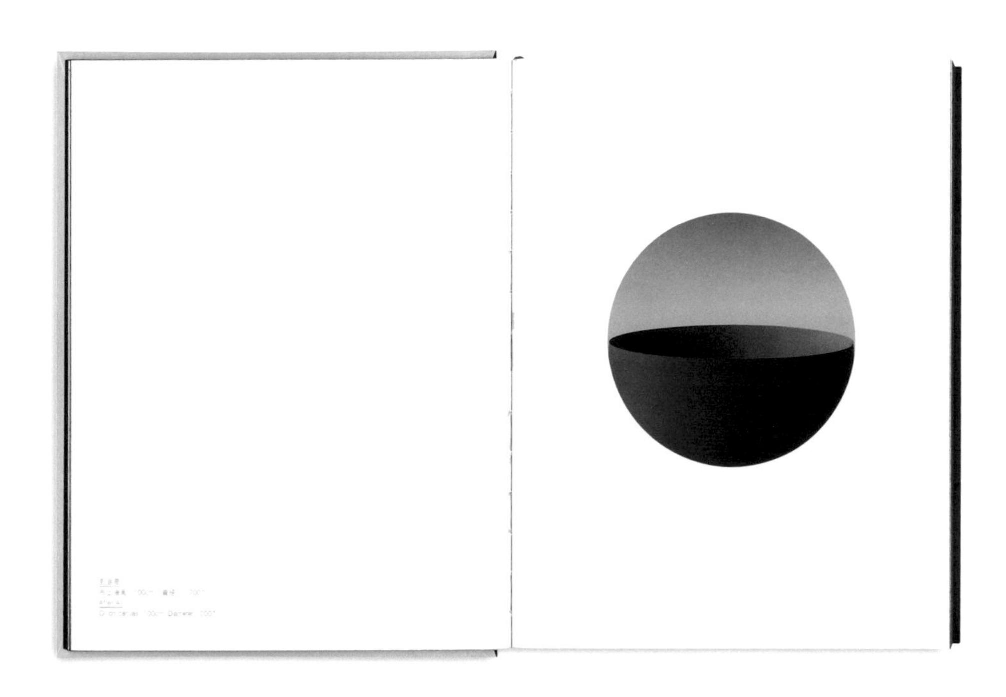

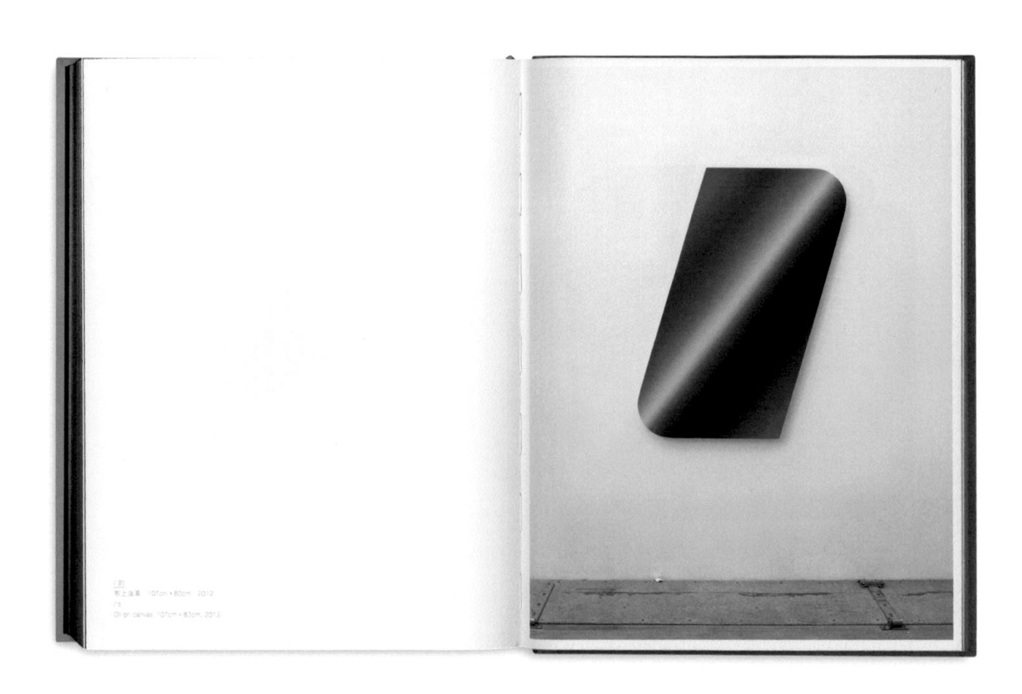
布上油画 107cm × 83cm 2012
Oil on canvas, 107cm × 83cm, 2012

评论 报道

陈文骥的画和一种静悄悄的革命
栗宪庭
《美术》 1987年第10期 P66-P67
《陈文骥》 人民美术出版社 2006 P226-P231
99艺术网 2012-8-21
http://magazine.99ys.com/periodical/per_157/article--701--3004_1.shtml
中央美术学院美术馆网站
http://museum.cafa.com.cn/cn/other/otherdetails_ot_71

前言
朱乃正
《陈文骥画人体》 四川美术出版社 1988

灵魂不肯离去
吕胜中
《当代学院艺术》 1994.1 P24
《陈文骥》 人民美术出版社 2006 P222-P225

关注 状态 ——陈文骥的绘画
梅墨生
《精神的逍遥——梅墨生美术论评集》 安徽美术出版社 1997 P376-P378
雅昌艺术家网 2013-09-09
http://artist.artron.net/20130909/n101364.html

窗外的风景与自己的房间——关于陈文骥的油画创作
冯博一
《陈文骥》 四合苑画廊 1999 P6-P11
《中国艺术》 2000.1 P36-P37
中央美术学院艺术资讯网 2010-11-11
http://www.cafa.com.cn/c/?t=832223
雅昌艺术家网 2010-3-31
http://artist.artron.net/20100331/n101345.html

个人体验与时代转变——陈文骥艺术变化的内在根据
宋晓霞
《美术研究》 2000.1 P25-P29

现实公路
马芝安（美国）
《世界都市》 2000.1-2 P126

个人体验与时代转变——漫谈陈文骥的艺术
宋晓霞
《中国当代画家丛书（二）》 四川美术出版社 2003 P156
《陈文骥》 人民美术出版社 2006 P212-P221
中央美术学院艺术资讯网 2010-11-11
http://www.cafa.com.cn/c/?t=832222

中国当代艺术数据库
http://www.artlinkart.com/cn/article/overview/7aahsum

一笔灰色——陈文骥的油画
冯博一
《陈文骥》 人民美术出版社 2006 P212-P221
《艺术家》 2006.11 总378 P472-P475
《今艺术》 2006.12 P206-P207
中国当代艺术数据库
http://www.artlinkart.com/cn/article/overview/1f6dxvs

作为态度的绘画——陈文骥的艺术
殷双喜
《陈文骥》 人民美术出版社 2006 P194-P201
《美术》 2008.10 P84-P89
雅昌艺术网 2010-12-02
http://comment.artron.net/20101202/n137481.html

在"美好的陷阱"中让生命永远充实——陈文骥与《红色领巾》
刘淳
《中国油画名作100讲》 百花文艺出版社 2006 P218-P220
中央美术学院美术馆网站
http://museum.cafa.com.cn/cn/other/otherdetails_ot_70

彼时彼刻，他在场
蒋岳红
《缤纷家居》 2006.6 P46-P47
雅昌艺术网 2008.01.11
http://news.artron.net/20080111/n40646.html

心象
尹吉男
《心象》 韩国·学古斋画廊 2007

艺随心走——谈陈文骥艺术道路
徐聪耘
雅昌艺术网 2008-5-14
http://news.artron.net/20080514/n45000.html

奏鸣的曲——陈文骥先生的艺术特色
尹吉男
《陈文骥》 东站画廊 2009
尹吉男的博客 2012-09-26
http://blog.sina.com.cn/u/1412401652

边缘的游戏——有感于陈文骥的新作
田佳
《陈文骥》 东站画廊 2009
《北京商报》 2011.12.2 C2版
中国当代艺术数据库
http://www.artlinkart.com/cn/article/overview/dbebxCm

識真山

傅山《奉祝碩公曹先生六十歲序》（圖一）這十二條屏把我感動了（以下簡稱《十二屏》），所以，基本上一個多月以來圍着這件東西在轉，想到的問題比較多，也不一定就能夠到傅山的點上，但是，傅山這個點確實激發了我對中國書法有一個新的感悟。

我的第一個感悟是齊白石《最後的輝煌》中的最後一張畫，《花神》（圖二），那張畫我看到的時候激動了好幾天，這是一九八八年，第二次是二〇〇六年，在天津博物館第一次看到了遼博藏的董源《夏景山口待渡》（圖三），這是我第二次激動，第三次激動就是這次看《十二條屏》，我想從以下幾個方面去講這幾個問題：行書和楷書之間的關係、線和點之間

圖二 齊白石《花神》

二〇一六年十二月十五日，薩本介先生在北京畫院做了一次講座，他用最本真的心面對傅山三百多年前的書法，彷彿打開了一扇時空之門，與傅山進行了一次直達心靈的對話。

圖三 董源《夏景山口待渡》

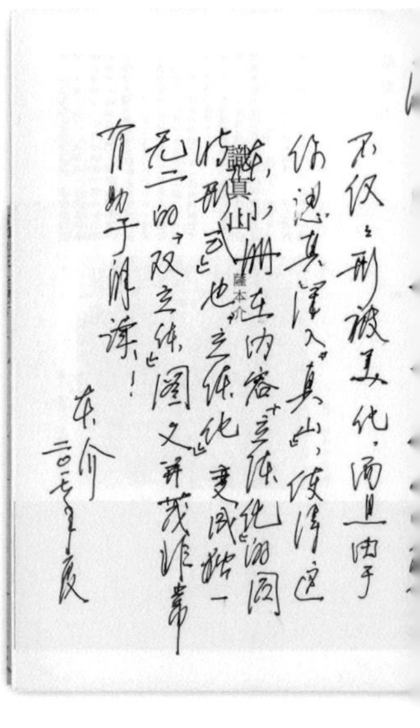

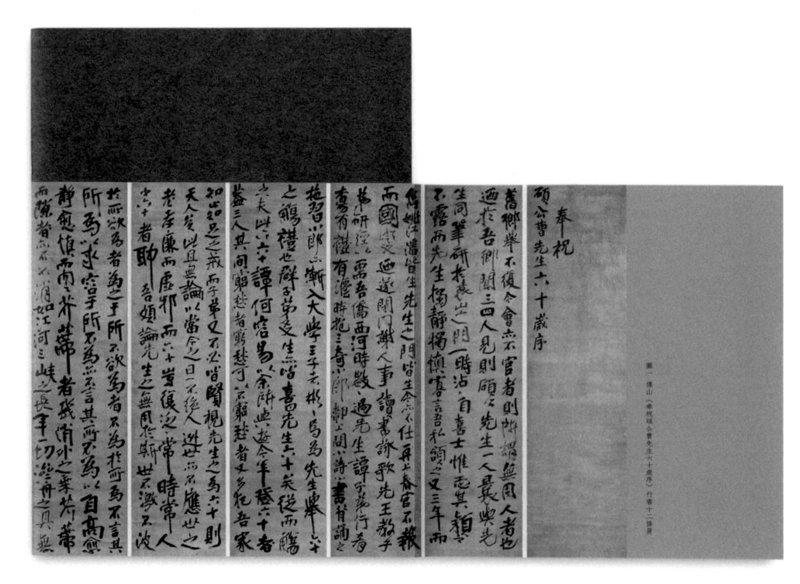

奉祝碩公曹先生六十歲序

圖一 傅山《奉祝碩公曹先生六十歲序》行書十二條屏

個過程，傅山這《十二條屏》給我的一個重要啟示，就是做事情要回零，也就是說，要打破以前的規律，重新考慮，重新審核，重新把碎片組合成一個新的東西。這是傅山給我的強烈的感受，最強烈的一點，我一開始是從回零這個概念接受這個事實，我覺得有太多地方回零了，幾乎是字字回零，筆筆回零，完全是重新開始。這個非常了得，所以他就從形體上破了歷史的慣性。就比如我一樣，我本來是個好人，去年八月八號，突然一下半身不遂了，這以後，你就要適應新的情況，你回零就不能走路了，一直到現在一瘸一拐還能上個六樓，還能夠走到這裡來做一個講座，我覺得這就是一種回零。人生如果每一個階段都回零，把前面的東西清理一下，然後根據前面的經驗，重新打碎再重新組合，這個觀念傅山在寫字上做到了，我覺得很了不起，因為一個文化人習慣性的動作，習慣性的思維，已經進入生活上的東西，他已經很習慣了，為什麼還要打破，為什麼還要重新組合，這個很了不

萨本介 **识真山**

作者自印本，北京，2017

240mm×170mm×2mm，

22 页

平装，骑马钉

1000 册

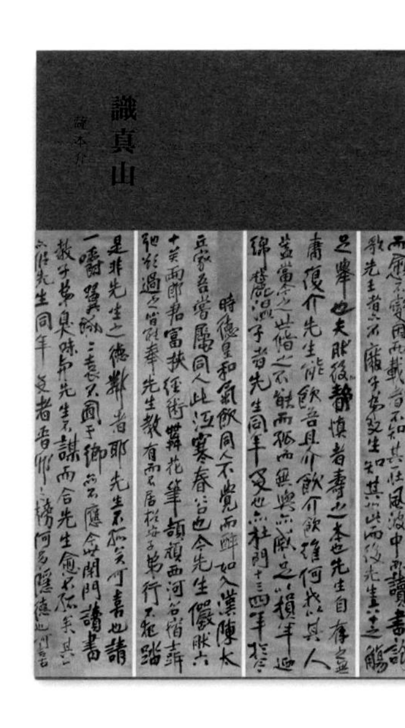

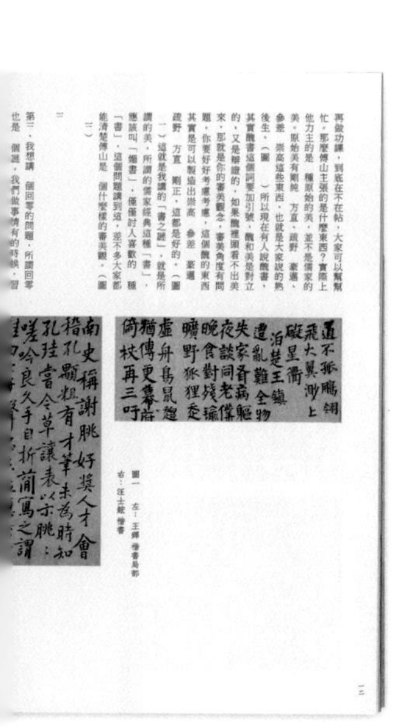

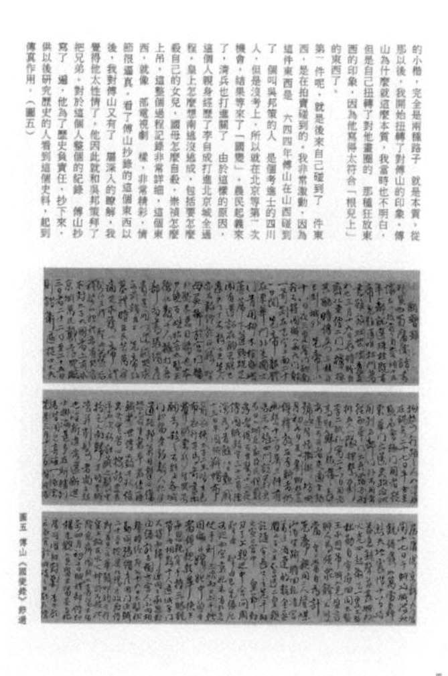

邬建安

邬建安

中国民族摄影艺术出版社，

北京，2018

306mm×236mm×26mm，

288 页

布面精装，外裹护封

2000 册

前言：茅为清

文章：吕胜中、阎安、

苏源熙、唐冠科

访谈：巫鸿 / 邬建安

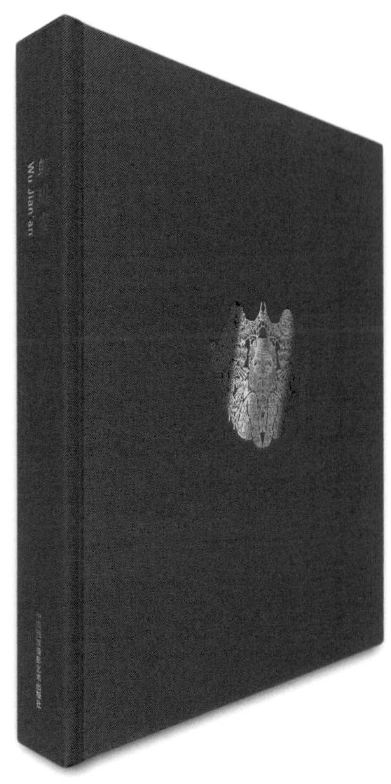

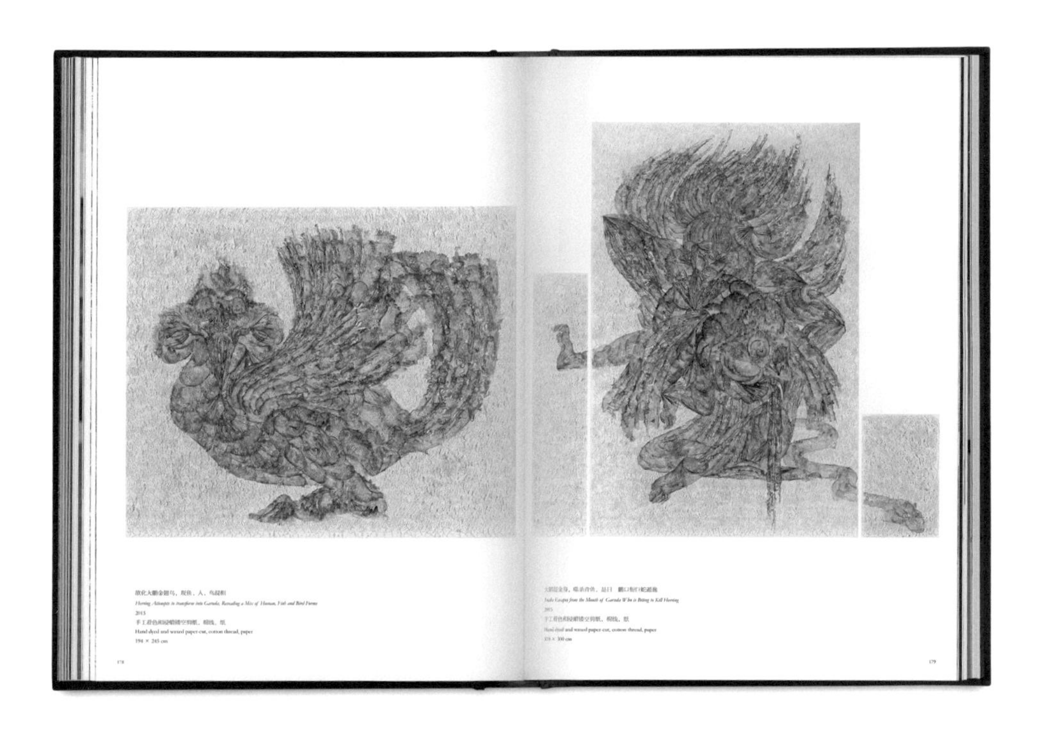
Herring Attempts to transform into Garuda, Revealing a Mix of Human, Fish and Bird Forms
2015
Hand dyed and waxed paper-cut, cotton thread, paper
194 × 245 cm
Hand dyed and waxed paper-cut, cotton thread, paper

36 Strokes, 2016
Marble, acrylic

北岛 **此刻**

地平线画廊，巴黎，2018

244mm×175mm×19mm，

136 页

布面精装

1000 册

文章：徐冰、李陀、北岛

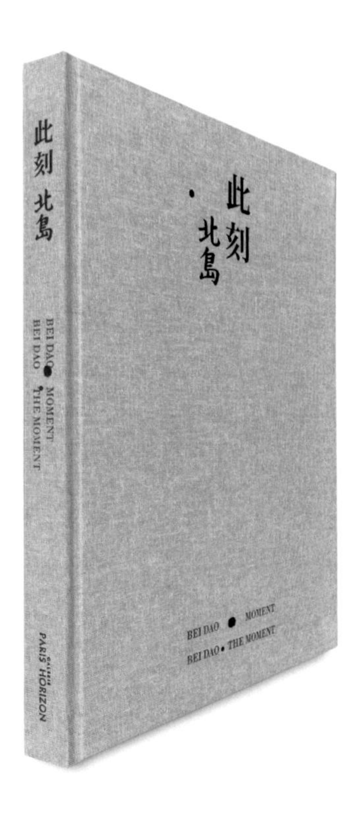

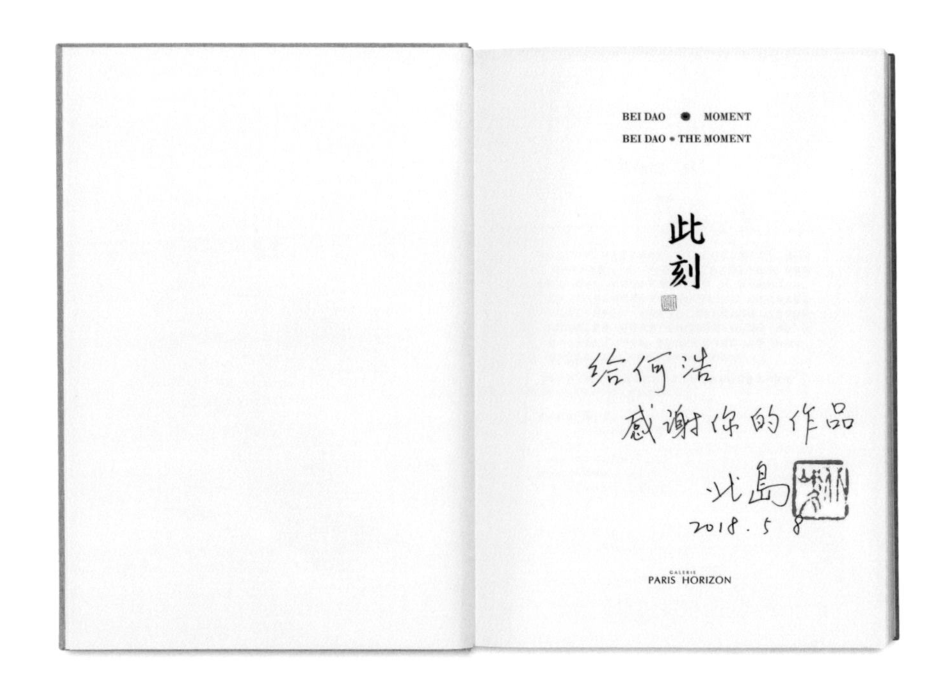

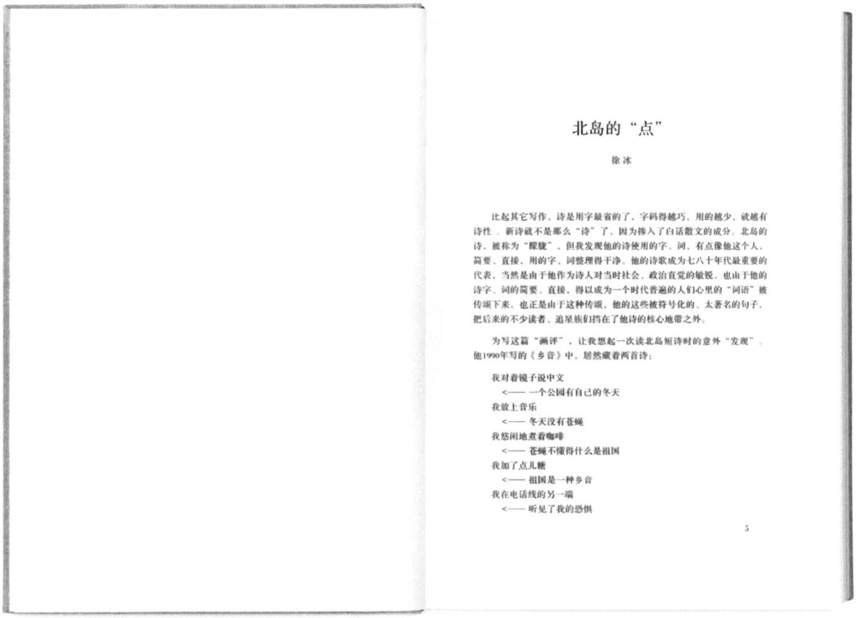

北岛的“点”

徐冰

比起其它写作，诗是用字最省的了，字码得越巧，用的越少，就越有诗性。新诗就不是那么“诗”了，因为掺入了白话散文的成分。北岛的诗，被称为“朦胧”，但我发现他的诗使用的字、词，有点像他这个人，简要、直接，用的字、词整理得干净。他的诗歌成为七八十年代最重要的代表，当然是由于他作为诗人对当时社会、政治直觉的敏锐，也由于他的诗字、词的简要、直接，得以成为一个时代普遍的人们心里的“词语”被传颂下来。也正是由于这种传颂，他的这些被符号化的、太著名的句子，把后来的不少读者、追星族们挡在了他诗的核心地带之外。

为写这篇“画评”，让我想起一次读北岛短诗时的意外“发现”。他1990年写的《乡音》中，居然藏着两首诗：

我对着镜子说中文
　<—— 一个公园有自己的冬天
我放上音乐
　<—— 冬天没有苍蝇
我悠闲地煮着咖啡
　<—— 苍蝇不懂得什么是祖国
我加了点儿糖
　<—— 祖国是一种乡音
我在电话线的另一端
　<—— 听见了我的恐惧

5

Les « points » de Bei Dao

Xu Bing

Comparée aux autres formes d'écriture, la poésie est la plus économe en mots, plus ils sont placés ingénieusement, moins ils sont employés, et plus ils sont poétiques. Or la nouvelle poésie, n'est plus vraiment « poétique », corrompue qu'elle est par l'incorporation d'éléments venus de la prose en langue parlée. On a qualifié la poésie de Bei Dao d' « obscure », mais j'ai constaté que les mots qu'il emploie, mono ou polysyllabiques, sont un peu comme lui: concis, directs, mis en ordre. Si ses poèmes sont devenus les plus représentatifs de la création poétique des années soixante-dix et quatre-vingt, c'est bien sûr par l'acuité de son intuition de poète sur la société et la politique, mais aussi, justement, c'est par cette concision, ce côté direct des mots, qu'ils ont pu devenir le « vocabulaire » disant l'état d'esprit commun à une époque et qu'ils se sont largement répandus. Et c'est en raison de cette appropriation que ces vers du poète, trop célèbres, devenus symboles, se sont retrouvés immobilisés en dehors du noyau de sa poésie par de nombreux fans ou simples lecteurs.

En écrivant cette « critique picturale », j'ai repensé à une « découverte » fortuite, faite à la lecture de courts poèmes de Bei Dao. « Accents du terroir », composé en 1990, à ma grande surprise, contenait en fait deux poèmes :

Je parle chinois devant le miroir
　<— *Un square a son propre hiver*
je mets de la musique
　<— *il n'y a pas de mouches en hiver*
pensivement je fais du café
　<— *une mouche ne comprend pas le mot patrie*
j'ajoute un peu de sucre
　<— *la patrie cet accent indéfinissable*
à l'autre bout de la ligne
　<— *j'entends ma propre peur*

8

Si l'on dépouille le poème de l'une ou de l'autre des deux parties, à la lecture, cela ne ressemble à rien, tandis que, lorsque les deux parties s'entrelacent, l'une occultant l'autre, à la lecture, on est pris, un peu comme avec le langage secret, dans des effets de complémentarité et d'interprétation mutuelle, comme si elles se répondaient l'une l'autre, mais avec une tentative autre : celle de créer de nouveaux territoires.

J'étais ravi d'avoir découvert quelque chose que ni les critiques de poésie et ni Bei Dao lui-même n'avaient repérée. Je comptais, la prochaine fois que je le verrai, lui demander confirmation, afin qu'il se réjouisse de sa trouvaille. Et je l'ai fait, un jour, Bei Dao a répondu « Mais c'était voulu, cette disposition décalée. » Et voilà comment ma « découverte » avait été ramenée à zéro. Ce bon vieux Bei Dao, qui semblait à première vue simple et obtus, en fait cultivait en douce le mystère.

En 2012, il est tombé malade, sa vivacité de pensée et ses facultés langagières s'en trouvèrent à l'arrêt, il n'avait plus assez de concentration pour taquiner l'écriture. Selon ce qu'il raconte, à ce moment-là, la régression la plus manifeste a concerné l'anglais (je pense qu'il faut lier cela au fait qu'il l'a étudié sur le tard et par obligation) , le chinois, sa langue maternelle, venait après (car appris dans son jeune âge). Le reste concerne des phénomènes substantiels liés à la physiologie, c'est-à-dire des gestes corporels : saisir le pinceau et les traces laissées par l'outil au contact du papier, et c'est ainsi que sont apparues des œuvres de l'« art» de l'image—des marques corporelles, en un instant particulier de la vie d'un écrivain.

Un jour, il me montra son téléphone portable: il peignait des tableaux. C'était les premiers qu'il avait exécutés juste après sa maladie; ils étaient composés de lignes régulières et denses, on aurait dit une empreinte digitale, celle du destin. Pourquoi ces peintures avec "lignes" avant celles avec "points"? On peut expliquer ainsi, grosso-modo, les pensées qui étaient les siennes quand il s'est lancé dans cette pratique : puisque je ne peux pas écrire, je vais peindre, et il faut que mes créations ressemblent à des tableaux, or pour cela, dans l'acte de peindre mon pinceau doit se déplacer. D'où les lignes avant les points. Dans son texte, Li Tuo analyse l'affirmation de Kandinsky selon laquelle « la ligne et la forme sont le prolongement du point ». Or pour Bei Dao, la ligne s'est rétractée en un point, il s'en suit que le message est enfoui plus profondément, que la suspension du « langage » est plus courte, le travail d' «énonciation» est des plus élémentaires, voilà qui lui ressemble davantage. Il y a quelques jours comme je lui demandais où étaient les tableaux composés de lignes, il a répondu : « Je ne les retrouve plus » On voit bien qu'il ne s'intéresse pas à ces œuvres-là, qu'il leur préfère celles

9

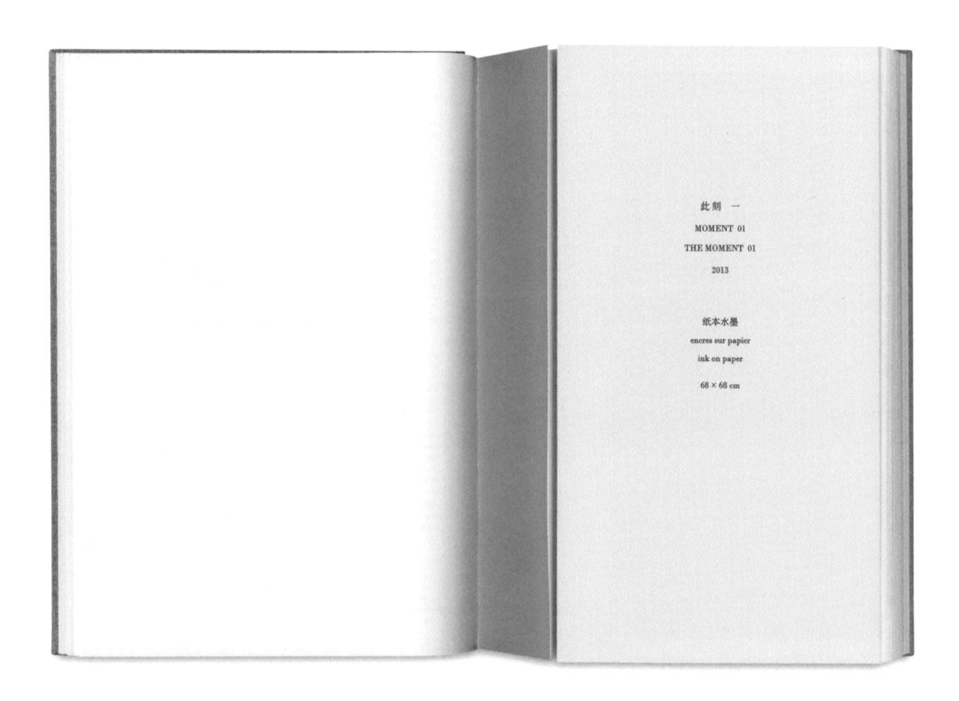
此刻 一
MOMENT 01
THE MOMENT 01
2013
纸本水墨
encres sur papier
ink on paper
68 × 68 cm

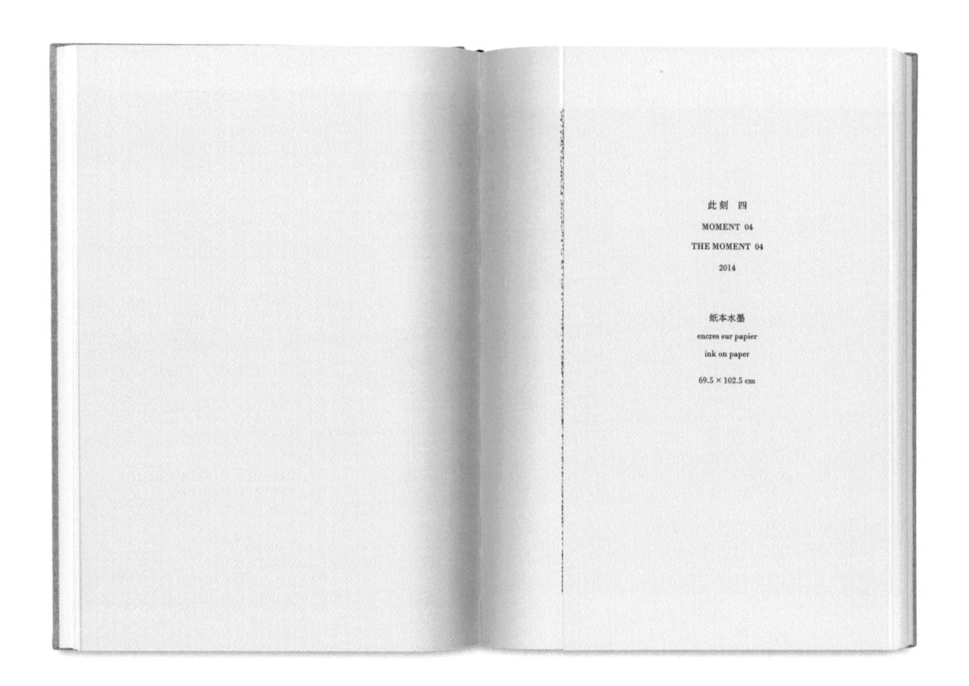
此刻 四
MOMENT 04
THE MOMENT 04
2014
纸本水墨
encres sur papier
ink on paper
69.5 × 102.5 cm

《此刻》

我的书籍设计总是从触摸作品开始的，而对作品的理解又往往始于对作者的认识。关于北岛，作家查建英在一篇回忆《今天》的文章结尾处这样写道：“严冬腊月，穿一件接过两截袖子的旧棉猴，站在北大食堂附近一面破墙前，光线不好，我的脸几乎要贴到那些被糨糊刷得凹凸不平的廉价油印纸上。

……

告诉你吧，世界，

我—不—相—信！

纵使你脚下有一千名挑战者，

那就把我算作第一千零一名。[1]

那时候没有粉丝这个词。那时候写出这样诗句的人，是我们的英雄。”这段文字是我所喜爱的，寥寥几笔勾勒出了北岛其人和他的时代。但若据此认为北岛就是那沉睡铁屋中激昂的呐喊者则未免简单刻板，我所认识的北岛，其实是个“少年”，那个《皇帝的新装》里径直说出自己眼中所见的少年，他作品中的张力正是来自少年的天然率性与苟且、昏庸的成人世界间所产生的落差，而由这一认识所生发出的清澈澄明之感，构成了我后来为北岛设计的所有书的美学基调。

《此刻》是北岛的第一本画集。2012年，北岛罹患中风，所幸救治及时而无性命之虞。祸福相依，康复期间北岛偶然拿起画笔，本是解忧破闷舒展身心之举，却无意间开启了他的另一种“写作”。

北岛的这些画作，在我看来有一个最重要的特点，它们是完全指向“内部”的。一方面，这些密密匝匝宛若繁星的墨点是心像和时间的轨迹，点随心变全无事先的预构，每一墨点即是“此刻”；另一方面，不同于专业艺术家，这些画作在创作之初，作者没有任何使之未来公共化的期许，是纯粹私人性的。

1. 查建英原文中此处为《雨夜》节选。

造册伊始，我跟北岛有一个不约而同的共识，这本画集从形态上讲应该更像一本“文集”而非“画册”。概念很对，但现实的情况是，如何用文字书的小开本来展示这些构成密集又饱含细微变化的画作？作品“向内”的特质又怎样用书籍来回应呢？

为此，我们首先从本来就数量有限的画作中再次甄选24幅，每幅以数字命名，从《此刻 一》至《此刻 二四》，以此暗合时间的流转。随后我把这24件作品在书中设置成为了24个折页，如同24通信件，打开才能阅读。在这里，“打开”既是心理过程亦是功能的解决之道——画页尺幅因此放大一倍，细微之处得以毕现。此外，我为每个折页的外部满纸敷上了淡淡一层暖灰色，一则以此强化了“内”与“外”的关系，二则灰色隐隐渗入背面，使得画面更为深醇蕴藉，含蓄隽永。

因为友谊，北岛赠我一幅他的画作，我很珍视，装裱起来挂在家中墙上，朝夕相对至今转眼已是两年。在这期间，偶尔我也会再次翻翻《此刻》这本小书。确如徐冰在书里文中所说，北岛的画似乎更适于平放在桌上观看，它其实更像书页。把书页挂起来当图看，看到的更多是版式、段落的图形变化，而当把一幅画在面前展开铺平时，其中的每个“点”都化为了浅诉低吟，给人以无限的幻想和猜测。这些画因为书又化作了诗。

人名索引

代后记 关于何浩的《写设计》

巫 鸿

看到《写设计：何浩艺术书籍设计2003—2018》校样，我有两个惊讶——虽然我自认对何浩的设计项目比较熟悉而且一直欣赏它们的当代感和文化气息。一个惊讶是这些项目对中国当代艺术的卷入程度之深，可以说是在一条脉搏上律动，分享着彼此的养分和生命。我甚至感到我在看一部本世纪以来的中国当代艺术史，因其特殊视点和材料而尤为可贵。另一个惊讶是我们二人互动之频繁，甚至超出我的记忆。翻到索引页一下查到 13 个出版物，多数是我策划的展览图录，少数包括我写的文章。我忽然感到我所面对的，是何浩创造的一条潜在线索，在近二十年中把我和艺术家、画廊与美术馆，和中国当代艺术联结在一起。我相信有这种感觉的不会只是我一人。

最使我记忆深刻的是我们的第一次合作——2004 年的《荣荣和映里：蜕》展览图录。此展在北京 798 大窑炉车间举行，以“蝉蜕”的意象把一个庞大的工业废墟转化成当代艺术场地。当展览本身业已消失 16 年，当大窑炉车间的废墟早已被闪亮的尤伦斯当代艺术中心掩埋，是何浩设计的画册给予了这个展览和它的废墟场地以持续的生命，甚至永生。

从设计上讲这也是我所有出版物中最具“展览性”的一个，把作品、文字和空间了无痕迹地融合在一本书中。它的视觉冲击力迅速而强烈：记得当我和纽约的亚洲协会美术馆商讨举办全球出版界大亨布雷克的私人美术藏品展览的时候，展览图录几稿设计他都不满意。我于是把何浩的《蜕》带到讨论现场。他眼睛顿时发亮，说“我要这个设计师”。很可惜这个展览最后因故未能举办，但他的反应使我难以忘怀。

2020 年 5 月于普林斯顿高研院

巫 鸿 美术史家、批评家、策展人，芝加哥大学教授

书籍态度 Attitude of Publishing

“何浩：书籍态度”展览现场，宁波美术馆，2013

徐冰
版画

“何浩：一个人的群展”展览现场，香港中文大学，2015

靳尚谊《向维米尔致意》

“何浩：一个人的群展”展览现场，香港中文大学，2015

香港中文大学主题演讲“平面设计的当代转向”暨与靳埭强对谈，2015

跋

何 浩

我的书籍设计历程从2003年开始。2003年是不同寻常的一年——“非典”就发生在那年。对于中国当代艺术而言，这同样是个值得被书写一笔的年份——798工厂在这一年正式作为“艺术区”进入了公众视野，这种确立并非是政府或某个机构的赋予，而是暗流涌动之后真正的水到渠成。今天回头看去，2003年当可视为中国当代艺术从地下转为地上的节点之年，只是当时的我未必有此洞见。

我获得的第一个书籍设计委托是本很小的小书，而且是个公益项目——大家合力给一位一年前去世的友人——戴汉志（Hans van Dijk）自印出版一本纪念画册，属于有钱出钱有力出力的性质。虽然无利可图，但这个委托对于当时的我却可谓正逢其时。一方面，经过之前数年的设计实践与思考，我对那些总是围绕着视觉形式展开的设计话题已是意兴阑珊，很想能够寻找到某种依托来冲破这种狭隘与贫乏，进而为我想知道的诸如设计的构造、发展这些本源问题找到实践层面的答案；另一方面，当时开始可以少量接触到的当代艺术以其超越具体媒介和手段甚至“艺术”本身的开放姿态为我猛然打开了一个新空间，使我发现了一类正对自己兴趣的新内容，而当代艺术创作中所蕴含的观念和方法更是让人兴奋。当时无法想到，这种仅是出于直觉的对于内容的信仰和对创作本源问题的探寻欲望竟无意间指引了我之后十年的创作方向，也铺就了我后来所有作品的底色。

虽然艺术和设计看上去似乎有着天然的关联，我的创作进程确实也在两者的融合和转换中获得滋养得以加速，但随着设计实践的逐渐深入，我反而越来越清晰地意识到设计者和艺术家在本质上是有着不同属性和要求的两种人。虽然关于这个话题的讨论由来已久，但我仍然认为有必要从更深层将之厘清，这种厘清对于设计者该如何认知自己的工作意义重大。

如果把艺术创作和设计都假设为一场演出的话，那么艺术家所做的可视为行为艺术表演，而设计者则必须是一个真正的演员——他的表演其实是扮演，有着“自己”和“角色”双重身份——演员需要让自己悄然走进另一个完全不同的人的生活并融入其中，以此使得观众感同身受。这是一种非常微妙并且奇妙的创作状态，“角色”和“自己”既重合又分离，既交织在一起又互相映衬。任何时代都不乏凌驾于角色之上、显赫一时的明星和为之买票叫好的观众，但纵观历史，一个伟大演员的“自我”必定要通过“无我”来成就。 而作为一个演员的乐趣和价值，也正是来自这个放下自我进入角色的过程，得以有机会体味和演绎各种不一样的悲欢人生。 这种乐趣和价值，同样属于设计者。

记得 2003 年在做戴汉志那本书时，跟我一起工作的朋友还在为离开电脑时如果不把网断开会不会产生高额费用或感染病毒而踟蹰，转眼间互联网已经成为他最重要的创作平台。事实上，网络所带来的信息裂变在过去的这些年彻底改变了我们每一个人的生活，与信息传达和视觉沟通共生的平面设计在这样的环境下则无从躲闪地亟须被重新定义和考量。美国设计先驱保罗 · 兰德（Paul Rand）曾经说：“设计是关系，是形式与内容的关系。”那么时至今日，这一关系该是一种怎样的构建呢？在我看来，今天的设计已不再是可以孤立观看甚至独立存在的物质，而是成为了一种渗透性媒介，具有类似黏合剂抑或水泥的属性——水泥本无形，与飘浮在空气中的各种细土灰尘看上去并无二致，但它一旦与水结合搅拌，即可生成万千形态，同时坚不可摧。基于这样的认识，我相信当下设计的发生必须相融而非飘浮于所面对的内容，设计的能量也只有在与内容发生“化学反应”后才能得以真正地释放和挥发。今天形式与内容的关系，已不再是把一筐葡萄喷上水或榨成汁，而是要酿成酒。设计本身即是写作的一部分。

今天的中国正处在一个大家都迫切想要证明自己的时代。在这样的大背景下，设计自然也会表现得有些亢奋，“乾隆品位”重领一时风尚。但是这种与中国“前现代”发展阶段相映照的审美趣味却实非我的取向，甚至我会有意小心避开，以免产生强加于人的视觉戾气。我设计的书大多看上去平淡普通，即便有些项目由于内容和编辑的特殊需要而使得书籍形态必须做出相应的回应，整本书从外表看上去依然是平实朴素的。对于书籍这种媒介形式而言，我始终觉得平朴是其本色，难容涂抹。这就如同富于真知灼见的表述从来都应是循循善诱、娓娓道来而非高谈阔论甚至危言耸听，这点并不会因为表述的内容不同而有所改变。

我一直觉得自己是个很幸运的人，因为我的合作者都是这个时代中国最出色的艺术家，每一个项目对我而言都是一个绝佳的学习机会，我因为设计这些书而成长而充盈。我跟这些艺术家也从开始时的不认识、不熟悉到因为一本书而成为挚友、诤友，人生中还有什么样的经历能比这更美妙呢？2015 年初秋，我为书法篆刻家刘彦湖设计他的印集。这是一批尺寸巨大、刀痕极深的陶印，在刘彦湖的创作手记中，有个小标题名为“大刀深刻”，这四个字使我感触颇深，它让我联想起《庄子》中的一则寓言：郢地有个人鼻尖上粘了一个白点，只有蚊蝇的翅膀那么薄，但他却让匠石用斧子把这个白点削掉。匠石二话不说抄起利斧劈将下来，白点随风而落，而郢人的鼻尖却毫发无损。整个过程匠石看似漫不经心信手拈来，而郢人则站在那里若无其事泰然处之。宋元君知道了这件事，召见匠石说：“你也同样为我试试。”匠石言道：“我确实曾经能够砍削掉鼻尖上的小白点。但我可以搭档的伙伴已经死去很久了。”审慎、专注与不间断的实践磨砺，使我的设计似乎亦渐得“大刀深刻”之法，但没有这些委托者的胆识和极度信任，我何处下刀？又怎敢发力呢？好的设计从来都是 1+1=3 的结果。

今天，人类的阅读进入了千年一遇的新时代，“数字技术”“互联网”“微博”“微信”“电子书”……这些既关联又支离的新词语构成了一个全新的阅读图景。近些年，不断有人跟我讨论纸质书的前景并表示为之担忧。其实，如同当年摄影对于绘画的冲击不仅没能使绘画消亡，反而为之提供了新的发展机遇，使绘画变得更为开阔和更具活力——在经历了数千年对外部的观看之后，终于有机会借此开启对于自身内部的审视——从此格局巨变。纸质书亦然。面对当今日益多元的阅读媒介，纸质书也势必会从过去单一的记录和叙述功能中蜕变、升华，重新审视自身作为不灭的实体物质所需具备的恒久、隽永、质感、温度…… 而这些正是需要通过更具前瞻性与洞察力的编辑、设计才能得以实现。我们有幸能够站在这个千载难逢的转折点上并有机会参与其中，是多么让人兴奋的一件事啊！

历史车轮滚滚向前永不停息。2012 年底十八大召开，中国的方方面面随之发生深刻改变，当代艺术作为社会文化最敏感的组成部分自然也无法不做出由表及里的回应。我相信 2013 年必可视为中国当代艺术的一个新节点，虽然这里面所暗含的转折同十年前一样看似平静但其实汹涌。自 2014 年起，我的创作轨迹由于时代和自身的双重驱动开始拓展偏移，从单一的艺术书设计转为艺术书 / 学术书双径并行，而在两个领域之间跳跃、穿行所引出的“化学反应”更助推了我的加速前行。尽管如此，对于这本书而言，我仍然认为专注地描述一段历史，哪怕仅仅是其中的一个面向、一条注脚，也比笼统地呈现一个人的全部工作要有意义得多，因此我并未将近五年来艺术书之外的工作成果收入其中，那些由新格局催生的新作品、新思想也许留在未来的某个历史契机中与读者分享会更为适宜，那是我的下一个计划，也将是另一个全新的话题。

本书的主体部分由香港中文大学出版社 2014 年 11 月以中英双语（繁体中文/ 英文）出版，名为《一个中国独立设计者的当代艺术史——何浩书籍设计 2003—2013》（*A Designer's Decade of Contemporary Art in China: The Book Design 2003—2013*）。感谢老友张赤兵先生和香港中文大学出版社甘琦社长、编辑部主任林颖博士，是他们的热忱和不弃才使得这样的一部书稿得以“社会化”，书中的“创作手记”也是应中大社编辑部同事们的要求、在编辑林骁的鼓励和陪伴中完成的。关于这部分内容，此次出版略有调整，替换删除了部分图文，个中原因无须赘言，而我则更倾向于将这一变更视为某种讨论。我相信任何基于不同视角的主张，只要是真诚严肃的，都会在讨论中让事物的真相更趋明晰。单就此书而言，这种讨论本身即是其中意味的一部分。

这本书的书名《写设计》源自宋晓霞《设计也可以是一种写作》一文的启示，而编辑曾诚更针对我的创作方法将其与文化人类学中的“写文化”概念加以连接，提出了“写设计”这一书名。所谓“写文化”，简而言之就是“写”与“文化”的关系，即作者竭力用反省克制的态度客观地对研究对象加以描述，但由于作者同时背负着自己的历史文化背景，这使得其在文本书写中又总会微妙地浮现自身的政治世界，写作正是在这种“经验接近”和“经验远离”的不断交替中去趋近事物的真相。

相对于文化人类学的研究，艺术书籍的设计确实也具备类似的属性。对于一个艺术书籍的设计者而言，实际上站在前后两个艺术活动的交汇处——在前一个以艺术品原作为中心的艺术活动中，他是接受者；而在后一个以艺术书籍为中心的艺术活动中，他又是艺术家、创作者。当设计师在得到为艺术家创作书籍的委托之时，艺术作品已经完成并独立于艺术家而存在，成为了客观的审美对象。这一审美对象，正待接受者去进行审美欣赏而得以完成前一个艺术活动。但和一般的读者相比，设计师又是一名特殊的接受者，他的审美自由，要在对原作品的接受过程中受到一定限制，他必须比一般读者更主动地融入原作者的世界，通过对话，与原作者产生“理解交集”、达成“审美共识”，从而在接下来的书籍设计创作过程中，展现原作者体现在作品中的原意、内容，用书籍工艺与设计语言的形式，尽可能多地将它们在艺术书籍这一新作品中传递出来。

尽管设计师有可能与艺术家的教育和社会背景、审美观点和趣味等方面有着高度的相似，但毕竟不是艺术家本人，不可避免地会有许多的不同。设计师作为“特殊”读者，他必须让渡出一部分审美自由，对作品的“误解”“偏见”虽然不可能完全消除，但要受到一定的限制。正如解释学家施莱尔马赫所说：“理解一位作者要像作者理解自己一样好，甚至比他对本人的理解还要好。”

而对于作为原作品创作者的艺术家来说，在下一个以书籍为中心的艺术活动中，会再次面对自己的作品。这时，艺术家在设计师的引导下，转换成为自己作品的接受者、评论者。二者在对话中，有机会在设计中获得对作品的新的“理解交集”和“共同美感”。这种以书籍设计为中心的对话、交流，甚至会产生出原作者创作时没有意识到的内蕴，对原意有所超越，产生新的意涵，从而实现对于原作的超越。所谓“写设计”正是对这一复杂而微妙的创作状态最为凝练而传神的表达。

最后，我要再次郑重地感谢我的良师益友宋晓霞教授和曾诚先生，上面关于书名的提示只是成书过程中的一环而已，事实上他们对于此书所起的作用是决定性的，没有他们的支持与帮助，就不会有这本书的诞生；芝加哥大学巫鸿教授是我在设计上一直以来的密切合作者，书稿付梓之际，巫老师中断了自己的著述，拨冗为本书写下后记，他将此书视为中国当代艺术“一条潜在的线索”，为之赋予了设计之外更为深远的意义，让我感铭在心，溢于言表。

我的导师谭平教授和挚友蒋华博士在过去十几年中给予了我最为深切无私的扶持与关爱。无以为报，唯有将创作进行到底。

2020年6月

何 浩，生于北京，毕业于中央美术学院，获艺术学博士学位，现为中央美院设计学院副教授。著有《一个中国独立设计者的当代艺术史——何浩书籍设计 2003—2013》（香港中文大学出版社，2014）。

教育部哲学社会科学研究重大课题攻关项目资助，项目批准号：12JZD015